Y0-AAX-928

Methods in Enzymology

Volume 333

REGULATORS AND EFFECTORS OF SMALL GTPases

Part G

Ras Family II

METHODS IN ENZYMOLOGY

EDITORS-IN-CHIEF

John N. Abelson Melvin I. Simon

DIVISION OF BIOLOGY
CALIFORNIA INSTITUTE OF TECHNOLOGY
PASADENA, CALIFORNIA

FOUNDING EDITORS

Sidney P. Colowick and Nathan O. Kaplan

Methods in Enzymology

Volume 333

Regulators and Effectors of Small GTPases

Part G
Ras Family II

EDITED BY

W. E. Balch

THE SCRIPPS RESEARCH INSTITUTE
LA JOLLA, CALIFORNIA

Channing J. Der

LINEBERGER COMPREHENSIVE CANCER CENTER
THE UNIVERSITY OF NORTH CAROLINA AT CHAPEL HILL
CHAPEL HILL, NORTH CAROLINA

Alan Hall

UNIVERSITY COLLEGE LONDON, LONDON, ENGLAND

ACADEMIC PRESS
San Diego London Boston New York Sydney Tokyo Toronto

This book is printed on acid-free paper. ∞

Academic Press
A Harcourt Science and Technology Company
525 B Street, Suite 1900, San Diego, California 92101-4495, USA
http://www.academicpress.com

Academic Press
Harcourt Place, 32 Jamestown Road, London NW1 7BY, UK
http://www.academicpress.com

International Standard Book Number: 0-12-182234-6

PRINTED IN THE UNITED STATES OF AMERICA
01 02 03 04 05 06 07 SB 9 8 7 6 5 4 3 2 1

Table of Contents

CONTRIBUTORS TO VOLUME 333		ix
PREFACE		xv
VOLUMES IN SERIES		xvii

Section I. Cytoplasmic and Nuclear Signaling Analyses

1. Determining Involvement of Shc Proteins in Signaling Pathways	JOHN P. O'BRYAN	3
2. Assaying Activity of Individual Protein Kinases in Crude Tissue or Cellular Extracts	SAID A. GOUELI, KEVIN HSIAO, AND BASEM S. GOUELI	16
3. Recombinant Adenoviral Expression of Dominant-Negative Ras N17 Blocking Radiation-Induced Activation of Mitogen-Activated Protein Kinase Pathway	PAUL DENT, CRAIG LOGSDON, BARBARA NICKE, KRISTOFFER VALERIE, JULIE FARNSWORTH, RUPERT SCHMIDT-ULLRICH, AND DEAN B. REARDON	28
4. Ras Activation of Phosphatidylinositol 3-Kinase and Akt	PABLO RODRIGUEZ VICIANA AND JULIAN DOWNWARD	37
5. Assays for Monitoring p70 S6 Kinase and RSK Activation	JEFFREY MASUDA-ROBENS, VERA P. KRYMSKAYA, HONGWEI QI, AND MARGARET M. CHOU	45
6. Ras Activation of PAK Protein Kinases	ALBERT CHEN, YI TANG, YA ZHUO, QI WANG, ALBERT PAHK, AND JEFFREY FIELD	55
7. Ras Signaling to Transcription Activation: Analysis with GAL4 Fusion Proteins	GABRIELE FOOS, CHRISTINA K. GALANG, CHAO-FENG ZHENG, AND CRAIG A. HAUSER	61
8. Ras Regulation of NF-κB and Apoptosis	MARTY W. MAYO, JACQUELINE L. NORRIS, AND ALBERT S. BALDWIN	73

9. Ras Activation of NF-κB and Superoxide — CHUNMING DONG AND PASCAL J. GOLDSCHMIDT-CLERMONT — 88

10. Ras, Metastasis, and Matrix Metalloproteinase 9 — ERIC J. BERNHARD AND RUTH J. MUSCHEL — 96

11. Ras Regulation of Urokinase-Type Plasminogen Activator — ERNST LENGYEL, SABINE RIED, MARKUS M. HEISS, CLAUDIA JÄGER, MANFRED SCHMITT, AND HEIKE ALLGAYER — 105

12. Ras Regulation of Cyclin D1 Promoter — DEREK F. AMANATULLAH, BRIAN T. ZAFONTE, CHRISTOPHER ALBANESE, MAOFU FU, CYNTHIA MESSIERS, JOHN HASSELL, AND RICHARD G. PESTELL — 116

13. Ras Regulation of Cyclin-Dependent Immunoprecipitation Kinase Assays — BRIAN T. ZAFONTE, DEREK F. AMANATULLAH, DANIEL SAGE, LEONARD H. AUGENLICHT, AND RICHARD G. PESTELL — 127

14. STAT Proteins: Signal Tranducers and Activators of Transcription — JACQUELINE BROMBERG AND XIAOMIN CHEN — 138

15. Integrin Regulation of Receptor Tyrosine Kinase and G Protein-Coupled Receptor Signaling to Mitogen-Activated Protein Kinases — RUDOLPH L. JULIANO, ANDREW E. APLIN, ALAN K. HOWE, SARAH SHORT, JUNG WEON LEE, AND SURESH ALAHARI — 151

16. R-Ras Regulation of Integrin Function — PAUL E. HUGHES, BEAT OERTLI, JAEWON HAN, AND MARK H. GINSBERG — 163

17. Caveolin and Ras Function — ROBERT G. PARTON AND JOHN F. HANCOCK — 172

Section II. Biological Analyses

18. Analyses of M-Ras/R-Ras3 Signaling and Biology — LAWRENCE A. QUILLIAM, JOHN F. REBHUN, HUI ZONG, AND ARIEL F. CASTRO — 187

19. Analyses of TC21/R-Ras2 Signaling and Biological Activity — SUZANNE M. GRAHAM, KELLEY ROGERS-GRAHAM, CLAUDIA FIGUEROA, CHANNING J. DER, AND ANNE B. VOJTEK — 203

20. Characterization of Rheb Functions Using Yeast and Mammalian Systems — JUN URANO, CHAD ELLIS, GEOFFREY J. CLARK, AND FUYUHIKO TAMANOI — 217

21. Ras Regulation of Skeletal Muscle Differentiation and Gene Expression — NATALIA MITIN, MELISSA B. RAMOCKI, STEPHEN F. KONIECZNY, AND ELIZABETH J. TAPAROWSKY — 232

22. Induction of Senescence by Oncogenic Ras — IGNACIO PALMERO AND MANUEL SERRANO — 247

23. Ras and Rho Protein Induction of Motility and Invasion in T47D Breast Adenocarcinoma Cells — PATRICIA J. KEELY — 256

24. Ras Regulation of Vascular Endothelial Growth Factor and Angiogenesis — JANUSZ RAK AND ROBERT S. KERBEL — 267

25. Ras Regulation of Radioresistance in Cell Culture — ANJALI K. GUPTA, VINCENT J. BAKANAUSKAS, W. GILLIES MCKENNA, ERIC J. BERNHARD, AND RUTH J. MUSCHEL — 284

26. Paired Human Fibrosarcoma Cell Lines That Possess or Lack Endogenous Mutant N-*ras* Alleles as Experimental Model for Ras Signaling Pathways — SWATI GUPTA AND ERIC J. STANBRIDGE — 290

27. Orally Bioavailable Farnesyltransferase Inhibitors as Anticancer Agents in Transgenic and Xenograft Models — MING LIU, W. ROBERT BISHOP, LORETTA L. NIELSEN, MATTHEW S. BRYANT, AND PAUL KIRSCHMEIER — 306

28. Animal Models for Ras-Induced Metastasis — CRAIG P. WEBB AND GEORGE F. VANDE WOUDE — 318

Section III. Regulation of Guanine Nucleotide Association

29. Nonradioactive Determination of Ras–GTP Levels Using Activated Ras Interaction Assay — STEPHEN J. TAYLOR, ROSS J. RESNICK, AND DAVID SHALLOWAY — 333

30. Measurement of GTP-Bound Ras-Like GTPases by Activation-Specific Probes — MIRANDA VAN TRIEST, JOHAN DE ROOIJ, AND JOHANNES L. BOS — 343

31. Immunocytochemical Assay for Ras Activity — LARRY S. SHERMAN AND NANCY RATNER — 348

AUTHOR INDEX 357

SUBJECT INDEX 389

Contributors to Volume 333

Article numbers are in parentheses following the names of contributors.
Affiliations listed are current.

SURESH ALAHARI (15), *Department of Pharmacology, University of North Carolina, Chapel Hill, North Carolina 27599-7365*

CHRISTOPHER ALBANESE (12), *Division of Hormone-Dependent Tumor Biology, Comprehensive Cancer Center, Department of Developmental and Molecular Biology, Albert Einstein College of Medicine, Bronx, New York 10461*

HEIKE ALLGAYER (11), *Department of Surgery, Klinikum Grosshadern, Ludwig-Maximillians Universität München, München D-81675, Germany*

DEREK F. AMANATULLAH (12, 13), *Division of Hormone-Dependent Tumor Biology, Comprehensive Cancer Center, Department of Developmental and Molecular Biology, Albert Einstein College of Medicine, Bronx, New York 10461*

ANDREW E. APLIN (15), *Department of Pharmacology, University of North Carolina, Chapel Hill, North Carolina 27599-7365*

LEONARD H. AUGENLICHT (13), *Department of Oncology, Montefiore and Albert Einstein Medical Centers, Bronx, New York 10467*

VINCENT J. BAKANAUSKAS (25), *Department of Radiation Oncology, University of Pennsylvania School of Medicine, Philadelphia, Pennsylvania 19104*

ALBERT S. BALDWIN (8), *Lineberger Comprehensive Cancer Center, University of North Carolina, Chapel Hill, North Carolina 27599*

ERIC J. BERNHARD (10, 25), *Department of Radiation Oncology, University of Pennsylvania School of Medicine, Philadelphia, Pennsylvania 19104*

W. ROBERT BISHOP (27), *Biological Research-Oncology, Schering-Plough Research Institute, Kenilworth, New Jersey 07033-1300*

JOHANNES L. BOS (30), *Department of Physiological Chemistry and Centre for Biomedical Genetics, University Medical Centre Utrecht, Utrecht 3584 CG, The Netherlands*

JACQUELINE F. BROMBERG (14), *Memorial Sloan Kettering Cancer Center, New York, New York 10021*

MATTHEW S. BRYANT (27), *Biological Research-Oncology, Schering-Plough Research Institute, Kenilworth, New Jersey 07033-1300*

ARIEL F. CASTRO (18), *Department of Biochemistry and Molecular Biology, Indiana University School of Medicine, Indianapolis, Indiana 46202*

ALBERT CHEN (6), *Department of Pharmacology, University of Pennsylvania School of Medicine, Philadelphia, Pennsylvania 19104*

XIAOMIN CHEN (14), *The University of Texas M. D. Anderson Cancer Center, Houston, Texas 77030*

MARGARET M. CHOU (5), *Department of Cell and Developmental Biology, University of Pennsylvania School of Medicine, Philadelphia, Pennsylvania 19104*

GEOFFREY J. CLARK (20), *National Cancer Institute, National Institutes of Health, Rockville, Maryland 20850-3300*

PAUL DENT (3), *Department of Radiation Oncology, Medical College of Virginia, Virginia Commonwealth University, Richmond, Virginia 23298-0058*

CHANNING J. DER (19), *Lineberger Comprehensive Cancer Center, University of North Carolina at Chapel Hill, Chapel Hill, North Carolina 27599-7295*

JOHAN DE ROOIJ (30), *Department of Physiological Chemistry and Centre for Biomedical Genetics, University Medical Centre Utrecht, Utrecht 3584 CG, The Netherlands*

CHUNMING DONG (9), *The Heart and Lung Institute, Division of Cardiology, Department of Internal Medicine, The Ohio State University, College of Medicine and Public Health, Columbus, Ohio 43210*

JULIAN DOWNWARD (4), *Imperial Cancer Research Fund, London WC2A 3PX, United Kingdom*

CHAD ELLIS (20), *National Cancer Institute, National Institutes of Health, Rockville, Maryland 20850-3300*

JULIE FARNSWORTH (3), *Department of Radiation Oncology, Medical College of Virginia, Virginia Commonwealth University, Richmond, Virginia 23298-0058*

JEFFREY FIELD (6), *Department of Pharmacology, University of Pennsylvania School of Medicine, Philadelphia, Pennsylvania 19104*

CLAUDIA FIGUEROA (19), *Department of Biological Chemistry, University of Michigan, Ann Arbor, Michigan 48109-0606*

GABRIELE FOOS (7), *La Jolla Cancer Research Center, The Burnham Institute, La Jolla, California 92037*

MAOFU FU (12), *Division of Hormone-Dependent Tumor Biology, Comprehensive Cancer Center, Department of Developmental and Molecular Biology, Albert Einstein College of Medicine, Bronx, New York 10461*

CHRISTINA K. GALANG (7), *La Jolla Cancer Research Center, The Burnham Institute, La Jolla, California 92037*

MARK H. GINSBERG (16), *Department of Vascular Biology, The Scripps Research Institute, La Jolla, California 92030*

PASCAL J. GOLDSCHMIDT-CLERMONT (9), *The Heart and Lung Institute, Division of Cardiology, Department of Internal Medicine, The Ohio State University, College of Medicine and Public Health, Columbus, Ohio 43210*

BASEM S. GOUELI (2), *Mayo Medical School, Rochester, Minnesota 55901*

SAID A. GOUELI (2), *Signal Transduction Group, Research and Development, Promega Corp., Madison, Wisconsin 53711, and Department of Pathology and Laboratory Medicine, University of Wisconsin School of Medicine, Madison, Wisconsin 53711*

SUZANNE M. GRAHAM (19), *Zoological Institute, Zurich University, Zurich, Switzerland*

ANJALI K. GUPTA (25), *Department of Radiation Oncology, University of Pennsylvania School of Medicine, Philadelphia, Pennsylvania 19104*

SWATI GUPTA (26), *Department of Microbiology and Molecular Genetics, College of Medicine, University of California, Irvine, California 92697-4025*

JAEWON HAN (16), *Department of Vascular Biology, The Scripps Research Institute, La Jolla, California 92030*

JOHN F. HANCOCK (17), *Laboratory of Experimental Oncology, Department of Pathology, University of Queensland Medical School, Brisbane, Queensland 4006, Australia*

JOHN HASSELL (12), *Institute for Molecular Biology and Biotechnology, McMaster University, Hamilton, Ontario L8S 4K1, Canada*

CRAIG A. HAUSER (7), *La Jolla Cancer Research Center, The Burnham Institute, La Jolla, California 92037*

MARKUS M. HEISS (11), *Department of Surgery, Klinikum Grosshadern, Ludwig-Maximillians Universität München, München D-81675, Germany*

ALAN K. HOWE (15), *Department of Pharmacology, University of North Carolina, Chapel Hill, North Carolina 27599-7365*

KEVIN HSIAO (2), *Signal Transduction Group, Research and Development, Promega Corp., Madison, Wisconsin 53711*

PAUL E. HUGHES (16), *Department of Vascular Biology, The Scripps Research Institute, La Jolla, California 92030*

CLAUDIA JÄGER (11), *Department of Obstetrics and Gynecology, Klinikum rechts der Isar, Technische Universität München, München D-81675, Germany*

RUDOLPH L. JULIANO (15), *Department of Pharmacology, School of Medicine, University of North Carolina, Chapel Hill, North Carolina 27599-7365*

PATRICIA J. KEELY (23), *Department of Pharmacology, University of Wisconsin, Madison, Wisconsin 53706*

ROBERT S. KERBEL (24), *Department of Medical Biophysics, Division of Cancer Biology Research, Sunnybrook Health Science Centre, University of Toronto, Toronto, Ontario M6G 2M9, Canada*

PAUL KIRSCHMEIER (27), *Biological Research-Oncology, Schering-Plough Research Institute, Kenilworth, New Jersey 07033-1300*

STEPHEN F. KONIECZNY (21), *Department of Biological Sciences, Purdue University, West Lafayette, Indiana 47907-1392*

VERA P. KRYMSKAYA (5), *Department of Medicine, University of Pennsylvania School of Medicine, Philadelphia, Pennsylvania 19104*

JUNG WEON LEE (15), *Department of Pharmacology, University of North Carolina, Chapel Hill, North Carolina 27599-7365*

ERNST LENGYEL (11), *Department of Obstetrics, Gynecology, and Reproductive Sciences and Cancer Research Institute, University of California, San Francisco, California 94143-0875*

MING LIU (27), *Biological Research-Oncology, Schering-Plough Research Institute, Kenilworth, New Jersey 07033-1300*

CRAIG LOGSDON (3), *Department of Physiology, University of Michigan, Ann Arbor, Michigan 48109*

JEFFREY MASUDA-ROBENS (5), *Department of Pharmacology, University of Pennsylvania School of Medicine, Philadelphia, Pennsylvania 19104*

MARTY W. MAYO (8), *Department of Biochemistry and Molecular Genetics, University of Virginia School of Medicine, Charlottesville, Virginia 22903*

W. GILLIES MCKENNA (25), *Department of Radiation Oncology, University of Pennsylvania School of Medicine, Philadelphia, Pennsylvania 19104*

CYNTHIA MESSIERS (12), *Institute for Molecular Biology and Biotechnology, McMaster University, Hamilton, Ontario L8S 4K1, Canada*

NATALIA MITIN (21), *Department of Biological Sciences, Purdue University, West Lafayette, Indiana 47907-1392*

RUTH J. MUSCHEL (10, 25), *Department of Pathology and Laboratory Medicine, University of Pennsylvania School of Medicine, Philadelphia, Pennsylvania 19104*

BARBARA NICKE (3), *Department of Physiology, University of Michigan, Ann Arbor, Michigan 48109*

LORETTA L. NIELSEN (27), *Biological Research-Oncology, Schering-Plough Research Institute, Kenilworth, New Jersey 07033-1300*

JACQUELINE L. NORRIS (8), *Paradigm Genetics, Inc., Research Triangle Park, North Carolina 27709*

JOHN P. O'BRYAN (1), *Laboratory of Signal Transduction, National Institute of Environmental Health Sciences, National Institutes of Health, Research Triangle Park, North Carolina 27709*

BEAT OERTLI (16), *Kantonsspital Bruderholz, Bruderholz CH-4101, Switzerland*

ALBERT PAHK (6), *Department of Pharmacology, University of Pennsylvania School of Medicine, Philadelphia, Pennsylvania 19104*

IGNACIO PALMERO (22), *Department of Immunology and Oncology, National Center of Biotechnology, Madrid E-28049, Spain*

ROBERT G. PARTON (17), *Centre for Microscopy and Microanalysis, Department of Physiology and Pharmacology, and Institute of Molecular Bioscience, University of Queensland, Brisbane, Queensland 4072, Australia*

RICHARD G. PESTELL (12, 13), *Division of Hormone-Dependent Tumor Biology, Comprehensive Cancer Center, Department of Developmental and Molecular Biology, Albert Einstein College of Medicine, Bronx, New York 10461*

HONGWEI QI (5), *Department of Cell and Developmental Biology, University of Pennsylvania School of Medicine, Philadelphia, Pennsylvania 19104*

LAWRENCE A. QUILLIAM (18), *Department of Biochemistry and Molecular Biology, Indiana University School of Medicine, Indianapolis, Indiana 46202*

JANUSZ RAK (24), *Department of Medical Biophysics, Division of Cancer Biology Research, Sunnybrook Health Science Centre, University of Toronto, Toronto, Ontario M6G 2M9, Canada*

MELISSA B. RAMOCKI (21), *Department of Human Genetics, University of Chicago, Chicago, Illinois 60637*

NANCY RATNER (31), *Department of Cell Biology, Neurobiology, and Anatomy, University of Cincinnati College of Medicine, Cincinnati, Ohio 45267-0521*

DEAN B. REARDON (3), *Department of Radiation Oncology, Medical College of Virginia, Virginia Commonwealth University, Richmond, Virginia 23298-0058*

JOHN F. REBHUN (18), *Department of Biochemistry and Molecular Biology, Indiana University School of Medicine, Indianapolis, Indiana 46202*

ROSS J. RESNICK (29), *Department of Molecular Biology and Genetics, Cornell University, Ithaca, New York 94720*

SABINE RIED (11), *Department of Obstetrics and Gynecology, Klinikum rechts der Isar, Technische Universität München, München D-81675, Germany*

PABLO RODRIGUEZ-VICIANA (4), *University of California, San Francisco Cancer Research Institute, San Francisco, California 94115*

KELLEY ROGERS-GRAHAM (19), *Lineberger Comprehensive Cancer Center, University of North Carolina, Chapel Hill, North Carolina 27599-7295*

DANIEL SAGE (13), *Division of Hormone-Dependent Tumor Biology, Comprehensive Cancer Center, Department of Developmental and Molecular Biology, Albert Einstein College of Medicine, Bronx, New York 10461*

RUPERT SCHMIDT-ULLRICH (3), *Department of Radiation Oncology, Medical College of Virginia, Virginia Commonwealth University, Richmond, Virginia 23298-0058*

MANFRED SCHMITT (11), *Department of Obstetrics and Gynecology, Klinikum rechts der Isar, Technische Universität München, München D-81675, Germany*

MANUEL SERRANO (22), *Department of Immunology and Oncology, National Center of Biotechnology, Madrid E-28049, Spain*

DAVID SHALLOWAY (29), *Department of Molecular Biology and Genetics, Cornell University, Ithaca, New York 94720*

LARRY S. SHERMAN (31), *Department of Cell Biology, Neurobiology, and Anatomy, University of Cincinnati College of Medicine, Cincinnati, Ohio 45267-0521*

SARAH SHORT (15), *Department of Pharmacology, University of North Carolina, Chapel Hill, North Carolina 27599-7365*

ERIC J. STANBRIDGE (26), *Department of Microbiology and Molecular Genetics, College of Medicine, University of California, Irvine, California 92697-4025*

FUYUHIKO TAMANOI (20), *Department of Microbiology and Molecular Genetics, University of California, Los Angeles, California 90095-1489*

YI TANG (6), *Dupont Pharmaceutical Co., Glenolden Laboratories, Glenolden, Pennsylvania 19036*

ELIZABETH J. TAPAROWSKY (21), *Department of Biological Sciences, Purdue University, West Lafayette, Indiana 47907-1392*

STEPHEN J. TAYLOR (29), *Department of Molecular and Cell Biology, University of California, Berkeley, California 94720*

JUN URANO (20) *Department of Biochemistry and Biophysics, University of California, San Francisco, California 94143-0448*

KRISTOFFER VALERIE (3), *Department of Radiation Oncology, Medical College of Virginia, Virginia Commonwealth University, Richmond, Virginia 23298-0058*

GEORGE F. VANDE WOUDE (28), *Van Andel Research Institute, Grand Rapids, Michigan 49503*

MIRANDA VAN TRIEST (30), *Department of Physiological Chemistry and Centre for Biomedical Genetics, University Medical Centre Utrecht, Utrecht 3584 CG, The Netherlands*

ANNE B. VOJTEK (19), *Department of Biological Chemistry, University of Michigan, Ann Arbor, Michigan 48109-0606*

QI WANG (6), *Department of Pharmacology, University of Pennsylvania School of Medicine, Philadelphia, Pennsylvania 19104*

CRAIG P. WEBB (28), *Van Andel Research Institute, Grand Rapids, Michigan 49503*

BRIAN T. ZAFONTE (12, 13), *Division of Hormone-Dependent Tumor Biology, Comprehensive Cancer Center, Department of Developmental and Molecular Biology, Albert Einstein College of Medicine, Bronx, New York 10461*

CHAO-FENG ZHENG (7), *Novasite Pharmaceuticals, San Diego, California 92121*

YA ZHUO (6), *Department of Pharmacology, University of Pennsylvania School of Medicine, Philadelphia, Pennsylvania 19104*

HUI ZONG (18), *Department of Biochemistry and Molecular Biology, Indiana University School of Medicine, Indianapolis, Indiana 46202*

Preface

As with the Rho and Rab branches of the Ras superfamily of small GTPases, research interest in the Ras branch has continued to expand dramatically into new areas and to embrace new themes since the last *Methods in Enzymology* Volume 255 on Ras GTPases was published in 1995. First, the Ras branch has expanded beyond the original Ras, Rap, and Ral members. New members include M-Ras, Rheb, Rin, and Rit. Second, the signaling activities of Ras are much more diverse and complex than appreciated previously. In particular, while the Raf/MEK/ERK kinase cascade remains a key signaling pathway activated by Ras, it is now appreciated that an increasing number of non-Raf effectors also mediate Ras family protein function. Third, it is increasingly clear that the cellular functions regulated by Ras go beyond regulation of cell proliferation, and involve regulation of senescence and cell survival and induction of tumor cell invasion, metastasis, and angiogenesis. Fourth, another theme that has emerged is regulatory cross talk among Ras family proteins, including both GTPase signaling cascades that link signaling from one family member to another, as well as the use of shared regulators and effectors by different family members.

Concurrent with the expanded complexity of Ras family biology, biochemistry, and signaling have been the development and application of a wider array of methodology to study Ras family function. While some are simply improved methods to study old questions, many others involve novel approaches to study aspects of Ras family protein function not studied previously. In particular, the emerging application of techniques to study Ras regulation of gene and protein expression represents an important direction for current and future studies. Consequently, *Methods in Enzymology,* Volumes 332 and 333 cover many of the new techniques that have emerged during the past five years.

We are grateful for the efforts of all our colleagues who contributed to these volumes. We are indebted to them for sharing their expertise and experiences, as well as their time, in compiling this comprehensive series of chapters. In particular, we hope these volumes will provide valuable references and sources of information that will facilitate the efforts of newly incoming researchers to the study of the Ras family of small GTPases.

CHANNING J. DER
ALAN HALL
WILLIAM E. BALCH

METHODS IN ENZYMOLOGY

VOLUME I. Preparation and Assay of Enzymes
Edited by SIDNEY P. COLOWICK AND NATHAN O. KAPLAN

VOLUME II. Preparation and Assay of Enzymes
Edited by SIDNEY P. COLOWICK AND NATHAN O. KAPLAN

VOLUME III. Preparation and Assay of Substrates
Edited by SIDNEY P. COLOWICK AND NATHAN O. KAPLAN

VOLUME IV. Special Techniques for the Enzymologist
Edited by SIDNEY P. COLOWICK AND NATHAN O. KAPLAN

VOLUME V. Preparation and Assay of Enzymes
Edited by SIDNEY P. COLOWICK AND NATHAN O. KAPLAN

VOLUME VI. Preparation and Assay of Enzymes (*Continued*)
Preparation and Assay of Substrates
Special Techniques
Edited by SIDNEY P. COLOWICK AND NATHAN O. KAPLAN

VOLUME VII. Cumulative Subject Index
Edited by SIDNEY P. COLOWICK AND NATHAN O. KAPLAN

VOLUME VIII. Complex Carbohydrates
Edited by ELIZABETH F. NEUFELD AND VICTOR GINSBURG

VOLUME IX. Carbohydrate Metabolism
Edited by WILLIS A. WOOD

VOLUME X. Oxidation and Phosphorylation
Edited by RONALD W. ESTABROOK AND MAYNARD E. PULLMAN

VOLUME XI. Enzyme Structure
Edited by C. H. W. HIRS

VOLUME XII. Nucleic Acids (Parts A and B)
Edited by LAWRENCE GROSSMAN AND KIVIE MOLDAVE

VOLUME XIII. Citric Acid Cycle
Edited by J. M. LOWENSTEIN

VOLUME XIV. Lipids
Edited by J. M. LOWENSTEIN

VOLUME XV. Steroids and Terpenoids
Edited by RAYMOND B. CLAYTON

VOLUME XVI. Fast Reactions
Edited by KENNETH KUSTIN

VOLUME XVII. Metabolism of Amino Acids and Amines (Parts A and B)
Edited by HERBERT TABOR AND CELIA WHITE TABOR

VOLUME XVIII. Vitamins and Coenzymes (Parts A, B, and C)
Edited by DONALD B. MCCORMICK AND LEMUEL D. WRIGHT

VOLUME XIX. Proteolytic Enzymes
Edited by GERTRUDE E. PERLMANN AND LASZLO LORAND

VOLUME XX. Nucleic Acids and Protein Synthesis (Part C)
Edited by KIVIE MOLDAVE AND LAWRENCE GROSSMAN

VOLUME XXI. Nucleic Acids (Part D)
Edited by LAWRENCE GROSSMAN AND KIVIE MOLDAVE

VOLUME XXII. Enzyme Purification and Related Techniques
Edited by WILLIAM B. JAKOBY

VOLUME XXIII. Photosynthesis (Part A)
Edited by ANTHONY SAN PIETRO

VOLUME XXIV. Photosynthesis and Nitrogen Fixation (Part B)
Edited by ANTHONY SAN PIETRO

VOLUME XXV. Enzyme Structure (Part B)
Edited by C. H. W. HIRS AND SERGE N. TIMASHEFF

VOLUME XXVI. Enzyme Structure (Part C)
Edited by C. H. W. HIRS AND SERGE N. TIMASHEFF

VOLUME XXVII. Enzyme Structure (Part D)
Edited by C. H. W. HIRS AND SERGE N. TIMASHEFF

VOLUME XXVIII. Complex Carbohydrates (Part B)
Edited by VICTOR GINSBURG

VOLUME XXIX. Nucleic Acids and Protein Synthesis (Part E)
Edited by LAWRENCE GROSSMAN AND KIVIE MOLDAVE

VOLUME XXX. Nucleic Acids and Protein Synthesis (Part F)
Edited by KIVIE MOLDAVE AND LAWRENCE GROSSMAN

VOLUME XXXI. Biomembranes (Part A)
Edited by SIDNEY FLEISCHER AND LESTER PACKER

VOLUME XXXII. Biomembranes (Part B)
Edited by SIDNEY FLEISCHER AND LESTER PACKER

VOLUME XXXIII. Cumulative Subject Index Volumes I-XXX
Edited by MARTHA G. DENNIS AND EDWARD A. DENNIS

VOLUME XXXIV. Affinity Techniques (Enzyme Purification: Part B)
Edited by WILLIAM B. JAKOBY AND MEIR WILCHEK

VOLUME XXXV. Lipids (Part B)
Edited by JOHN M. LOWENSTEIN

VOLUME XXXVI. Hormone Action (Part A: Steroid Hormones)
Edited by BERT W. O'MALLEY AND JOEL G. HARDMAN

VOLUME XXXVII. Hormone Action (Part B: Peptide Hormones)
Edited by BERT W. O'MALLEY AND JOEL G. HARDMAN

VOLUME XXXVIII. Hormone Action (Part C: Cyclic Nucleotides)
Edited by JOEL G. HARDMAN AND BERT W. O'MALLEY

VOLUME XXXIX. Hormone Action (Part D: Isolated Cells, Tissues, and Organ Systems)
Edited by JOEL G. HARDMAN AND BERT W. O'MALLEY

VOLUME XL. Hormone Action (Part E: Nuclear Structure and Function)
Edited by BERT W. O'MALLEY AND JOEL G. HARDMAN

VOLUME XLI. Carbohydrate Metabolism (Part B)
Edited by W. A. WOOD

VOLUME XLII. Carbohydrate Metabolism (Part C)
Edited by W. A. WOOD

VOLUME XLIII. Antibiotics
Edited by JOHN H. HASH

VOLUME XLIV. Immobilized Enzymes
Edited by KLAUS MOSBACH

VOLUME XLV. Proteolytic Enzymes (Part B)
Edited by LASZLO LORAND

VOLUME XLVI. Affinity Labeling
Edited by WILLIAM B. JAKOBY AND MEIR WILCHEK

VOLUME XLVII. Enzyme Structure (Part E)
Edited by C. H. W. HIRS AND SERGE N. TIMASHEFF

VOLUME XLVIII. Enzyme Structure (Part F)
Edited by C. H. W. HIRS AND SERGE N. TIMASHEFF

VOLUME XLIX. Enzyme Structure (Part G)
Edited by C. H. W. HIRS AND SERGE N. TIMASHEFF

VOLUME L. Complex Carbohydrates (Part C)
Edited by VICTOR GINSBURG

VOLUME LI. Purine and Pyrimidine Nucleotide Metabolism
Edited by PATRICIA A. HOFFEE AND MARY ELLEN JONES

VOLUME LII. Biomembranes (Part C: Biological Oxidations)
Edited by SIDNEY FLEISCHER AND LESTER PACKER

VOLUME LIII. Biomembranes (Part D: Biological Oxidations)
Edited by SIDNEY FLEISCHER AND LESTER PACKER

VOLUME LIV. Biomembranes (Part E: Biological Oxidations)
Edited by SIDNEY FLEISCHER AND LESTER PACKER

VOLUME LV. Biomembranes (Part F: Bioenergetics)
Edited by SIDNEY FLEISCHER AND LESTER PACKER

VOLUME LVI. Biomembranes (Part G: Bioenergetics)
Edited by SIDNEY FLEISCHER AND LESTER PACKER

VOLUME LVII. Bioluminescence and Chemiluminescence
Edited by MARLENE A. DELUCA

VOLUME LVIII. Cell Culture
Edited by WILLIAM B. JAKOBY AND IRA PASTAN

VOLUME LIX. Nucleic Acids and Protein Synthesis (Part G)
Edited by KIVIE MOLDAVE AND LAWRENCE GROSSMAN

VOLUME LX. Nucleic Acids and Protein Synthesis (Part H)
Edited by KIVIE MOLDAVE AND LAWRENCE GROSSMAN

VOLUME 61. Enzyme Structure (Part H)
Edited by C. H. W. HIRS AND SERGE N. TIMASHEFF

VOLUME 62. Vitamins and Coenzymes (Part D)
Edited by DONALD B. MCCORMICK AND LEMUEL D. WRIGHT

VOLUME 63. Enzyme Kinetics and Mechanism (Part A: Initial Rate and Inhibitor Methods)
Edited by DANIEL L. PURICH

VOLUME 64. Enzyme Kinetics and Mechanism (Part B: Isotopic Probes and Complex Enzyme Systems)
Edited by DANIEL L. PURICH

VOLUME 65. Nucleic Acids (Part I)
Edited by LAWRENCE GROSSMAN AND KIVIE MOLDAVE

VOLUME 66. Vitamins and Coenzymes (Part E)
Edited by DONALD B. MCCORMICK AND LEMUEL D. WRIGHT

VOLUME 67. Vitamins and Coenzymes (Part F)
Edited by DONALD B. MCCORMICK AND LEMUEL D. WRIGHT

VOLUME 68. Recombinant DNA
Edited by RAY WU

VOLUME 69. Photosynthesis and Nitrogen Fixation (Part C)
Edited by ANTHONY SAN PIETRO

VOLUME 70. Immunochemical Techniques (Part A)
Edited by HELEN VAN VUNAKIS AND JOHN J. LANGONE

VOLUME 71. Lipids (Part C)
Edited by JOHN M. LOWENSTEIN

VOLUME 72. Lipids (Part D)
Edited by JOHN M. LOWENSTEIN

VOLUME 73. Immunochemical Techniques (Part B)
Edited by JOHN J. LANGONE AND HELEN VAN VUNAKIS

VOLUME 74. Immunochemical Techniques (Part C)
Edited by JOHN J. LANGONE AND HELEN VAN VUNAKIS

VOLUME 75. Cumulative Subject Index Volumes XXXI, XXXII, XXXIV–LX
Edited by EDWARD A. DENNIS AND MARTHA G. DENNIS

VOLUME 76. Hemoglobins
Edited by ERALDO ANTONINI, LUIGI ROSSI-BERNARDI, AND EMILIA CHIANCONE

VOLUME 77. Detoxication and Drug Metabolism
Edited by WILLIAM B. JAKOBY

VOLUME 78. Interferons (Part A)
Edited by SIDNEY PESTKA

VOLUME 79. Interferons (Part B)
Edited by SIDNEY PESTKA

VOLUME 80. Proteolytic Enzymes (Part C)
Edited by LASZLO LORAND

VOLUME 81. Biomembranes (Part H: Visual Pigments and Purple Membranes, I)
Edited by LESTER PACKER

VOLUME 82. Structural and Contractile Proteins (Part A: Extracellular Matrix)
Edited by LEON W. CUNNINGHAM AND DIXIE W. FREDERIKSEN

VOLUME 83. Complex Carbohydrates (Part D)
Edited by VICTOR GINSBURG

VOLUME 84. Immunochemical Techniques (Part D: Selected Immunoassays)
Edited by JOHN J. LANGONE AND HELEN VAN VUNAKIS

VOLUME 85. Structural and Contractile Proteins (Part B: The Contractile Apparatus and the Cytoskeleton)
Edited by DIXIE W. FREDERIKSEN AND LEON W. CUNNINGHAM

VOLUME 86. Prostaglandins and Arachidonate Metabolites
Edited by WILLIAM E. M. LANDS AND WILLIAM L. SMITH

VOLUME 87. Enzyme Kinetics and Mechanism (Part C: Intermediates, Stereochemistry, and Rate Studies)
Edited by DANIEL L. PURICH

VOLUME 88. Biomembranes (Part I: Visual Pigments and Purple Membranes, II)
Edited by LESTER PACKER

VOLUME 89. Carbohydrate Metabolism (Part D)
Edited by WILLIS A. WOOD

VOLUME 90. Carbohydrate Metabolism (Part E)
Edited by WILLIS A. WOOD

VOLUME 91. Enzyme Structure (Part I)
Edited by C. H. W. HIRS AND SERGE N. TIMASHEFF

VOLUME 92. Immunochemical Techniques (Part E: Monoclonal Antibodies and General Immunoassay Methods)
Edited by JOHN J. LANGONE AND HELEN VAN VUNAKIS

VOLUME 93. Immunochemical Techniques (Part F: Conventional Antibodies, Fc Receptors, and Cytotoxicity)
Edited by JOHN J. LANGONE AND HELEN VAN VUNAKIS

VOLUME 94. Polyamines
Edited by HERBERT TABOR AND CELIA WHITE TABOR

VOLUME 95. Cumulative Subject Index Volumes 61–74, 76–80
Edited by EDWARD A. DENNIS AND MARTHA G. DENNIS

VOLUME 96. Biomembranes [Part J: Membrane Biogenesis: Assembly and Targeting (General Methods; Eukaryotes)]
Edited by SIDNEY FLEISCHER AND BECCA FLEISCHER

VOLUME 97. Biomembranes [Part K: Membrane Biogenesis: Assembly and Targeting (Prokaryotes, Mitochondria, and Chloroplasts)]
Edited by SIDNEY FLEISCHER AND BECCA FLEISCHER

VOLUME 98. Biomembranes (Part L: Membrane Biogenesis: Processing and Recycling)
Edited by SIDNEY FLEISCHER AND BECCA FLEISCHER

VOLUME 99. Hormone Action (Part F: Protein Kinases)
Edited by JACKIE D. CORBIN AND JOEL G. HARDMAN

VOLUME 100. Recombinant DNA (Part B)
Edited by RAY WU, LAWRENCE GROSSMAN, AND KIVIE MOLDAVE

VOLUME 101. Recombinant DNA (Part C)
Edited by RAY WU, LAWRENCE GROSSMAN, AND KIVIE MOLDAVE

VOLUME 102. Hormone Action (Part G: Calmodulin and Calcium-Binding Proteins)
Edited by ANTHONY R. MEANS AND BERT W. O'MALLEY

VOLUME 103. Hormone Action (Part H: Neuroendocrine Peptides)
Edited by P. MICHAEL CONN

VOLUME 104. Enzyme Purification and Related Techniques (Part C)
Edited by WILLIAM B. JAKOBY

VOLUME 105. Oxygen Radicals in Biological Systems
Edited by LESTER PACKER

VOLUME 106. Posttranslational Modifications (Part A)
Edited by FINN WOLD AND KIVIE MOLDAVE

VOLUME 107. Posttranslational Modifications (Part B)
Edited by FINN WOLD AND KIVIE MOLDAVE

VOLUME 108. Immunochemical Techniques (Part G: Separation and Characterization of Lymphoid Cells)
Edited by GIOVANNI DI SABATO, JOHN J. LANGONE, AND HELEN VAN VUNAKIS

VOLUME 109. Hormone Action (Part I: Peptide Hormones)
Edited by LUTZ BIRNBAUMER AND BERT W. O'MALLEY

VOLUME 110. Steroids and Isoprenoids (Part A)
Edited by JOHN H. LAW AND HANS C. RILLING

VOLUME 111. Steroids and Isoprenoids (Part B)
Edited by JOHN H. LAW AND HANS C. RILLING

VOLUME 112. Drug and Enzyme Targeting (Part A)
Edited by KENNETH J. WIDDER AND RALPH GREEN

VOLUME 113. Glutamate, Glutamine, Glutathione, and Related Compounds
Edited by ALTON MEISTER

VOLUME 114. Diffraction Methods for Biological Macromolecules (Part A)
Edited by HAROLD W. WYCKOFF, C. H. W. HIRS, AND SERGE N. TIMASHEFF

VOLUME 115. Diffraction Methods for Biological Macromolecules (Part B)
Edited by HAROLD W. WYCKOFF, C. H. W. HIRS, AND SERGE N. TIMASHEFF

VOLUME 116. Immunochemical Techniques (Part H: Effectors and Mediators of Lymphoid Cell Functions)
Edited by GIOVANNI DI SABATO, JOHN J. LANGONE, AND HELEN VAN VUNAKIS

VOLUME 117. Enzyme Structure (Part J)
Edited by C. H. W. HIRS AND SERGE N. TIMASHEFF

VOLUME 118. Plant Molecular Biology
Edited by ARTHUR WEISSBACH AND HERBERT WEISSBACH

VOLUME 119. Interferons (Part C)
Edited by SIDNEY PESTKA

VOLUME 120. Cumulative Subject Index Volumes 81–94, 96–101

VOLUME 121. Immunochemical Techniques (Part I: Hybridoma Technology and Monoclonal Antibodies)
Edited by JOHN J. LANGONE AND HELEN VAN VUNAKIS

VOLUME 122. Vitamins and Coenzymes (Part G)
Edited by FRANK CHYTIL AND DONALD B. MCCORMICK

VOLUME 123. Vitamins and Coenzymes (Part H)
Edited by FRANK CHYTIL AND DONALD B. MCCORMICK

VOLUME 124. Hormone Action (Part J: Neuroendocrine Peptides)
Edited by P. MICHAEL CONN

VOLUME 125. Biomembranes (Part M: Transport in Bacteria, Mitochondria, and Chloroplasts: General Approaches and Transport Systems)
Edited by SIDNEY FLEISCHER AND BECCA FLEISCHER

VOLUME 126. Biomembranes (Part N: Transport in Bacteria, Mitochondria, and Chloroplasts: Protonmotive Force)
Edited by SIDNEY FLEISCHER AND BECCA FLEISCHER

VOLUME 127. Biomembranes (Part O: Protons and Water: Structure and Translocation)
Edited by LESTER PACKER

VOLUME 128. Plasma Lipoproteins (Part A: Preparation, Structure, and Molecular Biology)
Edited by JERE P. SEGREST AND JOHN J. ALBERS

VOLUME 129. Plasma Lipoproteins (Part B: Characterization, Cell Biology, and Metabolism)
Edited by JOHN J. ALBERS AND JERE P. SEGREST

VOLUME 130. Enzyme Structure (Part K)
Edited by C. H. W. HIRS AND SERGE N. TIMASHEFF

VOLUME 131. Enzyme Structure (Part L)
Edited by C. H. W. HIRS AND SERGE N. TIMASHEFF

VOLUME 132. Immunochemical Techniques (Part J: Phagocytosis and Cell-Mediated Cytotoxicity)
Edited by GIOVANNI DI SABATO AND JOHANNES EVERSE

VOLUME 133. Bioluminescence and Chemiluminescence (Part B)
Edited by MARLENE DELUCA AND WILLIAM D. MCELROY

VOLUME 134. Structural and Contractile Proteins (Part C: The Contractile Apparatus and the Cytoskeleton)
Edited by RICHARD B. VALLEE

VOLUME 135. Immobilized Enzymes and Cells (Part B)
Edited by KLAUS MOSBACH

VOLUME 136. Immobilized Enzymes and Cells (Part C)
Edited by KLAUS MOSBACH

VOLUME 137. Immobilized Enzymes and Cells (Part D)
Edited by KLAUS MOSBACH

VOLUME 138. Complex Carbohydrates (Part E)
Edited by VICTOR GINSBURG

VOLUME 139. Cellular Regulators (Part A: Calcium- and Calmodulin-Binding Proteins)
Edited by ANTHONY R. MEANS AND P. MICHAEL CONN

VOLUME 140. Cumulative Subject Index Volumes 102–119, 121–134

VOLUME 141. Cellular Regulators (Part B: Calcium and Lipids)
Edited by P. MICHAEL CONN AND ANTHONY R. MEANS

VOLUME 142. Metabolism of Aromatic Amino Acids and Amines
Edited by SEYMOUR KAUFMAN

VOLUME 143. Sulfur and Sulfur Amino Acids
Edited by WILLIAM B. JAKOBY AND OWEN GRIFFITH

VOLUME 144. Structural and Contractile Proteins (Part D: Extracellular Matrix)
Edited by LEON W. CUNNINGHAM

VOLUME 145. Structural and Contractile Proteins (Part E: Extracellular Matrix)
Edited by LEON W. CUNNINGHAM

VOLUME 146. Peptide Growth Factors (Part A)
Edited by DAVID BARNES AND DAVID A. SIRBASKU

VOLUME 147. Peptide Growth Factors (Part B)
Edited by DAVID BARNES AND DAVID A. SIRBASKU

VOLUME 148. Plant Cell Membranes
Edited by LESTER PACKER AND ROLAND DOUCE

VOLUME 149. Drug and Enzyme Targeting (Part B)
Edited by RALPH GREEN AND KENNETH J. WIDDER

VOLUME 150. Immunochemical Techniques (Part K: *In Vitro* Models of B and T Cell Functions and Lymphoid Cell Receptors)
Edited by GIOVANNI DI SABATO

VOLUME 151. Molecular Genetics of Mammalian Cells
Edited by MICHAEL M. GOTTESMAN

VOLUME 152. Guide to Molecular Cloning Techniques
Edited by SHELBY L. BERGER AND ALAN R. KIMMEL

VOLUME 153. Recombinant DNA (Part D)
Edited by RAY WU AND LAWRENCE GROSSMAN

VOLUME 154. Recombinant DNA (Part E)
Edited by RAY WU AND LAWRENCE GROSSMAN

VOLUME 155. Recombinant DNA (Part F)
Edited by RAY WU

VOLUME 156. Biomembranes (Part P: ATP-Driven Pumps and Related Transport: The Na, K-Pump)
Edited by SIDNEY FLEISCHER AND BECCA FLEISCHER

VOLUME 157. Biomembranes (Part Q: ATP-Driven Pumps and Related Transport: Calcium, Proton, and Potassium Pumps)
Edited by SIDNEY FLEISCHER AND BECCA FLEISCHER

VOLUME 158. Metalloproteins (Part A)
Edited by JAMES F. RIORDAN AND BERT L. VALLEE

VOLUME 159. Initiation and Termination of Cyclic Nucleotide Action
Edited by JACKIE D. CORBIN AND ROGER A. JOHNSON

VOLUME 160. Biomass (Part A: Cellulose and Hemicellulose)
Edited by WILLIS A. WOOD AND SCOTT T. KELLOGG

VOLUME 161. Biomass (Part B: Lignin, Pectin, and Chitin)
Edited by WILLIS A. WOOD AND SCOTT T. KELLOGG

VOLUME 162. Immunochemical Techniques (Part L: Chemotaxis and Inflammation)
Edited by GIOVANNI DI SABATO

VOLUME 163. Immunochemical Techniques (Part M: Chemotaxis and Inflammation)
Edited by GIOVANNI DI SABATO

VOLUME 164. Ribosomes
Edited by HARRY F. NOLLER, JR., AND KIVIE MOLDAVE

VOLUME 165. Microbial Toxins: Tools for Enzymology
Edited by SIDNEY HARSHMAN

VOLUME 166. Branched-Chain Amino Acids
Edited by ROBERT HARRIS AND JOHN R. SOKATCH

VOLUME 167. Cyanobacteria
Edited by LESTER PACKER AND ALEXANDER N. GLAZER

VOLUME 168. Hormone Action (Part K: Neuroendocrine Peptides)
Edited by P. MICHAEL CONN

VOLUME 169. Platelets: Receptors, Adhesion, Secretion (Part A)
Edited by JACEK HAWIGER

VOLUME 170. Nucleosomes
Edited by PAUL M. WASSARMAN AND ROGER D. KORNBERG

VOLUME 171. Biomembranes (Part R: Transport Theory: Cells and Model Membranes)
Edited by SIDNEY FLEISCHER AND BECCA FLEISCHER

VOLUME 172. Biomembranes (Part S: Transport: Membrane Isolation and Characterization)
Edited by SIDNEY FLEISCHER AND BECCA FLEISCHER

VOLUME 173. Biomembranes [Part T: Cellular and Subcellular Transport: Eukaryotic (Nonepithelial) Cells]
Edited by SIDNEY FLEISCHER AND BECCA FLEISCHER

VOLUME 174. Biomembranes [Part U: Cellular and Subcellular Transport: Eukaryotic (Nonepithelial) Cells]
Edited by SIDNEY FLEISCHER AND BECCA FLEISCHER

VOLUME 175. Cumulative Subject Index Volumes 135–139, 141–167

VOLUME 176. Nuclear Magnetic Resonance (Part A: Spectral Techniques and Dynamics)
Edited by NORMAN J. OPPENHEIMER AND THOMAS L. JAMES

VOLUME 177. Nuclear Magnetic Resonance (Part B: Structure and Mechanism)
Edited by NORMAN J. OPPENHEIMER AND THOMAS L. JAMES

VOLUME 178. Antibodies, Antigens, and Molecular Mimicry
Edited by JOHN J. LANGONE

VOLUME 179. Complex Carbohydrates (Part F)
Edited by VICTOR GINSBURG

VOLUME 180. RNA Processing (Part A: General Methods)
Edited by JAMES E. DAHLBERG AND JOHN N. ABELSON

VOLUME 181. RNA Processing (Part B: Specific Methods)
Edited by JAMES E. DAHLBERG AND JOHN N. ABELSON

VOLUME 182. Guide to Protein Purification
Edited by MURRAY P. DEUTSCHER

VOLUME 183. Molecular Evolution: Computer Analysis of Protein and Nucleic Acid Sequences
Edited by RUSSELL F. DOOLITTLE

VOLUME 184. Avidin-Biotin Technology
Edited by MEIR WILCHEK AND EDWARD A. BAYER

VOLUME 185. Gene Expression Technology
Edited by DAVID V. GOEDDEL

VOLUME 186. Oxygen Radicals in Biological Systems (Part B: Oxygen Radicals and Antioxidants)
Edited by LESTER PACKER AND ALEXANDER N. GLAZER

VOLUME 187. Arachidonate Related Lipid Mediators
Edited by ROBERT C. MURPHY AND FRANK A. FITZPATRICK

VOLUME 188. Hydrocarbons and Methylotrophy
Edited by MARY E. LIDSTROM

VOLUME 189. Retinoids (Part A: Molecular and Metabolic Aspects)
Edited by LESTER PACKER

VOLUME 190. Retinoids (Part B: Cell Differentiation and Clinical Applications)
Edited by LESTER PACKER

VOLUME 191. Biomembranes (Part V: Cellular and Subcellular Transport: Epithelial Cells)
Edited by SIDNEY FLEISCHER AND BECCA FLEISCHER

VOLUME 192. Biomembranes (Part W: Cellular and Subcellular Transport: Epithelial Cells)
Edited by SIDNEY FLEISCHER AND BECCA FLEISCHER

VOLUME 193. Mass Spectrometry
Edited by JAMES A. MCCLOSKEY

VOLUME 194. Guide to Yeast Genetics and Molecular Biology
Edited by CHRISTINE GUTHRIE AND GERALD R. FINK

VOLUME 195. Adenylyl Cyclase, G Proteins, and Guanylyl Cyclase
Edited by ROGER A. JOHNSON AND JACKIE D. CORBIN

VOLUME 196. Molecular Motors and the Cytoskeleton
Edited by RICHARD B. VALLEE

VOLUME 197. Phospholipases
Edited by EDWARD A. DENNIS

VOLUME 198. Peptide Growth Factors (Part C)
Edited by DAVID BARNES, J. P. MATHER, AND GORDON H. SATO

VOLUME 199. Cumulative Subject Index Volumes 168–174, 176–194

VOLUME 200. Protein Phosphorylation (Part A: Protein Kinases: Assays, Purification, Antibodies, Functional Analysis, Cloning, and Expression)
Edited by TONY HUNTER AND BARTHOLOMEW M. SEFTON

VOLUME 201. Protein Phosphorylation (Part B: Analysis of Protein Phosphorylation, Protein Kinase Inhibitors, and Protein Phosphatases)
Edited by TONY HUNTER AND BARTHOLOMEW M. SEFTON

VOLUME 202. Molecular Design and Modeling: Concepts and Applications (Part A: Proteins, Peptides, and Enzymes)
Edited by JOHN J. LANGONE

VOLUME 203. Molecular Design and Modeling: Concepts and Applications (Part B: Antibodies and Antigens, Nucleic Acids, Polysaccharides, and Drugs)
Edited by JOHN J. LANGONE

VOLUME 204. Bacterial Genetic Systems
Edited by JEFFREY H. MILLER

VOLUME 205. Metallobiochemistry (Part B: Metallothionein and Related Molecules)
Edited by JAMES F. RIORDAN AND BERT L. VALLEE

VOLUME 206. Cytochrome P450
Edited by MICHAEL R. WATERMAN AND ERIC F. JOHNSON

VOLUME 207. Ion Channels
Edited by BERNARDO RUDY AND LINDA E. IVERSON

VOLUME 208. Protein–DNA Interactions
Edited by ROBERT T. SAUER

VOLUME 209. Phospholipid Biosynthesis
Edited by EDWARD A. DENNIS AND DENNIS E. VANCE

VOLUME 210. Numerical Computer Methods
Edited by LUDWIG BRAND AND MICHAEL L. JOHNSON

VOLUME 211. DNA Structures (Part A: Synthesis and Physical Analysis of DNA)
Edited by DAVID M. J. LILLEY AND JAMES E. DAHLBERG

VOLUME 212. DNA Structures (Part B: Chemical and Electrophoretic Analysis of DNA)
Edited by DAVID M. J. LILLEY AND JAMES E. DAHLBERG

VOLUME 213. Carotenoids (Part A: Chemistry, Separation, Quantitation, and Antioxidation)
Edited by LESTER PACKER

Volume 214. Carotenoids (Part B: Metabolism, Genetics, and Biosynthesis)
Edited by Lester Packer

Volume 215. Platelets: Receptors, Adhesion, Secretion (Part B)
Edited by Jacek J. Hawiger

Volume 216. Recombinant DNA (Part G)
Edited by Ray Wu

Volume 217. Recombinant DNA (Part H)
Edited by Ray Wu

Volume 218. Recombinant DNA (Part I)
Edited by Ray Wu

Volume 219. Reconstitution of Intracellular Transport
Edited by James E. Rothman

Volume 220. Membrane Fusion Techniques (Part A)
Edited by Nejat Düzguünes

Volume 221. Membrane Fusion Techniques (Part B)
Edited by Nejat Düzgünes

Volume 222. Proteolytic Enzymes in Coagulation, Fibrinolysis, and Complement Activation (Part A: Mammalian Blood Coagulation Factors and Inhibitors)
Edited by Laszlo Lorand and Kenneth G. Mann

Volume 223. Proteolytic Enzymes in Coagulation, Fibrinolysis, and Complement Activation (Part B: Complement Activation, Fibrinolysis, and Nonmammalian Blood Coagulation Factors)
Edited by Laszlo Lorand and Kenneth G. Mann

Volume 224. Molecular Evolution: Producing the Biochemical Data
Edited by Elizabeth Anne Zimmer, Thomas J. White, Rebecca L. Cann, and Allan C. Wilson

Volume 225. Guide to Techniques in Mouse Development
Edited by Paul M. Wassarman and Melvin L. DePamphilis

Volume 226. Metallobiochemistry (Part C: Spectroscopic and Physical Methods for Probing Metal Ion Environments in Metalloenzymes and Metalloproteins)
Edited by James F. Riordan and Bert L. Vallee

Volume 227. Metallobiochemistry (Part D: Physical and Spectroscopic Methods for Probing Metal Ion Environments in Metalloproteins)
Edited by James F. Riordan and Bert L. Vallee

Volume 228. Aqueous Two-Phase Systems
Edited by Harry Walter and Göte Johansson

Volume 229. Cumulative Subject Index Volumes 195–198, 200–227

Volume 230. Guide to Techniques in Glycobiology
Edited by William J. Lennarz and Gerald W. Hart

VOLUME 231. Hemoglobins (Part B: Biochemical and Analytical Methods)
Edited by JOHANNES EVERSE, KIM D. VANDEGRIFF, AND ROBERT M. WINSLOW

VOLUME 232. Hemoglobins (Part C: Biophysical Methods)
Edited by JOHANNES EVERSE, KIM D. VANDEGRIFF, AND ROBERT M. WINSLOW

VOLUME 233. Oxygen Radicals in Biological Systems (Part C)
Edited by LESTER PACKER

VOLUME 234. Oxygen Radicals in Biological Systems (Part D)
Edited by LESTER PACKER

VOLUME 235. Bacterial Pathogenesis (Part A: Identification and Regulation of Virulence Factors)
Edited by VIRGINIA L. CLARK AND PATRIK M. BAVOIL

VOLUME 236. Bacterial Pathogenesis (Part B: Integration of Pathogenic Bacteria with Host Cells)
Edited by VIRGINIA L. CLARK AND PATRIK M. BAVOIL

VOLUME 237. Heterotrimeric G Proteins
Edited by RAVI IYENGAR

VOLUME 238. Heterotrimeric G-Protein Effectors
Edited by RAVI IYENGAR

VOLUME 239. Nuclear Magnetic Resonance (Part C)
Edited by THOMAS L. JAMES AND NORMAN J. OPPENHEIMER

VOLUME 240. Numerical Computer Methods (Part B)
Edited by MICHAEL L. JOHNSON AND LUDWIG BRAND

VOLUME 241. Retroviral Proteases
Edited by LAWRENCE C. KUO AND JULES A. SHAFER

VOLUME 242. Neoglycoconjugates (Part A)
Edited by Y. C. LEE AND REIKO T. LEE

VOLUME 243. Inorganic Microbial Sulfur Metabolism
Edited by HARRY D. PECK, JR., AND JEAN LEGALL

VOLUME 244. Proteolytic Enzymes: Serine and Cysteine Peptidases
Edited by ALAN J. BARRETT

VOLUME 245. Extracellular Matrix Components
Edited by E. RUOSLAHTI AND E. ENGVALL

VOLUME 246. Biochemical Spectroscopy
Edited by KENNETH SAUER

VOLUME 247. Neoglycoconjugates (Part B: Biomedical Applications)
Edited by Y. C. LEE AND REIKO T. LEE

VOLUME 248. Proteolytic Enzymes: Aspartic and Metallo Peptidases
Edited by ALAN J. BARRETT

VOLUME 249. Enzyme Kinetics and Mechanism (Part D: Developments in Enzyme Dynamics)
Edited by DANIEL L. PURICH

VOLUME 250. Lipid Modifications of Proteins
Edited by PATRICK J. CASEY AND JANICE E. BUSS

VOLUME 251. Biothiols (Part A: Monothiols and Dithiols, Protein Thiols, and Thiyl Radicals)
Edited by LESTER PACKER

VOLUME 252. Biothiols (Part B: Glutathione and Thioredoxin; Thiols in Signal Transduction and Gene Regulation)
Edited by LESTER PACKER

VOLUME 253. Adhesion of Microbial Pathogens
Edited by RON J. DOYLE AND ITZHAK OFEK

VOLUME 254. Oncogene Techniques
Edited by PETER K. VOGT AND INDER M. VERMA

VOLUME 255. Small GTPases and Their Regulators (Part A: Ras Family)
Edited by W. E. BALCH, CHANNING J. DER, AND ALAN HALL

VOLUME 256. Small GTPases and Their Regulators (Part B: Rho Family)
Edited by W. E. BALCH, CHANNING J. DER, AND ALAN HALL

VOLUME 257. Small GTPases and Their Regulators (Part C: Proteins Involved in Transport)
Edited by W. E. BALCH, CHANNING J. DER, AND ALAN HALL

VOLUME 258. Redox-Active Amino Acids in Biology
Edited by JUDITH P. KLINMAN

VOLUME 259. Energetics of Biological Macromolecules
Edited by MICHAEL L. JOHNSON AND GARY K. ACKERS

VOLUME 260. Mitochondrial Biogenesis and Genetics (Part A)
Edited by GIUSEPPE M. ATTARDI AND ANNE CHOMYN

VOLUME 261. Nuclear Magnetic Resonance and Nucleic Acids
Edited by THOMAS L. JAMES

VOLUME 262. DNA Replication
Edited by JUDITH L. CAMPBELL

VOLUME 263. Plasma Lipoproteins (Part C: Quantitation)
Edited by WILLIAM A. BRADLEY, SANDRA H. GIANTURCO, AND JERE P. SEGREST

VOLUME 264. Mitochondrial Biogenesis and Genetics (Part B)
Edited by GIUSEPPE M. ATTARDI AND ANNE CHOMYN

VOLUME 265. Cumulative Subject Index Volumes 228, 230–262

VOLUME 266. Computer Methods for Macromolecular Sequence Analysis
Edited by RUSSELL F. DOOLITTLE

VOLUME 267. Combinatorial Chemistry
Edited by JOHN N. ABELSON

VOLUME 268. Nitric Oxide (Part A: Sources and Detection of NO; NO Synthase)
Edited by LESTER PACKER

VOLUME 269. Nitric Oxide (Part B: Physiological and Pathological Processes)
Edited by LESTER PACKER

VOLUME 270. High Resolution Separation and Analysis of Biological Macromolecules (Part A: Fundamentals)
Edited by BARRY L. KARGER AND WILLIAM S. HANCOCK

VOLUME 271. High Resolution Separation and Analysis of Biological Macromolecules (Part B: Applications)
Edited by BARRY L. KARGER AND WILLIAM S. HANCOCK

VOLUME 272. Cytochrome P450 (Part B)
Edited by ERIC F. JOHNSON AND MICHAEL R. WATERMAN

VOLUME 273. RNA Polymerase and Associated Factors (Part A)
Edited by SANKAR ADHYA

VOLUME 274. RNA Polymerase and Associated Factors (Part B)
Edited by SANKAR ADHYA

VOLUME 275. Viral Polymerases and Related Proteins
Edited by LAWRENCE C. KUO, DAVID B. OLSEN, AND STEVEN S. CARROLL

VOLUME 276. Macromolecular Crystallography (Part A)
Edited by CHARLES W. CARTER, JR., AND ROBERT M. SWEET

VOLUME 277. Macromolecular Crystallography (Part B)
Edited by CHARLES W. CARTER, JR., AND ROBERT M. SWEET

VOLUME 278. Fluorescence Spectroscopy
Edited by LUDWIG BRAND AND MICHAEL L. JOHNSON

VOLUME 279. Vitamins and Coenzymes (Part I)
Edited by DONALD B. MCCORMICK, JOHN W. SUTTIE, AND CONRAD WAGNER

VOLUME 280. Vitamins and Coenzymes (Part J)
Edited by DONALD B. MCCORMICK, JOHN W. SUTTIE, AND CONRAD WAGNER

VOLUME 281. Vitamins and Coenzymes (Part K)
Edited by DONALD B. MCCORMICK, JOHN W. SUTTIE, AND CONRAD WAGNER

VOLUME 282. Vitamins and Coenzymes (Part L)
Edited by DONALD B. MCCORMICK, JOHN W. SUTTIE, AND CONRAD WAGNER

VOLUME 283. Cell Cycle Control
Edited by WILLIAM G. DUNPHY

VOLUME 284. Lipases (Part A: Biotechnology)
Edited by BYRON RUBIN AND EDWARD A. DENNIS

VOLUME 285. Cumulative Subject Index Volumes 263, 264, 266–284, 286–289

VOLUME 286. Lipases (Part B: Enzyme Characterization and Utilization)
Edited by BYRON RUBIN AND EDWARD A. DENNIS

VOLUME 287. Chemokines
Edited by RICHARD HORUK

VOLUME 288. Chemokine Receptors
Edited by RICHARD HORUK

VOLUME 289. Solid Phase Peptide Synthesis
Edited by GREGG B. FIELDS

VOLUME 290. Molecular Chaperones
Edited by GEORGE H. LORIMER AND THOMAS BALDWIN

VOLUME 291. Caged Compounds
Edited by GERARD MARRIOTT

VOLUME 292. ABC Transporters: Biochemical, Cellular, and Molecular Aspects
Edited by SURESH V. AMBUDKAR AND MICHAEL M. GOTTESMAN

VOLUME 293. Ion Channels (Part B)
Edited by P. MICHAEL CONN

VOLUME 294. Ion Channels (Part C)
Edited by P. MICHAEL CONN

VOLUME 295. Energetics of Biological Macromolecules (Part B)
Edited by GARY K. ACKERS AND MICHAEL L. JOHNSON

VOLUME 296. Neurotransmitter Transporters
Edited by SUSAN G. AMARA

VOLUME 297. Photosynthesis: Molecular Biology of Energy Capture
Edited by LEE MCINTOSH

VOLUME 298. Molecular Motors and the Cytoskeleton (Part B)
Edited by RICHARD B. VALLEE

VOLUME 299. Oxidants and Antioxidants (Part A)
Edited by LESTER PACKER

VOLUME 300. Oxidants and Antioxidants (Part B)
Edited by LESTER PACKER

VOLUME 301. Nitric Oxide: Biological and Antioxidant Activities (Part C)
Edited by LESTER PACKER

VOLUME 302. Green Fluorescent Protein
Edited by P. MICHAEL CONN

VOLUME 303. cDNA Preparation and Display
Edited by SHERMAN M. WEISSMAN

VOLUME 304. Chromatin
Edited by PAUL M. WASSARMAN AND ALAN P. WOLFFE

VOLUME 305. Bioluminescence and Chemiluminescence (Part C)
Edited by THOMAS O. BALDWIN AND MIRIAM M. ZIEGLER

VOLUME 306. Expression of Recombinant Genes in Eukaryotic Systems
Edited by JOSEPH C. GLORIOSO AND MARTIN C. SCHMIDT

VOLUME 307. Confocal Microscopy
Edited by P. MICHAEL CONN

VOLUME 308. Enzyme Kinetics and Mechanism (Part E: Energetics of Enzyme Catalysis)
Edited by DANIEL L. PURICH AND VERN L. SCHRAMM

VOLUME 309. Amyloid, Prions, and Other Protein Aggregates
Edited by RONALD WETZEL

VOLUME 310. Biofilms
Edited by RON J. DOYLE

VOLUME 311. Sphingolipid Metabolism and Cell Signaling (Part A)
Edited by ALFRED H. MERRILL, JR., AND YUSUF A. HANNUN

VOLUME 312. Sphingolipid Metabolism and Cell Signaling (Part B)
Edited by ALFRED H. MERRILL, JR., AND YUSUF A. HANNUN

VOLUME 313. Antisense Technology (Part A: General Methods, Methods of Delivery, and RNA Studies)
Edited by M. IAN PHILLIPS

VOLUME 314. Antisense Technology (Part B: Applications)
Edited by M. IAN PHILLIPS

VOLUME 315. Vertebrate Phototransduction and the Visual Cycle (Part A)
Edited by KRZYSZTOF PALCZEWSKI

VOLUME 316. Vertebrate Phototransduction and the Visual Cycle (Part B)
Edited by KRZYSZTOF PALCZEWSKI

VOLUME 317. RNA–Ligand Interactions (Part A: Structural Biology Methods)
Edited by DANIEL W. CELANDER AND JOHN N. ABELSON

VOLUME 318. RNA–Ligand Interactions (Part B: Molecular Biology Methods)
Edited by DANIEL W. CELANDER AND JOHN N. ABELSON

VOLUME 319. Singlet Oxygen, UV-A, and Ozone
Edited by LESTER PACKER AND HELMUT SIES

VOLUME 320. Cumulative Subject Index Volumes 290–319

VOLUME 321. Numerical Computer Methods (Part C)
Edited by MICHAEL L. JOHNSON AND LUDWIG BRAND

VOLUME 322. Apoptosis
Edited by JOHN C. REED

VOLUME 323. Energetics of Biological Macromolecules (Part C)
Edited by MICHAEL L. JOHNSON AND GARY K. ACKERS

VOLUME 324. Branched-Chain Amino Acids (Part B)
Edited by ROBERT A. HARRIS AND JOHN R. SOKATCH

VOLUME 325. Regulators and Effectors of Small GTPases (Part D: Rho Family)
Edited by W. E. BALCH, CHANNING J. DER, AND ALAN HALL

VOLUME 326. Applications of Chimeric Genes and Hybrid Proteins (Part A: Gene Expression and Protein Purification)
Edited by JEREMY THORNER, SCOTT D. EMR, AND JOHN N. ABELSON

VOLUME 327. Applications of Chimeric Genes and Hybrid Proteins (Part B: Cell Biology and Physiology)
Edited by JEREMY THORNER, SCOTT D. EMR, AND JOHN N. ABELSON

VOLUME 328. Applications of Chimeric Genes and Hybrid Proteins (Part C: Protein-Protein Interactions and Genomics)
Edited by JEREMY THORNER, SCOTT D. EMR, AND JOHN N. ABELSON

VOLUME 329. Regulators and Effectors of Small GTPases (Part E: GTPases Involved in Vesicular Traffic)
Edited by W. E. BALCH, CHANNING J. DER, AND ALAN HALL

VOLUME 330. Hyperthermophilic Enzymes (Part A)
Edited by MICHAEL W. W. ADAMS AND ROBERT M. KELLY

VOLUME 331. Hyperthermophilic Enzymes (Part B)
Edited by MICHAEL W. W. ADAMS AND ROBERT M. KELLY

VOLUME 332. Regulators and Effectors of Small GTPases (Part F: Ras Family I)
Edited by W. E. BALCH, CHANNING J. DER, AND ALAN HALL

VOLUME 333. Regulators and Effectors of Small GTPases (Part G: Ras Family II)
Edited by W. E. BALCH, CHANNING J. DER, AND ALAN HALL

VOLUME 334. Hyperthermophilic Enzymes (Part C)
Edited by MICHAEL W. W. ADAMS AND ROBERT M. KELLY

VOLUME 335. Flavonoids and Other Polyphenols (in preparation)
Edited by LESTER PACKER

VOLUME 336. Microbial Growth in Biofilms (Part A: Developmental and Molecular Biological Aspects) (in preparation)
Edited by RON J. DOYLE

VOLUME 337. Microbial Growth in Biofilms (Part B: Special Environments and Physicochemical Aspects) (in preparation)
Edited by RON J. DOYLE

VOLUME 338. Nuclear Magnetic Resonance of Biological Macromolecules (Part A) (in preparation)
Edited by THOMAS L. JAMES, VOLKER DÖTSCH, AND ULI SCHMITZ

VOLUME 339. Nuclear Magnetic Resonance of Biological Macromolecules (Part B) (in preparation)
Edited by THOMAS L. JAMES, VOLKER DÖTSCH, AND ULI SCHMITZ

VOLUME 340. Drug-Nucleic Acid Interactions (in preparation)
Edited by JONATHAN B. CHAIRES AND MICHAEL J. WARING

VOLUME 341. Ribonucleases (Part A) (in preparation)
Edited by ALLEN W. NICHOLSON

VOLUME 342. Ribonucleases (Part B) (in preparation)
Edited by ALLEN W. NICHOLSON

VOLUME 343. G Protein Pathways (Part A: Receptors) (in preparation)
Edited by RAVI IYENGAR AND JOHN D. HILDEBRANDT

VOLUME 344. G Protein Pathways (Part B: G Proteins and Their Regulators) (in preparation)
Edited by RAVI IYENGAR AND JOHN D. HILDEBRANDT

VOLUME 345. G Protein Pathways (Part C: Effector Mechanisms) (in preparation)
Edited by RAVI IYENGAR AND JOHN D. HILDEBRANDT

Section I

Cytoplasmic and Nuclear Signaling Analyses

[1] Determining Involvement of Shc Proteins in Signaling Pathways

By JOHN P. O'BRYAN

Shc proteins are integral components in the action of a wide variety of receptors including receptor tyrosine kinases (RTKs), G protein-coupled receptors (GPCRs), immunoglobulin receptors, and integrins.[1] Activation of each of these receptors can lead to the recruitment of Shc proteins, culminating in their tyrosine phosphorylation. Activated, that is, tyrosine phosphorylated, Shc recruits the Grb2:Sos complex, which in turn activates the Ras signal transduction pathway through stimulation of nucleotide exchange on Ras. However, Shc proteins are also thought to possess additional functions.[1] Indeed, results suggest that Shc proteins may play an important role in the response of cells to oxidative stress and the initiation of apoptosis as a part of this response.[2] This finding coupled with the identification of multiple Shc family members, each with distinct expression patterns, suggests that this family of signaling proteins plays a central role in the function of many cell types.[3–6] In this chapter, several methods for examining the involvement of Shc proteins in various signaling pathways are discussed.

Overview of Shc Family Members

To date, three mammalian *Shc* genes have been identified: *ShcA, ShcB (Sck),* and *ShcC (N-Shc/Rai).*[3–6] All three *Shc* genes encode proteins that are highly related in sequence and structure, consisting of a carboxy-terminal Src homology 2 (SH2) domain, a central effector region rich in proline and glycine residues and containing two distinct sites for tyrosine phosphorylation (CH1), and an amino-terminal phosphotyrosine-binding (PTB) domain (Fig. 1). Although both the SH2

[1] L. Bonfini, E. Migliaccio, G. Pelicci, L. Lanfrancone, and P. G. Pelicci, *Trends Biochem. Sci.* **21,** 259 (1996).

[2] E. Migliaccio, M. Giorgio, S. Mele, G. Pelicci, P. Reboldi, P. P. Pandolfi, L. Lanfrancone, and P. G. Pelicci, *Nature* (*London*) **402,** 309 (1999).

[3] W. M. Kavanaugh and L. T. Williams, *Science* **266,** 1862 (1994).

[4] T. Nakamura, R. Sanokawa, Y. Sasaki, D. Ayusawa, M. Oishi, and N. Mori, *Oncogene* **13,** 1111 (1996).

[5] J. P. O'Bryan, Z. Songyang, L. Cantley, C. Der, and T. Pawson, *Proc. Natl. Acad. Sci. U.S.A.* **93,** 2729 (1996).

[6] G. Pelicci, L. Dente, A. De Giuseppe, B. Verducci-Galletti, S. Giuli, S. Mele, C. Vetriani, M. Giorgio, P. P. Pandolfi, G. Cesareni, and P. G. Pelicci, *Oncogene* **13,** 633 (1996).

0076-6879/00 $35.00

and PTB domains of all three family members are highly similar (68 and 78%, respectively), the central effector region (CH1) is less well conserved. There are, however, three regions of the CH1 domain that are highly conserved in mammalian Shc family members. First, the sequence Tyr-Val-Asn-(Thr/Ile/Val) is conserved in all three mammalian family members and represents a major site of tyrosine phosphorylation. Second, a more amino-terminal sequence of Tyr-Tyr-Asn-(Ser/Asp) also represents a prominent site of phosphorylation.[7] Interestingly, both sites conform to consensus Grb2-binding sites and, indeed, both bind Grb2 or Grb2-related family members. In contrast to the more carboxy-terminal tyrosine phosphorylation site, there are a number of additional amino acids surrounding the amino-terminal phosphorylation site that are also conserved between Shc family members, suggesting that these residues play an important role in Shc function through the recognition of effector proteins.[5] This notion is further strengthened by the fact that the amino-terminal tyrosine phosphorylation site is also conserved in *Drosophila* Shc.[8]

In addition to the well-conserved tyrosine phosphorylation sites, there is a third region of the CH1 domain conserved in all three mammalian Shc family members. This sequence, Asp-Leu-Phe-Asp-Met-(Lys/Arg)-Pro-Phe-Glu-Asp-Ala-Leu, has been mapped as the binding site for adaptins.[9] As their name suggests, members of this class of proteins function as adaptors that link the endocytic machinery of the clathrin-coated pit with integral membrane proteins.[10] Although this finding suggests a potential role for Shc proteins in endocytosis, there has not been any definitive proof of this hypothesis. Furthermore, this region is only weakly conserved in *Drosophila* Shc.[8]

Mammalian *Shc* genes encode a complex series of proteins. *ShcA* encodes three proteins termed $p46^{ShcA}$, $p52^{ShcA}$, and $p66^{ShcA}$ (Figs. 1 and 2). All three isoforms have a PTB domain, a CH1 domain, and an SH2 domain; however, the PTB domain in $p46^{ShcA}$ lacks a critical helix important for forming high-affinity contacts with the phosphopeptide ligand.[11] Thus, although the $p46^{ShcA}$ PTB domain does bind phosphopeptides, this truncated PTB domain appears to have a lower affinity for phosphopeptide ligand as compared with the PTB present in $p52^{ShcA}$.[12] $p66^{ShcA}$ possesses at the amino terminus an additional proline-rich extension that is thought

[7] P. van der Geer, S. Wiley, G. D. Gish, and T. Pawson, *Curr. Biol.* **6,** 1435 (1996).

[8] K.-M. V. Lai, J. P. Olivier, G. Gish, M. Henkemeyer, J. McGlade, T. Pawson, *Mol. Cell. Biol.* **15,** 4810 (1995).

[9] Y. Okabayashi, Y. Sugimoto, N. F. Totty, J. Hsuan, Y. Kido, K. Sakaguchi, I. Gout, M. D. Waterfield, and M. Kasuga, *J. Biol. Chem.* **271,** 5265 (1996).

[10] D. A. Lewin and I. Mellman, *Biochim. Biophys. Acta* **1401,** 129 (1998).

[11] M. M. Zhou, K. S. Ravichandran, E. T. Olejniczak, A. M. Petros, R. P. Meadows, M. Sattler, J. E. Harlan, W. S. Wade, S. J. Burakoff, and S. W. Fesik, *Nature (London)* **378,** 584 (1995).

[12] M. M. Zhou, J. E. Harlan, W. S. Wade, S. Crosby, K. S. Ravichandran, S. J. Burakoff, and S. W. Fesik, *J. Biol. Chem.* **270,** 31119 (1995).

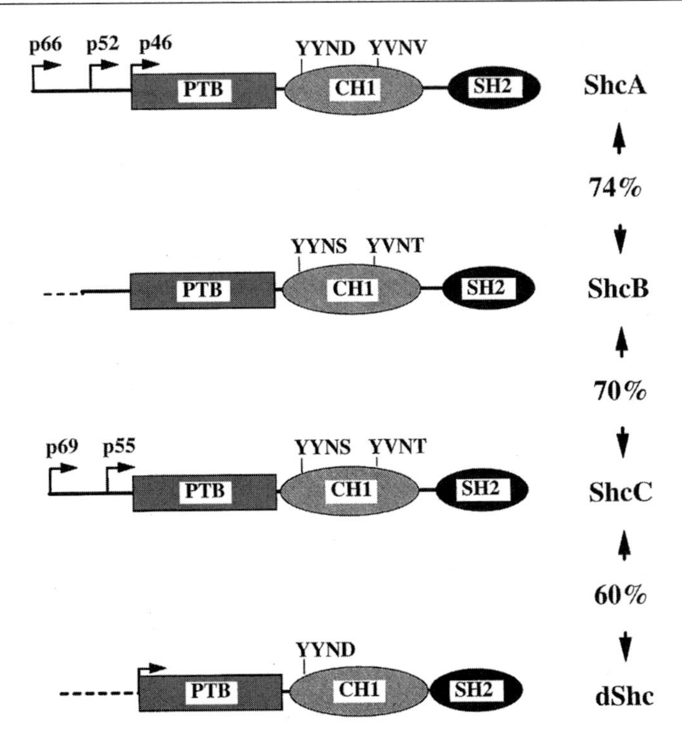

FIG. 1. Shc family of proteins. The percent similarity of the various family members is indicated. dShc, *Drosophila* Shc.[8]

to bind proteins containing Src homology 3 (SH3) domains.[13] Whether the presence of this extension alters the affinity of the PTB domain for tyrosine phosphorylated substrates is not known; however, $p66^{ShcA}$ does complex with the activated epidermal growth factor receptor (EGFR) after growth factor stimulation.[13,14] Although most data indicate that $p46^{ShcA}$ and $p52^{ShcA}$ are involved in activation of the Ras–MAPK (mitogen-activated protein kinase) signal transduction pathway, evidence suggests that $p66^{ShcA}$ may play an antagonistic role in the regulation of Ras activation.[13,14] In addition, targeted deletion of $p66^{ShcA}$ indicates that this isoform is important in the response of cells to oxidative stress.[2]

Similar to *ShcA,* the *ShcC* gene encodes multiple protein isoforms termed $p55^{ShcC}$ and $p69^{ShcC}$. There is no $p46^{ShcA}$ equivalent because the internal initiating

[13] E. Migliaccio, S. Mele, A. Salcini, G. Pelicci, K.-M. V. Lai, G. Superti-Furga, T. Pawson, P. P. Di Fiore, L. Lanfrancone, and P. G. Pelicci, *EMBO J.* **16,** 706 (1997).

[14] S. Okada, A. W. Kao, B. P. Ceresa, P. Blaikie, B. Margolis, and J. E. Pessin, *J. Biol. Chem.* **272,** 28042 (1997).

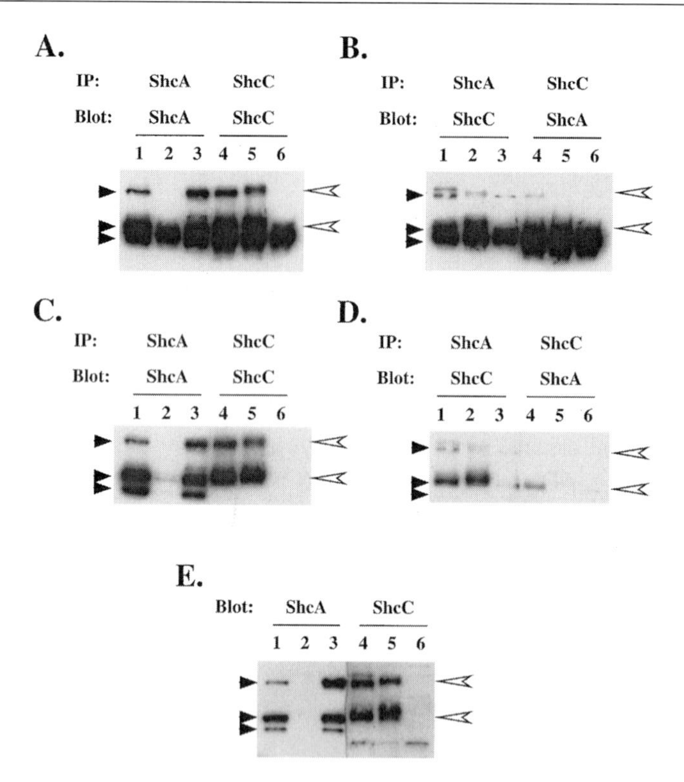

FIG. 2. Specificity of antibodies to Shc family of proteins. (A–D) Lysates (500 μg) from PFSK cells (lanes 1 and 4), newborn mouse brains (lanes 2 and 5), or A673 cells (lanes 3 and 6) were immunoprecipitated with either a commercially available ShcA antibody (Upstate Biotechnology; 5 μg per immunoprecipitation) or an ShcC antibody made in the laboratory of the author (2.5 μl of serum per immunoprecipitation). Immunoprecipitates were fractionated on an SDS–8% (w/v) polyacrylamide gel (Novex, San Diego, CA) and then analyzed by Western blot with the same antibodies. (A and B) Filters were probed with ShcA antibody (1 μg/ml) or ShcC antibody (1:500) as indicated, using anti-rabbit Ig–HRP as the detection reagent. Solid arrows denote the three ShcA isoforms. Open arrows denote the two ShcC isoforms. (C and D) Filters from (A) and (B) were stripped and reprobed with primary antibodies as described above but using protein A–HRP as the detection reagent. (E) Western blot of 25 μg of lysate from PFSK (lanes 1 and 4), newborn mouse brains (lanes 2 and 5), and A673 (lanes 3 and 6), probed as described in (C) and (D) with either ShcA or ShcC antibodies.

methionine present in ShcA is not conserved in ShcC. The two isoforms of ShcC appear to be equivalent to $p52^{ShcA}$ and $p66^{ShcA}$. In contrast to ShcA, which is widely expressed, ShcC expression is restricted to the brain.[4–6] As with ShcA, ShcC is thought to regulate the Ras–MAPK pathway.[4–6] The ability of $p69^{ShcC}$ to regulate stress-induced pathways has not as yet been investigated.

ShcB is another Shc family member similar in structure to both ShcA and ShcC (Fig. 1). Expression analysis of *ShcB* suggests that like *ShcC*, *ShcB* is more restricted in expression than *ShcA*[3,6,15]; however, analysis of the ShcB protein products has shown them to be elusive. To date there are no reports detailing the expression of ShcB proteins.

The absence of any enzymatic domain in Shc proteins has led to the classification of this family of proteins as scaffolding or adaptor proteins. Thus, the function of Shc proteins is to assemble multimeric protein complexes that in turn will stimulate signaling pathways such as the Ras signal transduction pathway.[16] As mentioned above, involvement of Shc in a signal transduction pathway has been inferred by the finding that Shc proteins become tyrosine phosphorylated after activation of that specific pathway. In addition, the use of dominant interfering Shc mutants as well as neutralizing Shc antibodies has led to the conclusion that Shc proteins function in distinct signaling pathways.

Analysis of Shc Proteins: Tyrosine Phosphorylation

To assess the involvement of Shc family members in a particular signaling pathway, it is necessary to ascertain whether Shc becomes tyrosine phosphorylated during activation of that pathway. This chapter focuses on the analysis of Shc in RTK signaling pathways, using the EGFR as an example. Although the same approach may be applied to other receptor-activated pathways, it is best to keep in mind that different parameters may need to be examined to ascertain whether Shc proteins are involved in a particular pathway. For example, length of stimulation with a growth factor/hormone, length of serum starvation, cell type, and so on, will need to be tested empirically.

In the case of EGF signaling, NIH 3T3 murine fibroblast cells as well as 293T human embryonic kidney cells have been used as model systems for analyzing Shc involvement. NIH 3T3 cells are an immortalized murine fibroblast cell line that has been previously described for use in signal transduction experiments.[17] 293T cells are human embryonic kidney epithelial cells that have been transformed by simian virus 40 (SV40) large T antigen, thereby allowing high-level expression of genes cloned into mammalian expression vectors that contain an SV40 origin of replication.[18] In addition, these cells are highly transfectable by the calcium phosphate precipitation method, as described in Dominant-Negative Shc Proteins (below). 293T cells are maintained in Dulbecco's modified Eagle's medium (DMEM) containing D-glucose (4500 mg/liter), $NaHCO_3$ (3.7 g/liter), and sodium pyruvate (110 mg/liter) supplemented with 10% (v/v) heat-inactivated fetal bovine

[15] T. Nakamura, S. Muraoka, R. Sanokawa, and N. Mori, *J. Biol. Chem.* **273,** 6960 (1998).

[16] M. Rozakis-Adcock, J. McGlade, G. Mbamalu, G. Pelicci, R. Daly, W. Li, A. Batzer, S. Thomas, J. Brugge, P. G. Pelicci, J. Schlessinger, and T. Pawson, *Nature (London)* **360,** 689 (1992).

[17] G. Clark, A. D. Cox, S. M. Graham, and C. J. Der, *Methods Enzymol.* **255,** 395 (1995).

[18] W. S. Pear, G. P. Nolan, M. L. Scott, and D. Baltimore, *Proc. Natl. Acad. Sci. U.S.A.* **90,** 8392 (1993).

serum. Medium may also be supplemented with penicillin (100 U/ml) and streptomycin (100 μg/ml) to prevent bacterial contamination. Cells are maintained at 37° in 10% CO_2. NIH 3T3 cells are grown in the same medium, except that 10% (v/v) calf serum is used in place of fetal bovine serum. We have found that calf serum obtained from Colorado Serum Company (Denver, CO) provides the best culturing conditions for NIH 3T3 cells.

Preparation of Cell Lysates for Analysis

1. If cells are to be transfected before analysis, see Dominant-Negative Shc Proteins (below) for transfection methods. For analysis of cell lines after stimulation, proceed to step 2.
2. Plate an equivalent numbers of cells ($1\text{–}5 \times 10^6$ cells) on 100-mm tissue culture plates and grow the cells to approximately 50–80% confluence.
3. Serum starve the cells overnight. In the case of NIH 3T3 cells, place the cells in medium (9 ml) containing 0.5% (v/v) calf serum and starve for no longer than 16 hr; otherwise, the cells begin to die. In the case of 293T cells, place the cells in medium (9 ml) lacking serum for at least 16 hr. Longer periods of starvation have been used for 293T cells; however, the cells appear to become less adherent with increased starvation.
4. On the next day, stimulate the cells with EGF at 100 ng/ml (100 μg/ml solution in 100 m*M* acetic acid stored at −20°; Upstate Biotechnology, Lake Placid, NY) for 5 min at 37°. The length of treatment should be determined empirically for a given stimulus. In the case of EGF, maximal Shc phosphorylation occurs within 1–5 min of growth factor addition.
5. Remove the medium and wash the cells once with phosphate-buffered saline (137 m*M* NaCl, 2.7 m*M* KCl, 8.1 m*M* Na_2HPO_4, 1.5 m*M* $KHPO_4$, pH 7.4). Normally, ice-cold phosphate-buffered saline (PBS) is used to wash the cells before lysis; however, 293T cells will detach from the tissue culture plates at temperatures below 37°. Therefore, when using 293T cells, the PBS must be warmed to 37° before washing the plates to prevent cell loss. In addition, caution must be taken not to dislodge cells with the PBS when washing because 293T cells are loosely adherent. For 100-mm plates, lyse the cells in 1 ml of ice-cold PLC-LB [50 m*M* HEPES (pH 7.5), 150 m*M* NaCl, 10% (v/v) glycerol, 1% (v/v) Triton X-100, 1 m*M* EGTA, 1.5 m*M* magnesium chloride, 100 m*M* sodium fluoride supplemented with 1 m*M* sodium orthovanadate, and appropriate protease inhibitors]. The preparation and use of vanadate as an inhibitor of protein tyrosine phosphatases have been previously described.[19] Transfer the lysates to fresh 1.5-ml centrifuge tubes.

[19] J. A. Gordon, *Methods Enzymol.* **201,** 477 (1991).

6. Incubate the lysates at 4° with gentle mixing on a nutator. Centrifuge the lysates for 10 min at 14,000 rpm [20,000 *g* in an Eppendorf (Hamburg, Germany) model 5417 microcentrifuge] at 4° to remove insoluble debris. Transfer the lysates to fresh 1.5-ml centrifuge tubes and determine protein concentrations by standard techniques.

To determine whether Shc proteins are involved in the signaling pathway of interest, the tyrosine phosphorylation status of the protein is determined. This analysis may be accomplished in several different but complementary approaches. We routinely immunoprecipitate total Shc proteins and then the immunoprecipitated proteins are subjected to Western blot analysis with anti-phosphotyrosine antibodies to determine the levels of tyrosine phosphorylation. Alternatively, following metabolic labeling of cells with [^{32}P]-orthophosphate, Shc proteins can be immunoprecipitated and analyzed by autoradiography for the incorporation of radiolabel.

Immunoprecipitation of Shc Proteins

1. Aliquot equivalent amounts of protein from each cell lysate to 1.5-ml centrifuge tubes and bring to equal volume with PLC-LB as described above. We routinely use 0.5–2 mg of protein in a volume of 0.5–1 ml for each sample.

2. Samples can be precleared by adding 25–50 μl of protein A– or protein G–agarose beads (Sigma, St. Louis, MO) and incubating at 4° with gentle mixing for 30–60 min. Spin out the beads and transfer the lysates to fresh tubes.

3. Add primary antibody and incubate for 1 hr at 4° as described above. For immunoprecipitation of Shc proteins, it is necessary to decide which Shc family members will be examined. There are a number of commercially available antibodies to ShcA and ShcC as listed in Table I. Each antibody has different cross-reactivities with other Shc family members (see Table I and comments below). Thus, caution must be exercised in the interpretation of results with a given antibody. For immunoprecipitation of tyrosine-phosphorylated proteins, we routinely use PY20 anti-phosphotyrosine antibody (Transduction Laboratories, Lexington, KY).

4. Add secondary affinity reagent and incubate for an additional 1 hr at 4° with gentle mixing as described above. There are a number of secondary affinity reagents that may be used at this step, including anti-immunoglobulin, protein A, or protein G linked to agarose or Sepharose beads. Protein A and protein G beads are routinely used for Shc and phosphotyrosine immunoprecipitates, respectively.

5. Pellet the immune complexes by centrifugation at 14,000 rpm for 3 min at 4°. Rinse the complexes extensively to remove residual lysate from the pellet. When using lysates from 293T cells transfected with SV40-based expression vectors, it

TABLE I
COMMERCIALLY AVAILABLE ANTIBODIES TO Shc FAMILY MEMBERS

Manufacturer	Antibody source	Antigen	Cross-reactivity[a,b]
Upstate Biotechnology	Protein A–purified polyclonal	hShcA (aa 366–473)[c]	ShcC[d]
Transduction Laboratories	Rabbit polyclonal	hShcA (aa 359–473)	?
	Mouse monoclonal IgG_1	hShcA (aa 359–473)	?
	Mouse monoclonal IgG_1	mShcC (aa 239–374)	?
Santa Cruz Biotechnology	Rabbit polyclonal	hShcA (aa 454–473)	ShcC
	Mouse monoclonal IgG_1	hShcA (aa 366–473)	—
	Rabbit polyclonal	hShcA (aa 366–473)	ShcC
	Goat polyclonal	hShcA (aa 2–20)	—
	Goat polyclonal	hShcC (aa 2–20)	—

[a] Cross-reactivity with different Shc family members was provided by the manufacturer.
[b] Cross-reactivity with ShcB is not known. However, given the difficulties encountered in detecting ShcB, reactivity with these antibodies in not likely to be due to ShcB.
[c] h, Human; m, mouse.
[d] Although not listed by the manufacturer, this antibody does cross-react with ShcC both in Western blot analysis and immunoprecipitations (Fig. 2).

is important to wash the immune complex at least three times with 0.5–1 ml of PLC-LB per wash. Because these cells vastly overexpress proteins from SV40-based expression vectors, care must be taken to remove any residual lysate that may result in a signal in the Western blot analysis. We have found that even with three washes, there may still be a signal due to residual lysate remaining in the immune complex if the protein is highly overexpressed. Thus, we recommend extensive washing.

6. Remove residual supernatant from the immune complexes and then resuspend the beads in 50 μl of 2× Laemmli sample buffer. Boil the sample for 10 min, cool on ice, and then fractionate half of the sample on a sodium dodecyl sulfate (SDS)–8% (w/v) polyacrylamide gel.

7. Transfer the proteins to Immobilon-P filters (Millipore, Danvers, MA). After transfer, rinse each membrane briefly in H_2O or TBST [137 m*M* NaCl, 2.7 m*M* KCl, 25 m*M* Tris-HCl (pH 7.4), 0.1% (v/v) Tween 20]. Block the filters in TBST supplemented with either 3% (w/v) nonfat dry milk if using Shc antibodies or with 3% (w/v) bovine serum albumin if using anti-phosphotyrosine antibodies in the Western blot analysis.

8. Add primary antibody and incubate at room temperature for 1 hr or overnight at 4°. Rinse three times with TBST (5 min for each wash). Add secondary antibody and incubate for an additional 1 hr at room temperature. We routinely use horseradish peroxidase linked to immunoglobulin (either anti-mouse or anti-rabbit;

Amersham-Pharmacia, Piscataway, NJ). Because p46^{ShcA}, p52^{ShcA}, and p55^{ShcC} migrate close to the heavy chain of the antibody used in the immunoprecipitation, it is often difficult to distinguish the signal from these isoforms versus the signal from the immunoglobulin heavy chain background (Fig. 2A). An alternative approach to decrease the heavy chain signal is to use protein A as the detection reagent. This reagent can be purchased linked to a number of different compounds for visualization [e.g., ^{125}I, horseradish peroxidase (HRP), alkaline phosphatase (AP), biotin]. We have successfully used protein A–HRP to visualize Shc in immunoprecipitation–Western experiments (Fig. 2C–E).

9. Rinse the blots five times with TBST (at least 5 min per wash). The HRP-labeled secondary antibody is detected with one of a number of commercially available reagents such as ECL (enhanced chemiluminescence) reagents (Amersham, Arlington Heights, IL) or SuperSignal reagents (Pierce, Rockford, IL).

Comments. The available antibodies to different Shc family members exhibit different degrees of cross-reactivity (Fig. 2 and Table I). Thus, it is important to be able to distinguish when a positive signal with a particular antibody is due to the specific Shc family member of interest or rather is due to cross-reactivity of that antibody with another Shc family member. This point is illustrated in the immunoprecipitation and Western blot analyses shown in Fig. 2. An ShcC antibody developed in the laboratory of the author, along with a commercially available ShcA antibody from Upstate Biotechnology, have been used to test the specificities of the two antibodies. Both antibodies were raised against the highly conserved SH2 domain of the respective Shc family member. Although both antibodies work well in immunoprecipitation and Western blot analysis (Fig. 2), the ShcA antibody exhibits a significant degree of cross-reactivity with ShcC in immunoprecipitation experiments. For example, newborn mouse brain lysates contain abundant ShcC proteins but low or undetectable ShcA (Fig. 2E, compare lanes 2 and 5). However, when an ShcA immunoprecipitate of a mouse brain lysate is blotted with the ShcC antibody, there is a significant signal, suggesting that the ShcA antibody recognizes both ShcA and ShcC (Fig. 2D, lane 2). In contrast, if an ShcC immunoprecipitate from A673 cells, which lack ShcC expression but have abundant ShcA (Fig. 2E, lanes 3 and 6), is blotted with ShcA antibodies, little or no signal is detected, indicating that the ShcC antibody does not cross-react with ShcA in immunoprecipitations (Fig. 2D, lane 6).

The specificities of the Shc antibodies are further illustrated by results from immunoprecipitation and Western blot analysis of a cell line that expresses both ShcA and ShcC (PFSK; Fig. 2E, lanes 1 and 4). When an ShcA immunoprecipitate is blotted with ShcC antibodies, only p55^{ShcC} and p69^{ShcA} are detected. If these signals were due to cross-reactivity of the ShcC antibody with ShcA in the Western blot analysis, then p46^{ShcA} would have been detected and the ShcC Western blot of the ShcA immunoprecipitate from A673 cells (Fig. 2D, lane 3) would have been positive. Thus, these results indicate that the ShcC antibody is specific for ShcC whereas the ShcA antibody cross-reacts with ShcC, at least in immunoprecipitation experiments.

Also illustrated in Fig. 2 is the difference in results obtained when using anti-rabbit HRP (Fig. 2A and B) versus protein A–HRP (Fig. 2C and D) as the secondary detection reagent. In Fig. 2A and B, the heavy chain band obscures $p46^{ShcA}$, $p52^{ShcA}$, and $p55^{ShcC}$. On shorter exposures, however, the individual Shc isoforms can be distinguished from the immunoglobulin heavy chain (data not shown). The results with protein A–HRP are much cleaner. As seen in Fig. 2C and D, there is no heavy chain signal, thus allowing for clear detection of all the Shc isoforms.

Dominant-Negative Shc Proteins

Dominant-negative proteins are mutant versions of a protein that interfere with the normal function of that protein in a dominant manner.[20] The classic example of a dominant-negative protein is the Asn-17 mutant of the Ras GTPase.[21] This mutant Ras allele prevents association of the endogenous protein with its guanine nucleotide exchange factor, thereby precluding activation of Ras. Thus, another approach for implicating Shc proteins in a particular signaling pathway is to utilize Shc dominant-negative mutants. The rationale behind this approach is that the mutant Shc protein will inhibit the signaling pathway though one of two mechanisms: (1) disrupting the interaction of the receptor or upstream activator with endogenous Shc proteins, or (2) blocking the interaction of Shc proteins with their downstream target(s). These two mechanisms are not necessarily mutually exclusive. Numerous groups have utilized this approach to implicate Shc proteins in the action of a variety of proteins (see Table II on page 356).[22–38] In each case, a particular mutant Shc

[20] I. Herskowitz, *Nature* (*London*) **329,** 219 (1987).

[21] L. A. Feig and G. M. Cooper, *Mol. Cell. Biol.* **8,** 3235 (1988).

[22] C. T. Baldari, G. Pelicci, S. M. M. Di, E. Milia, S. Giuli, P. G. Pelicci, and J. L. Telford, *Oncogene* **10,** 1141 (1995).

[23] P. A. Blaikie, E. Fournier, S. M. Dilworth, D. Birnbaum, J. P. Borg, and B. Margolis, *J. Biol. Chem.* **272,** 20671 (1997).

[24] L. R. Collins, W. A. Ricketts, L. Yeh, and D. Cheresh, *J. Cell Biol.* **147,** 1561 (1999).

[25] E. Fournier, P. Blaikie, O. Rosnet, B. Margolis, D. Birnbaum, and J. P. Borg, *Oncogene* **18,** 507 (1999).

[26] N. Gotoh, K. Muroya, S. Hattori, S. Nakamura, K. Chida, M. Shibuya, *Oncogene* **11,** 2525 (1995).

[27] N. Gotoh, A. Tojo, M. Shibuya, *EMBO J.* **15,** 6197 (1996).

[27a] N. Gotoh, M. Toyoda, and M. Shibuya, *Mol. Cell. Biol.* **17,** 1824 (1997).

[28] R. J. Hill, S. Zozulya, Y.-L. Lu, P. W. Hollenbach, B. Joyce-Shaikh, J. Bogenberger, and M. L. Gishizky, *Cell Growth Differ.* **7,** 1125 (1996).

[29] K. Li, R. Shao, and M.-C. Hung, *Oncogene* **18,** 2617 (1999).

protein was used to block a biochemical or biological effect. In addition, there are a number of biologic and biochemical end points with which the efficacy of these dominant negatives may be measured including DNA synthesis, transcription, cell growth, differentiation, and transformation. This section provides a detailed protocol on the use of Shc dominant-negative proteins to examine EGF-induced activation of transcription. This protocol can be adapted for use with agents other than EGF.

293T cells[31,37] or variations thereof[30] have been used to examine Shc function by the dominant-negative approach. The advantage of this cell system is that the cells have high transfection efficiencies (>50%) and that foreign genes present on expression vectors containing an SV40 origin of replication are expressed at high levels because of the presence of the SV40 large T antigen. These methods are presented below.

1. Rinse stock plates of 293T cells with PBS and then briefly (<2–3 min) trypsinize the cells with 0.05% (w/v) trypsin–0.53 m*M* tetrasodium EDTA in Hanks' balanced salt solution (HBSS) without magnesium or calcium (Life Technologies, Gaithersburg, MD).
2. Resuspend the cells in complete medium to inactive the tyrpsin and then count the cells.
3. Plate 3 ml of cells at a density of 2×10^5 cells per well in six-well tissue culture plates.
4. On the next day, transfect the cells by the calcium phosphate precipitation method. This method provides a rapid, reliable, and cost-effective means of introducing DNA into these cells with efficiencies of greater than 50%. For reporter assays, we have utilized the Gal4-Elk-1 reporter system to measure the effect of the ShcC dominant negatives on the activation of the Ras pathway by EGF.[31] There

[30] M. J. Lorenzo, G. D. Gish, C. Houghton, T. J. Stonehouse, T. Pawson, B. A. J. Ponder, and D. P. Smith, *Oncogene* **14,** 763 (1997).

[31] J. P. O'Bryan, Q. T. Lambert, C. J. Der, *J. Biol. Chem.* **273,** 20431 (1998).

[32] S. Raffioni, D. Thomas, E. D. Foehr, L. M. Thompson, and R. A. Bradshaw, *Proc. Natl. Acad. Sci. U.S.A.* **96,** 7178 (1999).

[33] W. A. Ricketts, D. W. Rose, S. Shoelson, and J. M. Olefsky, *J. Biol. Chem.* **271,** 26165 (1996).

[34] S. Roche, J. McGlade, M. Jones, G. Gish, T. Pawson, and S. A. Courtneidge, *EMBO J.* **15,** 4940 (1996)

[35] T. Sasaoka, H. Ishihara, T. Sawa, M. Ishiki, H. Morioka, T. Imamura, I. Usui, Y. Takata, and M. Kobayashi, *J. Biol. Chem.* **271,** 20082 (1996).

[36] L. E. Stevenson, K. S. Ravichandran, and A. R. Frackelton, *Cell Growth Differ.* **10,** 61 (1999).

[37] D. Thomas and R. A. Bradshaw, *J. Biol. Chem.* **272,** 22293 (1997).

[38] K. K. Wary, F. Mainiero, S. J. Isakoff, E. E. Marcantonio, and F. G. Giancotti, *Cell* **87,** 733 (1996).

are numerous reporter systems available with which to measure activation of additional signaling cascades.[39] However, we have had success with only the Gal4-Elk-1 system. The other reporters that we have tested in 293T cells appear to have high levels of background activity, which may be due to the presence of the SV40 large T antigen. For each well, mix 0.5–1 μg of each ShcC expression construct along with 0.5 μg of the Gal4-Elk-1 plasmid, 2.5 μg of the Gal4-luciferase reporter, and 1–1.5 μg of calf thymus DNA (Boehringer Mannheim, Indianapolis, IN), as carrier, yielding a total DNA amount of 5 μg in a volume of 112.5 μl of H_2O. Add 12.5 μl of 2.5 m*M* $CaCl_2$ dropwise with gentle shaking. Transfer this solution to an equivalent volume of 2× HBS (280 m*M* NaCl, 1.5 m*M* Na_2HPO_4, 12 m*M* dextrose, 50 m*M* HEPES, pH 7.05) with dropwise addition from a Pipetman (Gilson, Middleton, WI) while gently mixing the tube contents. Incubate the mixture for 20–30 min at room temperature, after which a fine white precipitate should be visible. Because multiple wells are transfected, a "master mix" containing the components common to all the samples is set up. This mix is then aliquoted to individual tubes and the DNAs specific to each transfection are added. This approach limits variability in luciferase activity due to errors in pipetting reporters between samples.

5. Gently mix the solution with a Pipetman, add DNA precipitate dropwise to the cells, and then gently rock the plates back and forth several times to mix. Let the cells incubate with the DNA precipitate for 3–4 hr. Do not allow longer incubations, as cell viability will decrease with increased incubation times. Aspirate medium containing DNA and then replace with complete medium. Do not rinse the plates as this will dislodge the cells. In addition, there is no need to glycerol shock the cells, given the high efficiency of transfection.

6. On the next day, remove the medium and replace with serum-free medium (2–3 ml per well). The cells should be starved for at least 16 hr.

7. On the next day, add EGF to a final concentration of 100 ng/ml and incubate for 4–6 hr at 37° in a tissue culture incubator. Note that the length of EGF stimulation is greater than the 5 min used for examining Shc phosphorylation. This is because the effects of Shc activation, that is, increased gene transcription, need to be converted into transcriptional changes in luciferase message leading to alteration in the protein level as measured with the luciferase assay. Thus, to detect changes in luciferase activity after growth factor addition, sufficient time must be allowed to manifest these changes in protein levels.

8. Rinse the cells with 2 ml of warmed PBS per well, being careful not to dislodge the cells. Rinse only two six-well plates at a time to minimize cell loss due to detachment from the plates.

9. Add 250 μl of luciferase lysis buffer (Analytical Luminescence, San Diego, CA) to each well and incubate the plates at 4° for 30–45 min.

[39] C. A. Hauser, C. J. Der, and A. D. Cox, *Methods Enzymol.* **238,** 271 (1994).

10. Transfer the lysates to 1.5-ml centrifuge tubes and pellet insoluble debris by centrifugation for 2 min at 14,000 rpm at 4°.

11. For luciferase assays, we routinely use 5–10 μl of lysate for analysis. Samples are analyzed on a luminometer to detect differences in luciferase activity. Both the Monolight 2000 luminometer (Analytical Luminescence) and the MLX microtiter plate luminometer (Dynex Technologies, Chantilly, VA), with developing reagents from Analytical Luminescence, have been used successfully. Follow the manufacturer's recommendations when using different reagents or machines.

Comments. The above-described protocols provide a starting point for analyzing the importance of Shc proteins in the action of a particular stimulus. It is important to keep in mind that the parameters described above have been optimized for analysis of Shc in EGF signaling. Thus, different stimuli may require different parameters that need to be determined empirically.

Furthermore, there are additional approaches for analyzing the involvement of Shc proteins in signaling pathways. Microinjection of cells with neutralizing antibody or dominant-negative expression constructs has been used by several groups to examine the role of Shc in signaling pathways.[33,34,40] Methods for these techniques have been described elsewhere.[41,42]

The finding that p66^{ShcA} may regulate the oxidative stress response and life span has further strengthened the notion that Shc proteins may be involved in signaling pathways other than Ras activation.[2] In addition, Shc proteins undergo serine and threonine phosphorylation in addition to tyrosine phosphorylation.[13,14] Thus, it is important to keep in mind that Shc may play an important role in signaling pathways in the absence of tyrosine phosphorylation or activation of Ras. The above described protocols provide a starting point for examination of Shc involvement in a signaling pathway as defined by tyrosine phosphorylation of Shc and activation of the Ras–Raf–MAPK pathway; however, they can be adapted for use with different experimental end points such as determination of serine and threonine phosphorylation or analysis of different reporters in different cell types.

Acknowledgments

I thank Drs. Dave Armstrong, Fernando Ribeiro-Neto, and Robert Mohney for reading the manuscript.

[40] T. Sasaoka, D. W. Rose, B. H. Jhun, A. R. Saltiel, B. Draznin, and J. M. Olefsky, *J. Biol. Chem.* **269,** 13689 (1994).

[41] D. Bar-Sagi, *Methods Enzymol.* **255,** 436 (1995).

[42] K. Kovary, *Methods Enzymol.* **254,** 445 (1995).

[2] Assaying Activity of Individual Protein Kinases in Crude Tissue or Cellular Extracts*

By SAID A. GOUELI, KEVIN HSIAO, and BASEM S. GOUELI

I. Introduction

Protein kinases and phosphatases play an important role in a variety of cellular functions such as cell growth, development, and gene expression.[1] It is estimated that about 2–3% of the genes in the entire genome of a eukaryotic cell may encode protein kinases and as many as 5% of human genes may encode protein kinases and phosphatases.[2] The fact that these protein kinases and phosphatases have multiple substrates *in vivo* may explain their diverse physiological functions.[1–3] Thus, it is of considerable interest to develop an assay system that is specific for certain protein kinases and simple enough for general use by investigators.

The availability of peptides that serve as specific substrates for certain protein kinases made it possible to determine the activity of a specific protein kinase in a tissue or cellular extract with minimal interference from other enzymes.[4] A widely used method to monitor the phosphopeptide product is the negatively charged phosphocellulose P-81 method, which requires the substrate to contain at least two or three basic amino acids, because the binding is based on electrostatic interaction.[5] The inclusion of basic amino acid residues, however, may alter the specificity of the substrate[6] and may give variable results depending on the sequence of the peptide, and in some instances, incomplete binding of phosphopeptides to the filters has been observed.[7] Because the binding of the phosphorylated proteins or peptides to the P-81 filter is electrostatic in nature, the washing protocol also may cause variability in the results and gentle washing is required to minimize the loss of the filter-bound peptide. In addition, any positively charged proteins (other than the phosphopeptide product) that are phosphorylated by protein kinase(s), including the autophosphorylated enzyme in the tissue extract, will also bind to the P-81 filters.[8] Thus, an assay method that can eliminate these pitfalls would offer

* U. S. Patent 6,066,462.

[1] T. Hunter, *Cell* **80,** 225 (1995).

[2] M. J. Hubbard and P. Cohen, *Trends Biochem. Sci.* **18,** 172 (1993).

[3] A. Levitzki and A. Gazit, *Science* **267,** 1782 (1995).

[4] B. E. Kemp and R. B. Pearson, *Methods Enzymol.* **200,** 121 (1991).

[5] J. E. Casnellie, *Methods Enzymol.* **200,** 115 (1991).

[6] L. J. Cisek and J. L. Corden, *Methods Enzymol.* **200,** 301 (1991).

[7] R. Toomik, P. Kman, and L. Engstorm, *Anal. Biochem.* **204,** 311 (1992).

[8] O. Hvalby, H. C. Hemmings, Jr., O. Paulsen, A. J. Czernik, A. C. Nairn, J. M. Godfraind, V. Jensen, M. Raastad, J. F. Storm, P. Andersen, and P. Greengard, *Proc. Natl. Acad. Sci. U.S.A.* **91,** 4761 (1994).

0076-6879/00 $35.00

important advantages over the existing methodologies. Toward this goal, we developed an assay system that has been successfully used to specifically determine the activity of an individual enzyme in crude tissue or cellular extract, circumvents the pitfalls associated with the phosphocellulose method, combines the attributes of simplicity and high sensitivity, and is amenable to both low- and high-throughput scales.[9,10]

II. Principle of Assay System

The assay system is based on the high affinity and selective binding of biotin to streptavidin (K_α of $10^{-14}M$). Thus, when biotinylated derivative of a selective peptide substrate is phosphorylated by the cognate protein kinase, the phosphorylated/biotinylated product can be separated from both free ATP and endogenously phosphorylated proteins that are nonbiotinylated, using a streptavidin-linked matrix. The only phosphorylated product that binds to the matrix is the phosphoform of the biotinylated peptide. The excess free [γ-^{32}P]ATP can be readily removed by a simple washing procedure (i.e., five to seven washes for 1–4 min each). The matrix-bound phosphopeptide is dried and the ^{32}P incorporated into the peptide substrate is quantified with a PhosphorImager (Molecular Dynamics, Sunnyvale, CA), by autoradiography, or with a liquid scintillation counter.[9]

III. Materials and Methods

A. Materials

1. Peptide Synthesis. The biotin-modified peptides are synthesized on a peptide synthesizer, using established solid-phase peptide procedures. The C_6-biotin moiety is coupled before cleavage of the peptide from the resin, and the peptide is purified by reversed-phase high-performance liquid chromatography (HPLC). The identity of the biotinylated peptides is confirmed by quantitative amino acid analysis and fast atom bombardment (FAB) mass spectrometry and their purity is confirmed by HPLC, using two solvent systems. These peptides are commercially available from Promega (Madison, WI), or they can be custom synthesized by several peptide synthesis companies.

2. Peptide Substrates. Peptide substrates specific for various protein kinases are selected on the basis of preferable and specific consensus sequences for each protein kinase (PK)[4] and synthesized in a biotinylated form by a vendor. Amino acid sequences of selective peptide substrates and the appropriate activators for some protein kinases are listed in Table I.

[9] B. S. Goueli, K. Hsiao, A. Tereba, and S. A. Goueli, *Anal. Biochem.* **225,** 10 (1995).
[10] S. A. Goueli, K. Hsiao, and C. Ruzicka, *Promega Notes* **64,** 2 (1997).

TABLE I
SELECTIVE BIOTINYLATED PEPTIDE SUBSTRATES FOR VARIOUS PROTEIN KINASES AND THEIR APPROPRIATE ACTIVATORS

Protein kinase	Sequence	Activator
PKA	Biotin-C_6-Leu-Arg-Ala-Ser-Leu-Gly	cAMP
PKC	Biotin-C_6-Ala-Ala-Lys-Ile-Gln-Ala-Ser-Phe-Arg-Gly-His-Met-Ala-Arg-Lys-Lys	Ca^{2+}/DAG/PS
Cdc2	Biotin-C_6-Pro-Lys-Thr-Pro-Lys-Lys-Ala-Lys-Lys-Leu	Cyclin
DNA-PK	Biotin-C_6-Glu-Pro-Pro-Leu-Ser-Gln-Glu- Ala-Phe-Ala-Asp-Leu-Trp-Lys-Lys	dsDNA
CK-1	Biotin-C_6-Asp-Asp-Asp-Glu-Glu-Ser-Ile-Thr-Arg-Arg	
CK-2	Biotin-C_6-Arg-Arg-Arg-Glu-Glu-Glu-Thr-Glu-Glu-Glu	Polyamines
CaM KII	Biotin-C_6-Lys-Lys-Ala-Leu-Arg-Arg-Gln-Glu-Thr-Val-Asp-Ala-Leu	Ca^{2+}/calmodulin
Tyrosine kinases (EGFR, IR, Src, etc.)	Two proprietary peptides (Promega)	

3. Purified Enzymes. Protein kinase A (PKA), PKC, cdc2, casein kinase 1 (CK-1), CK-2, epidermal growth factor receptor (EGFR), calcium and calmodulin-dependent protein kinase II (CaM KII), and DNA-dependent protein kinase (DNA-PK) are available through several commercial sources.

4. Buffers and Solutions.

Basic extraction buffer (used to extract all protein kinases listed in this chapter unless otherwise specified): 25 m*M* Tris-HCl (pH 7.4), 0.5 m*M* EDTA, 0.5 m*M* EGTA, 10 m*M* 2-mercaptoethanol, leupeptin (1 μg/ml), aprotinin (1 μg/ml), and 0.5 m*M* phenylmethylsulfonyl fluoride (PMSF). Store at 4° or, for up to 6 months, at −20°. *Note:* Just before use, add 0.5 ml of PMSF stock solution (100 m*M* PMSF in 100% ethanol) per 100 ml of extraction buffer. For PKC, extraction buffer should also contain 0.05% (v/v) Triton X-100

Basal kinase reaction 5× buffer: This buffer [100 mM Tris-HCl (pH 7.4), 50 m*M* $MgCl_2$] is used to assay protein kinases unless otherwise specified, that is, when additional ingredients such as activators or inhibitors are required

Protein kinase dilution 5× buffer: Basal kinase reaction buffer containing bovine serum albumin (BSA, 0.5 mg/ml) is recommended to dilute enzyme preparations before use. For PKC, a 0.05% (v/v) Triton X-100 should be included for enhanced enzyme stability

PKA activation 5× solution: Basal buffer plus 25 μ*M* cAMP

Protein Kinase C Activation. PKC activation requires two buffer solutions: coactivation and activation buffers.

PKC coactivation 5× solution: 1.25 m*M* EGTA, 2 m*M* $CaCl_2$, pH 7.4
PKC activation 5× buffer: 100 m*M* Tris-HCl (pH 7.4), 50 m*M* $MgCl_2$, L-α-phosphatidyl-L-serine (PS, 1.6 mg/ml), 1,2-dioleoyl-*sn*-glycerol ($C_{18:1}$, *cis*-9, (DAG, 0.16 mg/ml) and is prepared as follows.

PROTOCOL FOR MAKING 5× ACTIVATION BUFFER

1. Prepare DAG at 5 mg/ml in chloroform.
2. Prepare PS at 10 mg/ml in chloroform.
3. Pipette 0.16 ml of PS and 0.032 ml of DAG into an 18 × 150 mm tube.
4. Remove chloroform by using N_2 (gaseous); this takes 5–10 min.
Note: Steps 1 to 4 are performed in an ice bath.
5. Add 1.0 ml of 5× basal kinase reaction buffer, and then sonicate with a VC500 sonicator (Sonics and Materials, Danbury, CT). Use duty cycle 20 on a scale of 1–100 and set the output control dial to 4 with the use of the microtip. Sonicate, on ice, four times, eight pulses each. After each eight pulses, put the tube on wet ice for 15 sec.
6. This buffer should be prepared just 1 hr (on wet ice) before use; store at −70° if desired. *Note:* The final concentration of PS is 1.6 mg/ml, and the final concentration of DAG is 0.16 mg/ml.

PKC control 5× buffer: 100 m*M* Tris-HCl (pH 7.5), 50 m*M* $MgCl_2$
Termination buffer: 7.5 *M* guanidine hydrochloride in H_2O
Wash solutions: 2 *M* NaCl; 2 *M* NaCl in 1% (v/v) H_3PO_4
Other reagents: The peptide inhibitor of PKA (PKI), the myristoylated peptide inhibitor of PKC, and streptavidin-linked membranes (SAM^2 biotin capture membrane) (1.25 × 1.15 cm) are obtained from Promega. All other reagents are of high research grade and are obtained from Sigma (St. Louis, MO)

B. Methods

1. Determination of Protein Kinase Activity. The following protocols for PKA and PKC are described in detail to illustrate the utility and versatility of the assay system. Other enzymes can be assayed similarly, using selective biotinylated peptide substrates and appropriate activators (see Table I) under their optimal assay conditions.

a. PREPARATION OF CELLULAR OR TISSUE EXTRACTS

1. Precool the appropriate homogenizer and extraction buffer to 0 to 4°.
2. Tissue samples: Homogenize 1 g of tissue in 5 ml of cold extraction buffer with a cold homogenizer (e.g., a Polytron homogenizer).

3. Cultured cells: Wash 5×10^6–1×10^7 cells with phosphate-buffered saline (PBS) (5 ml per 100-mm dish) and remove the buffer completely. Suspend the cells in 0.5 ml of cold extraction buffer and homogenize with a cold homogenizer (e.g., a Dounce homogenizer).

4. Centrifuge the lysate for 5 min at 4° at 14,000 *g* in a microcentrifuge and save the supernatant. Crude extracts should be assayed the same day they are prepared to retain maximal activity and obtain optimal results.

b. Determination of Protein Kinase A Enzymatic Activity

1. Prepare the ATP mix as follows.

Component	Final per reaction (μl)	20 Reactions (μl)
ATP, 0.5 m*M*	5.00	100
[γ-^{32}P]ATP (3000 Ci/mmol), 10 μCi/μl	0.05	1

2. Prepare reaction mix in 0.5- to 1.5-ml microcentrifuge tubes as shown in Table II (a reaction without substrate should also be performed to determine background counts).

3. Prepare appropriate dilutions of the enzyme samples in enzyme dilution buffer and place at 0°. We recommend preparing and testing crude lysate samples undiluted and serially diluted 2- to 16-fold (a 1000-fold dilution is recommended for purified enzyme).

TABLE II
Experimental Design for Protein Kinase A Assay

Component	Final volume for number of reactions (μl)	
	5 Reactions	20 Reactions
PKA assay buffer (5×)	5	100
cAMP, 0.025 m*M*[a]	5	100
PKA biotinylated peptide substrate, 0.5 m*M*[b,c]	5	100
[γ-^{32}P]ATP mix	5	100

[a] This component is not required if the PKA catalytic subunit is the source of activity; replace with 5 μl of deionized water. Also replace with deionized water when measuring the basal activity of PKA.

[b] Final concentration is 100 μ*M;* other concentrations may be used but should not exceed 200 μ*M*.

[c] Larger volumes may be spotted; however, if more than 15 μl is to be spotted, separate the squares first to prevent cross-contamination. Do not exceed 30 μl per square. (Minor seepage of liquid onto adjacent squares does not cause contamination as the biotinylated peptide is rapidly immobilized to the SAM^2 membrane before liquid migration is complete.) The linear capacity of the membrane is 1.3 nmol/10 μl of terminated reaction volume.

4. Mix gently and preincubate the reaction mix (step 2 of this section) at 30° for 1–5 min.

5. Initiate the reaction by adding 5 μl of the enzyme sample to the reactants. The total reaction volume will be 25 μl. Incubate the reaction at 30° for 5 min. (Other time points and temperatures may be tested if desired.)

6. Terminate the reaction by adding 12.5 μl of termination buffer to each reaction; mix well. This solution is stable at 4° for at least 24 hr but can be kept at room temperature during processing.

7. Spot 10 μl from each terminated reaction onto a prenumbered SAM^2 membrane square. After all samples have been spotted, follow the wash and rinse steps as described below. Save the reaction tubes to be used for standards as shown below (step 11 of this section).

8. Place the SAM^2 membrane squares (1.25 × 1.15 cm) containing samples from the preceding step 7 into a washing container. Wash, using an orbital platform shaker set on low, or by occasional manual shaking as follows: Wash once for 30 sec with 200 ml of 2 *M* NaCl; wash three times for 2 min, each with 200 ml of 2 *M* NaCl; wash four times for 2 min, each with 200 ml of 2 *M* NaCl in 1% (w/v) H_3PO_4; wash twice for 30 sec each, with 100 ml of deionized water. *Note:* The total wash time is <20 min.

9. Dry the SAM^2 membrane squares on a piece of aluminum foil under a heat lamp for 5–10 min or air dry at room temperature for 30–60 min. (If the SAM^2 membrane has been washed with 95% (v/v) ethanol, shorten the drying time to 2–5 min under a heat lamp or 10–15 min at room temperature.)

10. Determine total counts for calculation of the specific activity of [γ-^{32}P]ATP as follows: Remove 5-μl aliquots from any two reaction tubes from step 7 of this section and spot onto individual SAM^2 membrane squares or Whatman (Clifton, NJ) 3-mm filter disks. For this step, dry without washing. After analysis use these results to calculate the specific activity of [γ-^{32}P]ATP as shown below. If 5 μl is not available from a single tube, combine the contents of several tubes for this step.

11. Analysis by scintillation counting: If still connected, separate the SAM^2 membrane squares with samples (from steps 9 and 10 of this section), using forceps, scissors, or a razor blade, and place squares or 3-mm filter disks into individual scintillation vials. Add scintillation fluid to the vials and count.

12. PhosphorImager analysis: Alternatively, the SAM^2 membrane may remain intact and the intact SAM^2 membrane may be analyzed with a PhosphorImager.

c. CALCULATION OF SPECIFIC ACTIVITY OF [γ-^{32}P]ATP

$$\text{Specific activity of } [\gamma\text{-}^{32}\text{P}]\text{ATP (in cpm/pmol of ATP)} = \frac{(37.5/5)(X)}{2500}$$

where 37.5 is the sum of the reaction volume (25 μl) plus the termination buffer volume (12.5 μl), 5 is the volume in microliters of the samples from step 10 of the preceding section, X is the average counts per minute of the 5-μl samples from step 11 of the preceding section, and 2500 is the number of picomoles of ATP in the reaction.

d. Calculation of Protein Kinase A Enzyme Activity

Enzyme specific activity (in pmol ATP/min/μg of protein)

$$= \frac{(\text{cpm}_{\text{reaction with substrate}} - \text{cpm}_{\text{reaction without substrate}})(37.5)}{(10)(\text{time}_{\text{min}})(\text{amount of protien in reaction}_{\mu\text{g}})(\text{specific activity of } [\gamma\text{-}^{32}\text{P}]\text{ATP})}$$

where 37.5 is the sum of the reaction volume (25 μl) plus the termination buffer volume (12.5 μl) and 10 is the volume in microliters of the sample from step 4 of Section III,B,1,b.

2. Determination of Protein Kinase C Enzymatic Activity

a. Preparation of Tissue or Cell Samples for Protein Kinase C Assay. Prepare crude extracts as described above. Pass the supernatant over a 1-ml column of DEAE-cellulose that has been preequilibrated in extraction buffer. Wash the column with 5 ml of extraction buffer, and then elute the PKC-containing fraction with 5 ml of extraction buffer containing 200 m*M* NaCl. Extracts should be assayed the same day they are prepared to retain maximal activity and obtain optimal results.

b. Protein Kinase C Assay Protocol

1. Prepare the ATP mix as described for PKA.
2. Prepare the reaction mix in 0.5- to 1.5-ml microcentrifuge tubes as shown in Tables III and Table IV.
3. Prepare appropriate dilutions of the enzyme samples to be tested, using enzyme dilution buffer. We recommend preparing and testing crude lysate samples undiluted and serially diluted 2- to 16-fold. Purified enzyme preparations may require greater dilution.

TABLE III
Protein Kinase C Reaction in Presence of Phospholipids (Activated Reaction)

Component	Final volume (μl) for:	
	1 Reaction	20 Reactions
PKC coactivation 5× buffer	5	100
PKC activation 5× buffer	5	100
PKC biotinylated peptide substrate, 0.5 m*M*	5	100
[γ-^{32}P]ATP mix	5	100

TABLE IV
PROTEIN KINASE C REACTION IN ABSENCE OF PHOSPHOLIPIDS (CONTROL REACTION)

Component	Final volume (μl) for:	
	1 Reaction	20 Reactions
PKC coactivation 5× buffer	5	100
PKC control 5× buffer	5	100
PKC biotinylated peptide substrate, 0.5 m*M*	5	100
[γ-^{32}P]ATP mix	5	100

4. Mix gently and preincubate the reaction mix at 30° for 1–5 min.
5. Initiate the reaction by adding 5 μl of the enzyme sample to the reactants. The total reaction volume will be 25 μl.
6. Incubate the reaction at 30° for 5 min. (Other time points and temperatures may be tested if desired.)
7. Terminate the reaction by adding 12.5 μl of termination buffer to each reaction; mix well. This solution is stable at 4° for at least 24 hr but can be kept at room temperature during processing.
8. Repeat steps 8–11 of Section III,B,1,b.

c. CALCULATION OF PROTEIN KINASE C ENZYME ACTIVITY. The enzymatic activity of PKC can be determined by subtracting the activity of the enzyme in the absence of phospholipids (control buffer) from that of the enzyme in the presence of phospholipids (activation buffer).

Enzyme activity (in pmol ATP/min/μg of protein)

$$= \frac{(\text{cpm}_{\text{reaction with phospholipids}} - \text{cpm}_{\text{reaction without phospholipids}})(37.5)}{(10)(\text{time}_{\text{min}})(\text{amount of protein in reaction}_{\mu g})(\text{specific activity of } [\gamma\text{-}^{32}\text{P}]\text{ATP})}$$

where 37.5 is the sum of the reaction volume (25 μl) plus the termination buffer volume (12.5 μl) and 10 is the volume (in microliters) of the sample.

IV. Results and Discussion

As mentioned above, the basic principle of this assay system is to use the biotinylated form of the selective peptide substrate for any protein kinase and [γ-^{32}P]ATP, to specifically assay for the kinase activity of the enzyme. To illustrate the utility of this method, we show results obtained with PKA and PKC as prototypes. Kinase activity of other kinases can be easily carried out, using the appropriate biotinylated peptide substrates and activators.

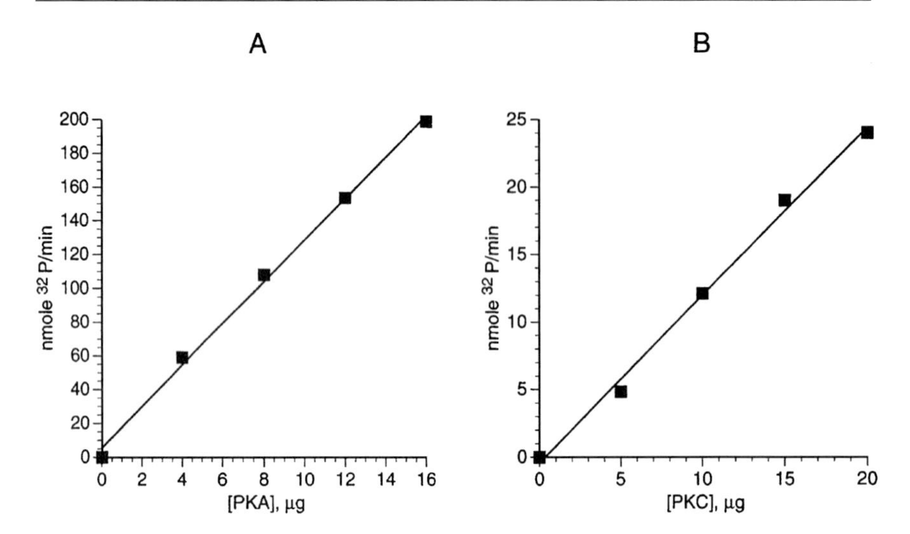

FIG. 1. Effect of enzyme concentration on the protein kinase activity of PKA (A) and PKC (B). The protein kinase activity was determined in the presence of a 200 μM concentration of the corresponding biotinylated peptide substrates and appropriate kinase activators as described in Section III,B.

When the kinase activity of PKA was assayed in the presence of increasing concentrations of the peptide substrate, a hyperbolic response curve that is typical of Michaelis–Menten saturation behavior was obtained. Maximal activity was obtained with 150–200 μM substrate, yielding a K_m value of 12 μM and a V_{max} value of 15 μmol/min/mg. Similar values (K_m of 10 μM and V_{max} of 16 μmol/min/mg of enzyme)[5] have been reported for PKA, using Kemptide as substrate. The response curve for PKC, using the peptide substrate neurogranin$_{(28-43)}$, showed a maximum activity of PKC at 100–150 μM peptide substrate. We have also observed that the basal activity of PKC (no phospholipids) was as low as the background (no enzyme added). This was an indication that all the activity of PKC was fully dependent on the presence of activators. It is noteworthy that the PKC activity determined here reflects those representing the conventional isoforms of PKC (α, β, and γ), which require the presence of all three activators (calcium, DAG, and PS). Other classes of PKC isoforms, such as the novel δ, ε, θ, η, and μ isoforms, require DAG and PS but do not require calcium for activation, and finally, the atypical isoforms λ and ξ require only PS for activation.[11]

Using saturating concentrations of biotinylated peptide substrates for PKA and PKC (200 μM), we were able to demonstrate a linear response in the activity of PKA and PKC by increasing the amount of each enzyme in their corresponding reactions (Fig. 1A and B). We obtained a linear response in the activity of PKA

[11] J. Hofmann, *FASEB J.* **11,** 649 (1997).

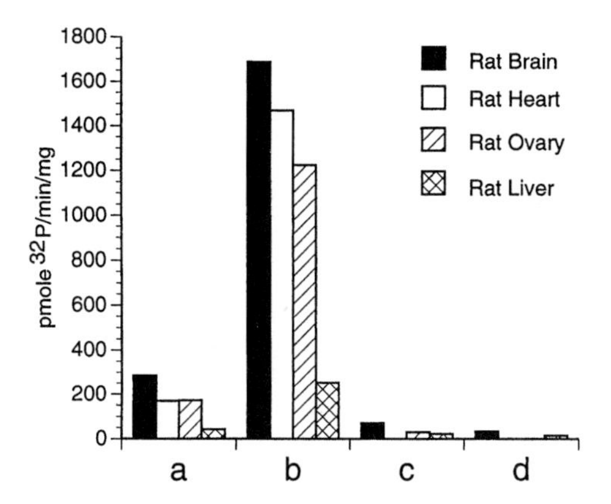

FIG. 2. The protein kinase activity of PKA in extracts of rat brain, heart, ovary, and liver was determined as described in Methods, using a 200 μM concentration of the biotinylated peptide substrate under several conditions. (a) None; (b) plus 5 μM cAMP; (c) plus 10 μM PKA inhibitor; (d) plus 5 μM cAMP and 10 μM PKA inhibitor.

with as low as 16 ng of pure PKA (Fig. 1A) and 20 ng of PKC (Fig. 1B). The addition of a 10 μM concentration of the PKA inhibitor (PKI) or a 100 μM concentration of the myristoylated peptide inhibitor of PKC drastically reduced the amount of ^{32}P incorporated into peptide substrates by more than 90%. These results confirm that phosphorylation of each peptide by the cognate kinase is selective. To further demonstrate the utility of this assay in determining the kinase activity of enzymes in tissue extract, the kinase activity of PKA and PKC was assayed in extracts of various rat tissues under the optimal conditions for both enzymes. The basal activity of PKA (in the absence of cAMP) was significantly low in all tissues examined (Fig. 2a). The addition of cAMP (5 μM) increased the activity of the enzyme by 6- to 9-fold (Fig. 2b). The addition of the PKI resulted in a remarkable inhibition of the basal as well as the activatable kinase activity (more than 90%), thus confirming that the phosphate incorporation was catalyzed by PKA (Fig. 2c and d). It is apparent that the remarkable low background observed in our assay attests to the fact that only the biotinylated phosphopeptide binds to the membranes. Thus the results obtained represent the true value for the ^{32}P that is incorporated into the peptide substrate and not in any additional proteins present in the extract. Similarly, PKC activity in extracts of various rat tissues was determined. The activity of the enzyme was stimulated by phospholipids (Fig. 3a vs. 3b) and inhibited by the addition of a 100 μM concentration of the myristoylated PKC inhibitor (Fig. 3c and d). It is noteworthy that the basal activity of PKC in various tissue extracts determined by this method (in the absence of activators) was as low as the background level (no enzyme added) and thus significantly high fold stimulation was achieved.

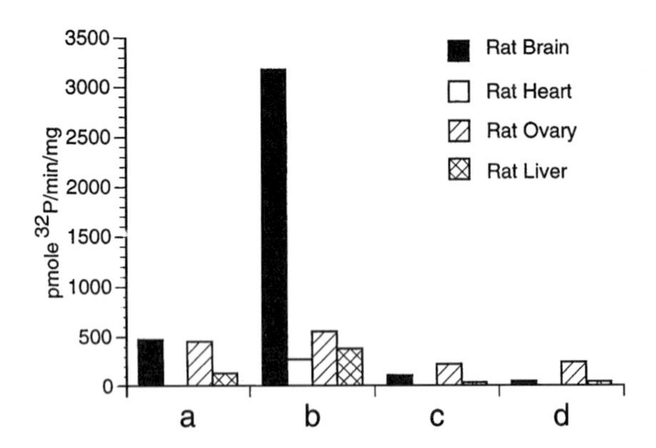

FIG. 3. The protein kinase activity of PKC in extracts of rat brain, heart, ovary, and liver was determined as described in Section III,B, using a 200 μM concentration of the biotinylated peptide substrate. (a) None; (b) plus phospholipids; (c) plus 100 μM myristoylated PKC inhibitor; (d) plus phospholipids and inhibitor.

An important consideration regarding this assay is that it can be easily scaled up to a high-throughput format by using high-capacity streptavidin-linked sheets that are fitted into the 96-well plate format. These plates are available from Promega. They offer several advantages that make them useful in pharmaceutical research for high-throughput drug screening assays of inhibitors/activators of protein kinases. The availability of equipment that facilitated complete automation of this assay, such as automatic plate handling using robotics, automatic washers, and fully automated liquid scintillation counters, provided the opportunity for efficient high-throughput analysis. We have demonstrated the feasibility of this system in a high-throughput format by using commercially available automatic washers and liquid scintillation counters of 96- and 384-well plates [Wallac (Gaithersburg, MD) MicroBeta Trilux and Packard (Downers Grove, IL) TopCount MicroPlate liquid scintillation counter.[10]

V. Conclusion

We have developed a novel protein kinase assay system that uses an innovative approach to accurately measure the kinase activity of various protein kinases and circumvents the pitfalls of existing methodologies. The assay offers the following advantages that make it unique and the method of choice for determining the kinase activity of various enzymes in tissue or cellular extracts.

Versatility and ease of use: The kinase activity of individual protein kinase can be quantified even in the presence of a mixture of other protein kinases, as is the case when working with tissue or cellular extracts, and the assay can be completed in 10–15 min after reaction completion.

Low background and high sensitivity: Because only the phosphorylated peptide (and not other phosphorylated proteins present in cellular or tissue extracts) binds to the matrix, the background obtained is significantly lower than that observed with other methods and the results generated represent a true estimate of enzymatic activity in the extract. This will allow the investigator to detect the activity of as little as a few femtomoles to picomoles of enzyme.

Correct consensus peptide sequence: There is no need to alter the optimal substrate recognition sequence in order for it to be used as substrate for the corresponding enzyme in this assay, as is the case with other assay systems such as the P-81 method. The only requirement is the addition of a biotin moiety to the N terminus amino group of the peptide, which, on the basis of our studies, does not affect its specificity for the enzyme. More recently, the utility of our approach has been elegantly demonstrated in the development of an assay to determine the kinase activity of I-κB kinase (IKK), using a biotinylated I-κBα-derived peptide substrate.[12]

Reliability and quantification of results: Because the binding of biotin to streptavidin is the strongest known noncovalent biological interaction (K_a of $10^{15} M^{-1}$), the phosphorylated biotinylated peptide strongly binds to the streptavidin-linked disk and is not affected by wide-ranging conditions. These include extremes of pH (2.0–10.0), temperature, organic solvents, ionic and nonionic detergents (SDS, CHAPS, Triton X-100, Tween 20, or Tween 80), and other denaturing agents (5 *M* guanidine hydrochloride and 2 *M* urea).[13] Therefore, the peptide is not likely to be washed off during the washing procedure as is the case when the peptide substrate is bound to filters via weak electrostatic binding (e.g., phosphocellulose filter assay).[7]

High binding capacity: The binding capacity of the disk for the biotinylated peptide substrate is high (a minimum of 2.5 nmol/disk). This allows the use of high concentrations (up to 1 m*M*) of the peptide substrates under optimal enzyme assay conditions and maximizes the signal-to-noise ratio.

Adaptability to high-throughput assay format: The assay has been successfully adapted to a high-throughput 96-well plate format. The 96-well plates offer a tremendous advantage[10] for scientists in pharmaceutical research or large-scale clinical studies to screen for activators and inhibitors of protein kinases.

[12] D. Wisniewski, P. LoGrasso, J. Calaycay, and A. Marcy, *Anal. Biochem.* **274,** 220 (1999).

[13] D. Savage, G. Mattson, S. Desai, G. Nielander, S. Morgensen, and E. Conklin, "Avidin–Biotin Chemistry: A Handbook." Pierce Chemical Co., Rockland, Illinois, 1992.

[3] Recombinant Adenoviral Expression of Dominant-Negative Ras N17 Blocking Radiation-Induced Activation of Mitogen-Activated Protein Kinase Pathway

By PAUL DENT, CRAIG LOGSDON, BARBARA NICKE, KRISTOFFER VALERIE, JULIE FARNSWORTH, RUPERT SCHMIDT-ULLRICH, and DEAN B. REARDON

Generation of recombinant adenoviruses to express a variety of proteins has become a method of choice for many investigators in the field of signal transduction. Using this technique, near 100% infection efficiencies can be obtained. In this chapter, we describe methodologies to make a recombinant dominant-negative Ras N17 adenovirus by two related homologous recombination techniques: (1) in eukaryotes (human 293 cells) and (2) in prokaryotes (*Escherichia coli*).

Ionizing radiation has been shown to activate multiple signaling pathways within cells *in vitro,* which can lead to either increased cell death or increased proliferation depending on the cell type, the radiation dose, and the culture conditions.[1–9] A novel cellular target for ionizing radiation has been shown to be the epidermal growth factor receptor (EGFR, also called ErbB1), which is activated in response to irradiation of several carcinoma cell types.[1,7–9] Radiation exposure, via the EGFR, can activate the extracellular signal-regulated kinase–mitogen-activated protein kinase (ERK/MAPK) pathway to a level similar to that observed by physiologic (~0.1 n*M*) EGF concentrations.[7–9] Radiation has also been shown to increase the amount of GTP associated with Ras.[10] The data presented in this chapter describe

[1] S. Carter, K. L. Auer, M. Birrer, P. B. Fisher, R. Scmidtt-Ulrich, K.Valerie, R. Mikkelsen, and P. Dent, *Oncogene* **16,** 2787 (1998).

[2] Z. Xia, M. Dickens, J. Raingeaud, R. J. Davis, and M. E. Greenberg, *Science* **270,** 1326 (1995).

[3] S. J. Chmura, H. J. Mauceri, S. Advani, R. Heimann, E. Nodzenski, J. Quintans, D. W. Kufe, and R. R. Weichselbaum, *Cancer Res.* **57,** 4340 (1997).

[4] P. Santana, L. A. Pena, A. Haimovitz-Friedman, S. Martin, D. Green, M. McLoughlin, C. Cordon-Cardo, E. H. Schuchman, Z. Fuks, and R. Kolesnik, *Cell* **86,** 189 (1996).

[5] C. Rosette and M. Karin, *Science* **274,** 1194 (1996).

[6] A. Haimovitz-Friedman, Radiation-induced signal transduction and stress response. *Radiat. Res.* **150,** S102 (1998).

[7] R. K. Schmidt-Ullrich, R. B. Mikkelsen, P. Dent, D. G. Todd, K. Valerie, B. D. Kavanagh, J. N. Contessa, W. K. Rorrer, and P. B. Chen, *Oncogene* **15,** 1191 (1997).

[8] B. D. Kavanagh, P. Dent, R. K. Schmidt-Ullrich, P. Chen, and R. B. Mikkelsen, *Radiat. Res.* **149,** 579 (1998).

[9] P. Dent, D. B. Reardon, J. P. Park, G. Bowers, C. Logsdon, K. Valerie, and R. Schmidt-Ullrich, *Mol. Biol. Cell* **10,** 4231 (1999).

[10] S. Suy, W. B. Anderson, P. Dent, E. Chang, and U. Kasid, *Oncogene* **15,** 53-61 (1997).

METHODS IN ENZYMOLOGY, VOL. 333
0076-6879/00 $35.00

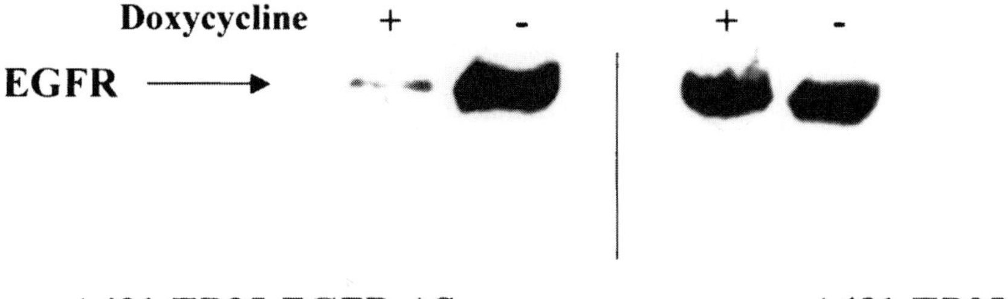

FIG. 1. Treatment of EGFR-antisense cells with doxycycline reduces expression of EGFR. EGFR-antisense cells were cultured as described in Materials and Methods and were treated with doxycycline (1 μg/ml) for 48 hr prior to experimentation. After 48 hr, cells were lysed and equal total protein amounts (100 μg) were immunoprecipitated, followed by immunoblotting with the same antibody to determine the expression of EGFR. Exposure was for 60 sec. A representative experiment ($n = 3$) is shown.

the generation of a recombinant adenovirus to express dominant-negative Ras N17 and the effect this gene product has on radiation-induced ERK–MAPK activity.

Materials and Methods

Generation of A431-TR25-EGFR-Antisense Cells

EGFR-CD533 is the wild-type EGFR with the COOH-terminal 533 amino acids deleted, previously shown to be a dominant-negative EGFR molecule, inhibiting EGFR function. The squamous vulval carcinoma cell line A431-TR25-EGFR-antisense has been generated as described for MDA-TR15-EGFR-CD533,[11,12] using the CD533 construct in the antisense orientation. In this chapter these cells are referred to as EGFR-antisense cells. Treatment of EGFR-antisense cells with doxycycline (1 μg/ml) for 48 hr induces antisense EGFR, and reduces expression of full-length wild-type EGFR protein by >100-fold (Fig. 1).

Culture of EGFR-Antisense Cells

Cells are cultured in RPMI 1640 supplemented with 5% (v/v) fetal calf serum at 37° in 95% (v/v) air/5% (v/v) CO_2.[1,11,12] Cells are plated at 2.5×10^6 cells per100-mm plate, in 5 ml of medium. For radiation-induced activation of protein kinases, cells are cultured for 4 days in this medium, and for 2 hr prior to irradiation are cultured in serum-reduced RPMI medium [0.5% (v/v) fetal calf serum].

[11] J. N. Contessa, D. B. Reardon, D. Todd, P. Dent, R. B. Mikkelsen, K. Valerie, G. D. Bower, and R. K. Schmidt-Ullrich, *Clin. Cancer Res.* **5,** 405 (1999).

[12] D. B. Reardon, J. N. Contessa, R. B. Mikkelsen, K. Valerie, C. Amir, P. Dent, and R. K. Schmidt-Ullrich, *Oncogene* **18,** 4756 (1999).

Recombinant Adenoviral Vectors: Generation and Infection in Vitro

Recombination in 293 Cells. A recombinant adenovirus expressing a dominant-negative Ras has been generated by cloning the human H-Ras cDNA with a serine-to-asparagine substitution at amino acid position 17 (gift of L. Feig, Tufts University, Boston, MA) into the multiple cloning site of the vector pAD.CMV-Link.1 (gift of K. J. Fisher, University of Pennsylvania, Philadelphia, PA). The p21-H-rasN17 cDNA is isolated from the plasmid pXVR by digestion with *Bam*HI and *Bgl*II; the resulting full-length cDNA is blunt ended with the Klenow fragment of *E. coli* DNA polymerase I and subcloned into the *Eco*RV site of the pAD.CMV-Link vector (pAD.CMV-Link.1.RasN17). pAD.CMV-Link.1.RasN17 and the pJM17 vector (Microbix Biosystems, Toronto, Ontario, Canada) are co-transfected into 80% confluent 293 cells by the $CaPO_4$–DNA coprecipitation technique. After 8–10 days, the death of the 293 cells indicates that a new recombinant virus encoding the p21-H-RasN17 protein has been generated. The viral DNA is then extracted from the 293 cells and confirmed by the Southern blot technique, using the Ras N17 cDNA as a probe. A single clone of recombinant adenovirus has been isolated through serial dilution, using a plaque assay. The expression of the recombinant adenovirus is performed as previously described with 293 cells, and the virus is subsequently concentrated on a cesium chloride gradient.[13,14] The concentration of the recombinant adenovirus is assessed on the basis of the absorbency at 260 nm and by a limiting dilution plaque assay. For controls, we utilized an emtpy adenovirus (pAD.CMV), which is isolated by cotransfection of the pJM17 vector with pAD.CMV.Link.1, not possessing an inserted gene. All purified viral stocks possess $\sim 10^{11}$ plaque-forming units (PFU)/ml. This procedure is similar to that described in Refs. 13–17.

Recombination in Escherichia coli. We have generated recombinant adenoviruses in bacteria, using a novel methodology. In this procedure, the full-length recombinant adenovirus genome is cloned in a plasmid, flanked by a rare cutter (*Pac*I)restriction site, and is generated by using a recombination-proficient *E. coli* strain (BJ5183) with the genotype *recBC sbcBC* (18, 19). We have developed a novel transfer plasmid, using pZero2.1 (InVitrogen, San Diego, CA) and a plasmid containing the 35-kbp adenoviral genome pTG-CMV (kindly provided by M. P. Wymann and S. B. Verca, University of Fribourg, Switzerland). Digestion of pZero2.1 with *Afl*III and*Stu*I is followed by insertion of a linker containing *Pac*I and *Bgl*II sites, forming the construct pZero-link. Digestion of pZero-link and pTG-CMV with *Pac*I, followed by fragment purification and annealing, produces the plasmid pZeroTG-CMV. The p21-H-rasN17 cDNA is isolated from the plasmid pXVR by digestion with *Bam*HI and *Bgl*II; the resulting full-length cDNA

[13] B. Nicke, M. J. Tseng, M. Fenrich, and C. Logsdon, *Am. J. Physiol.* **276,** G499 (1999).
[14] K. Valerie and A. Singhal, *Mutat. Res.* **336,** 91 (1995).

is 3′ blunt ended and subcloned into the *Bam*HI site of pZero TG-CMV (pZeroTG-CMV-RasN17). Colonies are selected with kanamycin. Recombination is achieved in the recombination-proficient *E. coli* strain BJ5183. The pZeroTG-CMV-RasN17 plasmid (500 ng) is digested with *Pac*I and *Bgl*II; the pTG-CMV plasmid is cut with *Cla*I (1 μg), followed by cotransformation of BJ5183 cells. Each DNA is added to BJ5183 competent cells and allowed to sit on ice for at least 30 min. Heat shock for 80 sec, and then put back on ice for 2 min. Add 250 ml of SOC medium and incubate at 37° for 1 hr. Plate out 150 ml of the transformation medium onto LB–glucose plates. Grow overnight at 37°. *Important*: Do not overgrow the plate; satellite colonies will form and must be avoided when picking colonies. After colonies have grown, pick about eight and grow all in 5 ml of LB–ampicillin. Harvest in the afternoon, allowing the bacteria to grow at least 6 hr. We use the Bio-Rad (Hercules, CA) minipreparation kit to prepare plasmid DNA. *Note*: If it is not possible to prepare DNA on the same day as the harvest, make sure to spin and pour off the LB, and then put the pellet into a −80° freezer to store for later use. Prepare BJ DNA using the Bio-Rad minipreparation kit. [*Note*: The Promega (Madison, WI) Wizard kit does not produce clean enough DNA for future transformation into XL1 Blue cells.] The Bio-Rad protocol is changed slightly.

1. Add 300 ml of cell resuspension solution to the bacterial pellet and resuspend the pellet with a pipette, or vortex.
2. Add 400 ml of cell lysis solution with gentle rocking. Allow the solution to sit at room temperature until the lysate is clear.
3. Add 400 ml of the neutralization solution, again with gentle agitation, followed by chilling on ice for 5 min
4. Clarify the solution by centrifugation at room temperature, discarding the pellet.
5. Add 200 ml of the binding matrix to the supernatant from the spin.
6. Follow the protocol as directed for wash steps.
7. Add 100 ml of TE to the binding matrix and filter. Spin and collect the DNA in a fresh tube.
8. Transform 5 ml into 50 ml of XL I Blue cells.
9. Heat shock for 42 sec followed by placement on ice for 2 min.
10. Add 250 ml of SOC medium and incubate at 37° for 1 hr.
11. Plate out 150 ml onto LB–glucose plates. Grow overnight at 37°.

Pick five colonies the following morning and prepare each separately with a Promega Wizard preparation kit. Digest each DNA, along with control empty pTG-CMV, with *Eco*RI to confirm recombinant plasmid pTG-CMV-RasN17. The pTG-CMV-RasN17 DNA is transfected into 80% confluent 293 cells, using the $CaPO_4$–DNA coprecipitation technique. Cells are overlaid with agarose and plaques form in 7–10 days. Plaques are isolated, the virus from each is expanded, and protein expression of Ras N17 is determined by Western blotting. The concentration of the recombinant adenovirus is assessed on the basis of the absorbency at 260 nm and by a limiting dilution plauqe assay. Recombinant adenoviruses are stored in small portions (~1 ml) at −80°. A reduction in titer is observed after more than one freeze–thaw cycle.

Infection of Cells with Recombinant Adenovirus. Prior to infection, the cell number is determined. For infection, cells are incubated in a minimal volume of serum-free medium for the plate size, for example, 3 ml for a 100-mm dish. To this medium, the appropriate amount of recombinant adenovirus is added to give the required multiplicity of infection (MOI). Cells are gently rocked for 4 hr at 37° in an incubator. At this time, medium can be replaced with serum-containing medium, or the original medium can be diluted with medium containing 2 × serum. EGFR-antisense cells are infected with dominant-negative Ras N17 adenovirus *in vitro* (MOI of 100), and incubated at 37° for an additional 24 hr. To assess expression, we performed Western immunoblots 24 hr after infection (Fig. 2, inset).

Treatment of Cells with Drugs, Neutralizing Antibody, Ionizing Radiation, and Cell Lysis

Treatment with U0126 is from a 100 m*M* stock solution (2 μ*M* final) and the maximal concentration of vehicle (dimethyl sulfoxide, DMSO) in medium is 0.02% (v/v). Cells are irradiated by a ^{60}Co source at a dose rate of 1.1Gy/min.[1,9] Cells are maintained at 37° throughout the experiment, except during the irradiation itself. Zero time is designated as the time point at which exposure to radiation ceases. After irradiation, cells are incubated for the specified times, followed by aspiration of medium and snap freezing at −70° on dry ice. Cells are lysed in 1 ml of ice-cold buffer A [25 m*M* HEPES (pH 7.4 at 4°), 5 m*M* EDTA, 5 m*M* EGTA, 5 m*M* benzamidine, 1 m*M* phenylmethylsulfonyl fluoride (PMSF), 1 μ*M* microcystin-LR, 0.5 m*M* sodium orthovanadate, 0.5 m*M* sodium pyrophosphate, 0.05% (w/v) sodium deoxycholate, 1% (v/v) Triton X-100, 0.1% (v/v) 2-mercaptoethanol],with trituration using a P1000 pipette. Lysates are clarified by centrifugation (4°). Immunoprecipitations from cell lysates are performed as in Refs. 1, 9, 11, and 12.

Assay of ERK–MAPK Activity

Immunoprecipitates are incubated (final volume, 50 μl) with 50 μl of buffer[1,9] containing 0.2 m*M* [γ-^{32}P]ATP (5000 cpm/pmol), 1 μ*M* microcystin-LR, and myelin basic protein (MBP, 0.5 mg/ml), which initiates reactions. After 20 min, 40-μl samples of the reaction mixtures are spotted onto a 2-cm circle of P81 paper (Whatman, Maidstone, UK) and immediately placed into 180 m*M* phosphoric acid. Papers are washed four times (10 min each) with phosphoric acid, and once with acetone, and ^{32}P incorporation into MBP is quantified by liquid scintillation

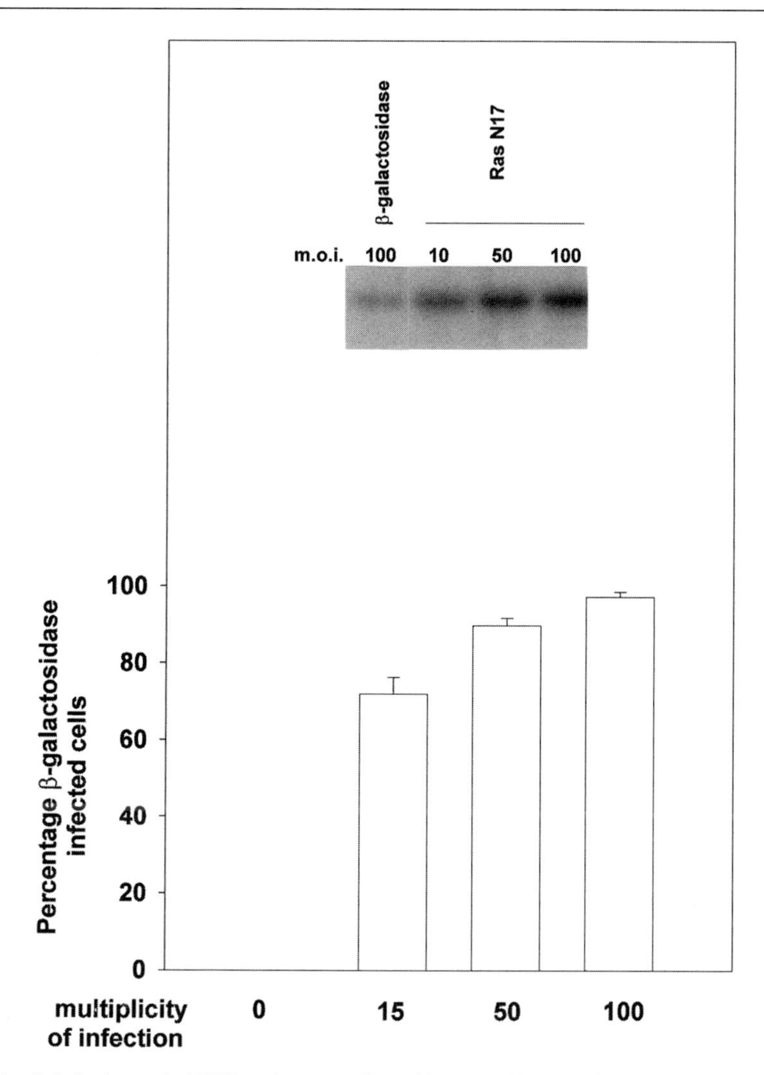

FIG. 2. Infection of EGFR-antisense cells with recombinant adenoviruses to express β-galactosidase or dominant-negative Ras N17. EGFR-antisense cells were infected at increasing multiplicities of infection of recombinant adenovirus to express β-galactosidase.Twenty-four hours after infection, cells were stained with X-Gal *in situ* and the percentage infected cells determined by counting blue cells. *Inset:* EGFR-antisense cells were infected with a recombinant adenovirus to express dominant-negative Ras N17.Twenty-four hours after infection, cells were lysed and portions (~100 μg) from each plate were subjected to SDS–PAGE and immunoblotting versus Ras using a monoclonal antibody (Y13-238); a representative experiment is shown (n = 5). Exposure time: 1 min.

spectroscopy. Preimmune controls are performed to ensure MBP phosphorylation is dependent on specific immunoprecipitation of ERK–MAPK.

Sodium Dodecyl Sulfate–Polyacrylamide Gel Electrophoresis and Western Blotting

Cells are irradiated and at specified time points/treatments medium is aspirated and the plates are snap frozen. Cells are lysed with homogenization buffer and subjected to immunoprecipitation. Immunoprecipitates are solubilized with 100 μl of 5× sodium dodecyl sulfate–polyacrylamide gel electrophoresis (SDS–PAGE) sample buffer, diluted to 250 μl with distilled water, and placed in a 100° dry bath for 15 min. One hundred-microliter aliquots of each time point are subjected to SDS–PAGE on 8% (v/v) gels. Gels are transferred to nitrocellulose and Western blotting, using specific antibodies, is performed as indicated. Blots are developed by enhanced chemiluminescence (ECL; Amersham, Arlington Heights, IL).

Results

Generation of A431-TR25-EGFR-Antisense Cells

The squamous vulval carcinoma cell line A431-TR25-EGFR-antisense was generated as described for MDA-TR15-EGFR-CD533,[11,12] using the CD533 construct in the antisense orientation. In this chapter, these cells are hereafter referred to as EGFR-antisense cells. Treatment of EGFR-antisense cells with doxycycline (1 μg/ml) for 24–48 hr induces antisens EGFR, and reduces expression of full-length wild-type EGFR protein by >100-fold (Fig. 1).

Generation of Recombinant Adenovirus to Express Dominant-Negative Ras N17

A recombinant adenovirus was generated as described in Materials and Methods.[13–17] To assess viral infection efficiency, we also generated a recombinant adenovirus to express β-galactosidase[14] and performed experiments in which cells were infected at increasing MOIs. At an MOI of 15, ~70% of the EGFR-antisense cells were infected (Fig. 2). At an MOI of 50, ~90% of the cells were infected. Because of this, we used an MOI of 100 in further studies with the recombinant Ras N17adenovirus. Cells were infected and the expression of Ras was monitored 24 hr later via Western blotting (Fig. 2, inset). Because we observed high expression of Ras N17 relative to endogenous wild-type Ras, we next determined whether Ras N17 could blunt ERK–MAPK activation by radiation.

[15] R. A. Gabbay, C. Sutherland, L. Gnudi, B. B. Kahn, R. M. O'Brien, D. K. Granner, and J. S. Flier, *J. Biol. Chem.* **271,** 1890 (1996).

[16] L. Gnudi, E. U. Frevert, K. L. Houseknecht, P. Erhardt, and B. B. Kahn, *Mol. Endocrinol.* **11,** 67 (1997).

[17] P. Kometiani, J. Li, L. Gnudi, B. B. Kahn, A. Askari, and Z. Xie, *J. Biol. Chem.* **273,** 15249 (1998).

Radiation Induction of Immediate Primary and Secondary Activations of EGFR and ERK–MAPK Pathway in EGFR-Antisense Carcinoma Cells

The ability of radiation to modulate EGFR and ERK–MAPK activity was investigated in EGFR-antisense cells. Radiation caused immediate primary activation of the EGFR and the ERK–MAPK pathway (0–10 min) followed by a later secondary activation (90–240 min) in EGFR-antisense cells prior to antisense induction. Inhibition of EGFR function by induction of antisense EGFR mRNA reduced EGFR protein levels and abolished activation of the EGFR (Fig. 3, inset). Furthermore, inhibition of EGFR function by induction of antisense EGFR mRNA also completely blocked the ability of radiation to activate ERK–MAPK (Fig. 3).

The ability of the growth factors, via the EGFR, to activate ERK–MAPK is known to be dependent on signaling through the Ras protooncogene; however, a direct role for Ras has not been definitively proved following radiation exposure.[10,18–20] Expression of dominant–negative Ras N17 blocked activation of ERK–MAPK by radiation in EGFR-antisense cells (Fig. 3). Incubation of cells with a specific inhibitor of MEK1/2, U0126, also blunted the ability of radiation to activate ERK–MAPK. These data demonstrate that radiation increased ERK–MAPK activity in carcinoma cells via an EGFR/Ras-dependent mechanism. To determine whether inhibition of ERK–MAPK signaling altered the ability of radiation to cause cell death (apoptosis), EGFR-antisense cells were irradiated (2 Gy) in the presence of the MEK 1/2 inhibitor U0126 and cell viability was determined 24 hr after exposure by terminal uridyl-nucleotide end labeling (TUNEL) of DNA. Radiation exposure increased the level of apoptosis from 4 to 6% without any additional treatment. Treatment of cells with U0126 increased apoptosis from 4 to 6%. However, combined irradiation with U0126 significantly increased apoptosis above either radiation-alone or treatment-alone values to 12% ($p < 0.05$). Our data argue that inhibition of the EGFR–ERK–MAPK pathway enhances the ability of radiation to kill carcinoma cells.

Thus radiation causes short immediate primary activation (0–5 min) and prolonged secondary activation (90–240 min) of the EGFR. Radiation also caused primary and secondary activation of the ERK–MAPK pathway, which were both dependent on EGFR function as judged by the ability of either antisense EGFR mRNA ordominant-negative EGFR-CD533 to block activation of both EGFR and ERK–MAPK. Expression of dominant-negative Ras N17 blocked the ability of radiation to alter ERK–MAPK pathway activity, which argues that radiation utilizes similar mechanisms to stimulate these signaling pathways as do natural ligands of

[18] K. Auer, J. Contessa, S. Brenz-Verca, L. Pirola, S. Rusconi, G. Cooper, A. Abo, M. Wymann, R. J. Davis, M. Birrer, and P. Dent, *Mol. Biol. Cell* **9,** 561 (1998).

[19] C. Charter, E. Degryse, M. Gantzer, A. Dieterle, A. Pavirani, and M. Mehtali, *J. Virol.* **70,** 4805 (1996).

[20] U. Kasid, S. Suy, P. Dent, T. Whiteside, and T. W. Sturgill, *Nature (London)* **382,** 316 (1996).

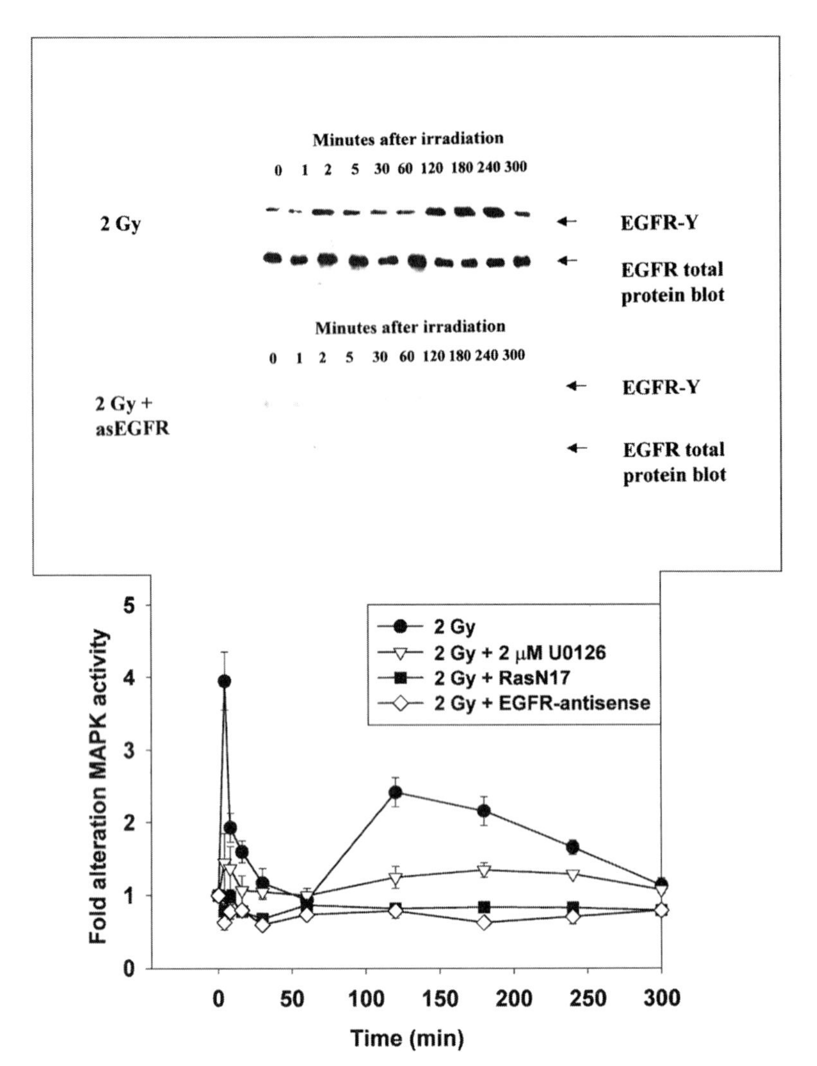

FIG. 3. Radiation-induced activation of ERK–MAPK in EGFR-antisense cells is blocked by expression of antisense EGFR mRNA, dominant-negative Ras N17, or incubation of cells with MEK1/2inhibitor U0126. EGFR-antisense cells were cultured, treated with doxycycline or infected with adenovirus as described in Materials and Methods. Cells were irradiated (2 Gy) and ERK–MAPK activity was determined over 0–300 min as described in Materials and Methods. Cells were lysed and portions (~100 μg) from each plate were used to immunoprecipitate ERK–MAPK followed by immune complex kinase assays as described in Materials and Methods. ERK–MAPK activity data are shown as fold increases in ^{32}P incorporation into MBP substrate, and are normalized to activity at time 0 from the means ± SEM of four independent experiments. *Inset:* EGFR-antisense cells were cultured, and were treated with doxycycline as described in Materials and Methods. Cells were irradiated (2 Gy) and the tyrosine phosphorylation of the EGFR was determined over 0–300 min as described in Materials and Methods. Cells were lysed and portions (~100 μg) from each plate were used to immunoprecipitate EGFR, followed by SDS–PAGE and immunoblotting versus either EGFR or phosphotyrosine (active) EGFR (EGFR-phosphotyrosine); a representative experiment is shown (n = 4). Exposure time: 30 sec.

EGFR. Several studies have suggested that activation of the ERK–MAPK pathway represents a cytoprotective signal.[1,21–23] Inhibition of ERK–MAPK function by the MEK 1/2 inhibitor U0126 during irradiation enhanced radiation-induced apoptosis. These data suggest that ionizing radiation may exert a self-limiting effect on its ability to kill and to reduce the proliferation of tumor cell ERK–MAPK activity, which in turn will lead to both increased proliferation of tumor cells as well as other cytoprotective responses.

Acknowledgments

Funding for this work was to P.D. from a fellowship from the V Foundation, a PHS grant (R01DK52835), a grant from the Jeffres Research Fund (J-464), and a grant from the Department of Defense (BC98-0148); to R.S.U. from PHS grants (P01CA72955) and (R01CA65896).The authors also thank J. Trzaskos (DuPont) for providing U0126. Figures 1 and 3 were reproduced from Dent *et al., Mol. Biol. Cell* **10,** 2493 (1999) by kind permission of the Journal.

[21] D. W. Abbott and J. Holt, *J. Biol. Chem.* **274,** 2732 (1999).
[22] P. C. Gokhale, D. McRae, B. P. Monia, A. Bagg, A. Rahman, A. Dritschilo, and U. Kasid, *Antisense Nucleic Acid Drug Dev.* **9,** 191 (1991).
[23] J. A. Vrana, S. Grant, and P. Dent, *Radiat. Res.* **151,** 559 (1999).

[4] Ras Activation of Phosphatidylinositol 3-Kinase and Akt

By PABLO RODRIGUEZ-VICIANA and JULIAN DOWNWARD

Introduction

Ras proteins have been found to interact with a number of target enzymes in addition to Raf (see Ref. 1 for review). An important alternative effector for Ras is the family of type I phosphoinositide 3-OH (phosphatidylinositol) kinases (PI 3-kinases) which bind to and activate the catalytic subunit of the lipid kinase.[2,3] The ability of Ras to activate PI 3-kinase in this way is important for the promotion of transformation by activated *ras* oncogenes,[4] at least part of which is due to the

[1] C. J. Marshall, *Curr. Opin. Cell Biol.* **8,** 197 (1996).
[2] P. Rodriguez-Viciana, P. H. Warne, R. Dhand, B. Vanhaesebroeck, I. Gout, M. J. Fry, M. D. Waterfield, and J. Downward, *Nature* (*London*) **370,** 527 (1994).
[3] P. Rodriguez-Viciana, P. H. Warne, B. Vanhaesebroeck, M. D. Waterfield, and J. Downward, *EMBO J.* **15,** 2442 (1996).
[4] P. Rodriguez-Viciana, P. H. Warne, A. Khwaja, B. M. Marte, D. Pappin, P. Das, M. D. Waterfield, A. Ridley, and J. Downward, *Cell* **89,** 457 (1997).

0076-6879/00 $35.00

ability of the PI 3-kinase pathway, acting through Akt, to inhibit apoptosis.[5,6] In this chapter we discuss assays by which the interaction of Ras with PI 3-kinase can be explored *in vitro,* and also assays that reveal the effects of these interactions on downstream readouts in whole cells.

Ras Interaction with Phosphatidylinositol 3-Kinase *in Vitro*

The interaction between Ras proteins and PI 3-kinase can be studied *in vitro,* using either soluble PI 3-kinase and immobilized Ras [e.g., purified untagged Ras covalently coupled to Affi-Gel 10 beads or glutathione *S*-transferase (GST)–Ras bound to glutathione–agarose] or immobilized PI 3-kinase and soluble Ras. The catalytic p110α subunit of PI 3-kinase has proved difficult to express in bacteria; however, it can be expressed as a full-length active protein, either without any tags or as a GST fusion protein, by infection of Sf9 (*Spodoptera frugiporda* ovary) cells with a recombinant baculovirus. p110 can be expressed either alone or bound to the p85α regulatory subunit, if the Sf9 cells are coinfected with a p85α baculovirus. However, the easiest way to look at the Ras–PI 3-kinase interaction *in vitro* is to use the Ras-binding domain (RBD) of the p110 subunit, amino acids 133–314 in p110α, expressed as a GST fusion protein in bacteria.

The p110 RBD GST fusion protein can be used in pulldown assays of cell lysates to determine the activation state of Ras in the extracts; however, the use of the Ras-binding domain of Raf is preferable for this due to its higher binding affinity for Ras.[7] The interaction of purified Ras with p110 RBD–GST can also be adapted to a 96-well format enzyme-linked immunosorbent assay (ELISA) and used to identify inhibitors of the interaction of Ras with effectors. In this case the lower affinity of the interaction of Ras with PI 3-kinase is an advantage, and allows weak inhibitors to be identified that are unable to significantly inhibit the interaction of Ras with Raf (J. Downward and P. H. Warne, unpublished data, 1999).

Expression of p110α Ras-Binding Domain

A 40-ml overnight culture of bacteria carrying the pGEX plasmid (Pharmacia, Piscataway, NJ) expressing amino acids (aa) 133–314 of p110α is diluted 1 : 25 into 1 liter of L-broth containing ampicillin. When the culture reaches an OD_{600} of 0.4–0.6 (2–3 hr), expression of the fusion protein is induced by incubation with 0.4 m*M* isopropyl-β-D-thiogalactopyranoside (IPTG) for an additional 2–4 hr at 37°. Bacteria are harvested by centrifugation at 4000 rpm for 10 min at 4°. The

[5] A. Khwaja, P. Rodriguez-Viciana, S. Wennstrom, P. H. Warne, and J. Downward, *EMBO J.* **16,** 2783 (1997).

[6] B. M. Marte, P. Rodriguez-Viciana, S. Wennström, P. H. Warne, and J. Downward, *Curr. Biol.* **7,** 63 (1997).

[7] S. J. Taylor and D. Shalloway, *Curr. Biol.* **6,** 1621 (1996).

pellet is resuspended in 15 ml of cold GST lysis buffer [1% (v/v) Triton X-100, 1 mM EDTA, 1 mM benzamidine, and 1 mM dithiothreitol (DTT) in phosphate-buffered saline (PBS)]. The bacteria are lysed by sonication on ice (four times, 1 min each time) and cell debris is removed by centrifugation at 15,000 rpm for 30 min at 4°. The supernatant is incubated with 1 ml of a 50% (w/v) slurry of glutathione–agarose beads that have been thoroughly washed in PBS. After 1–2 hr of rotation at 4°, the beads are centrifuged at 2000 rpm for 1 min and washed five times in PBS–1 mM EDTA to remove unbound proteins. Purified protein can be checked by running 10 μl of packed beads in a 10% (w/v) gel by sodium dodecyl sulfate–polyacrylamide gel electrophoresis (SDS–PAGE). Beads can be stored in PBS at 4° for at least 1 week. For long-term storage it is preferable to keep the beads at −20° in PBS–50% (v/v) glycerol.

GDP/GTP Loading of Purified Ras

Add Ras protein to loading buffer: 20 mM HEPES (pH 7.5), 50 mM NaCl, 5 mM EDTA, and bovine serum albumin (BSA, 5 mg/ml) (e.g., 10 μl of Ras protein solution to 20 μl of loading buffer). Ras proteins are stored in the presence of $MgCl_2$ [e.g., PBS, 50% (v/v) glycerol, 5 mM $MgCl_2$], and so the volume of Ras protein should always be less than the volume of loading buffer in order to ensure that all Mg^{2+} is chelated. Add GDP, GTP, or nonhydrolyzable GTP analog (GTPγS or GMP–PNP) to at least a 10-fold excess over Ras, for example, 0.5–1 μl of 1 mM GDP/GTP per microgram of Ras. Incubate for 5 min at 37°. Stop the reaction by placing the tubes on ice and adding $MgCl_2$ to a final concentration of 15 mM (so that there is a final excess of Mg^{2+} over EDTA, e.g., 1/20 of the reaction volume of 300 mM $MgCl_2$).

Binding Reaction

Add 0.5–1 μg of GDP/GTP-loaded Ras to a tube containing 200 μl of PBS, 5 mM $MgCl_2$, 0.1–1% (v/v) Triton X-100, and 30 μl [1 : 3 (w/v) in PBS] of glutathione beads bound to the GST–p110–RBD. Rotate at 4° for 1–2 hr. Wash five times with PBS, 5 mM $MgCl_2$, 0.1% (v/v) Triton X-100, drain the beads, and add 1× SDS–PAGE sample buffer. Run the samples in a 12.5–15% (w/v) polyacrylamide gel, transfer to nitrocellulose or polyvinylidene difluoride (PVDF) membrane, and perform a Western blot with anti-Ras antibody (e.g., pan-Ras monoclonal antibody 4 from Calbiochem, La Jolla, CA). GTP-bound Ras should interact with p110 at least 20-fold more strongly than GDP-bound Ras.

Ras Activation of Phosphatidylinositol 3-Kinase *in Vitro*

As well as interacting with PI 3-kinase *in vitro,* Ras will also cause activation of the enzymatic activity of PI 3-kinase after this interaction. This can be seen using purified posttranslationally modified Ras and purified baculovirus-expressed p85α/p110α–GST, along with liposomes containing phosphoinositide. The activation of the lipid kinase activity by Ras is synergistic with the effects of tyrosine phosphopeptide binding to p85: the activation may be due to improved access of PI 3-kinase to its lipid substrate because of its interaction with Ras that is inserted into the liposomes as a result of its isoprenylation, or may be due to a conformational change, or to a mixture of the two (see Rodriguez-Viciana *et al.*[3]).

Reconstitution of Ras into Liposomes

Liposomes are made by sonication of dried lipids in kinase buffer [40 m*M* HEPES (pH 7.5), fatty acid-free BSA (1 mg/ml) 2 m*M* EGTA, 100 m*M* NaCl, 5 m*M* $MgCl_2$, 50 μ*M* ATP]. The final lipid concentration is 640 μ*M* phosphatidylethanolamine, 600 μ*M* phosphatidylserine, 280 μ*M* phosphatidylcholine, 60 μ*M* sphingomyelin, 60 μ*M* phosphatidylinositol 4,5-biphosphate [PI(4,5)P_2]. Sonicate on ice (four times, 20 sec each) with a 4-mm probe at 15-μm amplitude in an MSE Soniprep 150. Pure modified Ras is added to the liposomes at a final concentration of 1 μ*M* and incubated on ice for 20 min. The modified Ras can be obtained by the method described by Porfiri *et al.*[8] Liposomes are then collected by centrifugation at 350,000*g* for 20 min. The supernatant is removed and the liposome pellet is resonicated in the same volume of kinase buffer. To load Ras with nucleotide, the mixture is made 10 m*M* in EDTA, and then 20 μ*M* GTP or GDP is added. After incubation at 30° for 10 min, 15 m*M* $MgCl_2$ is added.

Phosphatidylinositol 3-Kinase Assays in Liposomes

To the liposome mixture made as described above, either with or without Ras, 10 ng of p85α/GST–p110α and 1 μCi of [γ-^{32}P]ATP per assay are added. Tyrosine phosphopeptides can also be added at this point. The total assay volume is 25 μl, a 50% dilution of the liposome preparation. Assays are mixed continuously for 20 min at 20°; the reaction is then stopped by adding 400 μl of chloroform–methanol (1 : 2, v/v) and 100 μl of 2.4 *M* HCl. After vortexing and centrifugation (15,000*g*, 3 minutes), the lower chloroform layer is put into fresh tubes and dried down in a Speed-Vac. The residue is dissolved in 25 μl of chloroform and run on a silica thin-layer chromatography (TLC) plate in chloroform–methanol–acetone–acetic acid–water (40 : 13 : 15 : 12 : 7, by volume). The phosphatidylinositol triphosphate (PIP_3) spot is quantitated with a Molecular Dynamics (Sunnyvale, CA) PhosphorImager. See the next section for a fuller description of the TLC analysis. It should be possible to see a 3- to 5-fold activation of PI 3-kinase-induced PIP_3 generation in the presence of GTP-bound Ras.

[8] E. Porfiri, T. Evans, G. E. Bollag, R. Clark, and J. F. Hancock, *Methods Enzymol.* **255,** 13 (1995).

Ras Activation of Phosphatidylinositol 3-Kinase Activity in Intact Cells

To analyze the cellular levels of the lipid products of PI 3-kinase, it is necessary to metabolically label the cells with ortho[^{32}P]phosphate and identify and quantify the levels of PI(3,4)P_2 and phosphatidylinositol 3,4,5-triphosphate [PI(3,4,5)P_3] by high-performance liquid chromatography (HPLC). However, if an HPLC system is not available, a simpler and often sufficiently informative alternative is to perform TLC of the total lipid extract. Under the conditions described below, the TLC plate will resolve PI(3,4,5)P_3 as an individual spot (or band), but identification of PI(3,4)P_2 is not possible in this system. The PI(3,4,5)P_3 runs with a low mobility on the TLC plate, running at about 0.27 times the speed of the front, in an area that is predominantly free of other bands. When strong elevation of PIP_3 is occurring it should be readily detected by TLC alone; more subtle changes are likely to require HPLC analysis.

Ras activation of PI 3-kinase activity in intact cells can be looked at by measuring the levels of the lipid products of PI 3-kinase in the cell after expression of activated forms of Ras. This can be done by retroviral infection if a high-titer Ras retrovirus is available. The coexpression of PI 3-kinase with Ras (either the p110 subunit by itself or both the p110 and p85 subunits) will increase the signal-to-noise ratio and greatly increase the magnitude of the effects. This is most easily achieved by cotransfection. Because of the efficiency of transfection and the high levels of expression achieved, COS cells (either COS1 or COS7) are a convenient mammalian cell line with which to do this assay. COS cells can be successfully transfected in a variety of ways. For its efficiency and ease we routinely use lipofection with LipofectAMINE Plus (GIBCO-BRL, Gaithersburg, MD). Transfection of a 6-cm tissue culture dish is sufficient to perform HPLC analysis of the total cell lipids. If only TLC analysis is going to be performed, transfection in six-well plates is sufficient and reduces the amount of radioactivity involved.

The following protocol is for transfection of six-well plates. For transfection of 6-cm dishes all the volumes and numbers should be multiplied by two.

COS Transfection

Seed 1.5–2 × 10^5 COS cells per well the day before transfection. The next day (day 1), in 1.5-ml tubes, add 100 μl of Dulbecco's modified Eagle's medium without any serum or antibiotics (DMEM*). Add a total of 1 μg of plasmid(s) per tube. Cotransfection of Ras and wild-type p110 at a ratio of 1:1 (0.5 μg each) works well and is routinely used. The amount of plasmid per transfection should always be the same and made to 1 μg with empty vector when necessary. Add 6 μl of Plus reagent, mix, and let stand at room temperature for 15 min. Add 100 μl of DMEM* to which 4 μl of LipofectAMINE has been added. The use of a master mix is recommended; for example, for 10 transfections, add 44 μl of

LipofectAMINE to 1.1 ml of DMEM* and add 104 μl of this mix per tube. Mix and let stand for 15 min. Add the 200 μl of liposome mix to cells that have been washed once and left with 800 μl of DMEM*. After 3 hr, aspirate off and add 2 ml/well of DMEM–10% (v/v) fetal calf serum. The next day (day 2), cells can be serum starved if appropriate by leaving them overnight in DMEM–fatty acid-free BSA (1 mg/ml). On day 3 perform labeling.

^{32}P Labeling and Lipid Analysis

Two days after transfection, wash the cells twice with phosphate-free DMEM–fatty acid-free BSA (1 mg/ml) and leave for 1 hr. Replace with the same medium at 0.75 ml per well for a six-well plate or at 1.5 ml for a 6-cm dish. Using appropriate shielding and precautions for millicurie quantities of ^{32}P, add ortho[^{32}P]phosphate at 100 μCi/ml. Label the cells for 1.5–4 hr, typically 3 hr.

Prior to extraction of lipids, prepare in advance clean upper and lower phases (cUP and cLP). Make a solution of chloroform–methanol–1 *M* HCl, 25 m*M* EDTA, and 5 m*M* tetrabutylammonium hydrogen sulfate (TBAS; 8 : 4 : 3, by volume). Vortex and centrifuge. The resulting lower and upper phases are referred to as the clean upper phase (cUP) and clean lower phase (cLP), respectively. To avoid dripping when pipetting any chloroform phase it is always advisable to "seal" the tip by prepipetting first from a tube containing chloroform. Wash the cells with PBS and add 0.5 ml of ice-cold 1 *M* HCl–5 m*M* TBAS. Scrape the cells and transfer then to a 15-ml polypropylene or glass tube (tube 1) (Polybrene will melt in the presence of chloroform). Add 2 ml of methanol–chloroform (1 : 2, v/v) containing phosphoinositides (10 μg/ml; Sigma, St. Louis, MO) to act as carriers. Vortex vigorously and centrifuge for 2 min at 3000 rpm. Transfer the lower phase to a new 15-ml tube (tube 2) containing 1 ml of clean upper phase (cUP). Add 1.5 ml of clean lower phase (cLP) to tube 1 to reextract the aqueous upper phase. Vortex and centrifuge tubes 1 and 2. Transfer and combine the lower phases of tubes 1 and 2 in a clean tube (tube 3). Samples can be stored at −20° at this stage. Dry in a Speed-Vac vacuum concentrator.

Thin-Layer Chromatography Analysis

The large increase in PIP_3 seen in COS cells transfected with p110 and Ras can be easily seen on a TLC plate as discussed above. If HPLC analysis is going to be performed this step is optional. If a 6-cm dish has been used, one-tenth of the chloroform phase is dried separately and used for analysis by TLC. If the transfection is performed on a six-well dish, dry the entire chloroform phase to be used for TLC.

Resuspend the dried sample in 50 μl of chloroform. Load half on a silica gel 60 TLC plate (e.g., Whatman, Clifton, NJ) that has been sprayed with a 1% potassium oxalate solution and allowed to dry. Run for 4 hr in a TLC tank preequilibrated with chloroform–water–methanol–acetone–acetic acid (40 : 7 : 13 : 15 : 12, by volume) that has been freshly prepared. Let dry and expose by autoradiography or on a PhosphorImager.

Deacylation and High-Performance Liquid Chromatography Analysis

Phosphoinositides are analyzed by anion-exchange HPLC as deacylated, glycerophosphoinositides. Standards can be produced by phosphorylating PI(4)P and $PI(4,5)P_2$ with purified PI 3-kinase and $[\gamma\text{-}^{32}P]ATP$ and processing the samples as described below. Resuspend dried lipids in 200 μl of 33% methylamine solution (e.g., Fluka, Ronkonkoma, NY) by vigorous vortexing. Incubate for 30 min at 53°. Because methylamine is volatile it is advisable to use screw-top tubes to prevent the tubes from snapping open. Alternatively, put some weights on top of the tubes during the incubation. Let them air dry in a fume hood for several hours and finish drying in Speed-Vac. Add 0.5 ml of water and 0.6 ml of butanol–petroleum ether–ethyl formate (20 : 4 : 1, v/v/v). Vortex vigorously and centrifuge briefly. Transfer the upper butanol phase to a clean 1.5-ml tube and reextract with 0.5 ml of water. Vortex, centrifuge, and combine the two lower aqueous phases. The sample is now ready for injection and can be stored until use at −20°.

Inject half the sample (0.5 ml) into a Spherisorb S5SAX HPLC column (Waters, Milford, MA) and elute at 1 ml/min with a gradient of Milli-Q water (A) versus 1.25 *M* NaH_2PO_4, pH 3.8 (B), with the following program: $t=0$, $\%B=0$; $t=5$, $\%B=0$; $t=65$, $\%B=28$; $t=70$, $\%B=40$; $t=75$, $\%B=100$; $t=81$, $\%B=0$; $t=150$, $\%B=0$. Fractions are collected every 0.5 min and counted by Cerenkov emission on a scintillation counter. It is necessary to start collecting only after 45 min of the gradient described above. $PI(3,4)P_2$ elutes after 55 min of this gradient, followed four to six fractions later by $PI(4,5)P_2$. Another 25–30 fractions later, $PI(3,4,5)P_3$ is eluted, at 72 min. Different runs can be standardized to the level of $PI(4,5)P_2$. The performance of the column changes with the number of times it has been run, and so the elution times listed here are only a rough guide.

Ras Activation of Akt Activity in Intact Cells

Through its ability to activate PI 3-kinase, Ras is also able to stimulate the activity of the protein serine/threonine kinase Akt, also known as PKB.[6] When PI 3-kinase-produced lipids increase in the cell, Akt becomes localized to the plasma membrane as a result of these lipids binding to its pleckstrin homology domain. At the membrane, Akt is phosphorylated by two upstream kinases, PDK1 acting on T308 and PDK2 acting on S473. These phosphorylations result in stimulation of the protein kinase activity of Akt. The upstream kinases PDK1 and PDK2 are also stimulated by PI 3-kinase-generated lipids. For a review see Marte and Downward.[9]

[9] B. M. Marte and J. Downward, *Trends Biochem. Sci.* **22,** 355 (1997).

The simplest way to see the activation of Akt by Ras in intact cells is to use COS cell cotransfection assays, similar to those described above. Epitope-tagged Akt is transfected along with Ras, and then the kinase activity of the transfected Akt is measured in immune complexes, using histone H2B as substrate.

Kinase Assays

Hemagglutinin (HA)-tagged Akt is transfected into COS7 cells, with or without activated Ras, using LipofectAMINE (GIBCO-BRL) and the protocols described above. Typically, 0.5 μg of HA-tagged Akt in pcDNA3 can be used, along with 0.5 μg of V12 Ras in pSG5. Forty-eight hours after transfection, cells are serum starved overnight in DMEM plus 0.5% (v/v) fetal calf serum. Cells are lysed in 1 ml of lysis buffer on ice for 5 min. [Lysis buffer contains 50 m*M* HEPES (pH 7.5), 100 m*M* NaCl, 1% (v/v) Triton X-100, 1 m*M* EDTA, 5 m*M* NaF, and 0.1 m*M* $Na_3\ VO_4$.]

After centrifugation in a microcentrifuge at 4° for 5 min at 14,000*g*, HA–Akt is immunoprecipitated from the lysate with 2 μg of 12CA5 antibody for 1 hr at 4°, followed by tumbling with 20 μl of protein G–Sepharose for 30 min. The immunoprecipitates are washed with two 1-ml volumes of PBS–0.1% (v/v) Triton X-100–0.5 *M* NaCl, then with two 1-ml volumes of PBS–0.1% (v/v) Triton X-100, and then with 1 ml of kinase buffer [50 m*M* Tris (pH 7.5), 10 m*M* $MgCl_2$, 1 m*M* DTT]. Kinase assays are performed as in Burgering and Coffer,[10] except that histone H2B is used as a substrate. Briefly, 25 μl of kinase mix [50 m*M* Tris (pH 7.5), 10 m*M* $MgCl_2$, 1 m*M* DTT, histone 2B (0.1 mg/ml; Boehringer Mannheim, Indianapolis, IN) 1 μ*M* protein kinase A inhibitor (Sigma), 20 μ*M* cold ATP, and [γ-^{32}P]ATP (100 μCi/ml)] is added to each washed immunoprecipitate. These are then shaken at room temperature for 30 min. The reaction is then terminated by the addition of 25 μl of 2× gel sample buffer. The samples are run on a 15% (w/v) SDS–polyacrylamide gel. The incorporation of ^{32}P label into H2B at 18 kDa can be determined by scanning the gel in a PhosphorImager.

To check even expression and immunoprecipitation of HA–Akt, Western blots with a polyclonal antibody against the carboxy terminus of Akt (New England BioLabs, Beverly, MA) can be carried out on these samples. If necessary, the kinase activity can be normalized against the amount of Akt protein present as quantified by enhanced chemifluorescence detection and a STORM system scanner. This can be done on the same gel as the H2B analysis: Akt runs at 55 kDa, and so the upper half of the gel can be analyzed by Western blot for Akt, while the lower half is analyzed by scanning for quantitation of ^{32}P incorporation in H2B.

In these assays, cotransfection with V12 Ras typically activates Akt by about 20-fold without significant changes in expression levels.

[10] B. M. T. Burgering and P. J. Coffer, *Nature* (*London*) **376,** 599 (1995).

[5] Assays for Monitoring p70 S6 Kinase and RSK Activation

By JEFFREY MASUDA-ROBENS, VERA P. KRYMSKAYA, HONGWEI QI, and MARGARET M. CHOU

Introduction

A hallmark of the mitogenic response is an increase in protein synthesis. This increase arises from the modulation of multiple components of the translational machinery, including phosphorylation of the S6 polypeptide of the 40S ribosomal subunit.[1] Given the universality of S6 phosphorylation during proliferation, there was much impetus for identifying S6 kinases and characterizing their mechanism of activation. This led to the purification and cloning of two S6 kinases, p70 S6 kinase (p70S6k) and p90^{RSK} (more commonly referred to as RSK).[2–5] Since their initial identification, it has become evident that p70S6k is in fact the relevant S6 kinase in mammalian cells, whereas the role of RSK in S6 phosphorylation may be restricted to *Xenopus laevis* oocytes.[6–8] Nevertheless, RSK possesses an important role in the mitogenic response in mammalian cells, regulating gene expression through phosphorylation of proteins such as c-Fos, serum response factor (SRF), cAMP-response element-binding protein (CREB), and histone H3.[9–12]

Much has been learned about the signaling pathways that mediate activation of p70S6k and RSK. One common activator of both enzymes is p21$^{ras.}$ However, p21^{ras} utilizes distinct downstream effectors to elicit their activation. In the case of

[1] S. Ferrari and G. Thomas, *Crit. Rev. Biochem. Mol. Biol.* **29,** 385 (1994).

[2] P. Banerjee, M. F. Ahmad, J. R. Grove, C. Kozlosky, D. J. Price, and J. Avruch, *Proc. Natl. Acad. Sci. U.S.A.* **87,** 8550 (1990).

[3] D. A. Alcorta, C. M. Crews, L. J. Sweet, L. Bankston, S. W. Jones, and R. L. Erikson, *Mol. Cell. Biol.* **9,** 3850 (1989).

[4] S. W. Jones, E. Erikson, J. Blenis, J. L. Maller, and R. L. Erikson, *Proc. Natl. Acad. Sci. U.S.A.* **85,** 3377 (1988).

[5] S. C. Kozma, S. Ferrari, P. Bassand, M. Siegmann, N. Totty, and G. Thomas, *Proc. Natl. Acad. Sci. U.S.A.* **87,** 7365 (1990).

[6] J. Blenis, J. Chung, E. Erikson, D. A. Alcorta, and R. L. Erikson, *Cell Growth Differ.* **2,** 279 (1991).

[7] E. Erikson and J. L. Maller, *Proc. Natl. Acad. Sci. U.S.A.* **82,** 742 (1985).

[8] E. Erikson and J. L. Maller, *J. Biol. Chem.* **261,** 350 (1986).

[9] R. H. Chen, C. Abate, and J. Blenis, *Proc. Natl. Acad. Sci. U.S.A.* **90,** 10952 (1993).

[10] V. M. Rivera, C. K. Miranti, R. P. Misra, D. D. Ginty, R. H. Chen, J. Blenis, and M. E. Greenberg, *Mol. Cell. Biol.* **13,** 6260 (1993).

[11] P. Sassone-Corsi, C. A. Mizzen, P. Cheung, C. Crosio, L. Monaco, S. Jacquot, A. Hanauer, and C. D. Allis, *Science* **285,** 886 (1999).

[12] J. Xing, D. D. Ginty, and M. E. Greenberg, *Science* **273,** 959 (1996).

0076-6879/00 $35.00

p70S6k, activation is mediated by phosphatidylinositol 3-kinase (PI3K).[13,14] PI3K functions in part by stimulating p70S6k-activating kinases and also, most likely, by activating the Rho family GTPases Cdc42 and Rac.[15–17] Although the precise mechanism by which Cdc42 and Rac stimulate p70S6k remains unclear, they may function to relocalize p70S6k, juxtaposing it with its upstream kinases. Consistent with this model, Cdc42 and p70S6k have been shown to coimmunoprecipitate *in vivo.*[16]

In contrast, RSK activation by $p21^{ras}$ is mediated by the Raf → MEK → MAPK kinase module (MEK, MAPK/ErK kinase; MAPK, mitogen-activated protein kinase). MAPK kinase directly phosphorylates RSK on two sites, leading to activation of the N-terminal of RSK's two kinase domains.[18–21]

The development of specific antibodies against p70S6k and RSK has allowed monitoring of their activation *in vivo.* The next section details assays for measuring their activity, including preparation of recombinant substrate and a cotransfection protocol for examining activation by Ras superfamily GTPases.

Materials and Methods

Generation of Antisera

Rabbit polyclonal antibodies are generated against the extreme N- or C-terminal 20 amino acids of rat p70S6k (αII isoform) (antibodies denoted anti-p70S6k-N2 and -C2, respectively). Peptides are synthesized with an N-terminal cysteine to allow cross-linking to keyhole limpet hemocyanin (KLH). Although both antisera immunoprecipitate p70S6k efficiently, only anti-p70S6k-C2 functions well on immunoblots. The C-terminal antiserum recognizes p70S6k of rat, murine, and human origin. Commercially available anti-p70S6k antibodies suitable for immunocomplex kinase assays are also used (Santa Cruz Biotechnology, Santa Cruz, CA).

For RSK, rabbit polyclonal antibodies are generated against full-length recombinant avian RSK1 [denoted anti-RSK(6)], or the extreme C-terminal 20 amino acids coupled to KLH. Both antisera function for immunoprecipitation and

[13] J. Chung, T. Grammer, K. P. Lemon, A. Kazlauskas, and J. Blenis, *Nature (London)* **370,** 71 (1994).

[14] B. Cheatham, C. J. Vlahos, L. Cheatham, L. Wang, J. Blenis, and C. R. Kahn, *Mol. Cell. Biol.* **14,** 4902 (1994).

[15] B. M. T. Burgering and P. J. Coffer, *Nature (London)* **376,** 599 (1995).

[16] M. M. Chou and J. Blenis, *Cell* **85,** 573 (1996).

[17] N. Pullen, P. B. Dennis, M. Andjelkovic, A. Dufner, S. C. Kozma, B. A. Hemmings, and G. Thomas, *Science* **279,** 707 (1998).

[18] C. Sutherland, D. G. Campbell, and P. Cohen, *Eur. J. Biochem.* **212,** 581 (1993).

[19] T. W. Sturgill, L. B. Ray, E. Erikson, and J. L. Maller, *Nature (London)* **334,** 715 (1988).

[20] T. L. Fisher and J. Blenis, *Mol. Cell. Biol.* **16,** 1212 (1996).

[21] K. N. Dalby, N. Morrice, F. B. Caudwell, J. Avruch, and P. Cohen, *J. Biol. Chem.* **273,** 1496 (1998).

immunoblotting, and recognize RSK1 of avian, murine, rat, and human origin; only data for anti-RSK(6) are shown below. Anti-RSK(6) is affinity purified for immunoblotting, and demonstrates cross-reactivity with RSK2.

Purification of Recombinant S6 Kinase Substrate

The residues of S6 that are phosphorylated by p70S6k and RSK are contained within the C-terminal 32 amino acids of the polypeptide (Fig. 1). This sequence is identical in human, murine, and rat S6. An oligonucleotide encoding this peptide is subcloned directionally into *Bam*HI/*Eco*RI-digested pGEX-3X (Pharmacia Piscataway, NJ); this construct is termed glutathione S-transferase (GST)–S6(32aa).

A 50-ml overnight culture of DH5α expressing GST–S6(32aa) is grown in Luria broth containing ampicillin (200 μg/ml) at 37°. The next morning, the culture is diluted into 1 liter of the same medium and grown for 2 hr. Isopropyl-β-D-thiogalactopyranoside (IPTG) is added to a final concentration of 0.1 m*M*, and the culture is grown for an additional 2–4 hr. A small aliquot (e.g., 1 ml) of bacterial culture is collected before and after IPTG induction and pelleted for 30 sec in a microcentrifuge. The supernatant is removed and the bacterial pellet

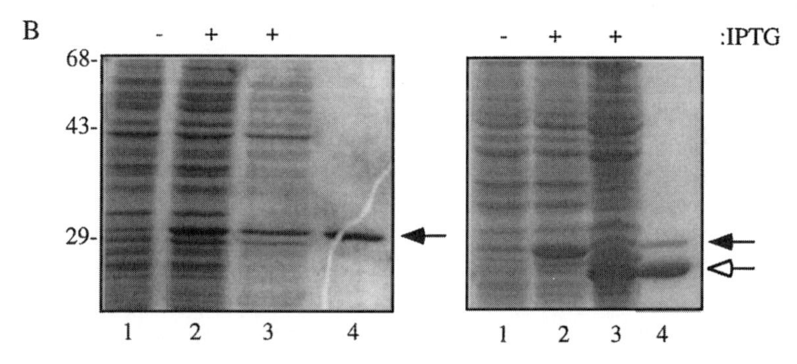

FIG. 1. Purification of GST–S6(32aa) substrate. (A) Sequence of the C-terminal 32 amino acids of S6 that are subcloned into pGEX-3X. Serine residues that are phosphorylated *in vivo* are outlined. (B) DH5α harboring S6(32aa)/pGEX-3X were grown at 37° for 2 hr, then induced with 0.1 m*M* IPTG for an additional 2 hr. A 1-ml aliquot of culture was collected before and after IPTG induction, and then boiled directly in sample buffer (lanes 1 and 2, respectively). The remaining culture was harvested, lysed by sonication, and the GST–S6(32aa) peptide purified by affinity chromatography using glutathione–Sepharose. GST–S6(32aa) was eluted and concentrated with Centricon filters. Lane 3, bacterial extract after sonication; lane 4, eluted GST–S6(32aa) peptide. Filled arrow, full-length GST–S6(32aa); open arrow, degradation product commonly observed. The left-hand and right-hand sides are two independent preparations; in the latter, significant cleavage of GST–S6(32aa) was observed on oversonication. Positions of the 29-, 43-, and 68-kDa markers are shown.

resuspended in 100–200 μl of 1× sample buffer [125 m*M* Tris(pH 6.8), 2% (w/v) sodium dodecyl sulfate (SDS), 5% (v/v) 2-mercaptoethanol (2-ME), 7.5% (v/v) glycerol, bromphenol blue] and boiled for 5 min. These "whole cell lysates" are later used as migration controls for the full-length GST–S6(32aa) polypeptide. This is particularly useful because the protein is susceptible to proteolytic cleavage, yielding a truncated product that migrates close to full-length GST–S6(32aa) but fails to serve as substrate (see notes below and Fig. 1).

The remainder of the bacterial culture is harvested by centrifugation at 5000 rpm for 10 min, and then resuspended in 10 ml of ice-cold lysis buffer [phosphate-buffered saline (PBS), 1% (v/v) Triton X-100, 100 m*M* EDTA] containing freshly added protease inhibitors phenylmethylsulfonyl fluoride (PMSF, 1 m*M*), aprotinin (1%, v/v), leupeptin (10 μg/ml), and pepstatin (0.7 μg/ml). Dithiothreitol (DTT) is specifically omitted, as its inclusion enhances proteolytic cleavage of the GST–S6(32aa) polypeptide. The resuspended bacteria are lysed by sonication on ice. Three 10- to 15-sec bursts with pulsing are used; oversonication enhances proteolytic cleavage (see Fig. 1). The bacterial lysate is then distributed into Eppendorf tubes and pelleted in a microcentrifuge at 4° for 10 min at full speed. The supernatants are combined and transferred to a 15-ml conical tube. Glutathione–Sepharose beads (Pharmacia; 1-ml bead volume) are added, and the sample is incubated for 2–3 hr at 4° with constant mixing. The beads are washed three times in lysis buffer, twice in high-salt buffer [20 m*M* Tris (pH 7.5), 1 *M* NaCl, 1% (v/v) Triton X-100], then once in PBS. All buffers are supplemented with freshly added protease inhibitors. Washes are performed by resuspending the beads in 10 ml of buffer, vortexing gently, and then pelleting at 2000–4000 *g* for 3 min. The entire washing procedure is conducted at 4° to minimize degradation.

The GST–S6(32aa) peptide is next eluted with ice-cold, freshly made elution buffer [50 m*M* Tris, 20 m*M* glutathione (reduced), 100 m*M* NaCl; final pH between 7.0 and 8.0]. Elutions are performed batchwise, such that the final combined eluate volume is 5 ml. At this point, the purified GST–S6(32aa) is no longer susceptible to degradation, and can be stored at −20° before proceeding to desalting of the sample, as described below.

Excess glutathione is removed from the GST–S6(32aa) preparation with PD10 columns (Pharmacia). Columns are equilibrated with 20 m*M* HEPES (pH 7.2)–10 m*M* $MgCl_2$ according to the manufacturer instructions. The GST–S6(32aa) is loaded onto the column, and then eluted with the same buffer. Yields are estimated by running a small fraction (e.g., 1/1000) of the total preparation on 12% SDS–polyacrylamide gels, followed by Coomassie staining. Purified bovine serum albumin (BSA) or ovalbumin is also run as a standard for quantitation. Typically, 3–6 mg of protein is obtained per liter of culture. The full-length GST–S6(32aa) polypeptide migrates at approximately 30 kDa; the proteolytically cleaved product (which fails to serve as a substrate for p70S6k and RSK) migrates at approximately 28 kDa (Fig. 1). The protein is concentrated with Centricon-10 filtration units

(Amicon, Danvers, MA) to a final concentration of 1–2 μg/μl, distributed into 50- to 100-μl aliquots, and stored at −20°. Repeated freeze–thawing does not adversely affect the substrate.

Cell Extract Preparation, Immunoprecipitation, and Kinase Assay Using GST–S6(32aa) as Substrate

NIH 3T3 cells are maintained in Dulbecco's modified Eagle's medium (DMEM) supplemented with 5% (v/v) calf serum, penicillin (100 U/ml)–streptomycin (0.1 mg/ml) (PS), and Glutamax (1%, v/v; GIBCO-BRL Gaithersburg, MD) (5% CM–DMEM). Prior to stimulation, NIH 3T3 cells are seeded to confluence, and then starved for 24–48 hr in 0.5% (v/v) fetal bovine serum (FBS)–DMEM to quiesce p70S6k and RSK activity. FBS is added to a final concentration of 20% (v/v) for various times to induce their activation. Human airway smooth muscle (HASM) cells are isolated and maintained in Ham's F12 medium supplemented with 10% (v/v) FBS.[22] Growth arrest is achieved by a 48-hr incubation in Ham's F12 medium containing 0.1% (w/v) BSA. Cells are stimulated with epidermal growth factor (EGF, 10 ng/ml) for various times.

After agonist stimulation, cells are washed twice with cold PBS, and then scraped on ice directly into cold cell lysis buffer [1 × KPO_4–EDTA buffer*, 5 m*M* EGTA (pH 7.2), 10 m*M* $MgCl_2$, 50 m*M* β-glycerophosphate, 0.5% (v/v) Nonidet P-40 (NP-40), 0.1% (v/v) Brij 35; final pH 7.28]. Sodium vanadate (1 m*M*), DTT (1 m*M*), PMSF (1 m*M*), leupeptin (0.5 μg/ml), pepstatin (0.7 μg/ml), and aprotinin (1%, v/v) are added fresh; approximately 1 ml of cell lysis buffer is used per 10-cm plate. Lysates are transferred to Eppendorf tubes, incubated for 10 min on ice, and then pelleted at full speed in a microcentrifuge for 10 min at 4°. Supernatants are collected and an aliquot removed for immunoblotting. The remainder of the supernatant is immunoprecipitated with antiserum against p70S6k or RSK. Approximately 5 μl of antiserum is used per 1×10^6–3×10^6 cells (i.e., one-fourth of a confluent 10-cm tissue culture dish). Alternatively, commercial antibodies are used for p70S6k (2 μg per 150 to 500 μg of lysate; Santa Cruz Biotechnology). Samples are incubated for 2 hr at 4° with constant mixing. Immunocomplexes are then precipitated with formalin-denatured *Staphylococcus aureus* (30–50 μl) or protein A–Sepharose (35 μl of a 1 : 1 slurry in PBS), and incubated for an additional 1–2 hr at 4° with constant mixing. Immunoprecipitates are washed once each in ice-cold buffer A [10 m*M* Tris (pH 7.2), 100 m*M* NaCl, 1 m*M* EDTA, 1% (v/v) NP-40, 0.5% (v/v) sodium deoxycholate; final pH 7.2], high-salt buffer [10 m*M* Tris (pH 7.2), 1 *M* NaCl, 0.1% (v/v) NP-40; final pH 7.2], ST [100 m*M*

[22] V. P. Krymskaya, R. B. Penn, M. J. Orsini, P. H. Scott, R. J. Plevin, T. R. Walker, A. J. Eszterhas, Y. Amrani, E. R. Chilvers, and R. A. Panettieri, *Am. J. Physiol.* **277,** L65 (1999).

*KPO_4–EDTA (10×): 0.07 *M* K_2HPO_4, 0.03 *M* KH_2PO_4, 10 m*M* EDTA; final pH 7.0.

Tris (pH 7.2), 150 mM NaCl], and 1.5× KB.** Washes are performed by pelleting for 1 min at full speed in a microcentrifuge at 4°, resuspending in 1 ml of the appropriate buffer, followed by gentle vortexing. All wash buffers are stored at 4°, and protease inhibitors and DTT are added to each just prior to use. After the final wash, pellets are resuspended in 20 μl of 1.5× KB. Kinase reactions are initiated by the addition of 10 μl of kinase assay cocktail (amounts given are per sample): 2 μg of GST–S6(32aa), 1.5 μl of 1 mM ATP, 1.0 μl of 22 μM protein kinase inhibitor (PKI; Sigma, St. Louis, MO), 10.0 μCi of [γ-^{32}p]ATP (3000 Ci/mmol), plus Milli-Q water to a final volume of 10 μl. Reactions are carried out for 5–10 min (see below) at 30°, and terminated by the addition of 10 μl of 4× sample buffer [500 mM Tris (pH 6.8), 8% (w/v) SDS, 20% (v/v) 2-ME, 30% (v/v) glycerol, bromphenol blue]. Samples are boiled for 5 min, and then fractionated on 12% SDS–polyacrylamide gels. Gels are Coomassie stained and subjected to autoradiography. GST–S6(32aa) phosphorylation is quantitated with a PhosphorImager (Molecular Dynamics, Sunnyvale, CA).

Optimal reaction times are determined for different batches of the S6 kinase antisera as follows: quiescent NIH 3T3 cells (two confluent 10-cm plates) are stimulated with 20% (v/v) FBS for 10 min, at which time activation of p70S6k and RSK is maximal. Cell extracts from the two plates are combined, and then immunoprecipitated with anti-p70S6k-N2 (40 μl) or anti-RSK(6) (20 μl). Immunoprecipitations and washes are conducted as described above. Pellets are resuspended in 120–160 μl of 1.5× kinase buffer, and reactions are initiated by the addition of the kinase assay cocktail (amount is scaled up on the basis of the number of time points to be tested). At various times, 30-μl aliquots are removed and boiled in 4× sample buffer. Under these assay conditions, reactions are linear for at least 14–20 min (Fig. 2). On the basis of these results, reaction times of 5 and 10 min are used for RSK and p70S6k, respectively. As seen in Fig. 3A, serum induces the potent and rapid activation of both p70S6k and RSK in NIH 3T3 cells.

Filter Kinase Assays

As an alternative to GST–S6(32aa), a commercially available peptide-based kinase assay kit is also used (Upstate Biotechnology, Lake Placid, NY). In the experiment shown in Fig. 3B, p70S6k is immunoprecipitated from HASM cells stimulated with EGF using anti-p70S6k (Santa Cruz Biotechnology). Immunoprecipitations and washes are performed as described above. However, instead of the final wash in 1.5× KB, two washes in assay dilution buffer are performed. Each pellet is then resuspended in 20 μl of assay dilution buffer, 10 μl of S6 kinase substrate peptide (AKRRRLSSLRA), 10 μl of protein kinase inhibitor cocktail

**Kinase buffer (KB, 10×): 0.2 M HEPES (pH 7.2), 0.1 M $MgCl_2$, BSA (1 mg/ml). Solution is stored at −20°. KB (1.5×) is prepared by diluting the 10× stock and supplementing it with 4.5 mM 2-ME; this solution is stored at 4° for short-term use (i.e., several weeks).

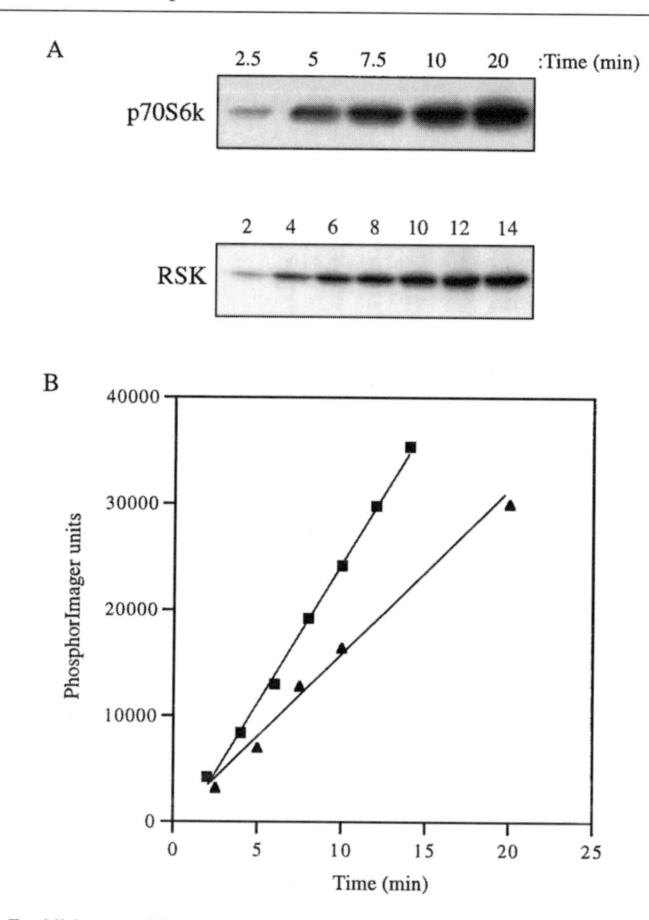

FIG. 2. Establishment of linear kinase assay conditions. NIH 3T3 cells were quiesced and then stimulated with 20% (v/v) FBS for 10 min, at which time maximal activation of p70S6k and RSK occurred. Extracts were prepared and immunoprecipitated with anti-p70S6k-N2 or anti-RSK(6) as described in Materials and Methods. The immunocomplexes were then incubated in kinase reaction cocktail, and at the indicated times aliquots were removed and boiled in sample buffer to terminate the reaction. Samples were fractionated by 12% SDS–PAGE, followed by autoradiography (A). The relative incorporation of ^{32}P was quantitated with a PhosphorImager (B). Triangles, anti-p70S6k-N2 immunoprecipitations; squares, anti-RSK(6) immunoprecipitations.

containing PKA inhibitor peptide, PKC inhibitor peptide, and calmodulin (CaM) kinase inhibitor. The reaction is initiated by adding to each sample 10 μl of magnesium/ATP cocktail containing 10 μCi of [γ-^{32}p]ATP (specific activity, 3000 Ci/mmol). After incubation at 30° for 10 min, a 25-μl aliquot of each sample is transferred onto p81 cation-exchange filters. The filters are washed three times for 5 min in 0.75% (w/v) H_3PO_4 and once for 3 min in 100% acetone. The radioactivity

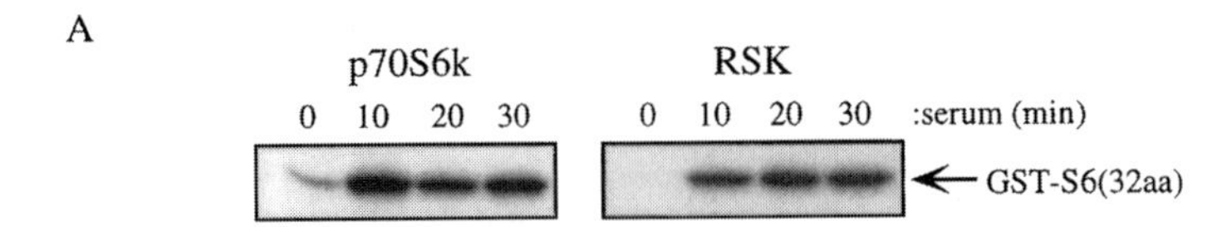

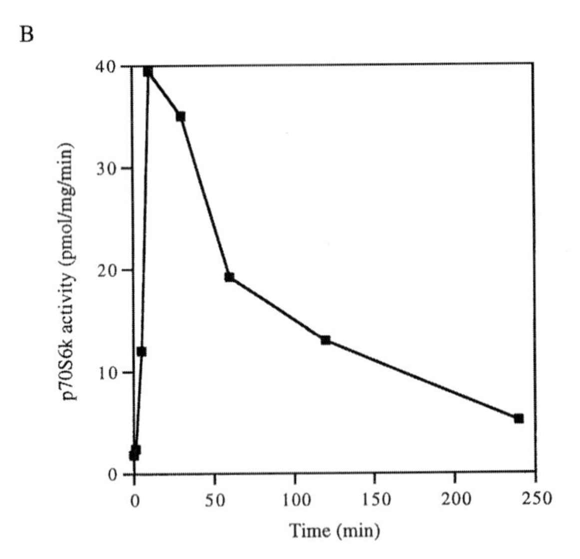

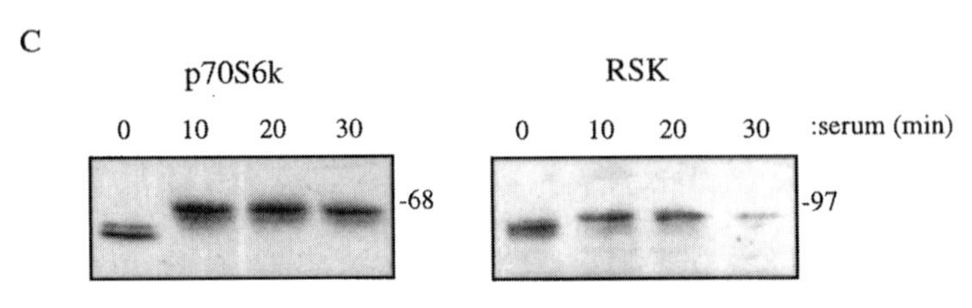

FIG. 3. Mitogen-dependent activation of p70S6k and RSK. (A) NIH 3T3 cells were quiesced for 48 hr, and then stimulated for 10, 20, or 30 min with 20% (v/v) FBS. Cell extracts were prepared and immunoprecipitated with anti-p70S6k(N2) or anti-RSK(6). Kinase assays were performed, using a 10-min reaction time for p70S6k, and a 5-min reaction time for RSK. Samples were fractionated on 12% SDS–polyacrylamide gels and subjected to autoradiography. (B) Filter kinase assays were also used to monitor activation of p70S6k. HASM cells were grown to confluence and then growth arrested for 48 hr. Cells were stimulated with EGF (10 ng/ml) for the indicated times. Cells were harvested and p70S6k immunoprecipitated with anti-p70S6k (Santa Cruz Biotechnology) as described in Materials and Methods. Kinase reactions were carried out for 10 min, using the peptide AKRRRLSSLRA as substrate. Aliquots were spotted onto p81 filters and washed. The radioactivity of each sample was determined by Cerenkov counting. The activity of p70 S6 kinase was calculated as ^{32}P incorporation into the substrate peptide minus nonspecific ^{32}P binding per milligram of total protein per minute. (C) Immunoblots were performed as an additional means of monitoring activation of the S6 kinases, and also to confirm that equal amounts of the enzymes were present in each lysate. The NIH 3T3 lysates prepared in (A) were fractionated on 8% SDS–polyacrylamide gels and probed with anti-p70S6k-C2 (1 : 5000 dilution) and anti-RSK(6) (affinity purified; 1 : 2000 dilution). Proteins were detected with horseradish peroxidase-linked anti-rabbit IgG, followed by enhanced chemiluminescence.

of each sample is determined by Cerenkov counting in a Beckman (Fullerton, CA) LS65000 scintillation counter. S6 kinase activity is measured by ^{32}P incorporation into the substrate peptide minus nonspecific ^{32}P binding per milligram of total protein per minute. Figure 3B demonstrates the potent and rapid activation of p70S6k by EGF in HASM cells, using this method.

Monitoring p70S6k and RSK Activation by Immunoblotting

In addition to measuring p70S6k and RSK activation by immunocomplex kinase assays, it is possible to monitor their activation by immunoblotting because both proteins exhibit shifts in their electrophoretic mobility due to multisite phosphorylation. In the case of p70S6k there are at least 10 basal and mitogen-induced sites, resulting in the production of four electrophoretically distinct bands (see Figs. 3 and 4).[17,23–27] While it is unclear which combination of phosphorylation events gives rise to each of the four bands, it is known that only the slowest mobility form is catalytically active (our unpublished observation, 1996). Optimal resolution of the phosphorylated forms is obtained by fractionating whole cell lysates (5–20 μg) on 8% SDS–polyacrylamide gels. Proteins are transferred to nitrocellulose membrane (Schleicher & Schuell, Keene, NH), and probed with anti-p70S6k-C2 overnight. Antibodies directed against the N terminus of p70S6k function poorly on immunoblots. Blots are then incubated with horseradish peroxidase-linked anti-rabbit IgG, followed by detection using enhanced chemiluminescence (ECL; Amersham, Arlington Heights, IL). In quiescent NIH 3T3 cells the majority of p70S6k is in the fastest migrating form; on serum stimulation, virtually all of the protein shifts to the slowest mobility form (Fig. 3C). In Fig. 4, the intermediate phosphorylated forms are also visible.

RSK also exhibits a retardation in its electrophoretic mobility on mitogenic activation, due to phosphorylation on at least four different sites *in vivo*.[18,20,21,28,29] However, its shift is not as dramatic as that of p70S6k (Fig. 3C).

[23] P. B. Dennis, N. Pullen, S. C. Kozma, and G. Thomas, *Mol. Cell. Biol.* **16,** 6242 (1996).

[24] N. K. Mukhopadhyay, D. J. Price, J. M. Kyriakis, S. Pelech, J. Sanghera, and J. Avruch, *J. Biol. Chem.* **267,** 3325 (1992).

[25] R. B. Pearson, P. B. Dennis, J. W. Han, N. A. Williamson, S. C. Kozma, R. E. H. Wettenhall, and G. Thomas, *EMBO J.* **14,** 5279 (1995).

[26] Q.-P. Weng, K. Andrabi, A. Klippel, M. T. Kozlowski, L. T. Williams, and J. Avruch, *Proc. Natl. Acad. Sci. U.S.A.* **92,** 5744 (1995).

[27] Q.-p. Weng, K. Andrabi, M. T. Kozlowski, J. R. Grove, and J. Avruch, *Mol. Cell. Biol.* **15,** 2333 (1995).

[28] C. J. Jensen, M. B. Buch, T. O. Krag, B. A. Hemmings, S. Gammeltoft, and M. Frodin, *J. Biol. Chem.* **274,** 27168 (1999).

[29] S. A. Richards, J. Fu, A. Romanelli, A. Shimamura, and J. Blenis, *Curr. Biol.* **9,** 810 (1999).

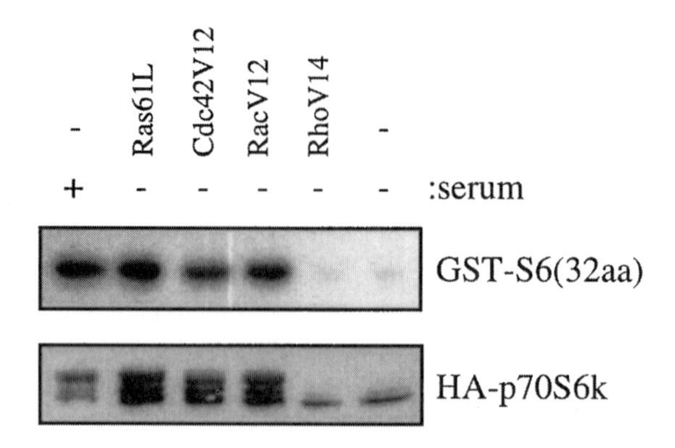

FIG. 4. Activation of p70S6k by selected Ras superfamily GTPases. NIH 3T3 cells were transiently cotransfected with HA–p70S6k/pBJ5 and the indicated GTPase, using LipofectAMINE. After transfection, cells were starved, then stimulated or not with 20% (v/v) FBS. The transfected p70S6k was immunoprecipitated with anti-HA antibody and subjected to kinase assays as described in Materials and Methods. One-half of the reaction was fractionated on 12% (w/v) gels and subjected to autoradiography (*top*); the second half of the reaction was fractionated on 8% gels and subjected to immunoblotting with anti-p70S6k-C2 antiserum (*bottom*).

Activation of p70S6k by Ras and Rho Family G Proteins

Hemagglutinin (HA) epitope-tagged p70S6k is subcloned into pBJ5 as previously described.[30] This construct is cotransfected with constitutively active forms of the Ras superfamily GTPases: Ras61L in pMT3, or Cdc42V12, RacV12, and RhoV14 in pEBG (which fuses the GTPases in frame with GST).[16] Cotransfection of p70S6k with HA- or Myc-tagged forms of the G proteins gives identical results (our unpublished observation).

NIH 3T3 cells are seeded at 3.0–3.4 $\times$ 10^5 per 35-mm dish the day prior to transfection (cells are approximately 80% confluent at the start of transfection). Transfections are performed with LipofectAMINE (Life Technologies, Gaithersburg, MD) according to manufacturer instructions. Briefly, 2 μg of total DNA and 6 μl of LipofectAMINE are used per sample. Typically, 0.2–0.4 μg of HA–p70S6k/pBJ5 is used per sample; constructs encoding the GTPases are added such that the total amount of DNA is 2 μg. After a 4- to 5-hr incubation, cells are allowed to recover overnight in 5% (v/v) CS–DMEM. The next morning, cells are starved in 0.5% (v/v) FBS–DMEM for 24–48 hr. Cells are harvested as described above, and HA–p70S6k is immunopurified with anti-HA (12CA5) antibody (1–2 μg per sample) and protein A–Sepharose. Kinase assays are performed as described above. Half the reaction is run on 12% SDS–polyacrylamide gels (for autoradiography), and the other half is run on 8% gels for subsequent immunoblotting

[30] L. Cheatham, M. Monfar, M. M. Chou, and J. Blenis, *Proc. Natl. Acad. Sci. U.S.A.* **92,** 11696 (1995).

with anti p70S6k-C2 (to confirm levels of protein and mobility shifts). Figure 4 demonstrates that activated alleles of Ras, Cdc42, and Rac 1, but not RhoA, are competent to stimulate p70S6k.

Concluding Remarks

The development of specific antibodies against p70S6k and RSK has allowed efficient monitoring of their activation *in vivo.* The availability of clean immunoblotting antibodies that detect mobility shifts are particularly useful for examining stimulation of transfected forms of p70S6k by candidate activators. Because p70S6k protein levels can vary from sample to sample during transfections, such antibodies provide an additional control by demonstrating p70S6k activation through changes in the ratio of the four phosphorylated bands (see Fig. 4).

Another important point is that the basal activity of p70S6k (i.e., under conditions of serum deprivation) can vary tremendously between different cell types. This seems to correlate with variations in the basal activity of P13K (our unpublished observation, 1997). In cell lines with low basal p70S6k activity, care must be taken to avoid overgrowth of cells as this leads to increased p70S6k activity under starved conditions. This is particularly important when testing activation of p70S6k by Cdc42 and Rac, as it becomes difficult to discern activation over the high basal activity.

[6] Ras Activation of PAK Protein Kinases

By ALBERT CHEN, YI TANG, YA ZHUO, QI WANG, ALBERT PAHK, and JEFFREY FIELD

Introduction

Because the small G protein Ras is mutationally activated in 20–30% of tumors, many laboratories have been tracing its downstream signals. Three major biological changes are promoted by Ras: cell proliferation, rearrangement of the actin cytoskeleton, and increases in cell survival. The first signal to be traced was the mitogen-activated protein (MAP) kinase cascade, consisting of the serine threonine kinases Raf → MEK (MAP kinase kinase) → ERK (extracellular singal-regulated kinase). This cascade plays a major role in the Ras proliferative signals.[1] Later, signals were identified through other effectors such as RalGDS (Ral guanine

[1] S. Campbell, R. Khosravi-Far, K. Rossman, G. Clark, and C. Der, *Oncogene* **17,** 1395 (1998).

0076-6879/00 $35.00

nucleotide dissociation stimulator) and phosphatidylinositol 3-kinase (PI 3-kinase). The signals through PI 3-kinase are the most well characterized. Through a PI 3-kinase to Akt signal, Ras promotes cell survival and through a PI 3-kinase to Rac signal, Ras mediates signals to the actin cytoskeleton to promote ruffling.[2,3] PAK protein kinases are direct targets of Rac and the related small G protein, Cdc42.[4]

The first evidence linking Ras to PAK (p21-activated kinase) was the demonstration that expression of kinase-deficient PAK mutant prevents Ras transformation in Rat-1 fibroblasts.[5] Kinase dead mutants often behave as dominant-negative mutants, presumably because they bind to key interacting proteins such as substrates, and block the endogenous enzymes from interacting productively. A single amino acid substitution at amino acid 299 (K299R) on PAK abolishes its kinase activity.[6] Later, Ras was shown to activate PAK in transfection assays of Rat-1 fibroblasts.[7] Ras activation of PAK was comparable to the levels observed by Rac. Raf itself did not activate PAK. However, activated PI 3-kinase and Ras^{C40}, an effector mutant that does not activate Raf but activates PI 3-kinase, both stimulated PAK. In addition, the PI 3-kinase-specific inhibitor LY294002 inhibited Ras and PI 3-kinase activation of PAK. This suggests that the signal was mediated by PI 3-kinase. Dominant-negative Rac and dominant-negative Cdc42 inhibited Ras and PI 3-kinase activation of PAK. Therefore, a Ras → PI 3-kinase → Rac/Cdc42 signal was traced. Surprisingly, the major downstream effector blocked by the dominant-negative PAK mutants was ERK, and not c-Jun N-terminal kinase (JNK) as had been predicted from the initial characterization of Rac and PAK.[5] Moreover, PAK was shown to transduce signals to ERK from other Rho family members.[7] Subsequently, direct phosphorylation of both Raf and MEK by PAK was reported at novel sites required for ERK signaling, providing a molecular mechanism for the signals from PAK to ERK.[8,9]

In this chapter, we describe protocols for analysis of Ras signaling to PAK protein kinases primarily through direct measurements of PAK activity. Figure 1 shows several gels using myelin basic protein (MBP) as the substrate. Not all cells

[2] P. Rodriguez-Viciana, P. H. Warne, A. Khwaja, B. M. Marte, D. Pappin, P. Das, M. D. Waterfield, A. Ridley, and J. Downward, *Cell* **89,** 457 (1997).

[3] A. Kauffmann-Zeh, P. Rodriguez-Viciana, E. Ulrich, C. Gilbert, P. Coffer, J. Downward, and G. Evan, *Nature* (*London*) **385,** 544 (1997).

[4] E. Manser, T. Leung, H. Salihuddin, Z.-S. Zhao, and L. Lim, *Nature* (*London*) **367,** 40 (1994).

[5] Y. Tang, Z. Chen, D. Ambrose, J. Liu, J. B. Gibbs, J. Chernoff, and J. Field, *Mol. Cell. Biol.* **17,** 4454 (1997).

[6] M. A. Sells, U. G. Knaus, S. Bagrodia, D. M. Ambrose, G. M. Bokoch, and J. Chernoff, *Curr. Biol.* **7,** 202 (1997).

[7] Y. Tang, J. Yu, and J. Field, *Mol. Cell. Biol.* **19,** 1881 (1999).

[8] J. A. Frost, H. Steen, P. Shapiro, T. Lewis, N. Ahn, P. E. Shaw, and M. H. Cobb, *EMBO J.* **16,** 6426 (1997).

[9] A. J. King, H. Sun, B. Diaz, D. Barnard, W. Miao, S. Bagrodia, and M. S. Marshall, *Nature* (*London*) **396,** 180 (1998).

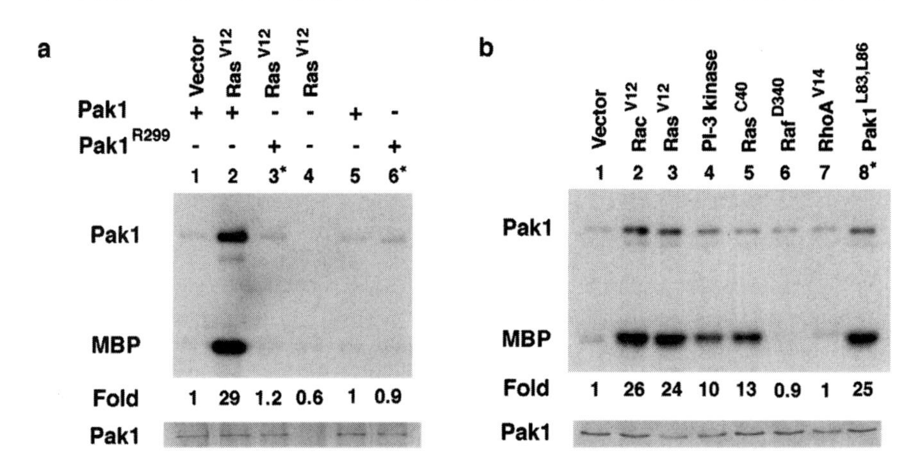

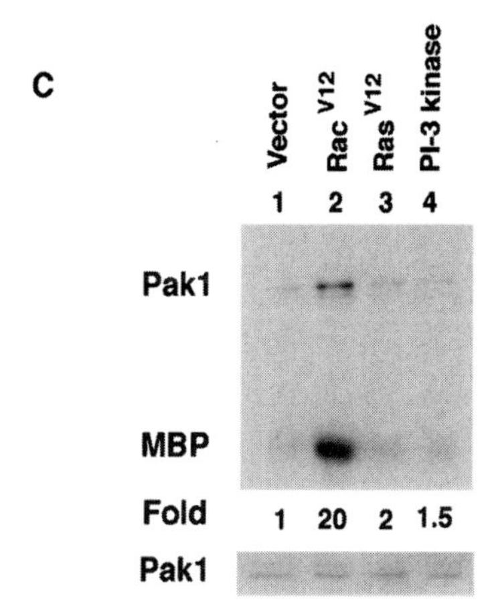

FIG. 1. Ras activation of PAK, using gel assays. Plasmids were transfected into either Rat-1 cells (a and b) or NIH 3T3 cells (c) and tested using myelin basic protein (MBP) as a substrate. The upper band is the PAK autophosphorylation band and the lower band is MBP. The radioactivity incorporated into MBP was quantitated on a PhosphorImager and expressed as fold activation compared with the vector control in lane 1. Beneath each kinase assay is a Western blot with antibody 9E10 for the transfected PAK. [Reprinted with permission from Y. Tang *et al.*[7]]

will support Ras signaling to PAK. We have found among fibroblast cell lines that Ras activates PAK in Rat-1 cells but not in NIH 3T3 cells (Fig. 1C). Interestingly, NIH 3T3 cells do not respond to PAK dominant-negative mutants either.[7] However, to date several other cell lines have supported Ras activation of PAK, including COS-7 cells and Schwann cells.[10]

Materials

Cell lines: Rat-1 cells are obtained from the Merck collection of strains and have been described elsewhere[11]; NIH 3T3 cells are a gift from S. Gutkind (NIH, Bethesda, MD)

Plasmids: cDNA expression plasmids utilizing the cytomegalovirus (CMV) promoter to express Myc-tagged PAK1, $PAK1^{R299}$, $PAK1^{L83, L86}$, and $PAK1^{L83,L86,R299}$ expressed from the plasmid pCMV6M (a modified version of pCMV5) have been described elsewhere[6]; Rac^{V12} and Ras^{C40} are gifts from D. Bar-Sagi, L. Van Aelst, and M. White[12]; Rac^{L61}, Raf^{D340}, and human K-Ras4B are gifts from C. Der. p110-C*aa*X, which consists of the p110 catalytic subunit of PI 3-kinase targeted to the membrane through fusion with the C*aa*X sequence of H-Ras, is as previously described[2]

Cell extract: Prepared from cells expressing PAK as described below

Antibodies: Antibody 9E10 for Myc-tagged constructs is available from Boehringer Mannheim (Indianapolis, IN) and antibodies for untagged PAK are available from Santa Cruz Biotechnology (Santa Cruz, CA)

Wash buffer: Phosphate-buffered saline (PBS)

Lysis buffer: 40 m*M* HEPES (pH 7.4), 1% (v/v) Nonidet P-40 (NP-40), 100 m*M* NaCl, 1 m*M* EDTA, 25 m*M* NaF, 1 m*M* sodium orthovanadate, leupeptin 10 μg/ml, aprotinin (10 μg/ml)

Phosphorylation buffer: 10 m*M* $MgCl_2$, 40 m*M* HEPES (pH 7.4)

Labeled ATP: $[\gamma\text{-}^{32}P]$ ATP (3000 Ci/mmol; 10 μCi/μl; Amersham, Arlington Heights, IL); it is important to use fresh isotope less than 1 week old

Unlabeled ATP: 1 m*M* stock solution (use 20 μ*M* final concentration)

Substrates:

Myelin basic protein (MBP; Sigma, St. Louis, MO): Stock solution, 10 mg/ml

Histone H4 (Boehringer Mannheim): Stock solution, ~1 mg/ml

$p47^{phox}$ peptide [peptide YRRNSVRF (amino acids 324–331 of $p47^{phox}$), synthesized by Research Genetics, Huntsville, AL]: Stock solution is

[10] Y. Tang, S. Marwaha, J. L. Rutkowski, G. I. Tennekoon, P. C. Phillips, and J. Field, *Proc. Natl. Acad. Sci. U.S.A.* **95,** 5139 (1998).

[11] N. E. Kohl, S. D. Mosser, S. J. deSolms, E. A. Giuliani, D. L. Pompliano, S. L. Graham, R. L. Smith, E. M. Scolnick, A. Oliff, and J. B. Gibbs, *Science* **260,** 1934 (1993).

[12] T. Joneson, M. McDonough, D. Bar-Sagi, and L. Van Aelst, *Science* **274,** 1374 (1996).

10 μg/μl, in 10% (v/v) acetic acid; dilute 1 : 100 into phosphorylation buffer and use 10 μl (1.0 μg) in assays

When using peptide substrates the following are needed.

Phosphocellulose columns: Phosphocellulose SpinZyme units (Pierce, Rockford, Il)
Phosphocellulose wash buffer (for $p47^{phox}$ assays only), ~75 m*M*: Dilute 4.14 ml of 85% (v/v) phosphoric acid into 1 liter of H_2O

When using protein substrates the following are needed.

Sodium dodecyl sulfate (SDS) gel equipment and supplies: Gel box, power supply, gel dryer, Coomassie blue stain, destain

Method

Cell Culture and Transfections

Cells are grown at 37° in 5% (v/v) CO_2 in high-glucose (4.5 g/liter) Mediatech Dulbecco's modified Eagle's medium (DMEM) purchased from Fisher Scientific (Pittsburgh, PA) supplemented with 10% (v/v) fetal bovine serum (Fisher), penicillin (100 U/ml), and streptomycin (100 μg/ml).

PAK assays are typically performed on PAK immunoprecitated from transfected cells. Transfection procedures are not described here in detail because there are many different methods depending on the cell lines used. To date, we have successfully used calcium phosphate and LipofectAMINE, and even generated stable cell lines expressing Myc-tagged PAK in these assays. Transient transfections are performed with 20 μg of total DNA (7 μg of each test DNA and 6 μg of PAK test DNA; the total DNA content in all transfections is brought to 20 μg, if necessary, with plasmid pUC19). We typically plate cells at a density of about 50% confluence on 60-mm-diameter dishes. We include Rac as a positive control and vector alone as a negative control for PAK activation. We also often use hyperactive PAK, $PAK^{L83,L86}$, and kinase dead PAK, PAK^{R299} as controls.

Cell Lysis and Extract Preparation

Cell lysates are prepared 12–48 hr posttransfection, depending on the procedure used for transfection. Each plate is washed twice with cold PBS and then lysed by addition of 1 ml of lysis buffer and incubated on a shaker at 4°. Samples are then centrifuged at full speed in a microcentrifuge for 20 min at 4° and the supernatants are collected and frozen until needed.

PAK Immunoprecipitations

Prior to the assay, about 100 μg of extracts from cells transfected with Myc-tagged PAK is mixed in a 1.5-ml microcentrifuge tube with ~3 μg of the anti-Myc antibody 9E10 and 30–40 μl of a 50% slurry of protein A beads and rotated for ~2 hr at 4°. Precipitates are washed three times with lysis buffer and kept on ice while the reaction is set up.

Kinase Assays

For convenience, a reaction mix is prepared, and the reactions are started by adding this mix to the immunoprecipitate. For each assay tube the reaction mix contains the following: substrate (5 μg of the myelin basic protein stock or 1 μg of the diluted peptide), 1 μl of 1 m*M* ATP (the final concentration is 20 μ*M*), and 0.5–1.0 μl (5–10 μCi) of [γ-^{32}P]ATP. Phosphorylation buffer is then added so that 20–25 μl is sufficient to deliver the correct quantity of each reagent. Reactions are started by addition of the appropriate amount of this reaction mix to the immunoprecipitated PAK. The samples are then incubated for 20 min at 22°.

For gel assays reactions are stopped by the addition of 1 part of 4× SDS gel sample buffer to three parts reaction, heated to 95°, and the products are resolved by SDS–12% (w/v) polyacrylamide gel electrophoresis and visualized by autoradiography. The gel is then stained with Coomassie blue, destained, and then dried on a gel dryer. Visualization of the gel is done by autoradiography at −80°. Exposure times can vary from 2 to 12 hr. Two major bands are usually seen. The upper band is the $p65^{PAK}$ autophosphorylation band and the lower band is the MBP band. The samples are quantitated by PhosphorImager analysis of the MBP band. Examples of kinase assays are shown in Fig. 1.

For p47 assays samples are quickly loaded onto phosphocellulose SpinZyme units (Pierce), centifuged in a microcentrifuge, and washed twice with phosphocellulose wash buffer, and then the filters are counted in a scintillation counter.[13]

Troubleshooting

Several problems are commonly encountered in these experiments. The first is the variation in transfection efficiencies of different cell lines and the steady reduction in transfection efficiency with continued cell culture. To minimize this, we do not culture cells for long periods of time and periodically test transfection efficiencies, using a marker such as green fluorescent protein (GFP) or β-galactosidase, and check expression of transfected genes on Western blots. Other problems that have been reported to us include the high backgrounds often seen when MBP is

[13] U. Knaus, S. Morris, H.-J. Dong, J. Chernoff, and G. Bokoch, *Science* **269,** 221 (1995).

used as a substrate. To avoid this, the pellets are briefly washed with phosphorylation buffer prior to loading on the gel. Others have begun switching to different substrates such as histone H4, which does not seem to require the washing step to maintain low backgrounds. We have also begun using the peptide substrate assay as described above. The advantage of using this substrate is that it can be isolated through a rapid filter-binding assay and then counted in a scintillation counter. The disadvantage of this assay is that without a gel the PAK autophosphorylation cannot be monitored.

Acknowledgments

Work in the laboratory is supported by grants to J. F. from the NIH (GM48241), the American Cancer Society, and the Neurofibromatosis Foundation. A. P. is supported by a fellowship from the HHMI.

[7] Ras Signaling to Transcription Activation: Analysis with GAL4 Fusion Proteins

By GABRIELE FOOS, CHRISTINA K. GALANG, CHAO-FENG ZHENG, and CRAIG A. HAUSER

Introduction

Activation of Ras signaling stimulates multiple downstream signaling cascades, many of which ultimately converge on the nucleus and broadly alter cellular gene expression. These widespread changes in gene expression are important for regulating normal growth, and when aberrantly regulated, for mediating oncogenesis. Many of the changes in gene expression mediated by Ras are the result of signaling through multiple parallel pathways resulting in activation of mitogen-activated protein (MAP) kinase family members.[1,2] These activated MAP kinases then translocate to the nucleus and phosphorylate transcription factors, which can greatly increase their *trans*-activation activity.[3] Reporter genes have been of great use in understanding both Ras signal transduction and its targets. There are several different methods utilizing reporter genes to analyze Ras signaling, and these approaches address distinct questions. One approach is to transiently contransfect oncogenic Ras and a reporter construct consisting of the promoter region of the gene fused to an easily assayed reporter coding sequence (e.g., luciferase). This

[1] M. H. Cobb, *Prog. Biophys. Mol. Biol.* **71,** 749 (1999).
[2] T. S. Lewis, P. S. Shapiro, and N. G. Ahn, *Adv. Cancer Res.* **74,** 49 (1998).
[3] R. Treisman, *Curr. Opin. Cell Biol.* **8,** 205 (1996).

0076-6879/00 $35.00

approach is useful for monitoring how the complex promoter elements of a particular gene together respond to Ras to regulate its transcription. Another reporter gene approach for analyzing how Ras alters the activity of a particular type of transcription factor involves transient cotransfections of Ras with a reporter construct consisting of synthetic binding sites for the transcription factor inserted into a minimal promoter–reporter. We have previously described the methods of this approach in detail,[4] and have used it to show that Ets transcription factors are targets of Ras signaling.[5] It is important to note that most transcription factors are part of large families that bind similar promoter sequences. Thus, unless one can observe "superactivation" by cotransfection of Ras and an expression construct for the specific transcription factor,[6] such reporters with synthetic binding sites may actually measure overall changes in the activity of the transcription factor family.

The type of reporter gene analysis described in this chapter utilizes fusion proteins between a heterologous DNA-binding domain from yeast GAL4, and the *trans*-activation domain of a specific transcription factor. The transcriptional activity of the hybrid transcriptional activator is monitored by cotransfection with a reporter gene containing multiple GAL4 DNA-binding sites in front of a minimal promoter and luciferase coding sequence. This approach allows analysis of Ras-induced changes in the *trans*-activation activity of a specific transcription factor, independent of its normal DNA-binding activity or the level of the endogenous factor expression. Mutational analysis of the *trans*-activation domain present in the fusion protein can be further used to define the residues involved in mediating Ras responsiveness. In addition, this type of reporter gene system is extremely useful as a downstream readout to study how different proteins or pharmacological agents alter specific Ras signal transduction pathways. For example, stimulation of the ERK (extracellular signal-regulated kinase), JNK (c-Jun N-terminal kinase), or p38 MAPK (mitogen-activated protein kinase) pathways have been monitored with GAL4–Elk, GAL4–Jun, or GAL4–CHOP reporters, respectively.[7–9]

General Considerations

Overall, this type of reporter gene analysis involves transient cotransfection of two or three different plasmids into an appropriate cell line. These plasmids are the expression construct for the GAL4 DNA-binding domain fused to a

[4] C. A. Hauser, J. K. Westwick, and L. A. Quilliam, *Methods Enzymol.* **255,** 412 (1995).

[5] C. K. Galang, C. J. Der, and C. A. Hauser, *Oncogene* **9,** 2913 (1994).

[6] B.-S. Yang, C. A. Hauser, G. Henkel, M. S. Colman, C. Van Beveren, K. J. Stacey, D. A. Hume, R. A. Maki, and M. C. Ostrowski, *Mol. Cell. Biol.* **16,** 538 (1996).

[7] R. Marais, J. Wynne, and R. Treisman, *Cell* **73,** 381 (1993).

[8] T. Smeal, M. Hibi, and M. Karin, *EMBO J.* **13,** 6006 (1994).

[9] X. Z. Wang and D. Ron, *Science* **272,** 1347 (1996).

trans-activation domain (e.g., GAL4–Elk), a reporter gene with multiple GAL4 DNA-binding sites preceding a minimal promoter fused to the luciferase coding sequence (e.g., pFR-Luc), and if desired, an expression construct for an activated signaling molecule [e.g., pZIP*ras*H(61L) which expresses oncogenic Ras]. If the altered signaling ultimately stimulates phosphorylation of the *trans*-activation domain in the hybrid transcription factor, then transcription of the luciferase reporter gene is activated. This approach is schematically shown in Fig. 1. Two days after transfection, the cells are harvested and the luciferase activity is then quantitated. While such assays are simple and rapid to perform, care must be taken in including appropriate controls. Thus, below we describe methods for inexpensive, high-throughput analysis of transcriptional signaling using GAL4 fusion proteins, and discuss the considerations for setting up and interpreting these assays.

Cell Type and Growth Conditions

The selection of a cell line to use depends on the specific regulation being analyzed. Different cell lines can have quite different signaling components, which can greatly affect the outcome of the GAL4 fusion activator assay. Because we have been using an NIH 3T3 cell model to study the role of transcriptional signaling in cellular transformation, we have utilized these cells for a number of studies. While the transfection efficiency of different cell lines can vary a great deal, nearly 100% of the cells that are transiently transfected receive all the plasmids in the transfection mix. Thus, even in poorly transfecting cell lines, it is reasonable to assume that every cell that expresses the luciferase reporter also is expressing the GAL4 fusion protein and the activated signaling protein. Cell lines that are difficult to transfect may require lipid-mediated transfection or electroporation. However, for NIH 3T3 cells, a substantial signal is obtained by the simple calcium phosphate method described below on cells plated in 12-well dishes. To avoid obscuring potential regulatory mechanisms by constantly stimulating the cells with serum growth factors, it is often useful after the transfection, to starve the cells for the final 24 hr of the assay. The composition and preparation of media and buffers used have been previously described in detail,[4] but the procedure below has been modified for 12-well dishes with NIH 3T3 cells.

NIH 3T3 cells are grown at 37° in a humidified 5% CO_2 incubator, in Dulbecco's modified Eagle's medium (DMEM) containing 10% (v/v) calf serum, 2 m*M* glutamine, and streptomycin and penicillin (100 units/ml). The day before transfection, split the cells to approximately 30% confluence, which is 1×10^5 cells/4-cm^2 well of a 12-well dish. For this split, cells from 10-cm stock dishes are removed with trypsin, resuspended by pipetting up and down several times with an equal volume of growth medium, and then suspended in growth medium at 1×10^5 cells/ml in 50-ml sterile tubes (a fairly dense 10-cm stock plate will yield about 70 wells). After thoroughly mixing the cell suspension by inversion,

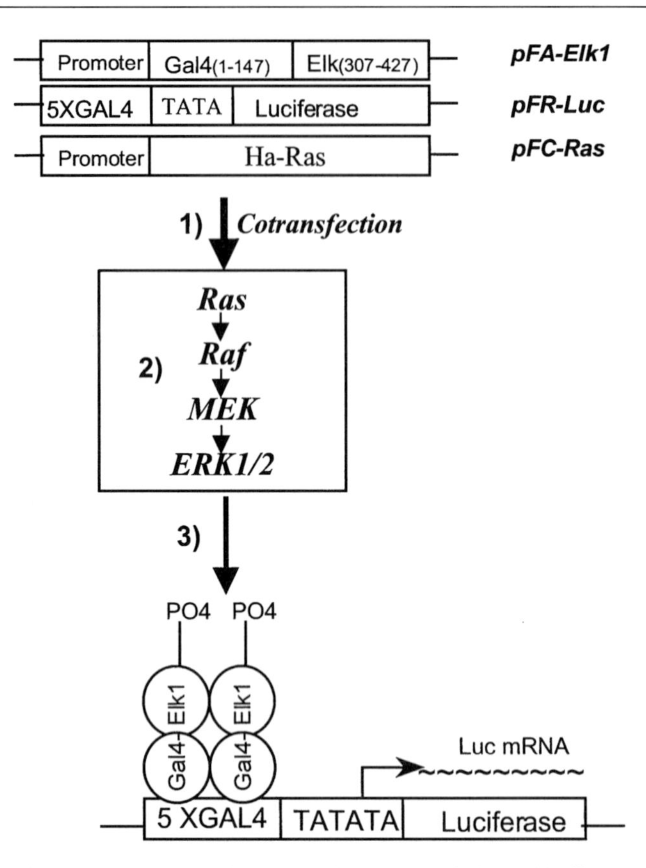

FIG. 1. Analysis of Ras activation of Elk-1 with GAL4–ELK1 fusion and luciferase reporter construct. *Top:* Schematic representations of the plasmids used. pFA-Elk1 expresses a fusion protein of GAL4 DNA-binding domain activation domain of the ternary factor Elk1. The plasmid pFC-Ras is an expression vector for oncogenic human Ha-Ras (61L). The pFR-Luc reporter plasmid expresses luciferase when GAL4-Elk1 fusion factor is activated in mammalian cells. (1) Cotransfection of the reporter plasmid (pFR-Luc), fusion *trans*-activator plasmid (pFA-Elk1), and the expression vector for the gene of interest (pFC-Ras) into mammalian cells such as NIH 3T3 or HeLa cells by calcium phosphate or lipid-mediated methods. (2) In the transfected cells, pFC-Ras expresses activated Ras protein, which causes the sequential activation of protein kinases Raf, MEK, and then the MAP kinases ERK1/2 in the cytoplasm. (3) Activated MAP kinase translocates into the nucleus and phosphorylates GAL4–Elk1, which is being constitutively expressed in the nucleus from pFA-Elk1. Phosphorylated GAL4–Elk1 then becomes transcriptionally active, stimulating the transcription of the luciferase gene on the pFR-Luc plasmid. Therefore, luciferace activity in transfected cells reflects the activation status of MAP kinases and the upstream components of the Ras pathway.

transfer 1 ml of cell suspension to each well of a 12-well plate. This large volume pooling of the cells prior to plating helps ensure a constant number of cells in each well. To attach the cells evenly, rock and swirl the plates well before placing them on a level incubator shelf. We have found that there is usually sufficient luciferase signal to transfect and assay even smaller wells (e.g., 24-well plates), but these have increased "edge effects" causing uneven cell plating and less reproducible transfections. On the other hand, it should be noted that for the standard assay, transfecting larger plates or dishes will not increase the luciferase signal, as a proportionally larger amount of cell lysis buffer will be needed to make cell extract.

Plasmids

Since the original observation that the DNA-binding and *trans*-activation activities of transcription factors can be found in separable structural domains was made with yeast GAL4,[10] the GAL4 DNA-binding domain/nuclear localization signal (residues 1–147) has been used to generate numerous chimeric transcription factors.[11] Reporter genes containing five GAL4-binding sites, the E1b minimal promoter,[7] and the luciferase coding sequence have also been widely used. The oncogenic Ras expression construct used in our experiments was pZIP*ras*H(61L).[12] It is possible to obtain specific GAL4 fusion transcription factors and the reporter gene from the many investigators who have reported their use, or purchase an assortment of them from Stratagene (La Jolla, CA). PathDetect Trans-Reporting systems (Stratagene) include mammalian expression constructs for GAL4 fusions with Elk1, Jun, CHOP, CREB, ATF2, and Fos. In addition, the GAL4-binding site-containing luciferase reporter gene (pFR-Luc) and several different expression constructs for activated components of the Ras signaling pathway (including the ones used in this work), the CMV-driven pFC-MEKK (MEKK residues 380–672), pFC-MEK1* (MEK1 S218/222E, Δ32–51), and pFC-PKA [the protein kinase A (PKA) catalytic subunit] can also be obtained from Stratagene.

Transient Transfection of NIH 3T3 Cells Grown in 12-Well Plates

Stock Solutions for Transfection

HBS (2×): 1.64% NaCl (w/v), 1.19% HEPES (w/v), 0.06% Na_2HPO_4 (w/v). Adjust to pH 7.12 with 1 *N* NaOH, filter sterilize, and store at 4°

$CaCl_2$ (2.5 *M*): To prepare a 10× stock, dissolve tissue culture-grade $CaCl_2$ in distilled H_2O, filter sterilize, and store at −20°

[10] J. Ma and M. Ptashne, *Cell* **48,** 847 (1987).

[11] I. Sadowski and M. Ptashne, *Nucleic Acids Res.* **17,** 7539 (1989).

[12] C. J. Der, B. Weissman, and M. J. MacDonald, *Oncogene* **3,** 105 (1988).

Calf thymus carrier DNA: Dried "highly polymerized" calf thymus DNA (Sigma, St. Louis, MO) is suspended in TE [10 m*M* Tris (pH 7.8), 0.1 m*M* EDTA] at 5 mg/ml and then incubated overnight at 60° to aid in resuspension. The DNA is then ethanol precipitated and resuspended as described above in sterile TE at about 4 mg/ml. The actual concentration of the slightly viscous carrier DNA is measured with a spectrophotometer, and aliquots are stored at 4°. We have formerly used herring testis DNA (Sigma), but several of the more recent batches of this product appear to be lower molecular weight DNA, and are 10-fold less efficient in potentiating transient transfection than the previous batches or the current calf thymus DNA

Plasmid DNAs: Prepare by Qiagen (Chatsworth, CA) columns as recommended by the manufacturer

Transfection Procedure

Three hours prior to transfection, draw off the medium and feed the cells with 1 ml of warm medium containing 10% (v/v) calf serum. For a typical transfection, add to a sterile microcentrifuge tube: 62.5 μl of 0.25 *M* $CaCl_2$ containing 0.5 μg of reporter construct (e.g., pFR-luc), 10–50 ng of GAL4 fusion construct (e.g., GAL4–Elk), 0.5 μg of oncogenic Ras expression construct [e.g., pZIP*ras*H(61L)], and calf thymus carrier DNA to a total of 5 μg of DNA. We do not typically make adjustments in the amount of calf thymus DNA for small differences in the amount of plasmid DNA, so a starting cocktail of 0.25 *M* $CaCl_2$ with 4.0 μg of carrier DNA and 0.5 μg of pFR-luc/62.5 μl is used. To another set of microcentrifuge tubes, add 62.5 μl of 2× HBS. Then, while vortexing the open microcentrifuge tube with the 2× HBS, add the contents of the DNA-containing tube dropwise to this tube over the course of about 10 sec. The dropwise addition is somewhat laborious, but makes transfection more reproducible. Enhanced reproducibility is particularly important for analysis with GAL4 fusion activators, with which we see substantially more assay variability than with standard reporter genes. This is the case even when the GAL4 fusion constructs have been properly diluted to assure accurate addition of the nanogram amounts of DNA used for transfection.

Let the precipitate stand for about 20 min. It should look cloudy and may be somewhat granular. Vortex each tube again prior to pipetting the precipitate onto the cells and then swirl the medium on the plate. For optimal precipitate formation, the components should be at room temperature. Mixing of the DNAs, precipitation, and addition of precipitate to the cells are routinely performed on the laboratory bench, without subsequent problems with contamination. Approximately 16 hr after transfection, the medium is drawn off, the cells are washed once with 1 ml of warm phosphate-buffered saline (PBS) containing 0.2 m*M* EGTA. The inclusion of EGTA in the wash is not essential with NIH 3T3 cells, but it can raise the transfection efficiency by reducing the calcium precipitate toxicity seen in more sensitive cell lines. After the wash, 1 ml of growth medium containing only 0.5% (v/v) calf serum is added to each well, and the cells are incubated for an additional 24 hr prior to harvest for the luciferase assay. This serum starvation is not always essential, but it can greatly reduce basal level signals in serum-sensitive pathways.

Luciferase Assay

Stock Solutions for Luciferase Assay

Cell lysis solution: 20 m*M* Tris (pH 7.8), 0.2% (v/v) Triton X-100, 1 m*M* dithiothreitol (DTT). In our hands, Tris works better than potassium-containing buffers in the lysis and assay buffers

ATP solution (250 m*M*): Dissolve ATP (Pharmacia, Piscataway, NJ) in H_2O. Adjust to pH 7.0 with concentrated NH_4OH. Store at −20°

Luciferin solution (10 m*M*): D-Luciferin-potassium salt (Analytical Luminescence Laboratory, San Diego, CA). Store at −20° protected from light

Luciferase assay mix: 20 m*M* Tris (pH 7.8), 9.3 m*M* ATP, 14.2 m*M* $MgSO_4$, 66 n*M* luciferin

Procedure

The following procedure has been developed for use with a microplate luminometer equipped with a 100 μl injector. We use a Berthold (Bad Wildbad, Germany) EG&G LB 96 P luminometer linked to a Macintosh computer. Although single-tube luminometers can be used, and they may be more sensitive to weak signals, there is a tremendous gain in throughput and a savings in labor and reagent costs when using a microplate luminometer. A microplate luminometer can read 96 assays unattended in about 30 min, and can provide the results as a Microsoft Excel file. Regarding assay cost, commercial luciferase assay kits such as that from Promega (Madison, WI) are excellent when high sensitivity and extended glow are needed. However, the assay described below works well with easily transfected cells, and is about 10-fold less expensive than commercial kits.

Wash cells twice with PBS. Thoroughly remove the PBS from the last wash by aspiration, and after keeping the 12-well dish tilted for a few seconds, aspirate each well again. Add 150 μl of cell lysis solution to each well. The cells will often detach and lyse by themselves, or after the lysis buffer-containing plate is hit sharply. If not, remove by scraping, and transfer the liquid to a microcentrifuge tube. Spin for 2 min at room temperature to pellet any debris. Transfer the supernatant to a new tube. These cell extracts may be assayed immediately or frozen at −70° for future assay. Transfer 100 μl of extract to a white microtiter plate (e.g., Corning Costar, Acton, MA). Include several wells containing just cell lysis buffer for background determination. Prepare luciferase assay mix (100 μl per assay plus a sufficient quantity to prime the injector and for blank wells) in a foil-wrapped tube. Wash and prime the luminometer injector as recommended by the manufacturer; we wash with 60 injections of H_2O followed by priming with 12 injections of luciferase assay mix. Failure to prime the injector with luciferase assay buffer will cause a loss of signal in the first several assay wells. The assay is carried out using a 100-μl injection of luciferase assay buffer, followed by a 10-sec integration with no delay. With our instrument, the background is about 120 relative light units (RLUs), and the signal is linear to more than 5 million RLUs.

Use of GAL4 Constructs to Analyze Ras Activation of Transcription Factor

We previously found that phosphorylation of Ets2 T72 is necessary for the large increase in Ets2 activity mediated by oncogenic Ras or Neu.[6,13] An example of using GAL4 constructs to analyze the effect of Ras on Ets2 *trans*-activation activity is shown in Fig. 2A. GAL4–Ets2 fusion constructs were made using a hemagglutinin (HA) epitope-tagged derivative of the Stratagene pFC-dbd vector (see below). We found that GAL4–Ets2(1–172) was strongly activated by oncogenic Ras in the reporter gene assay system. The Ets2 T72A mutation in this context caused a modest reduction of the basal *trans*-activation activity, but Ras-mediated activation was reduced to background levels seen with the empty GAL4 vector. Triple glycine substitution at the indicated residues of GAL4–Ets2(1–172) disrupted both the apparent basal *trans*-activation activity and Ras activation (Fig. 2A).

Titration experiments with GAL4–Ets2(1–172) plasmid, as well as with all the other GAL4 fusion activators that we have tested, consistently show that transfecting small amounts of the expression construct for the GAL4 hybrid transcription factor gives the best results for activation by Ras and other signaling molecules. In this example, only 20 ng of GAL4–Ets2(1–172) was cotransfected to obtain maximal activation with the pFR-luc reporter gene and the oncogenic Ras expression construct pZIP*ras*H(61L). If substantially more GAL4 construct is used (e.g., 100 ng), the basal level of expression of the pFR-luc reporter (that seen when cotransfected with the empty pZIP expression plasmid) is quite high, and coexpression of oncogenic Ras has little additional effect (data not shown). This is not a consequence of exceeding the linear range of the luciferase assay, but instead seems to involve titration of limiting cellular factors required for activation. While the amount of GAL4 construct used must be determined empirically, a general guide is to use between 5 and 100 ng and to use the least amount possible that gives a reliable basal signal.

[13] C. K. Galang, J. Garcia-Ramirez, P. A. Solski, J. K. Westwick, C. J. Der, N. N. Neznanov, R. G. Oshima, and C. A. Hauser, *J. Biol. Chem.* **271,** 7992 (1996).

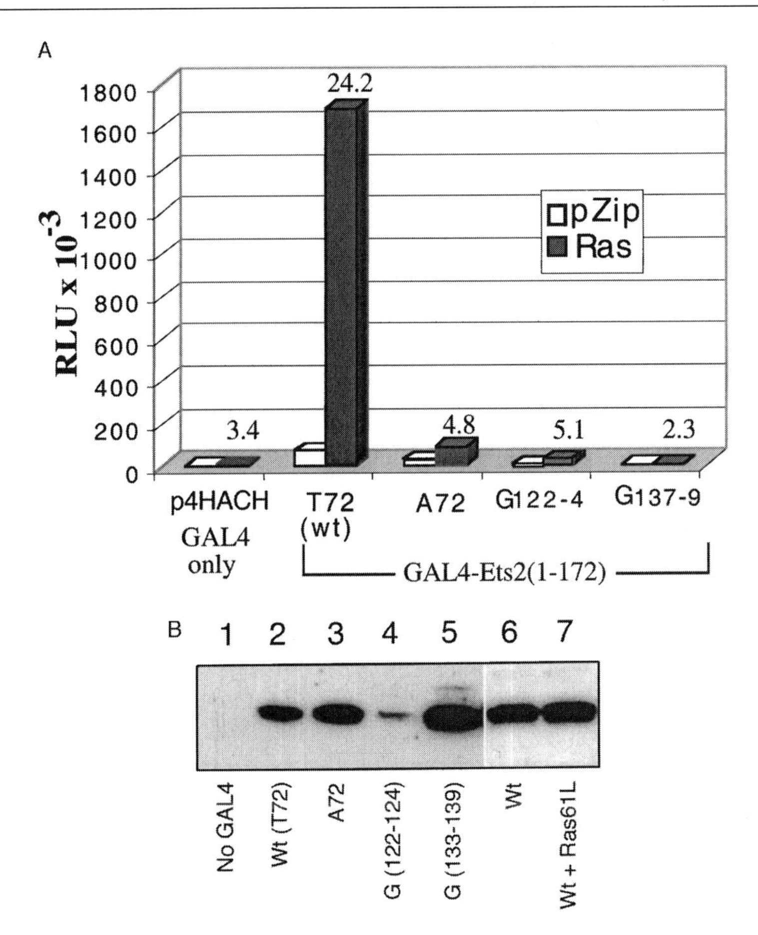

FIG. 2. Analysis of Ras-mediated *trans*-activation using GAL4–Ets2 constructs. (A) Results of a representative *trans*-activation assay. NIH 3T3 cells were transfected in 12-well dishes by the calcium phosphate method. Each well received 0.5 μg of pFR-Luc reporter, 0.5 μg of either pZIP*ras*H(61L) (solid columns) or the empty pZIP expression vector (open columns), 20 ng of the indicated GAL4–Ets2 construct, and 4 μg of calf thymus DNA. The cells were starved for 24 hr prior to harvest and assayed as described in text. The fold increase in luciferase activity mediated by Ras for each construct is shown above the solid columns. (B) Immunoblot of GAL4–Ets2 fusion proteins with an antibody directed against the HA epitope tag. Each well of a 12-well dish of NIH 3T3 cells was cotransfected with 0.5 μg of the indicated GAL4-Ets2(1–172) expression construct (except lane 1), and 0.5 μg of pZIP (lanes 1–6) or 0.5 μg of pZIP*ras*H(61L) (lane 7), using LipofectAMINE (GIBCO-BRL). The cells were then grown as in the reporter gene assay, harvested by scraping, washed, and resuspended in 20 μl of loading buffer. Then 10 μl of each extract was subjected to SDS–PAGE, transferred to nitrocellulose, probed with a monoclonal antibody against the HA epitope, and antibody binding visualized using Pierce Supersignal reagents.

Controls for *Trans*-Activation Assays

We have not observed significant transcriptional activity of the GAL4 DNA-binding domain (DBD) parental vector (e.g., pFC-dbd), but this is necessary to confirm in each cell line tested. An internal control plasmid is not included in our cotransfection assays; instead we rely on parallel control experiments and sufficient repetition in separate experiments to normalize for changes caused by experimental variation. Internal control plasmids can introduce new problems, as addition of a fourth cotransfected plasmid [e.g., cytomegatovirus (CMV)-β-Gal] can act as a transcriptional competitor, and can itself also be Ras responsive. If an internal control plasmid is desired, a non-Ras-responsive control reporter expressing *Renilla* luciferase (which can be separately be assayed from firefly luciferase) has been described.[14]

Typically, to generate reliable data, experiments are repeated three different times in duplicate. The duplicates should be separately precipitated and transfected DNAs, as this is the major source of variation within each experiment. There can be significant variation in the absolute value of luciferase signals obtained on different days, and thus the data from different experiments are often best combined when expressed as "fold activation" within each experiment by Ras or other signaling components being tested. This approach leads to good reproducibility. Other approaches to normalize experiments, such as expressing the data as relative light units per milligram of protein, are also used, but the majority of the variation seems to arise from different levels of transfection efficiency between experiments. We have not found it of value to normalize the luciferase data to the amount of protein in each extract, because the pooled plating, transfection, and direct lysis procedure described here rarely yields significant variation in extract protein content.

One important control for the experiment shown in Fig. 2A is to confirm that the abundance of the GAL4–Ets2 construct is not substantially altered either by expression of oncogenic Ras or by changes in protein stability caused by mutations in the Ets2 portion. As mentioned above, increasing the amount of GAL4 fusion activator can strongly increase reporter gene expression, and such a change could be misinterpreted as an Ras-mediated increase in *trans*-activation activity. One source of problems can be the use of a GAL4 fusion construct expression vector that contains a Ras-responsive promoter. Our experience is that simian virus 40 (SV40)-driven GAL4 construct vectors such as the original pFC-dbd (Stratagene) are not Ras responsive, whereas the CMV-driven pFC2-dbd (Stratagene), like other CMV-driven constructs, is quite Ras responsive (C. K. Galang and C. A. Hauser, unpublished data, 1999). To compare both the basal and Ras-induced *trans*-activation activity of various GAL4–Ets2 constructs, it is important to determine that truncations and mutations in Ets2 are not altering the hybrid protein stability.

[14] G. Behre, L. T. Smith, and D. G. Tenen, *BioTechniques* **26,** 24 (1999).

This presents a difficult technical problem, as the 20 ng of $CaPO_4$-transfected GAL4–Ets2 plasmid DNA transfected cells is expressed at levels well below the threshold of detection by immunoblotting. However, an estimate can be made of the relative protein expression and stability, using a parallel experiment of lipid-mediated transfection with substantially more plasmid DNA (e.g., 0.5 μg) under otherwise analogous conditions. This control is not ideal, but it allows a determination of whether there are large changes in the steady state levels of GAL4 fusion protein under related conditions.

Even after scaling up transfections, it can be difficult to obtain an immunoblot signal for the GAL4 fusion proteins. Two different commercial antibodies against the GAL4 DBD gave us weak signals with GAL4–Ets2 as well as other GAL4 fusions. Thus, we generated an HA epitope-tagged GAL4 DBD expression construct from pFC-dbd, with four HA tags at its N terminus. This SV40-driven construct, p4HACH, has the same cloning sites as pFC2-dbd, allowing a rapid transfer of Ets2 sequences to this context. *Trans*-activation studies showed that the p4HACH-Ets2 constructs behaved the same as the previously tested pFC-dbd-Ets2 constructs (G. Foos and C. A. Hauser, unpublished data, 1999). Immunoblot analysis of GAL4 construct expression, using an anti-HA monoclonal antibody, revealed that Ras was not inducing GAL4–Ets2(1–172) expression (Fig. 2B, lanes 6 and 7). In addition, this analysis showed that similar amounts of each GAL4–Ets2(1–172) protein were present under uninduced conditions, with the exception of the G(122–124) mutant (Fig. 2B, lane 4). The instability of this mutant protein, also observed in several other experiments, means that it cannot be concluded that the loss of basal and Ras-induced activity of the G(122–124) construct (Fig. 2A) was due to altered *trans*-activation activity. This example of an unstable fusion protein highlights the fact that without adequate controls, it cannot be assumed that the GAL4 fusion activator system provides a direct readout of *trans*-activation activity.

Use of GAL4 Constructs to Monitor Ras Signal Transduction

An example of an initial experiment to examine signaling pathway activation using a variety of GAL4 fusion activators in NIH 3T3 cells is shown in Table I. Some of these results are as expected, such as Ras and MEK activation of GAL4–Elk, and modest PKA activation of GAL4–CREB. It is informative to note the difference in the results obtained with the two similar GAL4–Elk constructs, as the only difference between them is that one is expressed using the SV40 promoter, and the other (denoted in Table I by "pFC2") from the CMV promoter. We have found that the huge increase in pFC2GAL4-Elk activity by Ras, MEKK, or MEK, is due to the combination of a substantial increase in the expression of the GAL4–Elk protein, as well as to increases in its transcriptional activation. We have further observed that MEKK even activates expression from the SV40 promoter, causing false readouts of increased *trans*-activation. This altered GAL4 fusion construct

TABLE I
EFFECTS OF COEXPRESSING ACTIVATED Ras PATHWAY COMPONENTS WITH GAL4 FUSION *trans*-ACTIVATORS AND GAL4-DEPENDENT REPORTER GENE

GAL4(1–147) fusion transcription activator[a]	Basal RLU[b]	Fold activation[c]			
		Ras	MEKK	MEK1	PKA
10 ng Elk (306–427)	48,167	19.4	43.6	9.7	0.1
10 ng Elk (306–427)pFC2	108,682	136.9	1,069.2	527.2	ND
40 ng Jun (1–223)	881	7.1	9.7	11.8	0.3
20 ng CREB (1–283)	1,899	0.6	15.0	0.2	3.7
10 ng VP-16 (401–479)	43,395	0.9	4.4	1.2	ND

[a]The amount of expression plasmid transfected into the NIH 3T3 cells and the residues of the *trans*-activation domain of each transcription factor included in the GAL4 fusion protein is indicated. The GAL4 expression constructs all use the SV40 promoter, except for Elk pFC2, which uses the CMV promoter.

[b]Basal RLU is the average number of RLUs (an arbitrary light unit) measured from NIH 3T3 cells transfected with the indicated fusion activator, pFR-luc reporter, and the empty expression construct pZIP.

[c]Fold activation is the average fold increase in reporter gene expression seen for each type of fusion activator, comparing luciferase levels seen on expression of the activated signaling component with that seen with empty expression vector pZIP. The amounts of expression plasmid used were 0.5 μg of pZIP or pZIP*ras*H(61L), 0.1 μg of pFC-MEKK, pFC- MEK1*, or pFC-PKA. ND, Not done.

expression complicates the analysis, and also highlights the need to measure protein levels. When working with activators such as Ras that can act across several signaling pathways, it is useful to find control GAL4 fusion activators that have significant basal *trans*-activation activity, but whose *trans*-activation activity is not affected by Ras signaling. This provides a valuable parallel negative control for Ras-mediated activation. The GAL4–VP16 fusion is the best such control that we have found, as its strong basal signal with the pFR-Luc reporter is not altered by coexpression of oncogenic Ras (Table I and data not shown). The finding that MEK1 expression can activate GAL4–Jun (Table I), likely reflects an autocrine loop similar to that reported with Raf activation of JNK.[15]

Both the specificity and magnitude of reporter gene response can be affected by the choice of the cell line assayed. These same plasmid constructs in HeLa cells showed more specificity when tested with the same activated MEKK, MEK1, and PKA constructs. In the HeLa cells, GAL4–Jun was activated by MEKK but not MEK, GAL4–Elk was activated by MEKK and MEK1, and GAL4–CREB

[15] S. A. McCarthy, M. L. Samuels, C. A. Pritchard, J. A. Abraham, and M. McMahon, *Genes Dev.* **9,** 1953 (1995).

was activated only by PKA.[16] It is interesting to note that while GAL4–c-Jun was not strongly activated by MEKK in NIH 3T3 cells, it was strongly activated in several other cell lines including HeLa, 293, and Chinese hamster ovary (CHO) cells (C.-F. Zheng, unpublished data, 1999). Overall, GAL4 fusion reporter gene systems are a powerful way to analyze signaling, but because of the complexity and cross-talk involved in Ras signaling,[17] one must not rely solely on this approach. Complementary approaches for monitoring downstream signaling, such as MAP kinase assays, are described elsewhere in this volume.

Acknowledgments

This work was supported by NIH grants CA63130 and CA74547 to C. A. H. G. F. is supported by a postdoctoral grant from the USAMRMC.

[16] L. Xu, T. Sanchez, and C.-F. Zheng, *Stratagies* **10,** 1 (1997).

[17] S. L. Campbell, R. Khosravi-Far, K. L. Rossman, G. J. Clark, and C. J. Der, *Oncogene* **17,** 1395 (1998).

[8] Ras Regulation of NF-κB and Apoptosis

By MARTY W. MAYO, JACQUELINE L. NORRIS, and ALBERT S. BALDWIN

I. Introduction

The Ras family consists of GTP-binding proteins that are responsible for regulating cellular proliferation, differentiation, and apoptosis.[1–4] Oncogenic Ras chronically binds GTP, resulting in the constitutive activation of downstream mitogen-activated protein kinase (MAPK), MAPK-like, and MAPK-independent signaling pathways.[3–5] Ras-dependent activation of these pathways, in turn, stimulates transcription factors that ultimately regulate gene expression and facilitate the transformation process. One of the transcription factors activated through Ras-dependent signal transduction pathways is the nuclear factor κB

[1] J. Downward, *Curr. Biol.* **7,** 258, (1997).

[2] R. Khosravi-Far and C. J. Der, *Cancer Metastasis Rev.* **13,** 67 (1994).

[3] J. Downward, *Curr. Opin. Genet. Dev.* **8,** 49 (1998).

[4] M. E. Katz and F. McCormick, *Curr. Opin. Genet. Dev.* **7,** 75 (1997).

[5] I. G. Macara, K. M. Lounsbury, S. A. Richards, C. McKiernan, and D. Bar-Sagi, *FASEB J.* **10,** 625 (1996).

 0076-6879/00 $35.00

(NF-κB).[6–8] NF-κB resides in the cytoplasm of unstimulated cells complexed with the inhibitor IκB.[9–10] In response to cellular stimuli, the IκB kinase (IKK) signalosome complex becomes activated and is responsible for phosphorylating the IκB proteins.[11] Once IκB becomes phosphorylated, it is targeted by the ubiquitin ligase, and subsequently degraded by the 26S proteasome.[11] The degradation of IκB liberates NF-κB, allowing the transcription factor to translocate to the nucleus, bind to *cis*-elements located within promoter regions, and upregulate gene expression.[9–10]

Our laboratory,[6,7,8,12] has demonstrated that the Ras oncoprotein can regulate the transcriptional activity of NF-κB by targeting the *trans*-activation domain of the p65 subunit of NF-κB. Moreover, the Ras oncoprotein requires the NF-κB transcription factor to initiate transformation,[8] in part because of the ability of NF-κB to suppress transformation-associated apoptosis.[13] We have shown that the H-Ras oncoprotein activates the transcriptional activity of NF-κB through a phosphoinoside 3-kinase (PI3K)- and Akt-dependent cell survival pathway.[14] Interestingly, H-Ras(V12) activates NF-κB through a mechanism involving both IKK and the ability of Akt to stimulate the *trans*-activation domain of the p65 subunit of NF-κB.[14] Thus, Ras stimulates NF-κB to upregulate gene products that are required to overcome Ras-induced apoptosis and potentiate proliferation. Consistent with Ras-induced proliferation, we and others[15,16] have shown that NF-κB can promote cell growth by controlling the transcription of the cyclin D1 gene. This is especially important because oncoproteins including Ras, the Ras-related GTPase, Rac, and the guanine nucleotide exchange factor family, Dbl, have been shown to upregulate cyclin D1 transcription and to induce deregulated cell cycle

[6] T. S. Finco and A. S. Baldwin, Jr., *J. Biol. Chem.* **268,** 17676 (1993).

[7] J. L. Norris and A. S. Baldwin, Jr., *J. Biol. Chem.* **274,** 13841 (1999).

[8] T. S. Finco, J. K. Westwick, J. L. Norris, A. A. Beg, C. J. Der, and A. S. Baldwin, *J. Biol. Chem.* **272,** 24113 (1997).

[9] S. Ghosh, M. J. May, and E. B. Kopp, *Annu. Rev. Immunol.* **16,** 225 (1998).

[10] A. S. Baldwin, Jr., *Annu. Rev. Immunol.* **14,** 649 (1996).

[11] E. Zandi and M. Karin, *Mol. Cell. Biol.* **19,** 4547 (1999).

[12] J. Y. Reuther, G. W. Reuther, D. Cortez, A. M. Pendergast, and A. S. Baldwin, Jr., *Genes Dev.* **12,** 968 (1998).

[13] M. W. Mayo, C. Y. Wang, P. C. Cogswell, K. S. Rogers-Graham, S. W. Lowe, C. J. Der, and A. S. Baldwin, *Science* **278,** 1812 (1997).

[14] L. V. Madrid, C. Y. Wang, D. C. Guttridge, A. J. Schottelius, A. S. Baldwin, and M. W. Mayo, *Mol. Cell. Biol.* **20,** 1626 (2000).

[15] M. Hinz, D. Krappmann, A. Eichten, A. Heder, C. Scheidereit, and M. Strauss, *Mol. Cell. Biol.* **19,** 2690 (1999).

[16] D. C. Guttridge, C. Albanese, J. Y. Reuther, R. G. Pestell, and A. S. Baldwin, *Mol. Cell. Biol.* **19,** 5785 (1999).

progression.[17,18] Moreover, Rac-induced upregulation of the cyclin D1 gene has been demonstrated to occur through an NF-κB-dependent manner.[17]

Because NF-κB is an inducible transcription factor, several molecular techniques have been developed to measure its activation in response to physiological and oncogenic stimuli. NF-κB DNA-binding activity in response to H-Ras(V12) expression can be measured by performing electrophoretic mobility shift assays (EMSAs), Western blot analysis for the presence of various nuclear NF-κB components, as well as transient transfection assays to measure Ras-responsive transcriptional activity of NF-κB. The purpose of this chapter is to describe methods commonly used to measure Ras-responsive activation of NF-κB, and to underscore techniques that illustrate the importance of NF-κB in Ras-induced apoptosis.

II. Analysis of NF-κB Transcriptional Activity

Transient transfection analysis is probably one of the quickest and most informative techniques performed to quantitate transcriptional activity. This method can be used to address whether the expression of a particular transgene can activate signal transduction pathways that stimulate a given transcription factor. NF-κB consists of a DNA-binding region and a *trans*-activation domain that interacts with core transcription factors, such as TATA-binding protein (TBP) and TBP-associated factors (TAFs), and coactivators, including CREB-binding protein (CBP) and p300.[19–21] This assay determines the activity of NF-κB because it requires both DNA binding and the assessment of the *trans*-activating potential of this transcription factor. Cells are transiently transfected with a plasmid encoding a particular transgene along with an NF-κB-responsive reporter. Two of the most commonly used reporter genes assayed for in transient transfections include the prokaryotic gene encoding chloramphenicol acetyltransferase (CAT), and the firefly luciferase (Luc) gene. For example, the 3×-κB-Luc reporter plasmid contains three NF-κB DNA-binding sites, originally identified in the MHC class I promoter, cloned upstream of a minimal TATA-containing promoter that drives the transcription of the luciferase gene. Thus, after transient transfection into cells, the reporter

[17] D. Joyce, B. Bouzahzah, M. Fu, C. Albanese, M. D'Amico, J. Steer, U. J. Klein, R. J. Lee, J. E. Segall, J. W. Westwick, C. J. Der, and R. G. Pestell, *J. Biol. Chem.* **274**, 25245 (1999).

[18] J. K. Westwick, R. J. Lee, Q. T. Lambert, M. Symons, R. G. Pestell, C. J. Der, and I. P. Whitehead, *J. Biol. Chem.* **273,** 16739 (1998).

[19] H. Zhong, R. E. Voll, and S. Ghosh, *Mol. Cell* **1,** 661 (1998).

[20] N. D. Perkins, L. K. Felzien, J. C. Betts, K. Leung, D. H. Beach, and G. J. Nabel, *Science* **275,** 523 (1997).

[21] M. E. Gerritsen, A. J. Williams, A. S. Neish, S. Moore, Y. Shi, and T. Collins, *Proc. Natl. Acad. Sci. U.S.A.* **94,** 2927 (1997).

gene product provides an accurate and quantitative measure of NF-κB-dependent gene expression.

A. *Transient Transfection of NIH 3T3 Cells*

There are several different methods to effectively transfect cells in culture. Some of the established protocols include calcium phosphate coprecipitation, electroporation, and DEAE-dextran transfection. However, there are many new commercially available reagents that are less toxic to cells and that generate more reproducible results than those mentioned above. These reagents include LipofectAMINE (Life Technologies, Rockville, MD), SuperFect (Qiagen, Valencia, CA), and GenePORTER (Gene Therapy Systems, San Diego, CA). In our hands, NIH 3T3 cells are effectively transfected with the SuperFect reagent. The advantages to using this method are that little plasmid DNA is required and that cells are transfected in the presence of serum, which helps to eliminate some of the toxicity associated with the transfection procedure. NIH 3T3 cells at 60% confluency (plated in six-well plates 18 hr prior to transfection) are transiently transfected using SuperFect reagent according to the manufacturer instructions. Plasmids, including 3×-κB-Luc, pCMV-LacZ, and either vector control plasmid (pCMV-5) or expression construct encoding H-Ras(V12) (1 μg each) are combined, diluted in 300 μl of serum- and antibiotic-free medium, and mixed with 15 μl of SuperFect reagent. Reagents are mixed and incubated for 10 min to allow SuperFect–DNA complexes to form. After 10 min, 2.7 ml of complete NIH 3T3 growth medium is added to the SuperFect–DNA complexes and mixed. NIH 3T3 cells are washed (once) with serum- and antibiotic-free medium and 1 ml of SuperFect–DNA complex is added to three wells of the six-well plate. Thus, one reaction mixture is enough for three wells and allows transfection experiments to be performed in triplicate. By including the pCMV-LacZ plasmid in the transfection reaction it is possible to normalize for efficiency of transfection. Detection of β-galactosidase-positive cells is described in Section V,A of this chapter. In general, cells are harvested 24 to 48 hr after transfection, depending on the optimal detection time for the reporter gene product assayed and also on the confluency of the cells. Note that NIH 3T3 cells that are confluent will display lower luciferase activity than cells actively proliferating.

B. *Assaying for Reporter Activity*

NIH 3T3 cells are harvested by washing the six-well plates (once) with ice-cold phosphate-buffered saline (PBS) and scraping the monolayer in 1 ml of ice-cold PBS, using a rubber policeman. Cell suspensions are transferred to a 1.5-ml microcentrifuge tube and pelleted after a 15-sec pulse spin. The luciferase enzyme is released from the cell by resuspending the cell pellet in 100 μl of reporter lysis buffer (Promega, Madison, WI) and incubating the cells on ice for

10 min. Cell debris is removed by centrifugation (12,000 rpm) for 10 min. Protein concentrations are determined with the Bio-Rad protein assay dye reagent (Bio-Rad, Hercules, CA). Protein concentrations are normalized to β-galactosidase activity using a β-Gal colorimetric assay (Promega, Madison, WI). To assay for luciferase activity, 100 μg of cell lysate is combined with ATP and the substrate luciferin in a luminometer (AutoLumat LB953; Berthold Analytical Instruments, Bad Wildbad, Germany). The luminometer injects 200 μl of luciferase assay buffer [25 m*M* glycylglycine (pH 7.8), 15 m*M* potassium phosphate (pH 7.8), 15 m*M* $MgSO_4$, 4 m*M* EGTA, 2 m*M* ATP, and 1 m*M* dithiothreitol (DTT)] and 200 μl of luciferin reagent [200 n*M* D-luciferin, 25 m*M* glycylglycine (pH 7.8)] into a tube containing the protein lysate (100 μg in a 100-μl volume, using reporter lysis buffer as a diluent). Luciferase and β-galactosidase activities are typically determined 24 hr posttransfection. Relative luciferase activities are determined and fold activities are calculated by normalizing values to total protein levels and to β-galactosidase enzyme levels.

An example of a transient transfection assay demonstrating the ability of H-Ras(V12) to stimulate the transcriptional activity of NF-κB is shown in Fig. 1. NIH 3T3 cells were transfected with either 3×-κB-Luc reporter construct or with 3×-mutκB-Luc, in which the κB elements have been mutated so that they no longer bind NF-κB. In addition, cells were transfected with either vector control or with expression vector encoding H-Ras(V12) protein. As shown in Fig. 1, the expression of H-Ras(V12) resulted in specific activation of NF-κB-dependent transcription.

III. Analysis of NF-κB DNA-Binding Activity

As described above, the transcriptional activity of NF-κB is controlled by signaling events that modulate nuclear translocation and DNA binding, as well as events that regulate the *trans*-activation function of various NF-κB components. The electrophoretic mobility shift assays (EMSAs) determine whether cellular stimuli activate nuclear translocation and DNA-binding activity of NF-κB. In the EMSA, proteins in nuclear extracts are analyzed for their ability to recognize a double-stranded DNA oligonucleotide corresponding to the NF-κB DNA-binding consensus site. DNA–protein complexes are resolved on a nondenaturing polyacrylamide gel. Gels are then dried and subjected to autoradiography. Unbound radiolabeled probe has a faster rate of migration than DNA bound by nuclear protein. Thus, radiolabeled DNA bound by nuclear protein results in a “shift.” The specificity of the shifted complexes can be determined by performing competition experiments in which unlabeled double-stranded competitor DNA is added in excess to the reaction mixture. Thus, an oligonucleotide that specifically binds NF-κB will compete for the nuclear proteins, while a nonspecific oligonucleotide will not. A more definitive analysis is to perform supershift experiments in which nuclear

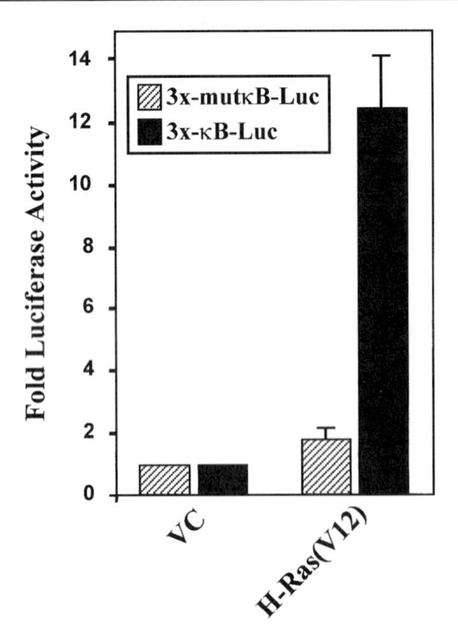

FIG. 1. Stimulation of NF-κB-dependent transcription by H-Ras(V12). NIH 3T3 cells were transiently cotransfected with the control plasmid (pCMV-Lac Z, 0.3 μg) and either an NF-κB-responsive reporter (3×-κB-Luc) or a mutant reporter (3×-mutκB-Luc, 0.3 μg each). In addition, cells were transfected with expression plasmids encoding H-Ras(V12) or empty vector control (VC, 0.3 μg each). Cell lysates were harvested 24 hr posttransfection and β-galactosidase and luciferase activities were assayed. The fold luciferase activity was determined by normalizing values to total protein levels and to β-galactosidase enzyme levels. Data are presented as fold activation, where the values obtained for the vector control were normalized to 1. Results represent means ± SD of three independent experiments performed in triplicate.

proteins are incubated in the presence of antibodies that specifically recognize various components of NF-κB. If a nuclear extract contains a particular subunit of NF-κB, the addition of antibody causes a further shift in the DNA–protein complex because of the added mass of the antibody (150 kDa). Therefore, while the EMSA is a quantitative measure of NF-κB DNA binding, supershift analysis provides a qualitative measure of which NF-κB subunits are being activated.

A. Isolation of Nuclear Extracts

Nuclear extracts can be isolated from cells and analyzed for the presence of NF-κB DNA binding after the addition of physiological stimuli [such as interleukin 1β (IL-1β), tumor necrosis factor (TNF), or platelet-derived growth factor (PDGF)] or in response to transgene expression. One disadvantage to analyzing NF-κB DNA-binding activity after transient transfection is that only cells that give high transfection efficiencies can be used. Some of the more commonly used cell lines for transient transfection and nuclear protein isolation include human embryonic kidney 293 cells and green monkey kidney COS cells. In our laboratory, we commonly analyze NF-κB DNA-binding activity in 293 cells after transient transfection of various expression constructs. Subconfluent 293 cells (5×10^6 cells per 100-mm dish) are transfected with either the vector control plasmid or with an expression vector encoding H-Ras(V12) (3 μg each), using 15 μl of SuperFect reagent as described in Section II,A. The efficiency of transfection can be analyzed in cells expressing the pCMV-LacZ (3 μg) plasmid by assaying for β-galactosidase activity (see description of this technique in Section V,A).

Nuclear extracts can be isolated from cells quickly and efficiently by a microisolation procedure. Briefly, 293 cells are washed (once) and scraped in 1 ml of ice-cold PBS and transferred into a 1.5-ml microcentrifuge tube. PBS is removed after a slow microcentifugation (2000 rpm for 5 min). Cell pellets are resuspended in 5 pellet volumes of CE buffer [10 m*M* HEPES (pH 7.6), 60 m*M* KCl, 1 m*M* EDTA, 1 m*M* DTT, 1 m*M* phenylmethylsulfonyl fluoride (PMSF), and 5 μg/ml each of aprotinin, leupeptin, and pepstatin] containing Nonidet P-40 (NP-40). The amount of NP-40 added to the CE buffer varies between cell line and ranges between 0.05 and 0.5% (v/v). We generally use 0.2% (v/v) NP-40 in CE buffer for lysis of 293 cells. Cells are lysed in CE buffer containing NP-40 by incubating the cells on ice for 3 min. Nuclei are pelleted by performing a slow centrifugation step (2000 rpm for 4 min) at 4°. After centrifugation, the CE buffer containing the cytoplasmic proteins is removed to a fresh microcentrifuge tube. To conserve protein–protein interactions, glycerol [20% (v/v) final] should be added to the CE buffer. Nuclei are washed to remove the nonionic detergent by resuspending nuclei in 5 pellet volumes of fresh CE buffer lacking NP-40. Carefully resuspend the nuclei in CE buffer by flicking the microcentrifuge tubes by hand. Do not vortex the samples because this will rupture the nuclei, releasing genomic DNA and greatly reducing the yield of nuclear protein. Nuclei are pelleted by centrifugation (2000 rpm for 4 min) at 4° and resuspended in 1 pellet volume of NE [20 m*M* Tris (pH 8.0), 420 m*M* NaCl, 1.5 m*M* $MgCl_2$, 0.2 m*M* EDTA, 25% (v/v) glycerol, 0.5 m*M* PMSF, and 5 μg/ml each of aprotinin, leupeptin, and pepstatin]. To achieve efficient recovery of nuclear proteins, the final salt concentration of the NE buffer needs to be 400 m*M* final. Therefore, the salt concentration is adjusted to 400 m*M* with 5 *M* NaCl, where the amount of salt to be added is equal to the pellet volume divided by 31.25. After the addition of NaCl, an additional pellet volume of NE is added and nuclei are further resuspended by vortexing the samples. Samples are incubated on ice for 10 min and centrifuged (14,000 rpm at 4°) for 10 min, and nuclear proteins are transferred to a fresh microcentrifuge tube. If the samples will be analyzed in multiple assays, store the nuclear proteins in multiple aliquots to avoid a loss of NF-κB DNA binding, which can result from multiple freeze–thaws. Both cytoplasmic and nuclear proteins can be stored at −70° for 2 to 3 months without a loss of activity. Protein concentrations are determined with the Bio-Rad protein assay dye reagent.

B. Radiolabeled Electrophoretic Mobility Shift Assay Probe for Detection of NF-κB DNA Binding

Normally, we use a 37-mer consisting of the NF-κB DNA-binding site and flanking regions from the MHC class I *H-2k^b^* gene (5′-CAGGGCTG-GGGATTCCCCATCTCCACAGTTTCACTTC-3′, binding site underlined). The oligonucl-eotide is annealed with its complementary strand (5′-GAAGTGAAACT-GTGG-3′) by boiling the two oligonucleotides (5 μg each) in the presence of 200 n*M* NaCl for 10 min. The oligonucleotides are allowed to anneal by slowly allowing the reaction to reach room temperature. The radiolabeled probe is generated by a random-primed labeling procedure in which 50 ng of double-stranded oligonucleotide is mixed with the labeling reaction mixture [1 m*M* each of dATP, dGTP, and dTTP, 0.1 m*M* dCTP, 1× Klenow buffer, 10 μl of [α-^{32}P]dCTP (3000 Ci/mmol), and 1 μl of Klenow enzyme in a 50-μl total volume]. The reaction is allowed to proceed for 30 min at 37°, after which 1 m*M* cold dCTP is added and the reaction is allowed to continue for an additional 15 min. Unincorporated α-^{32}P-labeled nucleotide is removed by loading the reaction mixture over a microspin Sephadex G-50 column (Amersham Pharmacia Biotech, Piscataway, NJ).

C. Electrophoretic Mobility Shift Assay for Detection of NF-κB DNA Binding

The binding reaction consists of nuclear protein (5–10 μg), nonspecific inhibitor poly(dI–dC) (Sigma, St. Louis, MO), binding buffer [50 m*M* NaCl, 10 m*M* Tris (pH 7.7), 0.5 m*M* EDTA, 10% (v/v) glycerol, and 1 m*M* DTT] and double-stranded ^{32}P-labeled probe (40,000 cpm) in a 20-μl total reaction volume. The binding buffer is typically made up as a 5× stock solution and can be stored at −20° for several months. NF-κB DNA binding is inhibited by salt concentrations that exceed 100 m*M*. For this reason, it is beneficial to make a 5× binding buffer that lacks NaCl. This is an important point to remember, because nuclear proteins isolated by the method described above are at 400 n*M* NaCl. Therefore, it is easy to exceed the optimum salt range for NF-κB DNA binding when a large volume of nuclear protein is needed. The use of a no-salt 5× binding buffer is an easy way to adjust for varying salt concentrations for different nuclear extracts.

DNA–protein complexes are resolved on a TGE (25 m*M* Tris, 190 m*M* glycine, 1 m*M* EDTA) 5% (w/v) nondenaturing polyacrylamide (30:1) gel containing 5% (v/v) glycerol. Gels are equilibrated by prerunning (18 mA) for 30 min and DNA–protein complexes are resolved by running gels at 22 mA for 2–2.5 hr. For antibody supershift assays and cold oligonucleotide competition experiments, nuclear extracts can be preincubated (10 min at room temperature) with either 1 μg of antiserum or with nonradioactive double-stranded oligonucleotide (10-, 50-, or 100-fold excess) before the addition of the radiolabeled EMSA probe. EMSA gels are dried and exposed to autoradiography overnight, using intensifying screens.

An example of an EMSA for H-Ras(V12)-induced NF-κB DNA binding is shown in Fig. 2. Transient transfection of a plasmid encoding H-Ras(V12) into 293

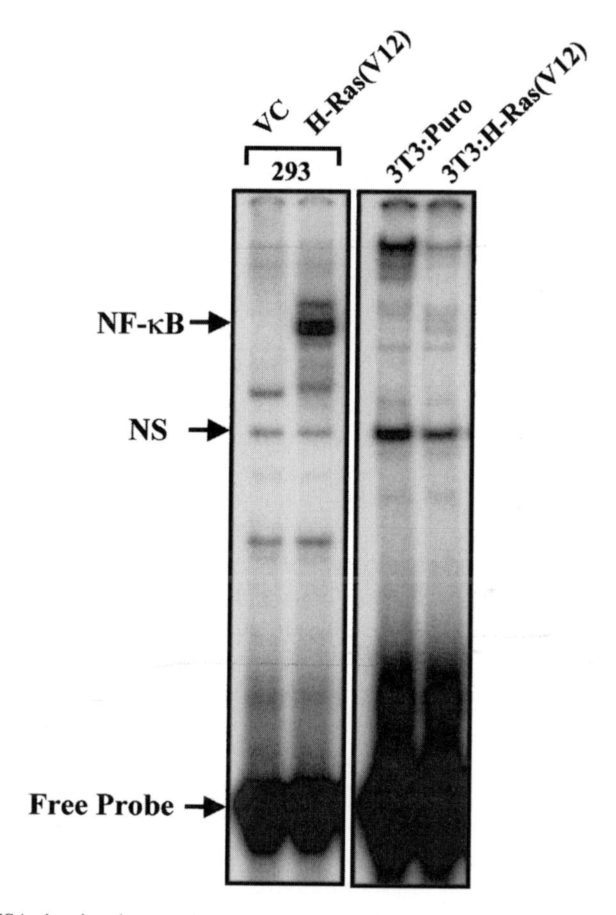

FIG. 2. EMSA showing that transient expression of H-Ras(V12) results in an increase in NF-κB DNA binding. Nuclear extracts were harvested from 293 cells transiently transfected with either an empty vector control plasmid (VC), or with an expression plasmid encoding H-Ras(V12) (3 μg each) for 24 hr. Nuclear extracts were also harvested from NIH 3T3 cells stably expressing either vector control (3T3 : Puro) or H-Ras(V12) [3T3 : H-Ras(V12)]. Nuclear extracts (8 μg) were incubated with a ^{32}P-labeled NF-κB-specific EMSA probe. DNA–protein complexes were resolved on a 10% (w/v) nondenaturing polyacrylamide TGE gel. The NF-κB–DNA complex is shown as well as the unbound "free" ^{32}P-labeled probe. A nonspecific (NS) complex is also observed in the EMSA analysis and demonstrates that equal amounts of nuclear proteins were analyzed between experimental groups.

cells results in an increase in NF-κB DNA-binding activity, as compared with cells transfected with empty vector control. As our laboratory has previously shown,[8] transformed NIH 3T3 cells stably expressing H-Ras(V12) fail to demonstrate an increase in NF-κB DNA binding when compared with parental cells expressing the vector control plasmid. This point is illustrated in Fig. 2, where NIH 3T3 cells stably expressing oncogenic Ras [3T3:H-Ras(V12)] display levels of NF-κB DNA binding similar to those of vector control cells (3T3:Puro), as detected by performing EMSA. These results indicate that transient expression of activated Ras stimulates signals that result in increased NF-κB DNA binding, whereas chronic H-Ras(V12) expression does not regulate NF-κB in the same manner.

IV. Analysis of Increased *Trans*-Activation Function of p65 Subunit of NF-κB

As is the case with many different transcription factors, NF-κB is regulated through signaling mechanisms that control nuclear translocation.[11] However, our laboratory has shown that oncoproteins can also upregulate the transcriptional activity of NF-κB through mechanisms that are independent of nuclear translocation events.[7,8,12] We have previously shown that H-Ras dramatically upregulates NF-κB-dependent transcription in H-Ras-transformed NIH 3T3 cells, and yet these transformed cells do not display an increase in NF-κB DNA binding when compared with vector control cells (Fig. 2). These results suggest that some oncoproteins are capable of upregulating NF-κB transcriptional activity without increasing the nuclear levels of this transcription factor. Thus, for example, H-Ras(V12) can upregulate NF-κB by targeting the *trans*-activation domain of the p65 subunit of NF-κB.[7,8] How can this increase in *trans*-activation potential be measured? This is achieved by cotransfecting cells with a plasmid encoding a Gal4–p65 fusion protein, in which sequences encoding the DNA-binding domain of the yeast transcription factor, Gal-4, have been joined with sequences encoding a *trans*-activation domain of the p65 subunit of NF-κB. Although there are three *trans*-activation (TA) domains located within the p65 subunit, we typically use sequences encoding the TA1 region, including amino acids 521–551. Gal4–p65 is cotransfected with a Gal4–luciferase (Gal4–Luc) reporter that contains four DNA-binding sites for Gal-4 and with expression constructs encoding various oncoproteins. Using this approach, it is possible to determine whether cellular signals induced by oncoproteins upregulate NF-κB transcriptional activity by specifically stimulating the *trans*-activation potential of the p65 subunit.

A. *Transient Cotransfection of Gal4–Luc and Gal4–p65 Plasmids into NIH 3T3 Cells*

NIH 3T3 cells at 60% confluency (plated in six-well plates 18 hr prior to transfection) are transiently transfected using SuperFect, following the same methodology described in Section II,A. Plasmids, including Gal4–Luc and Gal4–p65 (100 ng each) and either vector control plasmid (pCMV-5) or an expression construct encoding H-Ras(V12) (1 μg each), are combined, diluted in 300 μl of serum- and antibiotic-free medium, and mixed with 15 μl of SuperFect reagent. It is important to use limiting amounts of Gal4–p65 and Gal4–Luc in order to observe an induction after the transgene expression. Reagents are mixed, and incubated for 10 min to allow SuperFect–DNA complexes to form. Complete growth medium (2.7 ml) is added to the SuperFect–DNA complexes, mixed, and added to NIH 3T3 cells as described earlier. One reaction mixture is enough to perform the experiment in triplicate. Cells are harvested 24 to 48 hr after transfection and luciferase activities are analyzed. Gal4–Luc-mediated luciferase activity is analyzed in the same manner as previously described in Section II,B.

An example of a Gal4–p65/Gal4–Luc assay is shown in Fig. 3. NIH 3T3 cells either stably expressing the empty vector control (3T3:Puro) or oncogenic

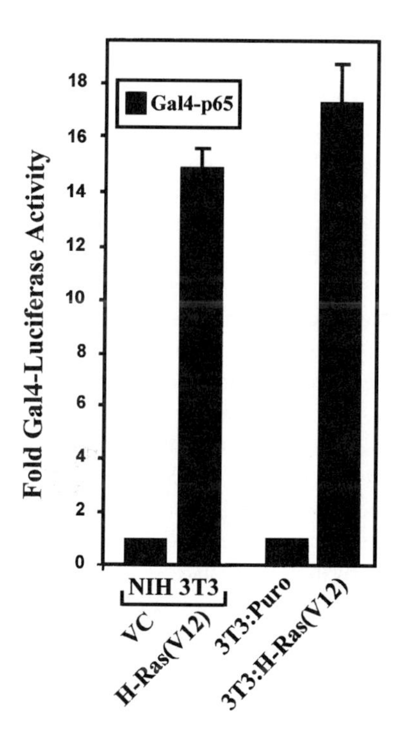

FIG. 3. H-Ras(V12) upregulates NF-κB by targeting the *trans*-activation domain of p65. NIH 3T3, 3T3 : Puro, and 3T3 : H-Ras(V12) cells were transiently cotransfected with the Gal4–Luc reporter and with plasmids encoding the Gal4–p65 fusion protein (100 ng each). In addition, NIH 3T3 cells were transfected with either empty plasmid vector (VC), or with plasmid encoding H-Ras(V12) (1 μg each). Luciferase activity was analyzed 48 hr posttransfection. Data shown represent the means ± SD of three independent experiments performed in triplicate.

H-Ras [3T3:H-Ras(V12)] were cotransfected with the 4×-Gal4 luciferase reporter, and with the Gal4–p65 expression construct. As shown in Fig. 3, H-Ras(V12)-transformed NIH 3T3 cells displayed significantly higher levels of p65 *trans*-activation potential compared with vector control cells. Collectively, these findings indicate that H-Ras(V12) stimulates the transcriptional activity of NF-κB by targeting basal levels of nuclear NF-κB and that this molecule effectively upregulates the *trans*-activation potential of this transcription factor.

V. Detection of H-Ras(V12)-Induced Cell Death after the Inactivation of NF-κB

We have previously demonstrated that oncogenic Ras stimulates the transcriptional activity of NF-κB in order to overcome H-Ras-induced apoptosis.[13] Apoptosis is an active regulatory response in which cells are programmed to commit suicide through a multistage process characterized by DNA fragmentation and morphologic cellular changes. Therefore, one of the apoptotic criteria commonly assayed for is the presence of endonucleolytically degraded genomic DNA. This can be accomplished by many different molecular techniques. In this section, we describe two of the more commonly used protocols to detect DNA fragmentation: DNA laddering by standard agarose gel electrophoresis, and deoxynucleotidyltransferase-mediated dUTP nick-end labeling (TUNEL) analysis. In addition, we describe another method, β-galactosidase expression assays, which we have successfully employed to rapidly screen signaling pathways involved in H-Ras(V12)-induced cell killing after the inactivation of NF-κB (13). β-Galactosidase expression assays rapidly determine the requirement of NF-κB, and other factors, in cell viability after coexpression of an oncoprotein. One disadvantage to this method is that a loss in cell viability does not necessarily indicate that cells are dying through apoptotic mechanisms, but because this method is easy and reproducible it makes an appropriate starting point. In this assay, cells are transiently cotransfected with a β-galactosidase reporter and with plasmids encoding H-Ras(V12) in the presence or absence of the NF-κB inhibitor, the super repressor-IκBα (SR-IκBα). The SR-IκBα acts as a dominant negative by blocking nuclear translocation and subsequently inhibiting NF-κB-regulated gene expression. Because coexpression of H-Ras(V12) and the SR-IκBα kills NIH 3T3 cells, this suggests that H-Ras(V12) activates NF-κB to transcriptionally upregulate genes that are required to overcome Ras-induced apoptosis. To definitively determine whether H-Ras(V12) requires NF-κB to overcome the induction of apoptosis, it is important to establish whether the coexpression of H-Ras(V12) and SR-IκBα induces programmed cell death. This can be accomplished by using cell lines that are inducible or by transiently expressing the appropriate transgenes and performing DNA laddering or TUNEL analysis.

A. Analysis of Cell Viability Using β-Galactosidase Expression Assays

NIH 3T3 cells are plated at 1×10^4 cells per 24-well plate 18 hr prior to transfection. Cells can be transfected with the SuperFect reagent as previously described. Briefly, one reaction is enough to transfect 12 wells. However, we typically analyze only three to six wells per group, depending on the experiment. Transfection reactions include the pCMV-LacZ plasmid (200 ng per reaction) and one of the following: (1) appropriate empty vector control plasmids (1 μg each), (2) the SR-IκBα (1 μg) and vector control (1 μg), (3) H-Ras(V12) and vector control (1 μg each), or (4) H-Ras(V12) and the SR-IκBα (1 μg each). SuperFect reactions are carried out as previously described in Section II,A; however, after the 10-min SuperFect–DNA incubation, the mixture is resuspended in 2 ml of complete medium rather than 2.7 ml. Next, 200 μl of SuperFect–DNA complexes is then added to each well of the 24-well plate. It is important to resuspend the SuperFect–DNA mixture periodically to ensure that the transfection reagent is effectively distributed. Cells are incubated with the SuperFect–DNA mixture for 2 hr, and then washed and fed with 2 ml of complete medium. Forty-eight hours posttransfection, medium is carefully removed and cells are fixed by adding 1 ml of 0.5% (v/v) glutaraldehyde for 10 min at room temperature. Cells are washed twice with ice-cold PBS containing 1 m*M* $MgCl_2$, and monolayers are covered with 300 μl of X-Gal working solution [5 m*M* $K_3Fe(CN)_6$, 5 m*M* $K_4Fe(CN)_6$, 1 m*M* $MgCl_2$ in PBS, containing 5-bromo-4-chloro-3-indolyl-β-D-galactopyranoside (X-Gal solution, 1 mg/ml), 0.01% (w/v) sodium dodecyl sulfate (SDS), and 0.01% (v/v) NP-40]. Twenty-four-well plates are covered in aluminum foil and incubated at 37° for 2–4 hr, after which β-galactosidase (β-Gal)-positive cells can be counted under the microscope. Each well of the 24-well plate is divided into 4 quadrants, allowing β-Gal-positive cells to be easily counted. Plates analyzed for β-galactosidase activity can be stored at 4° for up to 1 week. For example, because cells expressing the H-Ras(V12) oncoprotein require NF-κB for cell survival, then coexpression of H-Ras(V12) and the SR-IκBα leads to a loss of cell viability as determined by the decrease in the number of β-galactosidase-positive cells, as compared with control cells.

B. Isolation of Genomic DNA for Analysis of Apoptosis-Induced DNA Fragmentation

Approximately $1–4 \times 10^6$ cells are needed for the procedure. Cells are carefully washed once in ice-cold PBS, scraped into 1 ml of PBS, and transferred to a 1.5-ml microcentrifuge tube. Cells are incubated in ice-cold lysis buffer [20 m*M* Tris (pH 7.4), 10 m*M* EDTA, and 0.2% (v/v) TritonX-100] for 10 min on ice. Cells are centrifuged (10,000 rpm) for 10 min at 4° and supernatants are collected and placed in a fresh microcentifuge tube. During the isolation of fragmented DNA the slow centrifigation spin removes cell membrane debris and nonfragmented (intact) genomic DNA. However, fragmented DNAs are released from apoptotic cells and are collected in the centrifugation supernatant fraction. DNase-free RNase A (10 μl of a 500-μg/ml solution) is added to the supernatant and extracts are incubated at 37° for 1 hr, followed by the addition of proteinase K (100 μg/ml) and incubation at 50° for an additional 2 hr. DNAs are extracted twice with phenol–chloroform–isoamyl alchol (25 : 24 : 1, v/v/v) and once with chloroform–isoamyl alcohol (24 : 1, v/v), and precipitated by the addition of a 1 : 10 volume of sodium acetate (3 *M*), glycogen (5 μg/ml), and 2.5 volumes of ethanol. DNAs are resolved on a 1.5–2% (w/v) agarose gel containing ethidium bromide. Although this procedure does not isolate intact genomic DNA, an equivalent recovery of RNase-digested RNA demonstrates that equal numbers of cells were analyzed and this helps to normalize for nucleic acid loading.

An example of a DNA fragmentation ladder observed in Ras-induced apoptosis after the inactivation of NF-κB is shown in Fig. 4. Rat-1 cells stably expressing either the vector control (Rat-1:Hygro) or activated Ras [Rat-1:H-Ras(V12)] were infected with adenovirus encoding either the empty vector control (Ad-CMV) or the SR-IκBα (Ad-SRIκBα). Because greater than 95% of Rat-1 cells are

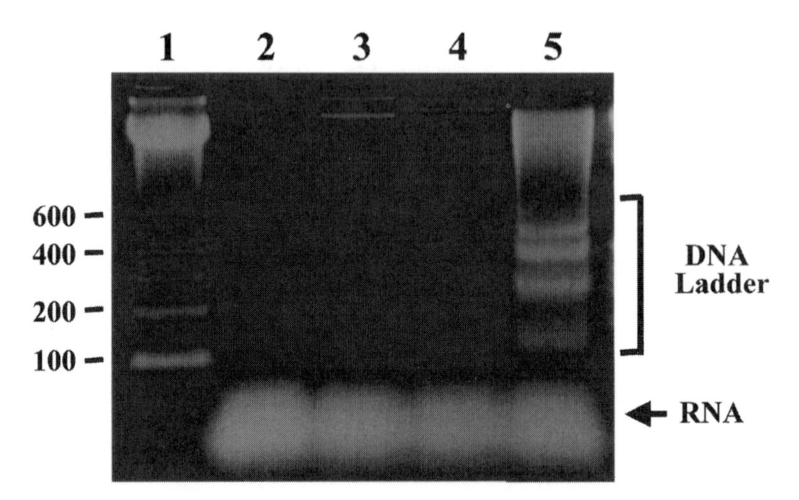

FIG. 4. H-Ras-transformed cells require NF-κB to overcome apoptosis. Rat-1 cells (1×10^6) stably expressing either the vector control plasmid (Rat-1 : Hygro) or oncogenic Ras [Rat-1 : H-Ras(V12)] were infected with adenovirus encoding either the NF-κB inhibitor (Ad-SRIκBα) or empty vector control (Ad-CMV) at 50 PFU/cell. DNAs were isolated 48 hr postinfection and run on a 2% (w/v) standard agarose gel. A 100-bp molecular weight marker was used to determine the approximate size of the fragmented DNAs. The nucleic acids shown are as follows: lane 1, a 100-bp DNA molecular weight marker; lane 2, DNAs isolated from Rat-1 : Hygro cells infected with Ad-CMV; lane 3, DNAs isolated from Rat-1 : Hygro cells infected with Ad-SRIκBα; lane 4, DNAs isolated from Rat-1 : H-Ras(V12) cells infected with Ad-CMV; lane 5, DNAs isolated from Rat-1 : H-Ras(V12) cells infected with Ad-SRIκBα. A 180-bp increment DNA ladder is shown. RNase A-digested RNA is shown and demonstrates that equal cell numbers were originally analyzed in the assay.

susceptible to adenovirus-mediated gene transfer, this approach makes it possible to analyze whole cell populations. As shown in Fig. 4 (lane 5), Rat-1:H-Ras(V12) cells demonstrated fragmented DNA ladder after adenovirus-mediated expression of the SR-IκBα. In contrast, expression of SR-IκBα did not induce apoptosis in vector control cells (Fig. 4, lane 3). These results are consistent with the requirement for NF-κB activation to overcome H-Ras(V12)-induced apoptosis.

C. Deoxynucleotidyltransferase-Mediated dUTP Nick-End Labeling Analysis

Cells (2.5×10^5 per six well plate) are plated on top of 22×22 mm sterile glass coverslips and allowed to adhere and grow 18 hr before the start of the experiment. Typically our laboratory analyzes for the presence of TUNEL-positive cells by performing the ApopTag procedure (Oncor, Gaithersburg, MD). Cells are carefully washed (once) and fixed for 15 min by incubating coverslips in 4% (w/v) paraformaldehyde in PBS (pH 7.4). Cells are washed (twice) with room temperature PBS. Samples are postfixed in ethanol–acetic acid (2 : 1, v/v) for 5 min at $-20°$. Slides are then drained and washed (twice) with room temperature PBS. Excess liquid is removed by gently turning coverslips on their edge and tapping onto a Kimwipe. Two drops of equilibration buffer ($1\times$) is then applied directly onto the coverslip. Once the equilibration buffer has been added, a fresh glass slide is placed over the coverslip containing the cells. The slide is then inverted so that the glass slide is on top and the samples are then placed in a humidified chamber for 5 min at room temperature. A humidifier can be made by taking a small plastic tray (4×5 inches), lining the bottom with several wet paper towels, and placing two disposable pipettes taped together with a 1.5-inch gap across the bottom of the tray for the slides to rest on. After the incubation, the coverslips are carefully removed from the slide and the excess liquid is removed. Fifty-four microliters of working strength TdT enzyme [reaction buffer plus terminal deoxynucleotidyltransferase (TdT) enzyme] is then added. Reapply the slide to the coverslip and incubate in a humidified chamber at 37° for 1 hr. Carefully remove the coverslip and place face up in a small container. Pipette prewarmed working strength stop/wash buffer (stop/wash buffer plus distilled water) and incubate for 30 min at 37°. Enough stop/wash buffer should be added to completely cover the samples. The slips need to be carefully rocked periodically (once every 10 min) to ensure equal distribution of the stop/wash buffer. The samples are removed from the stop/wash solution, the excess liquid is removed, and the slips are wash (three times) with PBS (3 min each). The excess liquid should be removed from the coverslips and 52 μl of working strength anti-digoxigenin–fluorescein (blocking solution plus anti-digoxigenin–fluorescein) is added to each coverslip. Reapply the slide and incubate in a humidified chamber for 30 min at room temperature. Remove the coverslip and wash the samples (three times) with PBS (5 min per wash) at room temperature. Mount the coverslips onto a fresh slide and visualize cells with a fluorescence microscope at a wavelength between 510 and 550 nm.

[9] Ras Activation of NF-κB and Superoxide

By CHUNMING DONG and PASCAL J. GOLDSCHMIDT-CLERMONT

p21*ras* (c-Ras) participates in various cellular processes, including proliferation, differentiation, apoptosis, and cytoskeletal organization.[1,2] Mutations in a *ras* allele that render it constitutively active have been described in approximately 30% of all human tumors, making it the most widely mutated human protooncogene. It appears that multiple pathways exist downstream of Ras: (1) Ras is known to activate both the Raf–MEK–ERK pathway and the MEKK–SEK–JNK pathway [MEK, MAP (mitogen-activated protein) kinase/ERK kinase; ERK, extracellular signal-regulated kinase; MEKK, MEK kinase; SEK, JNK kinase; JNK, c-Jun N-terminal kinase][3–6]; (2) Ras causes transformation via a pathway that includes the activation of Rac1, another small GTP-binding protein of the Rho family.[7] Activated Rac1 induces the production of reactive oxygen species (ROS), including superoxide anion ($O_2^{\cdot -}$)[8]; and (3) Ras has been shown to activate nuclear factor κB (NF-κB), which inhibits p53-mediated apoptosis in Ras-transformed NIH 3T3 fibroblasts, contributing to its oncogenic potential.[9] We have demonstrated that in HeLa cells Ras stimulates NF-κB activity via the activation of Rac1 and Rac1-induced ROS production, providing support for a sequential pathway linking Ras, Rac1, ROS production, and the activation of NF-κB.[10] Indeed, several lines of evidence indicate that extracellular stimuli induce the generation of ROS, which, in turn, either directly or indirectly via the activation of NF-κB, leads to the production of monocyte chemotactic protein 1 (MCP-1), which elicits direct migration of monocytes into inflammatory sites,

[1] M. S. Boguski and F. McCormick, *Nature* (*London*) **366,** 643 (1993).
[2] R. Khosravi-Far and C. J. Der, *Cancer Metastasis Rev.* **13,** 67 (1994).
[3] B. Derijard, M. Hibi, I. H. Wu, T. Barrett, B. Su, T. Deng, M. Karin, and R. J. Davis, *Cell* **76,** 1025 (1994).
[4] A. Minden, A. Lin, M. McMahon, C. Lange-Carter, B. Derijard, R. J. Davis, G. L. Johnson, and M. Karin, *Science* **266,** 1719 (1994).
[5] M. F. Olson, A. Ashworth, and A. Hall, *Science* **269,** 1270 (1995).
[6] P. Rodriguez-Viciana, P. H. Warne, A. Khwaja, B. M. Marte, D. Pappin, P. Das, M. D. Waterfield, A. Ridley, and J. Downward, *Cell* **89,** 457 (1997).
[7] R. G. Qiu, J. Chen, D. Kirn, F. McCormick, and M. Symons, *Nature* (*London*) **374,** 457 (1995).
[8] K. Irani and P. J. Goldschmidt-Clermont, *Biochem. Pharmacol.* **55,** 1339 (1998).
[9] M. W. Mayo, C. Y. Wang, P. C. Cogswell, K. S. Rogers-Graham, S. W. Lowe, C. J. Der, and A. S. J. Baldwin, *Science* **278,** 1812 (1997).
[10] D. J. Sulciner, K. Irani, Z. X. Yu, V. J. Ferrans, P. Goldschmidt-Clermont, and T. Finkel, *Mol. Cell. Biol.* **16,** 7115 (1996).

0076-6879/00 $35.00

playing an important role in inflammation and atherosclerosis.[11,12] Thus, it appears that Rac-induced superoxide production by NADPH oxidase can serve as a second messenger essential for signal transduction downstream of Ras, both in physiologic and nonphysiologic cells. In this chapter, we focus on the methodologies to detect Rac activation, subsequent superoxide production, and NF-κB activation.

The binding of Ras with GTP initiates the activation of signal transduction cascades, including the activation of ERK, JNK, and the association of Rac with GTP. The association of Rac with GTP triggers the clustering of a multiprotein complex, NADPH oxidase, which catalyzes the generation of superoxide in phagocytes as well as in nonphagocytic cells.[13–15] In phagocytes, the enzymatic activity of NADPH oxidase is provided by a flavocytochrome, cytochrome *b*-558, an integral membrane protein complex composed of two subunits, glycoprotein (gp) 91phox and p22phox.[8] In nonphagocytic cells, Mox1 (mitogenic oxidase-1) has been cloned and characterized as an analog of neutrophil gp91phox, and participates in the production of ROS together with p22phox.[16] While in phagocytes ROS production within phagolysosome plays a role in bactericidal activity, ROS produced in nonphagocytic cells appear to intersect with an array of intracellular pathways downstream of Ras and other signaling pathways. Furthermore, ROS can activate Ras through the activation of JAK2 (Janus protein kinase 2) and Fyn.[17] Thus, ROS not only serve as signaling molecules for activated Ras, but can also activate Ras in a positive feedback loop. These data point to a central role for Rac activation and ROS generation by NADPH oxidase in the regulation of growth, transformation, apoptosis, and senescence.[18,19]

NF-κB is a transcription factor that regulates genes involved in immune and inflammatory responses, cell proliferation, differentiation, and apoptosis.[20–22] There are five members in the NF-κB family: p50/NF-κB1, p52/NF-κB2, RelA/p65,

[11] B. S. Wung, J. J. Cheng, H. J. Hsieh, Y. J. Shyy, and D. L. Wang, *Circ. Res.* **81,** 1 (1997).
[12] M. Johnstone, A. J. Gearing, and K. M. Miller, *J. Neuroimmunol.* **93,** 182 (1999).
[13] P. Patriarca, R. Cramer, S. Moncalvo, F. Rossi, and D. Romeo, *Arch. Biochem. Biophys.* **45,** 255 (1971).
[14] D. Diekmann, A. Abo, C. Johnston, A. W. Segal, and A. Hall, *Science* **265,** 531 (1994).
[15] F. R. DeLeo and M. T. Quinn, *J. Leukoc. Biol.* **60,** 677 (1996).
[16] Y. A. Suh, R. S. Arnold, B. Lassegue, J. Shi, X. Xu, D. Sorescu, A. B. Chung, K. K. Griendling, and J. D. Lambeth, *Nature (London)* **401,** 79 (1999).
[17] J. Abe and B. C. Berk, *J. Biol. Chem.* **274,** 21003 (1999).
[18] T. Finkel, *J. Leukoc. Biol.* **65,** 337 (1999).
[19] K. Irani, Y. Xia, J. L. Zweier, S. J. Sollott, C. J. Der, E. R. Fearon, M. Sundaresan, T. Finkel, and P. J. Goldschmidt-Clermont, *Science* **275,** 1649 (1997).
[20] P. A. Baeuerle and T. Henkel, *Annu. Rev. Immunol.* **12,** 141 (1994).
[21] U. Siebenlist, G. Franzoso, and K. Brown *Annu. Rev. Cell Biol.* **10,** 405 (1994).
[22] A. S. J. Baldwin, *Annu. Rev. Immunol.* **14,** 649 (1996).

RelB, and c-Rel.[23–27] The Rel homology domain found in each of these proteins functions in DNA binding, dimerization between family members, and interaction with a family of inhibitory proteins known as I-κB, including I-κBα and I-κBβ, which bind to NF-κB and sequester it in the cytoplasm.[28] Stimulation of cells with inducers such as tumor necrosis factor α (TNF-α) and intersleukin 1 (IL-1) may activate Ras, which, in turn, either directly or indirectly through Rac activation and ROS production, causes I-κB phosphorylation, ubiquitination, and subsequent degradation, a process that reveals the nuclear localization signal of NF-κB, allowing for the translocation of NF-κB from the cytoplasm into the nucleus, where it affects the transcription of target genes.

Rac Activation Assay

Previously, Rac activation assay involved the use of the radioisotope ^{32}P. We have adopted a nonradioisotope protocol developed by Upstate Biotechnology, Inc. (Lake Placid, NY) to detect Rac activation. The assay is based on the principle that Rac–GTP binds p21-activated kinase 1 (PAK-1), while Rac–GDP is unable to associate with PAK-1. In cells, upon binding with Rac–GTP, PAK-1 becomes activated, and may regulate the dynamics of the actin cytoskeleton and gene transcription.

Method

1. Lyse cells with lysis buffer provided in the kit from Upstate Biotechnology.
2. Immediately add 5–10 μg (5–10 μl) of PAK-1–agarose suspension to 1 ml of cell lysate and gently rock the reaction mixture at 4° for 1 hr.
3. Collect the agarose beads by pulsing (5 sec in a microcentrifuge at 14,000 *g*) and discard the supernatant. Wash the beads three times with 0.5 ml of lysis buffer.
4. Resuspend the agarose beads in 30 μl of Laemmli sample buffer; boil for 5 min.
5. Perform sodium dodecyl sulfate–polyacrylamide gel electrophoresis (SDS–PAGE) with the supernatant and transfer the proteins to nitrocellulose membrane.

[23] G. P. Nolan, S. Ghosh, H. C. Liou, P. Tempst, and D. Baltimore, *Cell* **64,** 961 (1991).
[24] S. M. Ruben, P. J. Dillon, R. Schreck, T. Henkel, C. H. Chen, M. Maher, P. A. Baeuerle, and C. A. Rosen, *Science* **251,** 1490 (1991).
[25] R. M. Schmid, N. D. Perkins, C. S. Duckett, P. C. Andrews, and G. J. Nabel, *Nature (London)* **352,** 733 (1991).
[26] V. Bours, P. R. Burd, K. Brown, J. Villalobos, S. Park, R. P. Ryseck, R. Bravo, K. Kelly, and U. Siebenlist, *Mol. Cell. Biol.* **12,** 685 (1992).
[27] R. P. Ryseck, P. Bull, M. Takamiya, V. Bours, U. Siebenlist, P. Dobrzanski, and R. Bravo, *Mol. Cell. Biol.* **12,** 674 (1992).
[28] P. A. Baeuerle and D. Baltimore, *Cell* **87,** 13 (1996).

6. Block the blotted nitrocellulose in freshly prepared phosphate-buffered saline (PBS) with 0.1% (v/v) Tween 20, containing 3% (w/v) nonfat dry milk, for 1 hr.

7. Wash the nitrocellulose three times with PBS–0.1% (v/v) Tween 20, 10 min each.

8. Incubate the nitrocellulose with anti-Rac antibody (mouse monoclonal IgG_{2b}, 1 μg/ml) overnight at 4°.

9. Wash the nitrocellulose three times with PBS–0.1% (v/v) Tween 20, 10 min each.

10. Incubate the nitrocellulose with goat anti-mouse IgG for 30 min.

11. Wash the nitrocellulose three times with PBS–0.1% (v/v) Tween 20, 10 min each.

12. Develop by enhanced chemiluminescence (ECL; Amersham, Arlington Heights, IL).

Comments

Note 1: Cells transfected with adenovirus encoding RacV12 and RacN17 separately should be used as positive and negative controls, respectively. RacV12 is a mutant of Rac, which is unable to hydrolyze its bound nucleotide, and therefore is located in the GTP-bound conformation, and will bind PAK-1 constitutively. RacN17 is unable to exchange its bound nucleotide, but has preserved GTP triphosphatase activity. Hence, RacN17 is located in the GDP-bound conformation, and will sequestrate exchange factors for Rac, thereby preventing the interaction of endogenous Rac with PAK-1.

Note 2: An aliquot of cell lysate is used to measure protein concentration to normalize the amount of protein on loading into an SDS–polyacrylamide gel.

Assays for Oxygen-Free Radical Production

There are several methods to detect the production of oxygen free radicals within cells and tissues. These include spin trapping of hydroxyl radicals using 5,5-dimethyl-1-pyrroline *N*-oxide (DMPO) followed by detection by electron paramagnetic resonance (EPR) spectroscopy, microscopic detection of oxygen radicals using chemical probes, and other assays such as the deoxyribose assay, aromatic hydroxylation assay, pulse radiolysis, and bleomycin assay. We have routinely used the EPR spin trapping method, which can be applied to cells and is also appropriate for tissues, and provides a "signature" spectrum for superoxide. Another "workhorse" assay used by many laboratories is based on *in vivo* detection by dichlorofluorescein microfluorography to detect ROS production in cells. In the following sections, we describe detailed protocols for these two methods.

Measurement of Reactive Oxygen Species by Dichlorofluorescein

2′,7′-Dichlorofluorescein diacetate (DCF) is a unique fluorescence precursor that rapidly diffuses into cells whereupon cellular esterases cleave the acetate moiety, allowing for the accumulation of the membrane-impermeable form. Upon interaction with intracellular hydrogen peroxide, DCF is oxidized and produces a fluorescent moiety that can be conveniently detected by fluorescence microscopy, flow cytometry, and a CytoFluor multiwell plate reader (PerSeptive Biosystems/PE Biosystems, Foster City, CA), allowing for qualitative or quantitative assessment of intracellular ROS. The major ROS detected by DCF is H_2O_2. More important is the fact that the signal corresponds to an integration of all oxidants produced from the time DCF enters the cell, rather than a reflection of the concentration of ROS at a specific time point.

Methods

1. Dissolve 50 μg of 2′,7′-dichlorofluorescein diacetate (DCF-DA; Molecular Probes, Eugene, OR) in 40 μl of dimethyl sulfopide (DMSO).
2. Add 20 μl of DCF-DA to 10 ml of Hanks' balanced salt solution (HBSS).
3. Wash the cells/tissues three times with HBSS, 5 min each.
4. Incubate the cells/tissues with DCF-DA/HBSS at 37° for 15 min.
5. Wash the cells/tissues with HBSS three times, 5 min each.

The cells and tissues are now ready for fluorescence microscopic visualization. The cells can also be subjected to flow cytometric analysis or microplate assay, using a CytoFluor multiwell plate reader (CytoFluor II; PerSeptive Biosystems) at an excitation of 485/ ± 20 and emission of 530/ ± 25.

Comments

Note 1: Cells or tissues should contain low levels of autofluorescence.

Note 2: Tissues should be free of factors that may cause DCF clearance or extracellular leakage.

Note 3: Tissues should not contain color-quenching materials, such as hemoglobin, which may disturb digital fluorometry.

Note 4: DCF-DA should be prepared freshly before use, and should be maintained during the experiment in sealed, foil-wrapped vials to avoid air and light.

Electron Paramagnetic Resonance Spin Trapping

When oxygen free radicals, superoxide, or hydroxyl radicals are produced *in vivo,* they are rapidly metabolized by cellular antioxidant systems (superoxide dismutase, glutathione peroxidase, catalase, etc.), and have a short half-life, making direct *in vivo* measurement impossible. Thus, spin trapping of hydroxyl radical by DMPO combined with EPR spectroscopy has emerged as one of the most frequently used methods for the detection of superoxide.

Method

1. Collect cells, using a plastic scraper, in 1 ml of PBS; spin down at 1200 rpm.
2. Resuspend the cell pellet in 250 μl of PBS.
3. Add 25 μl of cell suspension to 175 μl of 100 m*M* DMPO in water.
4. Transfer the reaction mixture into a quartz flat cell (capacity, 200 μl).
5. Measure EPR spectra.

All measurements are carried out at room temperature on a Bruker (Billerica, MA) spectrometer ESP 300E, operating at microwave power, 20 mW; modulation frequency, 9.75 GHz. Other instrument settings are as follows: 100-G scan range, 81.92-msec time constant, 164-msec time conversion, 1.01-G amplitude modulation, and 2×10^{-4} gain receiver.

Comments

Note 1: The specificity of the assay can be enhanced with the use of inhibitors. The hydroxyl radical can arise either from superoxide or from H_2O_2. Addition of superoxide as a first control and of catalase as a second control helps confirm the nature of the hydroxyl radical precursor.

Note 2: DMPO stock solution should be prepared in the presence of activated charcoal in order to remove impurities that can produce ROS, and the charcoal is removed by decantation.

Assays for NF-κB Activation

Electrophoretic Mobility Shift Assays (EMSA)

Translocation of NF-κB from the cytosol to the nucleus leads to the accumulation of NF-κB proteins in the nucleus. Incubation of a ^{32}P-labeled oligonucleotide containing the DNA-binding site for NF-κB proteins (5′ -AGTTGAGGGGACTT-TCCCAGGC-3′; Promega, Madison, WI) with nuclear extracts that contain increased concentrations of NF-κB proteins results in binding of NF-κB proteins to the oligonucleotide. Such a binding results in retardation ("shift") of the electrophoretic mobility of the oligonucleotide on a nondenaturing polyacrylamide gel, which can be detected by autoradiography and serves as a marker of the presence of NF-κB in the nucleus.

Preparation of Nuclear Extracts

METHODS

1. Centrifuge cells at 750 *g* for 5 min at 4°.
2. Resuspend the cell pellets in ice-cold PBS, pH 7.4, and centrifuge again at 750 *g* for 5 min at 4°.

3. Resuspend the cell pellets in 300 μl of buffer containing 10 m*M* HEPES (pH 7.9), 10 m*M* KCl, 2 m*M* $MgCl_2$, 0.1 m*M* ethylenediaminetetraacetic acid (EDTA, sodium salt, pH 8.0), 0.2 m*M* NaF, 0.2 m*M* Na_3VO_4, 0.4 m*M* phenylmethylsulfonyl fluoride (PMSF), leupeptin, (0.3 mg/ml), and 1 m*M* dithiothreitol (DTT).

4. Incubate the suspension on ice for 20 min, and then add 20 μl of 10% (v/v) Nonidet P-40.

5. Vortex the suspension vigorously for 15 sec and centrifuge at 12,000 *g* for 5 min at 4°. This results in the pelleting of nuclear extracts.

6. Resuspend the nuclear pellet in 50 μl of buffer containing 50 m*M* HEPES (pH 7.9), 50 m*M* KCl, 300 m*M* NaCl, 0.1 m*M* EDTA, 0.2 m*M* NaF, 0.2 m*M* Na_3VO_4, 0.4 m*M* PMSF, leupeptin (0.3 mg/ml), 1 m*M* DTT, and 10% (v/v) sterile glycerol.

7. Incubate the nuclear lysates on a rocking platform for 30 min at 4°.

8. Centrifuge the nuclear lysates at 12,000 *g* for 10 min at 4°, and collect the supernatant.

9. Measure the protein concentration in the supernatant using the Bio-Rad assay method (Bradford protein assay reagent; Bio-Rad, Hercules, CA).

10. Aliquots of the extracts are used for the gel-shift assay, or stored at −80°.

COMMENTS

Note 1: Stock solutions of protease inhibitors PMSF, leupeptin, and DTT should be stored at −20°, and added fresh each time.

Note 2: Aliquots of nuclear extracts are either used for EMSA, or quick-frozen and stored at −80°. They should not be thawed more than twice.

DNA-Binding Reaction and Electrophoresis

METHOD

1. Incubate 5–10 μg of nuclear extracts with 60,000–80,000 cpm of labeled probe, in the presence of 40 m*M* HEPES (pH 7.9), 10% (v/v) glycerol, 1 m*M* $MgCl_2$, 0.1 m*M* DTT, bovine serum albumin (BSA, 300 μg/ml), and poly(dI–dC) · (dI–dC) (133 μg/ml) at room temperature for 20 min.

2. Run the samples on a nondenaturing 40% (w/v) acrylamide gel.

3. Heat and vacuum dry the gel, and detect the band shifts by autoradiography.

COMMENTS

Note 1: It is important to keep the volume of the buffer used to resuspend the nuclear pellet consistent in the DNA-binding reaction, as the high salts present in the buffer determine the binding kinetics of the protein to the nucleotide. Thus, strict control of the relative concentration of the salts by diluting the more concentrated samples using the same buffer should be performed.

Note 2: To further ascertain the specificity of NF-κB–DNA binding, reactions containing a 100-fold molar excess of unlabeled competitive or uncompetitive oligonucleotides and nuclear extracts are preincubated for 15 min, before adding the labeled probe. Competitive oligonucleotide should result in the disappearance of all NF-κB complexes, whereas uncompetitive oligonucleotide should not compete for binding.

Note 3: The transcription factor Rel family of genes is composed of several NF-κB protein subunits. To establish the composition of the band that is shifted, specific antibodies against p65, p50, p52, and c-Rel subunits (Santa Cruz Biotechnology Santa Cruz, CA) are incubated with the DNA-binding reactions for an additional 30 min prior to electrophoresis. When antibody binding results in an additional electrophoretic retardation, it is referred to as a "supershift," which increases the specificity of the assay for individual NF-κB subunits.

Trans-Activation Assays of NF-κB by Transfected Reporter Genes

Activation of NF-κB induces the transcription of NF-κB target genes. The transcriptional regulatory activity of NF-κB can be studied in cells transiently transfected with an NF-κB consensus DNA sequence upstream to a reporter gene, such as luciferase. The level of reporter gene expression in these transiently transfected cells reflects the transcription regulatory activity of NF-κB.

Transient Transfection. There are a variety of transient transfection protocols. We have routinely used the liposomal transfection reagent DOTAP (1,2-dioleoyl-3-trimethylammoniumpropane) (Boehringer Mannheim, Indianapolis, IN). Cells are transfected in 12-well plates with an NF-κB–luciferase reporter vector along with pSV-β-galactosidase control vector (SVβGal; Promega) to normalize the transfection efficiency.

METHOD. For one well with 50–60% confluent cells:

1. Dilute 1 μg of DNA to a final volume of 12.5 μl in HEPES buffer.
2. Dilute 7.5 μl of DOTAP to a final volume of 25 μl in HEPES buffer.
3. Mix the DNA and DOTAP for 15 min at room temperature.
4. Replace the culture medium with 1 ml of fresh medium containing the mixture from step 3 and culture for 6 hr.
5. Replace with normal medium and culture for a further 36–48 hr.

Luciferase Assay

METHOD. The measurement is performed with the luciferase assay system (Promega) in a Berthold Lumat LB 9507 luminometer.

1. Remove the cell growth medium and wash the cells twice with 1× PBS.
2. Lyse the cells with 200 μl of 1× reporter lysis buffer.
3. Scrape the cells, and transfer the cell lysates to a microcentrifuge tube; vortex for 10 sec.
4. Freeze the cells on dry ice for 20 min.

5. Thaw the cell extract and luciferase assay reagent in a water bath at room temperature for 30 min.

6. Centrifuge the cell extract at room temperature for 10–15 sec at 12,000 *g*, and transfer the supernatant to a new tube.

7. Mix 20 μl of cell extract and 100 μl of luciferase assay reagent.

8. Measure the luciferase activity for 10 sec, using a luminometer.

COMMENTS

Note 1: Cells transfected with a luciferase vector not regulated by NF-κB or a vector with a mutated NF-κB site should be used as a negative control.

Note 2: β-Galactosidase activity may be measured similarly with a Galactolight kit (Tropix, Bedford, MA); the measurement time is 5 sec.

Note 3: RLE cells exposed to TNF-α for 18 hr can be used as positive control.

We have reviewed the protocols to study Rac and NF-κB activation, as well as ROS production currently used in our laboratory. These methods will allow us to better understand the mechanisms by which Ras exerts its effects. They may also have broad use in any biological systems that involve the activation of Rac, NF-κB, and ROS.

[10] Ras, Metastasis, and Matrix Metalloproteinase 9

By ERIC J. BERNHARD and RUTH J. MUSCHEL

Introduction to Metastasis and Role of Metalloproteinases

ras oncogene activation has been shown to promote metastasis in many experimental tumor cell models (reviewed in Himelstein *et al.*[1]), and has been implicated in the metastatic progression of human cancers. The pathways that are responsible for induction of metastasis by *ras* are being elucidated in genetic studies that examine effectors of metastasis downstream of *ras*. Webb *et al.*[2] have shown using *ras* effector loop mutants[3] that tumorigenicity was retained by all *ras* mutants tested, but that only cells transformed by a mutant that activates the

[1] B. P. Himelstein, R. Canete-Soler, E. J. Bernhard, D. W. Dilks, and R. J. Muschel, *Invasion Metastasis* **14,** 246 (1994).

[2] C. P. Webb, L. Van Aelst, M. H. Wigler, and G. F. Woude, *Proc. Natl. Acad. U.S.A.* **95,** 8773 (1998).

[3] M. White, C. Nicolette, A. Minden, A. Polverino, L. van Aeist, M. Karin, and M. Wigler, *Cell* **80,** 533 (1995).

0076-6879/00 $35.00

Raf–MAP (mitogen-activated protein) kinase pathway (*ras*^{V12S35}) were capable of forming metastases. Kraemer *et al.* have demonstrated that c-*jun*-transformed cells are both tumorigenic and as metastatic in experimental metastasis assays as are *ras*-transformed cells.[4] These results imply that the capacity for metastasis induced by *ras* activation is mediated through AP-1 transcriptional activation. Furthermore, mice deficient in c-*fos* form only papillomas after skin exposure to carcinogens, not the invasive carcinomas seen in wild-type mice,[5] although other data implicate AP-1 even in papilloma formation.[6] The induction of metastatic behavior in cells expressing activated *ras* has been correlated with increased type IV matrix metalloproteinase (MMP) activity, and AP-1 is known to promote the expression of type IV MMPs.[7–9] Thus the picture that emerges is one that traces a direct link between *ras* expression and metastasis through the induction of MMP expression.

The process of metastasis involves breakdown and penetration of intercellular matrix and basement membranes (reviewed in Refs. 10 and 11). The basement membrane in particular is a barrier to tumor invasion, intravasation, and extravasation. Type IV collagen is the predominant collagen component of the basement membrane, and type IV matrix metalloproteinase activity may contribute significantly to the metastatic process by mediating its cleavage. In addition, MMPs have been implicated in the regulation of tumor cell proliferation both at the primary site and at the site of metastasis.[12] Other studies suggest that MMPs may also regulate endothelial cell growth within tumors,[13] thus influencing tumor angiogenesis.

It has been shown that the release of type IV collagenolytic activity by tumor cells correlates with the metastatic phenotype in *ras*-transformed rodent fibroblasts.[14] One of the enzymes implicated in type IV collagen breakdown is a 92-kDa

[4] M. Kraemer, R. Tournaire, V. Dejong, N. Montreau, D. Briane, C. Derbin, and B. Binetruy, *Cell Growth Differ.* **10,** 193 (1999).

[5] E. Saez, S. E. Rutberg, E. Mueller, H. Oppenheim, J. Smoluk, S. H. Yuspa, and B. M. Spiegelman, *Cell* **82,** 721 (1995).

[6] M. R. Young, J. J. Li, M. Rincon, R. A. Flavell, B. K. Sathyanarayana, R. Hunziker, and N. Colburn, *Proc. Natl. Acad. Sci. U.S.A.* **96,** 9827 (1999).

[7] B. P. Himelstein, E. J. Lee, H. Sato, M. Seiki, and R. J. Muschel, *Oncogene* **14,** 1995 (1997).

[8] M. Kirstein, L. Sanz, S. Quinones, J. Moscat, M. T. Diaz-Meco, and J. Saus, *J. Biol. Chem.* **271,** 18231 (1996).

[9] L. M. Matrisian, S. McDonnell, D. B. Miller, M. Navre, E. A. Seftor, and M. J. Hendrix, *Am. J. Med. Sci.* **302,** 157 (1991).

[10] L. Liotta, *Cancer Metastasis Rev.* **9,** 285 (1990).

[11] L. M. Coussens and Z. Werb, *Chem. Biol.* **3,** 895 (1996).

[12] A. F. Chambers and L. M. Matrisian, *J. Natl. Cancer Inst.* **89,** 1260 (1997).

[13] L. Lozonschi, M. Sunamura, M. Kobari, S. Egawa, L. Ding, and S. Matsuno, *Cancer Res.* **59,** 1252 (1999).

[14] S. Garbisa, A. Negro, T. Kalebic, R. Pozzatti, R. Muschel, U. Saffiotti, and L. A. Liotta, *Adv. Exp. Med. Biol.* **233,** 179 (1988).

type IV collagenase (gelatinase, MMP-9) expressed in a number of *ras*-transformed metastatic rodent tumor cells[15–17] and human tumors.[18–22] The 92-kDa gelatinase, interstitial (type I) collagenase (MMP-1), 72-kDa type IV collagenase (MMP-2), and stromelysin (transin) (MMP-3) belong to the metalloproteinase family of enzymes (reviewed in MacDougall and Matrisian[23]). The metalloproteinase family shares the functional characteristics of optimal activity at neutral pH and a requirement for divalent cations. In addition, they are inhibited by tissue inhibitors of metalloproteinases (TIMPs). Structurally, members of this family share extensive homology in several domains, including a metal-binding domain that mediates zinc binding and a conserved nine-amino acid sequence immediately adjacent to the cleavage site, which is lost on processing of the proenzyme to the mature form.[24] A subset of this family, the gelatinases, consists of MMP-2 and MMP-9. These are called gelatinases because of their efficient degradation of gelatin.

The role of the 92-kDa gelatinase in metastasis was previously examined using a panel of transformed rat embryo fibroblasts (REF) cells. REF transformants obtained by cotransfection with v-*myc* plus Ha-*ras,* or by transfection with Ha-*ras* alone, frequently form many metastases when injected into nude mice. The high metastatic potential of these cells is associated with type IV collagenolytic activity[14] and the release of a 92-kDa gelatinase.[16] A reduction in the metastatic potential of Ha-*ras*-transformed REF cells has been demonstrated on transfection with the E1A gene[25] or with a ribozyme targeting 92-kDa collagenase mRNA,[26] which results in a loss of, or marked reduction of, 92-kDa type IV collagenolytic activity. Similar findings were obtained in Ha-*ras*-transformed human

[15] M. Ballin, D. E. Gomez, C. C. Sinha, and U. P. Thorgeirsson, *Biochem. Biophys. Res. Commun.* **154,** 832 (1988).

[16] E. J. Bernhard, R. J. Muschel, and E. N. Hughes, *Cancer Res.* **50,** 3872 (1990).

[17] E. J. Bernhard, S. B. Gruber, and R. J. Muschel, *Proc. Natl. Acad. Sci. U.S.A.* **91,** 4293 (1994).

[18] C. Arkona and B. Wiederanders, *Biol. Chem.* **377,** 695 (1996).

[19] P. A. Forsyth, T. D. Laing, A. W. Gibson, N. B. Rewcastle, P. Brasher, G. Sutherland, R. N. Johnston, and D. R. Edwards, *J. Neurooncol.* **36,** 21 (1998).

[20] Y. Inoue, K. Abe, K. Obata, T. Yoshioka, G. Ohmura, K. Doh, K. Yamamoto, H. Hoshiai, and K. Noda, *J. Obstet. Gynaecol. Res.* **23,** 139 (1997).

[21] H. Iwata, S. Kobayashi, H. Iwase, A. Masaoka, N. Fujimoto, and Y. Okada, *Jpn. J. Cancer Res.* **87,** 602–11 (1996).

[22] A. E. Kossakowska, S.A. Huchcroft, S. J. Urbanski, and D. R. Edwards, *Br. J. Cancer* **73,** 1401 (1996).

[23] J. R. MacDougall and L. M. Matrisian, *Cancer Metastasis Rev.* **14,** 351 (1995).

[24] G. Murphy, H. Stanton, S. Cowell, G. Butler, V. Knauper, S. Atkinson, and J. Gavrilovic, *APMIS* **107,** 38 (1999).

[25] E. J. Bernhard, B. Hagner, C. Wong, I. Lubenski, and R. J. Muschel, *Int. J. Cancer* **60,** 718 (1995).

[26] G. Sehgal, J. Hua, E. J. Bernhard, I. Sehgal, T. C. Thompson, and R. J. Muschel, *Am. J. Pathol.* **152,** 591 (1998).

bronchial epithelial cells after E1A transfection.[27] Conversely, transfection of non-metastatic REF cells transformed with H-*ras* plus adenovirus E1A with a vector that results in constitutively expressed 92-kDa collagenase causes these cells to acquire metastatic capacity.[17]

To study the expression of collagenolytic enzymes in tumor cells, many investigators use gelatin zymography to examine gelatinase expression in cultured or explanted tumor cells. This assay yields a semiquantitative profile of gelatinolytic activity in a given cell line or tumor explant and identifies the molecular weight of the enzymes responsible for this activity. It is based on digestion by the enzyme of interest of substrate, in this case gelatin, embedded within a polyacrylamide gel. Enzyme activity is indicated by a band of clearing after developing and staining of the gel, indicating enzyme activity. As was noted above, both MMP-2 and -9 may exist in a latent form before processing of the N-terminal end occurs to release the inhibitory propeptide. Processing of the gelatinase gel also leads to gelatinase activation through poorly understood mechanisms. Hence the pro forms of both enzymes will appear as bands, MMP-9 at 92 kDa for human and rat and 105 kDa for mouse (the murine gene has an internal duplication leading to the increased size without any known functional alterations). During activation MMP-9 can be cleaved to 87-, 82-, and 67-kDa forms.[24, 28] The 67-kDa form can be confused with MMP-2 and Western blotting or immunoprecipitation may be required to identify the components of this band. The 82- and 67-kDa forms are the active forms. Thus a gelatinase gel that shows 92-kDa bands indicates that the MMP-9 protein was present, but not active, whereas bands at 82 kDa indicate that active enzyme was present in the sample. MMP-2 is 72 kDa in all tested species and its activated form is at 65/59 kDa and to a lesser extent at 45 kDa.

Gelatinase Zymography

Cell Culture and Conditioned Serum-Free Medium Collection

1. Grow cells to approximately 80% confluence.
2. Remove the serum-containing medium and wash the monolayer with three changes of serum-free medium.
3. Overlay the cells with 0.1 ml of serum-free medium per centimeter of culture dish or flask area.
4. Harvest the medium at the desired time (24 hr) after initiation of serum-free culture.
5. Spin the medium at 14,000 rpm in a microcentrifuge for 10 min at room temperature.

[27] G. I. Goldberg, S. M. Frisch, C. He, S. M. Wilhelm, R. Reich, and I. E. Collier, *Ann. N.Y. Acad. Sci.* **580,** 375 (1990).

[28] Y. Ogata, Y. Itoh, and H. Nagase, *J. Biol. Chem.* **270,** 18506 (1995).

6. Harvest the supernatant and freeze at −20 or −70° until ready for analysis. For samples to be analyzed multiple times, aliquot the samples to avoid repeated freezing and thawing. Samples may also be concentrated with Millipore (Danvers, MA) filters (e.g., Centricon 30) if the activity in the sample is low.

Gelatinase Gel Casting and Electrophoresis of Samples

Substrate polyacrylamide gel electrophoresis is carried out by a modification of the protocol of Heussen and Dowdle[29] with gelatin as the substrate embedded within a 7% (w/v) polyacrylamide (29 : 1, acrylamide:bisacrylamide) separating gel containing 0.1% (w/v) gelatin. This gel acts as both the protein separation medium and the substrate for the enzymes that are renatured after electrophoresis and become active. Enzyme activity is detected after staining as cleared bands on a blue background composed of the remaining undigested gelatin substrate (note that casein can be substituted for gelatin in order to detect stromelysin activity). In the case of gelatinolysis, the extent of digestion can be monitored by densitometry of processed gels and a linear range of gelatinase activity of approximately 50-fold can be determined by using serial dilutions of a standard enzyme preparation. The stacking gel, which is ovelayed on the fully polymerized separating gel, is a 4% (w/v) gel without gelatin. All reagents are made up in deionized water. Always use appropriate caution when handling acrylamide.

To cast separating gel mix together	To cast stacking gel mix together
1. Combine the following: 5.83 ml of 30% (29 : 1) acrylamide: bisacrylamide 9.4 ml of 1 *M* Tris-HCl (pH 8.8) 0.25 ml of 10% (w/v) SDS 2.5 ml of 1.0% (w/v) gelatin solution [made by warming gelatin granules (from porcine skin; Sigma, St. Louis, MO) in water until dissolved] 0.312 ml of 10% (w/v) ammonium persulfate (freshly made up) 8.41 ml of water 2. Filter the mixture through a 0.4- or 0.22-μm pore size filter 3. Add 6.25 μl of TEMED	1. Combine the following: 1.67 ml of 30% (29 : 1) acrylamide: bisacrylamide 3.15 ml of 0.5 *M* Tris-HCl (pH 6.8) 0.125 ml of 10% (w/v) SDS 0.5 ml of 10% (w/v) ammonium persulfate (freshly made up) 7.05 ml of water 2. Add 5 μl of *N,N,N',N'*-tetramethylethylenediamine (TEMED)

This will make up 25 ml of separating gel and 12.5 ml of stacking gel solution, which is enough for two 1.0-mm minigels or a single 0.75 mm × 10 cm × 15 cm

[29] C. Heussen and E. B. Dowdle, *Anal. Biochem.* **102,** 196 (1980).

gel. A thorough mixing of the gelatin with the acrylamide solution is essential. Insufficient mixing can result in gel failure due to poor staining. To avoid the formation of bubbles in the gel, the acrylamide can be stored for a limited time at room temperature. If stored at 4°, the gel solution should be degassed prior to polymerization. Gels are cast according to the instructions for the gel apparatus used. After polymerization of the stacking gel, carefully rinse out the wells with running buffer to remove unpolymerized acrylamide prior to loading samples.

1. The samples to be analyzed for gelatinolytic activity are thawed, mixed, and centrifuged to remove precipitated proteins.
2. Aliquots of the medium to be analyzed (20–40 μl) are mixed with a one-fourth volume of 4× Laemmli sample loading buffer (without reducing agents) and warmed to 37°. Recombinant MMP-2 or -9 can be purchased to use as positive controls (Oncogene Research Products, Cambridge, MA).
3. Equal volumes of sample are then loaded immediately onto the gel, which is run at 25-mA constant current until the bromphenol blue dye front reaches the bottom of the gel.
4. The gel is disassembled and rinsed three times (10 min each rinse) in 50 m*M* Tris, pH 7.4, containing 2% (v/v) Triton X-100 with gentle rocking.
5. The gel is then rinsed three times (5 min each rinse) in 50 m*M* Tris, pH 7.4, with gentle rocking.
6. The gel is incubated overnight in developing buffer at 37° without agitation Developing buffer consists of 50 m*M* Tris, pH 7.4, containing 0.2 *M* NaCl, 1% (v/v) Triton X-100, 0.02% (w/v) sodium azide, and 5 m*M* $CaCl_2$. Developing times can be altered to modify the sensitivity of the assay.
7. The gel is stained with 0.2% (w/v) Coomassie blue in 50% (v/v) methanol, 10% (v/v) acetic acid for 1 hr and destained in 20% (v/v) methanol, 10% (v/v) acetic acid until bands are easily seen.
8. Best results for archiving of gels are obtained if the gel is dried between sheets of clear cellulose for transillumination.

To determine the relationship between gelatinolytic activity and metastatic capacity, the metastatic efficiency of tumor cells must be determined. Two assays exist for this purpose: The spontaneous metastasis assay, in which metastasis from a subcutaneous or orthotopically implanted tumor is determined, and the experimental metastasis assay, in which the formation of metastatic colonies from an intravenous injection of tumor cells is determined. The former assay is frequently not possible because of growth of the primary metastasis to dimensions that attain the limits of National Institutes of Health (NIH) guidelines for animal use prior to the detection of metastases. Therefore only the experimental metastasis assay is discussed here.

Experimental Metastasis Assay

These experiments are carried out with 4- to 6-week-old NCr-*nu/nu* or BALB-*nu/nu* mice. The use of young nude mice permits evaluation of metastasis in the absence of immune response to the tumor cells.

1. Culture cells to log phase and split to low density 1–2 days prior to harvest for injection. At the time of harvest for injection, the majority of cells should not be in close contact with neighboring cells.
2. Harvest the cells with minimal trypsinization or with nonenzymatic cell dissociation medium. Place the cells in a 50-ml centrifuge tube with cold medium and centrifuge for 5 min at 500 *g* at 4°. Keep the cells on ice for all subsequent steps.
3. Count the cells, using a hemocytometer and trypan blue, to determine culture viability. If viability is less than 90% or if the cells are not in a single-cell suspension, discontinue the experiment and begin again at the first step (step 1 above).
4. Centrifuge at 500 *g* at 4° for 5 min and resuspend the cells in serum-free medium at a concentration that gives the desired cell number in 0.1–0.2 ml. Larger volumes can compromise murine circulation.
5. Immediately prior to injection, prepare a 1.0-ml tuberculin syringe with cell suspension by gently drawing cells into the syringe fitted with a 28- to 30-gauge needle.
6. Warm mouse tails with heated water (48° or below) until tail veins are distended and easily visible. Do not use higher temperatures as this can cause sloughing of the mouse tail skin.
7. Inject cells into the lateral tail vein slowly. If the needle is inside the vein, the injection can be accomplished with minimal pressure on the plunger, and the tail vein shows signs of clearing as the blood is replaced by the injection medium. If the needle is outside the vein, resistance will be felt and a subcutaneous bleb will appear. When an injection fails, another attempt may be tried on the contralateral vein or at a point proximal to the failed injection site.
8. Withdraw the needle and apply pressure to the injection site for a few seconds.
9. Monitor animals for immediate signs of distress indicating embolism, which can result from tumor cell aggregates, and thereafter three times a week for signs of cachexia or labored breathing. Cells that metastasize outside the lungs may show other signs or symptoms associated with metastasis to a particular site.
10. Kill the animals at the first sign of distress by anesthetic administration or CO_2 asphyxiation.
11. Open the thorax with a midline incision extending from the chin to the abdomen.
12. Dissect the neck, exposing the trachea.

13. Grasp the trachea with forceps and remove the lungs *en bloc* with the heart by dissecting through the trachea and along the vertebral column while gently lifting the heart/lung block. Cut the esophagus to release the block and lay on paper towels moistened with saline.

14. Insert an 18-gauge needle with a blunted tip (emery paper is best to remove the tip) into the tracheal orifice and clamp into place with light pressure from forceps, hemostats, or a suture tie.

15. Inject a 10% neutral buffered formalin solution into the lungs until they become fully inflated.

16. Remove the anterior rib cage with a pair of blunt scissors, taking care not to rupture the lungs.

17. Place the lungs in a 50-ml tube containing 10% neutral buffered formalin for further fixation.

18. Necropsy the mouse to document the presence of metastases occurring at other sites.

19. Count metastatic colonies in a double-blind fashion by examining the exterior surface of each lung lobe individually. Count metastases associated with the pericardium, mediastinum, sternum, and pleural cavity separately. A dissecting scope is helpful in scoring smaller metastatic nodules.

Visualizing the metastatic process at early times after tumor cell injection can yield clues to the metastatic process in different cell lines, and limitations to their metastatic efficiency. For this assay, a larger number of cells are injected to facilitate their identification, and the lungs of animals injected with tumor cells are examined prior to the establishment of visually detectable metastatic colonies.

In Situ Microscopy of Metastatic Cells in Lungs

To detect metastatic cells at early time points after injection and monitor their fate and location, tumor cells can be labeled by transfection with a vector expressing green fluorescent protein (GFP) such as pEGFP-C2 (Clontech, Palo Alto, CA)

1. Transfect cells with LipofecTAMINE (GIBCO-BRL, Gaithersburg, MD) followed by 2 weeks of selection (for pEGFP-C2 use Geneticin).

2. Sort GFP-positive cells in a flow cytometer, harvesting the 5% brightest cells and subclone this population. Retest the subclones for GFP expression by flow cytometry. This protocol should yield cells with constitutive high-level expression of GFP.

3. Inject 3×10^5 cells intravenously, as described above, in a volume of 100 μl for evaluation at 5–7 days postinjection. For examination at shorter intervals

(1 to 48 hr after injection) inject 2×10^6 cells intravenously in a volume of 200 μl. Particular care must be taken to avoid aggregates of cells and embolism in these concentrated cell preparations. Injections should also be done slowly.

4. Harvest lungs from mice at times ranging from 6 to 72 hr. Freeze the lungs in mounting medium such as Histo Prep (Fisher Scientific, Fair Lawn, NJ).

5. Cut 10-μm sections, using a cryostat, and fix in 2% (w/v) paraformaldehyde.

6. Stain section with 4′,6-diamidino-2-phenylindole (DAPI, 2.5 μg/ml) to identify cell nuclei.

7. Examine sections under epifluorescence at 535 nm (GFP) and 460 nm (DAPI) emission. Green fluorescent cells with a DAPI (blue)-stained nucleus represent tumor cells.

Examination of Tumor Cells in Lungs by Immunohistochemistry

1. After sectioning and fixation as described in step 4 of the preceding section, fix the cells again in an ethanol–acetic acid solution (2 : 1, v/v) for 5 min at −20°. Block the sections to prevent nonspecific binding by incubation for at least 1 hr at 4° in 5% (v/v) goat serum. Rinse the sections with Tris–NaCl buffer [1 *M* Tris, 140 *M* NaCl, 0.1% (v/v) Tween 20, pH 7.6 (rinsing buffer)].

2. Bind primary anti-GFP antibody overnight at 4° (we use a polyclonal rabbit anti-GFP antibody, diluted 1 : 1500; obtained from the University of Alberta, Calgary, Canada). Dip the slides 10 times in rinsing buffer.

3. Incubate sections with alkaline phosphatase–anti-rabbit IgG complex (PharMingen, San Diego, CA) at a 1 : 100 dilution for 1 hr at room temperature. Include levamisole (10 mg/ml; Sigma) to block any endogenous alkaline phosphatase activity. Dip the slides 10 times in rinsing buffer.

4. Stain anti-GFP-labeled cells with stable fast red/naphthol phosphate (Research Genetics, Huntsville, AL) chromogen to visualize the reaction.

5. Counterstain with methyl green for 10 min at room temperature.

6. Wash the slides in two changes of distilled water, dipping them each 10 times, followed by a 30-sec incubation in fresh distilled water.

7. Wash the slides in two changes of butanol, dipping them 10 times each, and then incubate them in butanol for 30 sec.

8. Carry out three incubations in xylene (2 min each). Mount the slides with a coverslip, using Permount.

[11] Ras Regulation of Urokinase-Type Plasminogen Activator

By ERNST LENGYEL, SABINE RIED, MARKUS M. HEISS, CLAUDIA JÄGER, MANFRED SCHMITT, and HEIKE ALLGAYER

Introduction and Background

Degradation of the extracellular matrix (ECM) is essential in many physiological and pathological processes, including involution of the prostate, cytotrophoblast implantation, wound healing, and tumor cell invasion, all of which require extensive extracellular matrix proteolysis.[1–3] Many different types of ECM-degrading enzymes have been implicated in these processes, including matrix metalloproteinases,[4] cysteine proteases (cathepsins B, L, and H),[5] caspases,[6] aspartyl proteases (cathepsin D),[7] and serine proteases.[8] The urokinase-type plasminogen activator (urokinase) is a serine protease[9] that converts the abundant zymogen plasminogen into the widely acting serine protease plasmin. Plasmin has a broad substrate specificity toward components of the extracellular matrix including laminin, vitronectin, and fibronectin.[10–12] Membrane attachment of urokinase to the urokinase receptor increases the rate of plasmin formation at the plasma membrane[13] and focuses proteolytic activity at the leading edge of the tumor.[14] Together, these proteolytic functions facilitate the migration of tumor cells through the extracellular matrix and basement membrane barriers. High urokinase expression

[1] M. S. Stack, S. M. Ellerbroek, and D. A. Fishman, *Int. J. Oncol* **12,** 569 (1999).
[2] G. A. Dekker and B. M. Sibai, *Am. J. Obstet. Gynecol.* **179,** 1359 (1998).
[3] C. Marschall, E. Lengyel, T. Nobutoh, E. Braungart, K. Douwes, A. Simon, V. Magdolen, U. Reuning, and K. Degitz, *J. Invest. Dermatol.* **113,** 69 (1999).
[4] M. J. Duffy and K. McCarthy, *Int. J. Oncol.* **12,** 1343 (1998).
[5] C. Thomssen, M. Schmitt, L. Goretzki, P. Oppelt, L. Pache, P. Dettmar, F. Jänicke, and H. Graeff, *Clin. Cancer Res.* **1,** 741 (1995).
[6] S. Kumar and P. A. Colussi, *Trends Biochem. Sci.* **24,** 1 (1999).
[7] H. Rochefort and E. Liaudet-Coopman, *APMIS* **107,** 86 (1999).
[8] U. Reuning, V. Magdolen, O. Wilhelm, K. Fischer, V. Lutz, H Graeff, and M. Schmitt, *Int. J. Oncol.* **13,** 893 (1998).
[9] T. C. Wun and E. Reich, *J. Biol. Chem.* **257,** 3276 (1982).
[10] K. Dano, P. A. Andreasen, J. Grondahl-Hansen, P. Kristensen, L. S. Nielsen, and L. Skriver, *Adv. Cancer Res.* **44,** 139 (1985).
[11] O. Wilhelm, R. Hafter, E. Coppenrath, M. Pflanz, M. Schmitt, and H. Graeff, *Cancer Res.* **48,** 3507 (1988).
[12] L. Liotta, R. Goldfarb, R. Brundage, G. Siegel, V. Terranova, and S. Garbisa, *Cancer Res.* **41,** 4629 (1981).
[13] V. Ellis, N. Behrendt, and K. Dano, *J. Biol. Chem.* **266,** 1257 (1991).
[14] F. Blasi, *Thromb. Haemost.* **82,** 298 (1999).

0076-6879/00 $35.00

is correlated with a poor prognosis for patients suffering from a variety of different types of cancer including breast, ovary, lung, and gastrointestinal cancer.[15–17]

Studies of the regulation of urokinase expression have shown that the urokinase gene is regulated at the transcriptional level.[18–21] The urokinase promoter contains functional binding sites for the transcription factors AP-1 (activator protein 1), PEA3 (polyoma virus enhancer activator 3), and NF-κB (nuclear factor κB),[21–23] which are important for both constitutive and regulated urokinase expression.[24] An additional important enhancer sequence of the urokinase promoter is a combined PEA3/AP-1 motif,[20,25] homologous to the polyomavirus enhancer A element,[26] that has been shown to be a target for transcriptional activation by oncogenes in several promoters.[27]

In this context there is good evidence[19,28–32] that members of the Ras family can induce urokinase gene expression, urokinase-dependent proteolysis, invasion, and metastasis in diverse tumor models. Ha-Ras induces urokinase promoter activity in a dose-dependent manner, as shown in a low urokinase-producing ovarian cancer cell line, OVCAR-3 (Fig. 1). A study by Jankun and colleagues[28] has shown an increased amount of urokinase bound to the receptors on fibroblasts transformed with the K-*ras* oncogene. This can be explained either as an induction

[15] W. Kuhn, L. Pache, B. Schmalfeldt, P. Dettmar, M. Schmitt, F. Jänicke, and H. Graeff, *Gynecol. Oncol.* **55,** 401 (1994).

[16] M. M. Heiss, R. Babic, H. Allgayer, K. U. Grützner, K. W. Jauch, U. Löhrs, and F. W. Schildberg, *J. Clin. Oncol.* **13,** 2084 (1995).

[17] P. F. M. Choong, M. Ferno, M. Akerman, H. Willen, E. Langstrom, P. Gustafson, T. Alvegard, and A. Rydholm, *Int. J. Cancer* **69,** 268 (1996).

[18] P. Verde, S. Boast, A. Franze, F. Robbiati, and F. Blasi, *Nucleic Acids Res.* **16,** 10699 (1988).

[19] E. Lengyel, E. Stepp, R. Gum, and D. Boyd, *J. Biol. Chem.* **270,** 23007 (1995).

[20] C. Nerlov, P. Rorth, F. Blasi, and M. Johnsen, *Oncogene* **6,** 1583 (1991).

[21] E. Lengyel, R. Gum, E. Stepp, J. Juarez, H. Wang, and D. Boyd, *J. Cell. Biochem.* **61,** 430 (1996).

[22] C. Nerlov, D. De Cesare, F. Pergola, A. Carracciolo, F. Blasi, M. Johnsen, and P. Verde, *EMBO J.* **11,** 4573 (1992).

[23] U. Reuning, O. Wihelm, T. Nishiguchi, L. Guerrini, F. Blasi, H. Graeff, and M. Schmitt, *Nucleic Acids Res.* **23,** 3887 (1995).

[24] P. Rorth, C. Nerlov, F. Blasi, and M. Johnsen, *Nucleic Acids Res.* **18,** 5009 (1990).

[25] S. Ried, C. Jäger, M. Jeffers, G. F. Vande Woude, H. Graeff, M. Schmitt, and E. Lengyel, *J. Biol. Chem.* **274,** 16377 (1999).

[26] M. E. Martin, J. Piette, M. Yaniv, W. J. Tang, and W. R. Folk, *Proc. Natl. Acad. Sci. U.S.A.* **85,** 5839 (1998).

[27] C. Wasylyk, P. Flores, A. Gutman, and B. Wasylyk, *EMBO J.* **8,** 3371 (1989).

[28] J. Jankun, V. Maher, and J. McCormick, *Cancer Res.* **51,** 1221 (1991).

[29] J. E. Testa, R. L. Medcalf, J. F. Cajot, W. D. Schleuning, and B. Sorbat, *Int. J. Cancer* **43,** 816 (1989).

[30] J. H. Axelrod, R. Reich, and R. Miskin, *Mol. Cell. Biol.* **9,** 2133 (1989).

[31] G. Brunner, J. Pohl, L. J. Erkell, A. Radler -Pohl, and V. Schirrmacher, *J. Cancer Res. Clin. Oncol.* **115,** 139 (1989).

[32] S. M. Bell, D. C. Connoly, N. J. Maihle, and J. L. Degen, *Mol. Cell. Biol.* **13,** 5888 (1993).

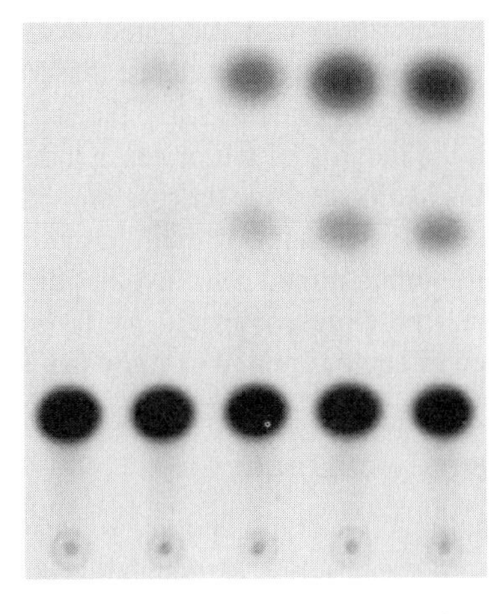

u-PAR CAT	-	+	+	+	+
pSV_0 CAT	+	-	-	-	-
c-Ha-ras (μg DNA)	+	-	0.3	1	5
pSV_2 neo	-	-	-	-	-

FIG. 1. Transient expression of Ha-Ras increases urokinase promoter activity in a dose-dependent manner. OVCAR-3 cells were transiently transfected with 10 μg of a CAT reporter driven by the wild-type urokinase promoter in the absence or presence of varying amounts of a vector bearing the Ha-Ras sequence. All transfections were performed in the presence of 5 μg of a β-galactosidase-expressing vector. After 5 hr, the medium was changed, and cells were cultured for an additional 36 hr. The cells were harvested and assayed for β-galactosidase activity. Cell extracts, corrected for differences in transfection efficiency, were incubated with [^{14}C]chloramphenicol for 8 hr. The mixture was extracted with ethyl acetate and subjected to thin-layer chromatography. The extent of conversion of [^{14}C]chloramphenicol to acetylated derivatives was determined with a PhosphorImager.

of urokinase-secretion by the K-*ras*-transformed fibroblasts or as elevated urokinase receptor expression, which therefore captures more of the protease. In either case, combined evidence suggests that *ras* transformation can lead to upregulation of both urokinase and its binding site, the consequence of this being an efficient induction of plasminogen-dependent proteolysis. In addition, our data have shown that the expression of the urokinase receptor, in at least a subpopulation of colon cancer patients, is regulated by a mutation-activated K-Ras.[33] In a study by Brunner *et al.*[31] it was demonstrated that Ha-Ras-transfected cells exhibited a significant increase in urokinase secretion as well as an increase in invasive [Matrigel (Becton Dickinson, Franklin Lakes, NJ) assay] and metastatic capacity (tail vein injection of mouse bladder carcinoma cells in C57BL/6 mice). This corroborates the notion that *ras* transfection not only induces urokinase gene

[33] H. Allgayer, H. Wang, S. Shirasawa, T. Sasazuki, and D. Boyd, *Br. J. Cancer* **80,** 1884 (1999).

expression, but also urokinase-mediated proteolysis, which contributes to the invasion phenomenon. There is evidence that Ras activates other protease systems such as matrix metalloproteinases[34,35] (see [10] in this volume[35a]), or cathepsins. Furthermore, the identity of the protease induced by Ras potentially depends on the signal transduction cascade that preferentially mediates the Ras induction.[36,37] Taken together, the data suggest that the induction of the invasive and metastatic potential of Ras-transformed cells involves different patterns of proteases, including urokinase. In the present article, we try to give an overview on the methods that have been or can be applied to study the regulation of urokinase by Ras.

Plasminogen Zymography

Zymography is a useful technique to quantify the activity of a plasminogen activator toward its substrate. Although it is not specific for urokinase, lysis zones at about 33 and 55 kDa represent the two-chain low molecular weight form of urokinase and the single-chain prourokinase, respectively. For interpretation of the results it is helpful to run recombinant single-chain and two-chain urokinase (American Diagnostica, Greenwich, CT) alongside the samples as a control, or to add antibodies against urokinase to be able to attribute the activity detected to urokinase. This assay, however, also detects tissue-type plasminogen activator (t-PA) with a molecular mass of 70 kDa and sometimes an additional band at 92 kDa that probably represents a urokinase/PAI-1 (plasminogen activator inhibitor 1) complex.

Because serum contains protease inhibitors that could distort the results of the assay, the medium on the cells must be changed to a medium lacking serum 24 hr prior to harvesting. At the time of harvest, the medium is removed and centrifuged at 14,000 rpm for 10 min (4°) to remove cell debris and may be stored at $-80°$ or used immediately. For the gel, conditioned medium normalized for equal cell numbers is denatured at 95°, for 5 min in a nonreducing 3× Laemmli buffer containing 62.5% Tris-HCl (pH 6.8), 6% (w/v) sodium dodecyl sulfate (SDS), 10% (v/v) glycerol, and 0.1% (w/v) bromphenol blue. The samples are electrophoresed in a 7.5% (w/v) SDS–polyacrylamide gel containing 0.1% (w/v) casein and plasminogen (5 μg/ml) (Roche, Nutley, NJ). The gel is incubated at

[34] R. Gum, E. Lengyel, J. Juarez, J. H. Chen, H. Sato, M. Seiki, and D. Boyd, *J. Biol. Chem.* **271,** 10672 (1996).

[35] J. A. Aguirre -Ghiso, P. Frankel, E. F. Farias, Z. Lu, H. Jiang, A. Olsen, E. B. de Kier Joffe, and D. A. Foster, *Oncogene* **18,** 4718 (1999).

[35a] E. J. Bernhard and R. J. Muschel, *Methods Enzymol.* **333,** Chap. 10, 2001 (this volume).

[36] M. Janulis, S. Silberman, A. Ambegaokar, S. Gutkind, and R. Schultz, *J. Biol. Chem.* **274,** 801 (1999).

[37] S. Silberman, M. Janulis, and R. Schultz, *J. Biol. Chem.* **272,** 5927 (1997).

room temperature for 2 hr in the presence of 2.5% (v/v) Triton X-100, 50 mM Tris-HCl (pH 7.5), and 0.05% (w/v) NaN_3 optional and washed three times for 5–10 min with sterile water to remove traces of Triton X-100. Subsequently, the gel is incubated for 16 hr at 37° in a buffer containing 10 mM $CaCl_2$, 0.15 M NaCl, and 100 mM Tris-HCl (pH 7.5) to renature proteins. The gel is stained for protein in a buffer containing 0.25% (w/v) Coomassie blue, 45.5% (v/v) methanol, and 1.1% (v/v) glacial acidic acid for 1 hr and destained for 1–2 hr in a buffer containing 10% (v/v) glacial acidic acid and 25% (v/v) methanol. Plasminogen-dependent proteolysis is detected as a white zone in a dark field.[30,38,39] To ensure that the activity is due to a plasminogen activator, the same samples are run in a gel without plasminogen, under the same conditions. If no activity is detected in the conditioned medium analyzed, it is worthwhile to concentrate the medium with MW10 spin columns (Centricon YM10; Amicon, Beverly, MA) and to use a high urokinase-producing cell line (e.g., UM-SCC-1)[21] as a positive control.

Laminin Degradation Mediated by Receptor-Bound Urokinase

The plasminogen assay described above is an experiment that gives good qualitative and quantitative results regarding the activity of total urokinase collected from conditioned media of tumor cells. However, in certain tumor types, plasminogen-dependent proteolysis is largely due to urokinase bound to the specific cell receptor (urokinase receptor).[40] To measure the enzymatic activity of receptor-bound urokinase an assay can be employed that uses laminin as a specific substrate for receptor-bound urokinase.[33,41,42]

Laminin is iodinated by a standard chloramine-T method.[41–43] Laminin (100 μg; Collaborative Research, Bedford, MA) is iodinated in 200 μl of phosphate-buffered saline (PBS), using 0.5 mCi of ^{125}I (ICN, Costa Mesa, CA) and 50 μl of chloramine-T (5 mg/ml) for 10 min at room temperature. The reaction is terminated (5 min) with 50 μl of sodium metabisulfite (12 mg/ml) solution. The next step is to separate radioactive laminin from unincorporated ^{125}I by extensive dialysis against PBS. Toward this end, the volume is expanded to 4 ml in PBS and dialyzed (dialysis tubing with pore size $<$10 kDa) against two rounds of PBS for 24 hr at 4°. An aliquot is counted for radioactivity. The dialysate containing the ^{125}I-labeled

[38] E. Lengyel, B. Singh, R. Gum, C. Nerlov, A. Sabichi, M. Birrer, and D. Boyd, *Oncogene* **11,** 2639 (1995).

[39] M. Niedbala and M. S. Picarella, *Blood* **79,** 678 (1989).

[40] L. Ossowski, *J. Cell Biol.* **107,** 2473 (1988).

[41] W. Schlechte, G. Murano, and D. Boyd, *Cancer Res.* **49,** 6064 (1989).

[42] D. Boyd, B. Ziober, S. Chakrabarty, and M. G. Brattain, *Cancer Res.* **49,** 816 (1989).

[43] W. Hunter, *in* "Handbook for Experimental Immunology" (D. M. Weir, ed.), p.1. Blackwell Press, Oxford, 1973.

laminin is filter sterilized and each well of a 24-well plate is coated with 5×10^6 cpm of radioactive laminin. The wells are then washed with serum-free medium and the plates are stored at 4° until required.

Cells are harvested nonenzymatically (3 m*M* EDTA–PBS) and 100,000 cells are plated on the ^{125}I labeled laminin-coated 24-well culture dishes to attach overnight. At 80–100% confluency cell surface urokinase receptors are saturated for 30 min with 5 n*M* urokinase (recombinant single-chain urokinase; American Diagnostica) and the cells are washed extensively with serum-free medium to remove unbound protease. Cells are then replenished with serum-free medium with, or without (as a negative control), plasminogen (10 μg/ml, final concentration). After various time points (20–150 min) at 37°, 50-μl aliquots of the culture supernatant are withdrawn and counted for γ radioactivity. Urokinase cleaves plasminogen to plasmin which then cleaves the glycoprotein laminin into the culture medium. Solubilized laminin represents the degraded glycoprotein and reflects receptor-bound urokinase activity of the cell line analyzed. If it is necessary to show the occurrence of proteolytic laminin fragments, the supernatants (2×10^4 dpm) are denatured in the presence of 2-mercaptoethanol and electrophoresed on a 7.5% (w/v) SDS polyacrylamide gel.[41] It needs to be emphasized that this assay does not necessarily reflect the invasive or metastatic capacity of cells and thus cannot replace invasion assays or animal models in the study of metastasis.

Magnetic Separation of Transfected and Nontransfected Cells

The objective, to prove up- or downregulation of the expression of a gene of interest at the protein level by a regulator, often implies the necessity to generate stably transfected cell lines, which is laborious, expensive, and time-consuming. An approach to circumvent this problem is the separation of cells with magnetic beads after transfection. The principle of the method is cotransfection of the gene of interest with an expression vector encoding a cell surface marker (a truncated CD4 molecule, pMACS4; Miltenyi Biotech, Auburn, CA), incubation with a magnetic bead-conjugated antibody specific for the CD4 protein, and selection for the transfected cells with a magnet (VarioMACS; Miltenyi Biotech).

Ha-*ras* and an expression vector that encodes the truncated CD4 molecule (pMACS4) are transfected in a 3 : 1 molar ratio, using the transfection protocol that gives the best result with the cell line employed. After 48 hr the cells are washed twice with PBS and harvested sterilely in 320 μl of degassed PBE buffer [PBS, 0.5% (w/v) bovine serum albumin (BSA), 5 m*M* EDTA], into a microcentrifuge tube, followed by an incubation step for 15 min at room temperature with 80 μl of a magnetic bead-conjugated antibody specific for the truncated CD4 protein (which is not cross-reactive with authentic human CD4). PBE is then added to the cells to reach a total volume of 2 ml and the cells are resuspended carefully to avoid clumping of cells. The VS$^+$ magnetic separation column is washed with 5 ml of

PBE and then the cell suspension is added to the column placed in the VarioMACS magnet. The flowthrough contains the untransfected cells, which can be used as a negative control. The retained material is washed four times (3 ml each) with PBE buffer to remove nonlabeled material. The separation column is then removed from the magnet, the transfected cells are pushed through the column with a piston, and the cells are collected. To increase purity, cells can be passed over a new column.

The cells can now be subjected to Western blotting or other protein assays or can be cultivated to generate stable cell clones. In our experience, this assay has given good results for urokinase and urokinase receptor regulation when using tumor cell lines of low-grade differentiation.[44]

Northern Blot

After determining an induction of urokinase by *ras* or any other oncogene or tumor suppressor at the functional and protein level, it might be worthwhile to investigate whether activation/inhibition is due to transcriptional regulation of gene expression. A first step toward answering this question is to do a Northern blot for urokinase mRNA, although this assay cannot differentiate between increased mRNA stability and increased mRNA transcription. RNA is prepared according to standard procedures [RNA-Clean (Hybaid, Franklin, MA), TRIzol (Life Technologies, Rockville, MD), etc.] and 10–20 μg of RNA is aliquoted into sterile tubes and dried in a vacuum centrifuge. The pellets are resolved in 20 μl of denaturation buffer containing 500 μl of formamide, 50 μl of 10× MOPS [360 m*M* morpholinepropanesulfonic acid (MOPS), 100 m*M* sodium acetate), 175 μl of 37% (v/v) formaldehyde, and 275 μl of diethylpyrocarbonate (DEPC)–H_2O and incubated at 65° for 15 min. Samples are cooled on ice and 5 μl of RNA-loading dye containing 35% (v/v) glycerol and 0.25% (w/v) bromphenol blue is added and mixed well. The samples are loaded on an agarose gel containing 1 g of agarose, 72.2 g of DEPC–H_2O, 10 ml of 10× MOPS, and 17.8 ml of 37% (v/v) formaldehyde. The gel is run in electrophoresis buffer containing 1× MOPS, 107 ml of 37% (v/v) formaldehyde, and up to 600 ml of DEPC–H_2O at 60 V for 2–3 hr with circulation of the buffer to avoid development of a pH gradient. Afterwards the gel is carefully washed in water for 5 min and incubated in 10× SSC buffer (1.5 *M* NaCl, 0.15 *M* sodium citrate) for 15 min. The RNA is then transferred from the gel to a nylon membrane (GeneScreen; Du Pont, Wilmington, DE) overnight according to standard procedures. The next day the membrane is incubated in 2× SSC for 10 min, dried at 80° for 3 hr, and cross-linked by UV light (1200 J/cm^2). Subsequently the membrane is premoistened in sterile water and prehybridized in 5 ml of Church buffer containing 7% (w/v) SDS, 0.25 *M* NaP_i (0.5 *M* $Na_2HPO_4 \cdot 2H_2O$, pH 7.2),

[44] H. Allgayer, H. Wang, Y. Wang, M. M. Heiss, R. Bauer, O. Nyormoi, and D. Boyd, *J. Biol. Chem.* **274,** 4702 (1999).

1 m*M* EDTA, and 0.1% (w/v) bovine albumin fraction at 65° for at least 1 hr in an hybridization oven.

The probe is generated by releasing a 1.5-kb fragment from the urokinase cDNA [American Type Culture Collection (ATCC), Manassas, VA] by cutting it with *Pst*I.[30,45,46] The fragment is denatured and labeled with the Prime-It-II random primer labeling kit (Stratagene, La Jolla, CA) with 5 μl of [α^{32}P]dCTP (3000 μCi/mmol). To remove free radioactivity the probe is purified through a MicroSpin G-25 column (Amersham Pharmacia, Piscataway, NJ). The probe is then diluted in 300 μl of TE buffer, and radioactivity is measured in a scintillation counter. Labeled urokinase cDNA probe (2.0×10^6 counts/ml) is denatured in 30% of the calculated probe volume with 2 *N* NaOH for 10 min and then added to the prehybridization solution. The membrane is hybridized at 65°, overnight. The next day the wash solution [0.04 *M* Na_2HPO_4, 1% (w/v) SDS] is heated to 65° and the membrane is washed three times with 10 ml of wash solution, each at 65° for 15 min. The final wash solution is removed and the excess liquid on the membrane is blotted on filter paper. Subsequently the membrane is covered with Saran Wrap and autoradiographed at −80° overnight.

Transient Transfection Reporter Assay to Determine Urokinase Reporter Activity

Elevation of mRNA amounts in a Northern blot can be explained either by an elevated transcriptional rate or enhanced mRNA stability. We and others have shown with nuclear run-on experiments that a high expression of urokinase is, at least in part, due to an increase in transcription.[21] Another step toward demonstrating transcriptional induction of urokinase, and delineating the promoter elements that mediate it, is transient transfection with a chloramphenicol acetyltransferase (CAT) or luciferase reporter gene. These are experiments that specifically address promoter activity.[18–22,25]

For transient transfections cells are plated on tissue culture dishes (diameter, 10 cm) or in six-well plates (diameter, 2 cm) in complete medium containing 10% (v/v) fetal calf serum (FCS). On the next day, cells should have grown to 60–70% confluency. The medium is then changed to fresh complete medium containing 10% (v/v) FCS. Cells are transiently cotransfected, using the best method for obtaining a high transfection efficiency, often calcium phosphate or a liposome-mediated method, with 1 μg (2-cm plates) or 5 μg (10-cm dishes) of urokinase–CAT reporter constructs fused to the human urokinase promoter[18] and 0.4 or 2 μg of a vector encoding Ha-Ras protein. To correct for transfection efficiencies the

[45] E. Lengyel, J. Klostergaard, and D. Boyd, *Biochim. Biophys. Acta* **1268,** 65 (1995).

[46] P. Verde, M. P. Stoppelli, P. Galeffi, P. Di Nocera, and F. Blasi, *Proc. Natl. Acad. Sci. U.S.A.* **81,** 4727 (1984).

transient transfection is performed in the presence of 1 or 4 μg of a β-galactosidase expression vector. We have also obtained good transfection efficiencies with a simple and reliable method using poly-L-ornithine (Sigma, St. Louis, MO).[47]

For the calcium phosphate method plasmid DNA and 62 μl of 2 *M* calcium chloride are added to sterile water (to 500 μl). A solution containing 15 m*M* HEPES (pH 7.1), 280 m*M* NaCl, and 1.5 m*M* Na_2HPO_4 is added drop by drop to plasmid DNA with continuous stirring. The tubes are incubated at room temperature for 30 min, vortexing every 10 min. The transfection solution is then added to the 60–70% confluent cells. After 6 hr the cells are rinsed twice with phosphate-buffered saline, changed to fresh 10% (v/v) FCS-containing medium, and cultured for an additional 36 hr. The cells are harvested in a buffer containing 40 m*M* Tris-HCl (pH 8.0), 1 m*M* EDTA, and 6 m*M* NaCl and lysed by repeated freeze (−70°) and thaw (37°) cycles in 0.25 *M* Tris-HCl (pH 7.8). Transfection efficiencies are determined by assaying for β-galactosidase activity. Lysate (10 μl) is added to 75 μl of a buffer containing 6.5 m*M* $NaPO_4$, 16.8 m*M* 2-mercaptoethanol, 0.15 m*M* $MgCl_2$, and 4.3 m*M* *o*-nitrophenylgalactopyranoside solution (1.3 mg/ml in 0.1 *M* sodium phosphate, store at −20°). The samples are vortexed and incubated for 1–4 hr at 37°. The reaction is stopped with 50 μl of 1 *M* Na_2CO_3 and the color change is measured at 405 nm in an enzyme-linked immunosorbent assay (ELISA) reader. Because Ha-Ras, but also other oncogens, has been shown to upregulate expression of AP-1 transcription factors, β-galactosidase or luciferase expression could possibly be affected, influencing normalization and creating misleading results. Alternatively, cell extracts could be normalized after determination of protein concentration.

After normalization for transfection efficiency, the lysates are heated at 72° for 10 min and spun at 14,000 rpm for 4 min to denature most of the proteins. The CAT enzyme itself is stable. CAT activity is measured by incubating cell lysates at 37° with 4 μ*M* [^{14}C]chloramphenicol (ICN) and a 1-mg/ml concentration of acetyl-coenzyme A (4 mg/ml in 0.25 *M* Tris-HCl, pH 7.8; Pharmacia, Uppsala, Sweden) (store at −80°; acetyl-CoA is sensitive to degradation) for 6 hr to overnight. If the activity is low, acetyl-CoA should be replenished to the reaction after 3–4 hr. The next day 800 μl of ethyl acetate is added to the mixture, which is then vortexed continuously for 30 sec and spun in a microcentrifuge at 14,000 rpm for 6 min. The organic top phase is transferred to a new tube and dried in a vacuum centrifuge for 45 min. The pellets are solubilized in 20 μl of ethyl acetate and vortexed well, and the acetylated products are plated on thin-layer chromatography plates (20 × 20 cm; Whatman, Clifton, NJ) and separated for 1.5 hr, using 95% (v/v) chloroform and 5% (v/v) methanol as the mobile phase. Reactions are visualized by autoradiography and radioactivity is quantified with a PhosphorImager (Molecular Dynamics, Sunnyvale, CA).

[47] M. A. Nead and D. J. McCance, *J. Invest. Dermatol.* **105,** 668 (1995).

Electrophoretic Mobility Shift Assay for Determining Transcription Factors Involved in Ras Regulation of Urokinase

The induction of urokinase gene by Ha-Ras is mediated[19] by a promoter region between bp −2109 and −1870. We and others have identified a combined AP-1/PEA3 motif (at bp −1967) and an AP-1 motif (at bp −1880) as important for oncogene- and t-PA-mediated-induction of the urokinase promoter.[19,20,22,38,48] To determine which transcription factors are involved in Ras-mediated urokinase regulation and to identify the individual transcription factors, mobility shift assays were employed. Nuclear extracts were prepared as described by Dignam *et al.*[49] Nuclear extract (7.5 μg) was incubated in a buffer containing 20 m*M* HEPES, 0.2 m*M* EDTA, 0.25 m*M* dithiothreitol, 50 m*M* NaCl, 10% (v/v) glycerol, and 1 μg of poly (dI-dC) at room temperature. The critical component of the incubation buffer is the salt concentration and sometimes it is necessary to try different NaCl concentrations or to substitute NaCl for KCl. To each reaction 5 fmol of a Klenow end-labeled [α-^{32}P]ATP-oligonucleotide corresponding to one of the aforementioned promoter motifs is added in the absence or presence of a 100-fold excess of unlabeled competitor oligonucleotide. Binding is allowed for 15 min. The mixture is electrophoresed in a 5% (w/v) polyacrylamide gel, using 0.5% (v/v) TBE (89 m*M* Tris, 89 m*M* boric acid, 1 m*M* EDTA) buffer, and the gel is dried and exposed overnight.

One strategy for specifically identifying bound transcription factors is the "supershift" technique, which employs specific antibodies to known transcription factors in an electrophoretic mobility shift assay (EMSA). The additional binding of the antibody results in a slower mobility of the retarded complex, which includes the transcription factor recognized by the specific antibody. In our study of hepatocyte growth factor (HGF)/RAS induction of urokinase gene expression, we found the best supershift results when coincubating the reaction mixture with a JunD antibody at 4° for 1 hr before running the gel. However, it may take some time to find the best conditions suited for antibody binding. Critical components are incubation time, temperature, and binding of the antibody. Some companies sell specific antibodies for supershift that are more concentrated, but even with these antibodies the attempt to supershift is sometimes unsuccessful. One reason for that might be that the antibody is directed against the DNA-binding part of the transcription factor, resulting at most in reduction of the binding signal, but not in a supershift. Another major reason is that the large antibody–protein–DNA complex can be easily disrupted in the strong electric field of the gel.

When analyzing the AP-1/PEA3 site at bp −1967 for regulation by oncogenes including *ras,* it might be worthwhile to analyze each part of the AP-1/PEA3 site

[48] D. D'Orazio, D. Besser, R. Marksitzer, C. Kunz, D. A. Hume, B. Kiefer, and Y. Nagamine, *Gene* **201,** 179 (1997).

[49] J. D. Dignam, R. M. Lebovitz, and R. G. Roeder, *Nucleic Acids Res.* **11,** 1475 (1983).

separately with oligonucleotides mutated either in the AP-1 or the PEA3 part of the binding site. Analysis of urokinase regulation by HGF/scatter factor (SF), shows that transcription factor binding to this promoter region involves the AP-1 part but not the PEA3 part of the motif at bp −1967. A remarkable difference between t-PA or H-Ras[19] and HGF/SF induction of the urokinase promoter is that the PEA3 part of this site is not required for HGF/SF induction, while it is necessary for induction by H-Ras.[25]

Conclusion

Overexpression or constitutive activation of *ras* genes induces urokinase-dependent enzymatic activity in a variety of cell lines analyzed. There has been intense effort to elucidate the signal transduction pathway downstream of Ras and the transcription factors that are activated besides AP-1, Ets, and NF-κB. Although it is well recognized that Ras can signal through the Rac-1, MEKK (MAPK/ERK kinase kinase), JNK (c-Jun N-terminal kinase) pathway,[50,51] or a separate phosphotidylinosetol (PI) 3-kinase-dependent pathway,[52] the data suggest that the induction of urokinase and also urokinase receptor by Ras is mediated mainly via the classic Ras–Raf– Erk1 pathway that transmits signals from Ras to AP-1.[19,21,53–55] Interestingly, this cascade has been demonstrated to mediate the upregulation of urokinase gene expression by several other potential oncogenes and growth factors, for example, HGF, vascular endothelial growth factor (VEGF), epidermal growth factor (EGF), insulin-like growth factor (IGF), basic fibroblast growth factor (bFGF), v-*mos, tpr-met,* or c-*erbB-2*.[38,55,55–57] Thus, this cascade is one of the central molecular pathways inducing an invasive and proliferative phenotype after transformation.

Besides its proteolytic activity, addition of urokinase to cells has revealed additional protein functions. Urokinase induces cell proliferation, migration, and

[50] A. Minden, A. Lin, M. McMahon, C. Lange-Carter, B. Derijard, R. J. Davis, G. L. Johnson, and M. Karin, *Science* **266,** 1719 (1994).

[51] M. Russell, C. A. Lange-Carter, and G. L. Johnson, *J. Biol. Chem.* **270,** 11757 (1995).

[52] P. Rodriguez-Viciana, P. H. Warne, R. Dhand, B. Vanhaesebroeck, I. Gout, M. J. Fry, M. D. Waterfield, and J. Downward, *Nature* (*London*) **370,** 527 (1994).

[53] J. P. Irigoyen, D. Besser, and Y. Nagamine, *J. Biol. Chem.* **272,** 1904 (1997).

[54] G. Cirillo, L. Casalino, D. Vallone, A. Caracciolo, D. De Cesare, and P. Verde, *Mol. Cell. Biol.* **19,** 6240 (1999).

[55] J. A. Ghiso, D. F. Alonso, E. F. Farias, D. E. Gomez, and E. B. de Kier Joffe, *Eur. J. Biochem.* **263,** 295 (1999).

[56] D. Besser, A. Bardelli, S. Didichenko, M. Thelen, P. Comoglio, C. Ponzetto, and Y. Nagamine, *Oncogene* **14,** 705 (1997).

[57] M. S. Pepper, K. Matsumoto, T. Nakamura, L. Orei, and R. Montesano, *J. Biol. Chem.* **267,** 20493 (1992).

cytoskeletal reorganization of different cell lines, and activates itself the ERK (extracellular signal-regulated kinase)–mitogen-activated protein kinase (MAPK) pathway.[53,58–60] The urokinase-dependent induction of these effects relies on its receptor-binding domain, but not on the proteolytic domain. While a direct involvement of Ras in these effects has not been shown yet, phosphorylation of focal adhesion kinase (FAK)[59] and ERK-1/2,[59,60] which are Ras effectors, has been reported. Therefore, it is conceivable that urokinase may also activate proteins of the Ras family.

In summary, constitutively active Ras has been shown to induce the expression of urokinase, urokinase receptor, and other invasion-related genes. Considering the synthesis of farnesyltransferase inhibitors that counter Ras activity, it is attractive to speculate that Ras inhibition could be an effective adjuvant strategy for preventing invasion and metastasis of tumors characterized by molecular aberrations involving the Ras pathway.

Acknowledgments

The work of E.L. and M.S. is supported by the Deutsche Forschungsgemeinschaft (DFG Le 889-4/1) and the Deutsche Krebshilfe (10-1197-Le 1). We thank Dr. D. Boyd (M.D. Anderson Cancer Center, Houston, TX) for continuous support.

[58] K. Fischer, V. Lutz, O. Wilhelm, M. Schmitt, H. Graeff, P. Heiss, T. Nishiguchi, N. Harbeck, H. Kessler, V. Luther, V. Magdolen, and U. Reuning, *FEBS Lett.* **438,** 101 (1998).
[59] H. Tang, D. M. Kerins, Q. Hao, T. Inagami, and D. E. Vaughan, *J. Biol. Chem.* **273,** 18268 (1998).
[60] D. H. D. Nguyen, I. M. Hussaini, and S. L. Gonias, *J. Biol. Chem.* **273,** 8502 (1998).

[12] Ras Regulation of Cyclin D1 Promoter

By DEREK F. AMANATULLAH, BRIAN T. ZAFONTE, CHRIS ALBANESE, MAOFU FU, CYNTHIA MESSIERS, JOHN HASSELL, and RICHARD G. PESTELL

Introduction

Cyclin D1, the regulatory subunit of cyclin-dependent kinases (CDK) 4 and 6, is required for, and capable of shortening, the G_1 phase of the cell cycle through the formation of holoenzyme complexes (cyclin–CDK) that phosphorylate retinoblastoma protein (pRB).[1,2] Ras transformation is inhibited by antisense to cyclin D1 mRNA.[3] Murine skin tumors induced by transgenic overexpression of Ras were reduced in number in the cyclin $D1^{-/-}$ background.[4] Together these studies suggest

0076-6879/00 $35.00

that cyclin D1 is required for Ras-induced transformation. The abundance of cyclin D1 is determined by transcriptional, translational, and posttranslational mechanisms. Cyclin D1 protein, which has a short half-life, is phosphorylated at Thr-268 by glycogen synthase kinase 3β (GSK-3β), which leads to polyubiquitin-targeted proteosome degradation.[5,6] The cyclin D1 promoter is directly induced by Ras, and it is thought that the transcriptional induction of cyclin D1 contributes to the increased abundance of cyclin D1 in transformed cells.[7,8] Subsequent analyses of many Ras superfamily members have identified transcriptional induction of the cyclin D1 gene as a common response. Indeed, analysis of the intracellular signaling pathways activated by Ras superfamily members provides evidence that the induction of cyclin D1 promoter activity correlates well with transforming function.[9–11] The purpose of this review is to describe the current understanding of the mechanisms by which Ras family proteins regulate the cyclin D1 promoter and to outline important methodological caveats relevant to the design of experiments analyzing Ras regulation of gene transcription.

Transforming mutants of $p21^{ras}$ induced the cyclin D1 promoter in JEG-3, Mv1.Lu, and CHO cell lines.[8] An AP-1 (activator protein 1) site at position –954 was required for full induction of the cyclin D1 promoter in JEG-3 trophoblastic cells. This site was also required for cyclin D1 promoter activation by the G protein-coupled mitogen angiotensin II (AII)[12] and by the p300-coactivator.[13] Electrophoretic mobility shift assays (EMSAs) using nuclear extracts from cultured

[1] R. G. Pestell, C. Albanese, A. T. Reutens, J. E. Segall, R. J. Lee, and A. Arnold, *Endocr. Rev.* **20,** 501 (1999).

[2] C. J. Sherr, *Cell* **79,** 551 (1994).

[3] J.-J. Liu, J.-R. Chao, M.-C. Jiang, S.-Y. Ng, J. J.-Y. Yen, and H.-F. Yang-Yen, *Mol. Cell. Biol.* **15,** 3654 (1995).

[4] A. I. Robles, M. L. Rodriguez-Puebla, A. B. Glick, C. Trempus, L. Hansen, P. Sicinski, R. W. Tennant, R. A. Weinbergh, S. H. Yuspa, and C. J. Conti, *Genes Dev.* **12,** 2469 (1998).

[5] J. A. Diehl, M. Cheng, M. F. Roussel, and C. J. Sherr, *Genes Dev.* **12,** 3499 (1998).

[6] J. A. Diehl, F. Zindy, and C. J. Sherr, *Genes Dev.* **11,** 957 (1997).

[7] H. Hendrik Gille and J. Downward, *J. Biol. Chem.* **274,** 22033 (1999).

[8] C. Albanese, J. Johnson, G. Watanabe, N. Eklund, D. Vu, A. Arnold, and R. G. Pestell, *J. Biol. Chem.* **270,** 23589 (1995).

[9] D. Joyce, B. Bouzahzah, M. Fu, C. Albanese, M. D'Amico, J. Steer, J. U. Klein, R. J. Lee, J. E. Segall, J. K. Westwick, C. J. Der, and R. G. Pestell, *J. Biol. Chem.* **274,** 25245 (1999).

[10] J. K. Westwick, R. J. Lee, Q. T. Lambert, M. Symons, R. G. Pestell, C. J. Der, and I. P. Whitehead, *J. Biol. Chem.* **273,** 16739 (1998).

[11] J. K. Westwick, Q. T. Lambert, G. J. Clark, M. Symons, L. Van Aelst, R. G. Pestell, and C. J. Der, *Mol. Cell. Biol.* **17,** 1324 (1997).

[12] G. Watanabe, R. J. Lee, C. Albanese, W. E. Rainey, D. Batlle, and R. G. Pestell, *J. Biol. Chem.* **271,** 22570 (1996).

[13] C. Albanese, M. D'Amico, A. T. Reutens, M. Fu, G. Watanabe, R. J. Lee, R. N. Kitsis, B. Henglein, M. Avantaggiati, K. Somasundaram, B. Thimmapaya, and R. G. Pestell, *J. Biol. Chem.* **274,** 34186 (1999).

cells and primary tissues have shown that several AP-1 proteins, c-Jun, Jun B, Jun D, and c-Fos, bound the cyclin D1 –954 region. AII induction of the cyclin D1 promoter, which required $p21^{ras}$, was associated with increased binding of AP-1 proteins to the cyclin D1 AP-1 site.[12] Site-directed mutagenesis of the AP-1-like sequences at position –954 abolished c-Jun- as well as $p21^{ras}$-dependent activation of cyclin D1 promoter activity.

The Dbl-related proteins (Tiam1, Bcr, Fgd1, Sos1, Sos2, RasGRF, Dbl, Vav, Ect2, Lbc, Lfc, Lsc, Ost, Dbs, Net1, and Tim) function as guanine nucleotide exchange factors (GEFs) and activators of the Rho family of Ras-related GTPases. Members of the Dbl family proteins stimulate the catalytic activities of c-Jun N-terminal kinase 1 (JNK1) and p38/Mpk2 as well as the transcriptional activation of c-Jun, implicating them in mitogen-activated protein kinase (MAPK)-mediated signaling pathways. Because Ect2 is a potently transforming Dbl family member that exhibits weak activation of JNK1 and c-Jun, and Lfc is a weakly transforming Dbl family member that exhibits strong activation of JNK1 and c-Jun, the transforming ability of the Dbl family does not always require JNK1 or c-Jun activation. The transcriptional induction of the cyclin D1 promoter, however, correlated directly with the transforming potential of the Dbl family members suggesting an important role for cyclin D1 in transformation by these proteins.[10]

The Ras-related proteins, Rac and Rho, also induce the cyclin D1 promoter.[9,11] Rac1 regulates several distinct pathways; no single Rac effect is necessary or sufficient for transformation. Rac1 is part of the NADPH oxidase complex, which induces the formation of reactive oxygen species and induces the cyclin D1 promoter in bovine tracheal myocytes.[14] Platelet-derived growth factor (PDGF) activates the extracellular signal-regulated kinase (ERK) pathway, which has been shown to induce DNA synthesis, increase cyclin D1 protein levels, and stimulate the transcription of the cyclin D1 promoter.[8,15–17] Dominant-negative Rac1 mutants inhibited PDGF-induced cyclin D1 transcription and the formation of radical oxygen species. Antioxidant treatment reduced both Rac1 and PDGF induction of the cyclin D1 promoter, inhibited cyclin D1 protein levels, but did not alter the effect of MEK1. Rac1 was therefore neither necessary nor sufficient to activate the ERK pathway, suggesting that Rac regulation of cyclin D1 is ERK independent.[14] Rac1 induction of the cyclin D1 promoter in NIH 3T3 cells involved an NF-κB signaling pathway and required both ATF-2 (activating transcription factor 2)

[14] K. Page, J. Li, J. A. Hodge, P. T. Liu, T. L. Vanden Hoek, L. B. Becker, R. G. Pestell, M. R. Rosner, and M. B. Hershenson, *J. Biol. Chem.* **274,** 22065 (1999).

[15] M. Ramakrishnan, N. L. Musa, J. Li, P. T. Liu, R. G. Pestell, and M. B. Hershenson, *Cell. Mol. Biol.* **18,** 736 (1998).

[16] J. N. Lavoie, G. L'Allemain, A. Brunet, R. Müller, and J. Pouysségur, *J. Biol. Chem.* **271,** 20608 (1996).

[17] G. Watanabe, A. Howe, R. J. Lee, C. Albanese, I.-W. Shu, A. N. Karnezis, L. Zon, J. Kyriakis, K. Rundell, and R. G. Pestell, *Proc. Natl. Acad. Sci. U.S.A.* **93,** 12861 (1996).

and NF-κB-binding sites (located at positions –54 and –39, respectively) of the cyclin D1 promoter. Activating Rac1 mutants increased binding and activity of RelA and NF-κB1.[9] Trans-dominant inhibitor mutants of NF-κB also reduced Rac1 activation of the cyclin D1 promoter. NF-κB, part of the Rel family (members of which dimerize and induce nuclear gene transcription), induces the cyclin D1 promoter[18] through both the NF-κB and ATF-2 sites.[9] This cross-coupling of AP-1 and NF-κB proteins at the ATF-2 site may serve as a point of biological integration.[9] It is likely that Rac1 transformation involves an NF-κB signaling pathway to cyclin D1.

Methods

The purpose of this part of the chapter is to detail the potential artifacts that may be encountered during analysis of promoter activity. Several different methods of gene transfer including calcium phosphate precipitation, electroporation, cationic liposomes, and gene gun may be used and are described elsewhere.[19] The type of gene transfer used is frequently dictated by the transfection efficiency for a given cell type. It is important to note that the method of gene transfer may independently regulate stress signaling pathways and or cell cycle events (below).

Vector Backbone Sequences

Several different types of reporter plasmids are available and the investigator should be aware of the potential limitations of the reporter system used. The luciferase reporter system is commonly used as a rapid and automated system for gene analyses. Because Ras activates a broad array of signaling pathways [ERK, p38, JNK, phosphatidylinositol (PI) 3-kinase (PI3-kinase), NF-κB], and plasmid DNA may contain sequences spuriously activated by Ras, it is important to examine the empty (promoterless) reporter plasmid itself for signaling responses. DNA sequences in the vector backbone of pUC plasmids have been identified as giving rise to spurious effects, and include the AP-1 site TGACACA.[20] This vector sequence independently confers responsiveness to UV irradiation, tissue plasminogen activator (TPA), induction by c-Jun, and repression by thyroid hormone[21] (reviewed in Ref. 22). This element can be removed as a 267-bp *Nde*I–*Eco*O109 fragment from

[18] D. C. Guttridge, C. Albanese, J. Y. Reuther, R. G. Pestell, and A. S. Baldwin, *Mol. Cell. Biol.* **19,** 5785 (1999).

[19] T. Sambrook, E. R. Fritsch, and T. Maniatis, "Molecular Cloning: A Laboratory Manual." Cold Spring Harbor Laboratory Press, Cold Spring Harbor, New York, 1989.

[20] C. H. Jonat, K.-K. Rahmsdorf, A. C. B. Cato, S. Gebel, H. Ponta, and P. Herrlich, *Cell* **62,** 1189 (1990).

[21] G. Lopez, F. Schaufele, P. Webb, J. M. Holloway, J. D. Baxter, and P. J. Kushner, *Mol. Cell. Biol.* **13,** 3042 (1993).

[22] P. J. Kushner, J. D. Baxter, K. G. Duncan, G. N. Lopez, F. Schaufele, R. M. Uht, P. Webb, and B. L. West, *Mol. Endocrinol.* **8,** 405 (1994).

the pUC backbone. These and other DNA sequences in the pBL2-derived plasmids have been removed from several current reporter systems such as pA_3LUC.[23]

Transfection Artifacts

Four main categories of potential artifacts may appear during experimentation if vector backbone sequences are a confounding variable. In the first type of experiment, the sequential deletion of a large promoter fragment cloned into the reporter vector results in the concomitant sequential induction of basal level reporter activity. In this circumstance the sequential promoter deletion results in closer proximity of the vector backbone enhancer to the gene promoter. These results may lead the investigator to falsely attribute the increase in basal activity to the loss of a repressor element.

In the second type of scenario, an investigator examining Ras signaling and nuclear receptor function may find that Ras activation of AP-1 activity leads to apparent synergistic activation of the unliganded receptor. This observation may be due to the presence of the AP-1 site in the vector backbone.

A third type of error is found in paradigms based on the use of heterologous reporter systems. In these assays a transcription factor is linked to the yeast GAL4 DNA-binding domain to form a chimeric transcription factor fusion protein. Reporter activity is assessed using a multimerized sequence (UAS, upstream activating sequence) linked to a reporter gene. The multimerized reporter sequence is activated on binding the GAL4-binding site. The presence of vector backbone sequences in the reporter may lead to spurious conclusions that the transcription factor fusion protein is activated by the Ras signal whereas, in fact, the induction occurs through activation of the vector backbone sequences.

A fourth type of error occurs when the investigator looking for an Ras response element deletes a basal level enhancer, which reduces activity below the reproducible detectable range of the reporter assay. If this candidate sequence is now cloned into the reporter system, which contains Ras-responsive sequences, the basal enhancer elements will be falsely ascribed Ras-responsive enhancer properties.

Other Reporter Vector Artifacts

Additional sequences that have been identified in pUC-based vectors include sequences resembling the CACCC transcription factor-binding site,[24] between the *Nde*I site and the polylinker region. Many of the current reporter plasmids are known to be spuriously activated by different signaling pathways and the information is conveniently accessed on line (*http://vectordb.atcg.com/vectordb/*

[23] W. M. Wood, M. Y. Kao, D. F. Gordon, and E. C. Ridgway, *J. Biol. Chem.* **264,** 14840 (1989).

[24] R. Schule, M. Muller, H. Otsuka-Murakami, and R. Renkawitz, *Nature (London)*. **332,** 87 (1988).

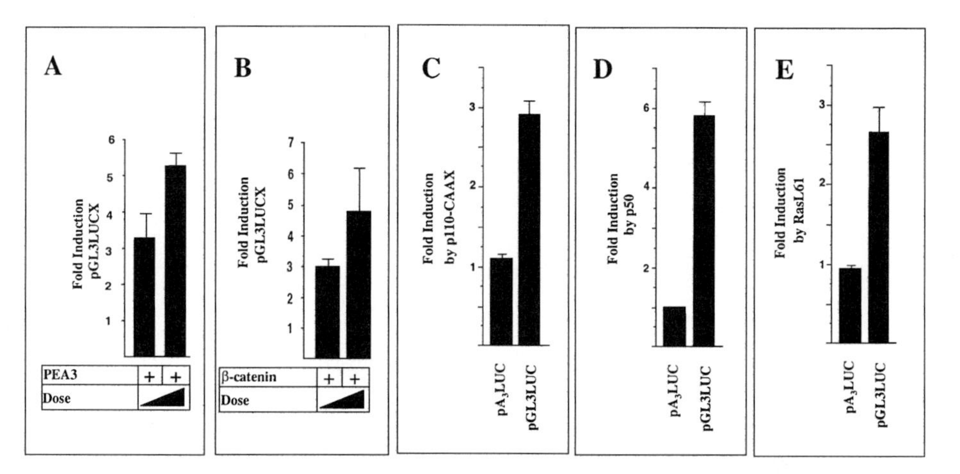

FIG. 1. Luciferase activity was determined from cells transfected into cultured mammalian cells. The empty expression vector luciferase reporter (pGL3basic, pA_3LUC) was transfected in conjunction with the mammalian expression vector encoding a transcription factor or activator of components of the Ras-signaling pathway. (A) COS cells were transfected with the empty reporter plasmids pGL3LUC and an expression plasmid for PEA3 at a molar ratio of either 1 : 0.33 or 1 : 0.67. (B) COS cells were transfected with pGL3LUC with increasing amounts of the effector plasmid for β-catenin Y33.[25a] The pGL3LUC empty reporter was induced 3- to 5-fold. (C) Chinese homster ovary (CHO) cells were transfected with the membrane-targeted p110-CAAX expression vector known to activate PI3-kinase signaling[25b] and the empty reporter plasmid (either pA_3LUC or pGL3LUC). The data represent means ± SEM for at least $N = 5$ separate transfections. (D and E) MCF7 cells were transfected with the empty expression vectors for pA_3LUC or pGL3LUC (2.4 μg) and 300–600 ng of effector plasmid for (D) the NF-κB protein (p50) and (E) RasL61.

vector.html). Thus, constructions such as pXP_2 derivatives[25] carry online warnings of cryptic response elements. Ras superfamily signaling can regulate NF-κB/AP-1 activity and T cell-specific transcription factor (TCF) signaling in a cell type-specific manner. Commercial vectors contain within their vector backbone sequences identical to the TCF site (TTG ATC TT; pGL3basic), and others resembling the NF-κB (GGC GAG TT; pGL3basic) or AP-1 site (TGA CAC T; pXP_2LUC). We are not aware of studies that have formally addressed the role of these vector sequences as was performed for the pUC AP-1-like element.[21] It is likely that several different factors may contribute to spurious activation of the plasmid vector backbones (Fig. 1[25a,b]), and the investigator should always assess the empty reporter vector in the analyses (Fig. 1). Some of the commercially available reporter vectors can be induced by overexpression of transcription factors including

[25] S. Nordeen, *BioTechniques* **6,** 454 (1988).

[25a] M. Shtutman, J. Zhurinsky, I. Simcha, C. Albanese, M. D'Amico, R. Pestell, and A. Ben-Ze'ev, *Proc. Natl. Acad. Sci. U.S.A.* **96,** 5522 (1999).

[25b] I. Matsumura, T. Kitamura, H. Wakao, H. Tanaka, K. Hashimoto, C. Albanese, J. Downward, R. G. Pestell, and Y. Kanakura, *EMBO J.* **18,** 1367 (1999).

PEA3 (polyomavirus enhancer activator 3),[26] β-catenin, and the NF-κB component p50 (Fig. 1). In addition, a membrane-targeted activator of the PI3-kinase pathway (p110-CAAX) induced the pGL3LUC reporter but not the pA_3LUC reporter vector. The mechanism responsible for this spurious activation is not known but underscore the importance of analyzing the effect of specific transcription factors on the empty reporter before analyzing the promoter for Ras responsiveness.

Normalization of Gene Reporter Activity

Internal Control Reporter Gene Artifacts

Regulation of Viral Promoters. To determine the effect of Ras, or the effect of a transfected expression vector of any sort, some form of normalization is required to compare between populations of transfected cells. Viral promoters driving additional reporter genes such as β-galactosidase, growth hormone, *Renilla* luciferase, or green fluorescent protein (GFP) may be cotransfected as a form of internal standardization. Because viral promoters driving the expression vectors are not inert and may be regulated by growth factors, cell cycle changes, integrin signaling, and hormones, the internal control plasmid must be carefully scrutinized before experiments commence. The Rous sarcoma virus (RSV) promoter is induced by p38 and ERK[27] and contains an inverted CAAT box and a hormonal response element.[28] These properties of the RSV promoter may confound data interpretation when used as an internal control. The cytomegalovirus (CMV) promoter is regulated by several kinases activated by Ras signaling pathways,[27] is induced by cAMP and butyrate, contains E2F and Sp1 sites (which contribute to cell cycle responsiveness in other promoters),[29] and is regulated by the PTEN, Akt kinase signaling pathway.[30] The induction of the CMV promoter by p38 (8-fold) and ERK (4-fold)[27] may lead to significant data misinterpretation if this promoter is used to drive internal control plasmid reporter genes. As these effects are cell type specific, it is important to consider the possibility that the internal control vectors may be regulated by the pathway being examined and assess this possibility early.

Cross-Squelching by Internal Control Viral Promoters. The yeast transcriptional activator GAL4 binds specific sites on DNA to activate transcription of adjacent genes.[31] Various GAL4 derivatives, which lack DNA-binding ability when transfected into yeast, have been observed to inhibit the GAL4 reporter.

[26] J. H. Xin, A. Cowie, P. Lachance, and J. A. Hassel, *Genes Dev.* **6,** 481 (1992).

[27] J. Raingeaud, A. J. Whitmarsh, T. Barrett, B. Dérijard, and R. J. Davis, *Mol. Cell. Biol.* **16,** 1247 (1996).

[28] F. Saatcioglu, T. Deng, and M. Karin, *Cell* **75,** 1095 (1993).

[29] J. R. Nevins, *Cell Growth Differ.* **9,** 585 (1998).

[30] X. Wu, K. Senechal, M. S. Neshat, Y. E. Whang, and C. L. Sawyers, *Proc. Natl. Acad. Sci. U.S.A.* **95,** 15587 (1998).

[31] G. Gill and M. Ptashne, *Nature* (*London*). **334,** 721 (1988).

This inhibition of gene expression is referred to as "squelching." Specific *trans*-activation factors differentially cross-squelch in a cell type- and dose-dependent manner.[32] This inhibition may be in part mediated by limiting amounts of coactivator proteins within the cells or recruitment of corepressors.[33,34] More powerful promoters will cross-compete weaker promoters in part by competing for limiting coactivators in the cell, and the cross-competition may be sequence specific. Thus, the viral promoters used as internal controls can cross-compete weaker promoters in a sequence-dependent manner and thereby lead to spurious mislocalization of a target response element.

Activation of Expression from SV40 Promoterless Constructs by Cotransfected Internal Control Plasmids. Some of the commonly used internal control plasmids (i.e., pCM5β-gal) have been shown to independently activate reporter genes.[35] Chloramphenicol acetyltransferase (CAT) expression of three vectors (pCATbasic, pBLCATERE, and pCATbTOR) was increased up to 300-fold by cotransfection with constructions including pCH110, or pCMV5β-gal. The effect was not observed with a construct not containing the simion virus 40 (SV40) *ori* (pCMVb). Removal of the SV40 *ori* 27-bp palindrome and large T antigen-binding site prevented activation of CAT expression. A similar effect was observed using a growth hormone (GH)-based reporter system, which was activated 20-fold when transfected with constructs containing the SV40 *ori*.[36] It has been proposed that the spurious activation of gene expression is due to recombination after internalization of transfected DNA, causing a sequence containing the SV40 *ori* to be transferred from plasmids containing it, to those that do not. This was thought to increase replication rates of the reporter construction with a consequent increase in gene reporter activity.[35] It would be therefore predicted that the method of "boosting" transfection efficiency by transfecting large T antigen[37] would lead to spurious activation of reporter gene activity by this mechanism in permissive cell types such as CV-1.[35]

Alternative Approaches to Normalization. Because of the risk that spurious effects of the promoters driving reporter genes used as internal standards may represent an additional independent variable, alternative methods are used. One approach is to perform multiplicate plate analysis. The variability between transfections with a constant amount of DNA into an equal number of cells at the same confluence is frequently small (less than 5% variability). As this variation is often less than the independent effect of a particular stimulus on the internal "control

[32] T. Bocquel, V. Kumar, C. Stricker, and H. Gronemeyer, *Nucleic Acids Res.* **17,** 2581 (1989).

[33] Y. Kamei, L. Xu, T. Heinzel, J. Torchia, R. Kurakawa, B. Gloss, S.-C. Lin, R. A. Heyman, D. W. Rose, C. K. Glass, and M. G. Rosenfeld, *Cell* **85,** 403 (1996).

[34] J. D. Chen and R. M. Evans, *Nature (London).* **377,** 454 (1995).

[35] A. Flint, S. Kaluz, M. Kaluzova, L. Sheldrick, P. Fisher, and K. Derecka, *BioTechniques* **27,** 728 (1999).

[36] M. Simoni and J. Grommoll, *J. Endocrinol. Invest.* **19,** 359 (1996).

[37] R. de Chasseval and J. P. de Villarty, *Nucleic Acids Res.* **20,** 245 (1992).

reporter," multiplicate plates with statistical analyses frequently remain a preferable alternative for controlling between transfected plates.

Artifacts Due to Enrichment for Cellular Transfection

The analysis of Ras signaling frequently leads the investigator to study the effects of Ras on cell cycle and apoptosis. Because sustained overexpression of components of the Ras signaling pathway in cultured cells may lead to adaptive alternations in the signaling pathway itself, increasing use is being made of cell sorting to isolate a pool of transiently transfected cells. Cotransfected marker plasmids include the GFP plasmids with retrieval of transfected cells based on green fluorescence, using fluorescence-activated cell sorting (FACS) selection. Other cell surface markers including CD19 or CD7 are also popular.[38–40] The CD4 plasmid has been used as a marker with subsequent selection based on magnetic-activated cell sorting (MACS).[41,42]

The introduction of DNA (pGL3basic) can induce a transient increase in G_0/G_1-phase cells that peaks at 8 hr in HeLa cells; and cellular apoptosis in many cell types.[43] The increase in apoptosis at 24 hr in cells transfected with pGL3basic (18%) was not observed with mock-transfected cells, suggesting the Ca_2PO_4 precipitation itself is not responsible for DNA-mediated apoptosis. Thus, these results contrast with previous studies in which Ca_2PO_4 treatment itself induced a G_1-phase arrest of NIH 3T3 cells.[44] The effects of pGL3basic DNA in HeLa cells were elicited in a p53-independent manner and are due, in part, to the entry of marker genes into the nucleus of transfected cells.[43] Because the nuclear translocation of plasmid DNA played an important role in apoptosis, it would be predicted that the subsequent development of membrane-localized GFP plasmids may reduce this problem.[45]

Materials

For the methods of transient expression analysis involving calcium phosphate precipitation, DEAE-dextran, electroporation, and liposome-mediated transfection

[38] C. Cayrol and E. K. Flemington, *EMBO J.* **15,** 2748 (1996).

[39] J. V. Frangioni, N. Moghal, A. Stuart-Tilley, B. G. Neel, and S. L. Alper, *J. Cell Sci.* **107,** 827 (1994).

[40] T. F. Tedder and C. M. Isaacs, *J. Immunol.* **143,** 712 (1989).

[41] G. Siebenkotten, U. Behrens-Jung, S. Miltenyi, K. Petry, and A. Radbruch, Employing surface markers for the selection of transfected cells. *In* "Cell Separation: Methods and Applications" (D. Recktenwald and A. Radbruch, eds.), p. 271. Marcel Dekker, New York, 1998.

[42] S. Miltenyi, W. Muller, W. Weichel, and A. Radbruch, *Cytometry* **11,** 231 (1990).

[43] A. Rodriguez and E. K. Flemington, *Anal. Biochem.* **272,** 171 (1999).

[44] J. Renzing and D. Lane, *Oncogene* **10,** 1865 (1995).

[45] R. F. Kalejta, T. Shenk, and A. J. Beavis, *Cytometry* **29,** 286 (1997).

the reader is referred to Ref. 19. A protocol for the commonly used Ca_2PO_4 transfection method is detailed below.

Transfection of Mammalian Cells with Calcium Phosphate

1. Plate cells at about 1×10^5 per 3 cm^2 of plate surface area (for a 12-well plate).
2. Allow the cells to adhere to the plate.
3. To 1.5-ml microcentrifuge tubes add the following:
   a. Luciferase reporter/vectors [it is typical to use molar ratios of 8 : 1 (or even less vector) to avoid cross-squelching].
   b. A volume of 120 μl of HEPES–HBS (HEPES-buffered saline: 137 m*M* NaCl, 5 m*M* KCl, 0.7 m*M* Na_2HPO_4, 6 m*M* dextrose, and 21 m*M* HEPES, to exactly pH 7.05; sterile filter) per 1 cm of well radius.
   c. $CaCl_2$ (2 *M*): Add dropwise 8.25 μl and mix while adding per 1 cm of well radius.
4. Incubate between 20 and 40 min at room temperature.
5. In a hood, add the contents of each microcentrifuge tube to the corresponding plated cells.
6. Swirl the plate.
7. Aspirate the medium after a 6-hr or overnight incubation (depending of $CaCl_2$ toxicity to cells).
8. Add medium to the cells and incubate the cells for another 18 to 30 hr.
9. Perform the luciferase assay.

Reporter Gene Assays

Several different types of reporter gene systems are used, including growth hormone (GH), chloramphenicol acetyltransferase (CAT), luciferase (firefly and *Renilla*) and green fluorescence protein (GFP). The commonly used luciferase assay is detailed below. A CMV–*Renilla* luciferase plasmid, pRL-CMV (Promega, Madison, WI), may be included to control for transfection efficiency with the caveats outlined above (Normalization of Gene Reporter Activity). *Renilla* luciferase activity is assayed on the same cell extracts according to the manufacturer instructions for the Dual-Luciferase reporter assay system. (Note that the ethanol used to wash single-injection machines must be completely removed to avoid spurious luminescence in the presence of firefly luciferin.)

Luciferase Assay

1. Prepare solutions:

   GME buffer (store at 4°): 25 m*M* glycylglycine (pH 7.8), 15 m*M* magnesium sulfate, and 4 m*M* EGTA in distilled water.

Extraction buffer (make fresh): 1 mM dithiothreitol and 1% (v/v) Triton X-100 in GME buffer.

Assay mix (always make fresh): 16.67 mM potassium phosphate (pH 7.8), 1 mM dithiothreitol, and 2.22 mM adenosine 5′- triphosphate in GME buffer.

Luciferin solution (make fresh; small leftover amounts can be stored at –20°): 250 μM luciferin and 12.5 mM dithiothereitol in GME buffer.

2. Lyse the cells:
   a. Aspirate the medium.
   b. Add 100 μl of extraction buffer for each 1 cm of well radius to each well.
   c. Shake, until cellular debris is visible (some cell types may need scraping).
3. Prepare sample:
   a. To each sample plus the blanks, add 300 μl of assay buffer.
   b. To each sample, add the cell extract (100 μl).
   c. Measure the integrated light output over 10 to 60 sec.

In this laboratory samples receive 100 μl of luciferin substrate solution (0.2 mM luciferin, 10 mM dithiothreitol in GME) by automatic injection in an AutoLumat LB 953 luminometer (Berthold, Bad Wildbad, Germany). Luciferase content is measured by calculating the light emitted during the initial 10 sec of the reaction and the values are expressed in arbitrary light units.[12,46] Statistical analyses are performed using the Mann–Whitney U test with significant differences established as $p < 0.05$.

Conclusion

The cyclin D1 promoter has been used as a molecular probe of the signal transduction pathways involved in Ras signaling. The valid assessment of promoter activation by Ras requires the use of a reporter system that is not itself responsive to components of the Ras signaling pathway. Because of the potential artifacts from vector backbone sequences and transcriptional readthrough, it is critical to perform control experiments with the empty reporter vector. The promoters used to drive "internal control vectors" are regulated by diverse signaling pathways and can spuriously cross-squelch weaker promoters being analyzed for responsive elements. The magnitude of the changes in the activity of viral promoters induced by proliferative signals may be dramatic. The magnitude of these spurious effects predicates the need for careful assessment of the relative benefits of these "controls" compared with comparisons deduced from replicate plates, in which the well-to-well variability may be minimal. Because the transcriptional induction of the cyclin D1 promoter by Ras family members correlates well with transforming capacity,

[46] R. G. Pestell, C. Albanese, A. Hollenberg, and J. L. Jameson, *J. Biol. Chem.* **269,** 31090 (1994).

analyses of the molecular mechanisms regulating cyclin D1 promoter activity will likely contribute to our understanding of transformation by Ras.

Acknowledgments

This work was supported in part by grants from the NIH (R01CA70897, RO1CA75503), and by awards from the Susan G. Komen Breast Cancer Foundation, Irma T. Hirschl, and the Pfeiffer Foundation (R.G.P.) and the Medical Research Council of Canada and by the Canadian Breast Cancer Research Institute (J.H.). Work conducted at the Albert Einstein College of Medicine was supported by Cancer Center Core National Institutes of Health grant 5-P30-CA13330-26.

[13] Ras Regulation of Cyclin-Dependent Immunoprecipitation Kinase Assays

By BRIAN T. ZAFONTE, DEREK F. AMANATULLAH, DANIEL SAGE, LEONARD H. AUGENLICHT, and RICHARD G. PESTELL

I. Introduction

Many of the known transforming characteristics of Ras appear to be the result of perturbations in normal cell cycle control. Mammalian systems have evolved an elaborate program for integrating signals and regulating specific molecular events governing progression through the cell cycle (Fig. 1). The cyclin-dependent kinase (CDK) holoenzymes consist of a catalytic subunit and a regulatory or cyclin subunit. Typically the abundance of the cyclin subunit is rate limiting and highly regulated within the mammalian cell. Phosphorylation of the cyclin–CDK heterodimeric complex by a CDK-activating kinase (CAK) activates the holoenzyme. Activated cyclin–CDK heterodimers promote G_1 to S-phase progression. Cyclin–CDK-mediated phosphorylation of the retinoblastoma tumor suppressor protein (pRB) is essential for passage through the G_1 restriction point.[1–4] CDK4 and CDK6 preferentially bind to the D-type cyclins. The D-type cyclins interact with pRB and the pRB-related proteins, p107 and p130. The immune precipitation assays and synthetic glutathione *S*-transferase (GST) pocket protein substrates described in this chapter are used to assess the activity of these holoenzyme.

[1] C. J. Sherr, *Science* **274,** 1672 (1996).
[2] R. A. Weinberg, *Cell* **81,** 323 (1995).
[3] T. Motokura and A. Arnold, *Curr. Opin. Genet. Dev.* **3,** 5 (1993).
[4] T. Hunter, *Curr. Opin. Genet. Dev.* **3,** 1 (1993).

0076-6879/00 $35.00

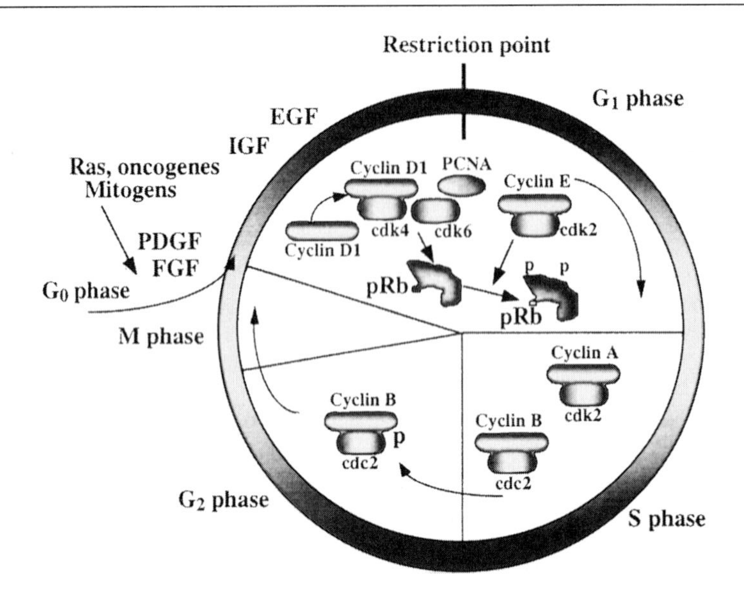

FIG. 1. The normal mammalian cell cycle involves sequential progression through the G_1, S, G_2, and M phases. A restriction point is proposed in the G_1 phase. After passage through the restriction point the cell is committed to completion of the cell cycle in the absence of serum.

Cyclin D1–CDK4/6 activity is maximal early in G_1 phase and the cyclin E–CDK2 complex is maximally active in late G_1/S phase. Overexpression of either cyclin D_1 or cyclin E decreases the time it takes the cell to complete the G_1 phase and enter S phase. Cyclin D1 and cyclin E appear to subserve at least partially overlapping functions with distinct target substrate specificities, as cyclin D1 induction of cell cycle progression appears to be pRB dependent, whereas cyclin E function is not. In cultured Rat-1 cells, cyclin D1, but not cyclin E, induced pRB phosphorylation.[5] Although pRB can be phosphorylated at an overlapping subset of sites by these two distinct kinases[6] and cyclin E binds and phosphorylates pRB in NIH 3T3 cells,[7] cyclin E enhancement of G_1 phase progression can occur independently of pRB.[8,9] Alternate cyclin E–CDK substrates were therefore sorted and identified, including NPAT protein (nuclear protein mapped to the *ATM* locus).[10] It is thought that phosphorylation of NPAT by cyclin E–CDK2 may promote S-phase

[5] D. Resnitzky and S. I. Reed, *Mol. Cell. Biol.* **15,** 3463 (1995).

[6] S. A. Ezhevsky, H. Nagahara, A. M. Vocero-Akbani, D. R. Gius, M. C. Wei, and S. F. Dowdy, *Proc. Natl. Acad. Sci. U.S.A.* **94,** 10699 (1997).

[7] B. L. Kelly, K. G. Wolfe, and J. M. Roberts, *Proc. Natl. Acad. Sci. U.S.A.* **95,** 2535 (1998).

[8] J. Lukas, T. Herzinger, K. Hansen, M. C. Moroni, D. Resnitzky, K. Helin, S. I. Reed, and J. Bartek, *Genes Dev.* **11,** 1479 (1997).

[9] D. Resnitzky, M. Gossen, H. Bujard, and S. I. Reed, *Mol. Cell. Biol.* **14,** 1669 (1994).

[10] J. Zhao, B. Dynlacht, T. Imai, T. Hori, and E. Harlow, *Genes Dev.* **12,** 456 (1998).

entry. Currently histone H1 is used as a convenient substrate for cyclin E–CDK2 *in vitro* kinase assays because pRB is a relatively poor substrate.

The activity of the cyclin–CDK holoenzyme is determined both by abundance of the protein subunits and by other regulatory proteins including CAK, the cyclin-dependent kinase inhibitors (CKI), and the CDK phosphatases (Cdc25). CAK, which is itself composed of CDK7 (MO15 protein) and cyclin H, phosphorylates the cyclin D1–CDK complex bound to pRB and is required for full kinase activity of cyclin D1.[1,11] The two CKI families, Cip/Kip and INK4, have previously been shown to inactivate the cyclin–CDK holoenzyme complexes.[1,2,12,13] The INK4a/ARF (ADP ribosylation factor) locus encodes two distinct gene products, p16INK4a and p19ARF, which are modulators of the pRB and p53 pathways, respectively.[14] Members of the Cip/Kip family include $p21^{Cip1}$, $p27^{Kip1}$, and $p57^{Kip2}$, whereas $p16^{INK4a}$, $p15^{INK4b}$, $p18^{INK4c}$, and $p19^{INK4d}$ comprise the INK4 inhibitor group. The INK4 CKIs specifically inhibit the catalytic domains of CDK4 and CDK6. The broader acting Cip/Kip "inhibitors" can function as either inhibitors or activators of CDK function. Historically, the Cip/Kip inhibitors were considered universal inhibitors that at high concentrations/signal intensities blocked cyclin D-, E-, and A-dependent kinase activities.[15,16] More recently the "activator" function of the Cip/Kip proteins has been examined. In immunodepletion studies $p21^{Cip1}$ was shown to function in a dose-dependent manner, inhibiting CDK activity only at high concentrations.[17] In $p21^{-/-}$, $p27^{-/-}$, or $p21^{-/-}/p27^{-/-}$ mouse embryo fibroblasts (MEFs), cyclin D–CDK complex assembly was impaired, resulting in a reduced cyclin D1–kinase activity that was restored by the addition of $p21^{Cip1}$ and/or $p27^{Kip1}$.[16] The cyclin–CDK complexes are also influenced by the Cdc25A phosphatase, which increases the activity of the cyclin D1–CDK4 holoenzyme.[18,19] The Cdc25 phosphatases can function as protooncogenes in transformation assays and activate the CDKs by removing their inhibitory phosphorylation of tyrosine and threonine residues.[20,21]

[11] H. Matsushime, D. E. Quelle, S. A. Shurtleff, M. Shibuya, C. J. Sherr, and J.-Y. Kato, *Mol. Cell. Biol.* **14,** 2066 (1994).

[12] A. T. Reutens, G. Watanabe, G. E. Shambaugh III, and R. G. Pestell, *Einstein Q. J. Biol. Med.* **14,** 3 (1997).

[13] C. J. Sherr and J. M. Roberts, *Genes Dev.* **9,** 1149 (1995).

[14] C. J. Sherr, *Genes Dev.* **12,** 2984 (1998).

[15] C. J. Sherr and J. M. Roberts, *Gene Dev.* **13,** 1501 (1999).

[16] M. Cheng, P. Olivier, J. A. Diehl, M. Fero, M. F. Roussel, J. M. Roberts, and C. J. Sherr, *EMBO J.* **18,** 1571 (1999).

[17] J. LaBaer, M. D. Garrett, L. F. Stevenson, J. M. Slingerland, C. Sandhu, H. S. Chou, A. Fattaey, and E. Harlow, *Genes Dev.* **11,** 847 (1997).

[18] K. Galaktionov, X. Chen, and D. Beach, *Nature (London)* **382,** 511 (1996).

[19] K. Galaktionov, C. Jessus, and D. Beach, *Genes Dev.* **9,** 1046 (1995).

[20] K. Galaktionov, A. K. Lee, J. Eckstein, G. Draetta, J. Meckler, M. Loda, and D. Beach, *Science* **269,** 1575 (1995).

[21] S. Jinno, K. Suto, A. Nagata, M. Igarashi, Y. Kanaoka, H. Nojima, and H. Okayama, *EMBO J.* **13,** 1549 (1994).

Ras signaling induces both growth inhibitory effects, which resemble cellular senescence,[22] and mitogenic, antiapoptotic, and cooperative transforming functions.[23] The effects of Ras on cellular components that contribute to CDK activity are complex and the reader is referred to reviews on this topic.[23–26] In brief, Ras induces cyclin D1 mRNA and promoter activity.[27–29] The abundance of cyclin D1 is rate limiting in Ras transformation in cultured cells[30] and in the skin, using transgenic mice.[31] The effects of Ras signals on CDK activity are in part determined by the intensity of the Ras signal, with sustained low-level induction inducing proliferation.[32] High-intensity signals induce p53[33] and $p21^{Cip1}$, inhibiting CDK activity,[34] and resulting in cell cycle arrest and premature cell senescence. In proliferating mouse fibroblasts expressing a high-intensity Raf-1 signal, $p21^{Cip1}$ was induced. Both $p21^{Cip1}$ and $p27^{Kip1}$ maintained their associations with CDK2/4 and cyclin D1 as the cell underwent cell cycle arrest.[25,31] In contrast with the induction of $p21^{Cip1}$, Raf-1 signaling repressed $p27^{Kip1}$ abundance and activity,[35] and Ras-induced ERKK (extracellular signal-regulated kinase kinase) activity directly phosphorylated $p27^{Kip1}$, abrogating its binding to cyclin–CDK complexes.[36] The cooperative functions of Ras are species specific. In rodent cells Ras induces focus formation without the aid of a cooperating oncogene within monolayers of either INK4a/ARF ($INKa^{-/-}$),[37] $p19ARF^{-/-}$,[38] or $p53^{-/-}$ mouse fibroblasts.[37] In human

[22] M. Serrano, A. W. Lin, M. E. McCurrach, D. Beach, and S. W. Lowe, *Cell* **88,** 593 (1997).

[23] A. C. Lloyd, *Curr. Opin. Genet. Dev.* **8,** 43 (1998).

[24] R. G. Pestell, C. Albanese, A. T. Reutens, J. E. Segall, R. J. Lee, and A. Arnold, *Endocr. Rev.* **20,** 501 (1999).

[25] E. Kerkhoff and U. R. Rapp, *Oncogene* **17,** 1457 (1998).

[26] X. Grana and E. P. Reddy, *Oncogene* **11,** 211 (1995).

[27] J. N. Lavoie, G. L'Allemain, A. Brunet, R. Müller, and J. Pouysségur, *J. Biol. Chem.* **271,** 20608 (1996).

[28] C. Albanese, J. Johnson, G. Watanabe, N. Eklund, D. Vu, A. Arnold, and R. G. Pestell, *J. Biol. Chem.* **270,** 23589 (1995).

[29] J. Filmus, A. I. Robles, W. Shi, M. J. Wong, L. L. Colombo, and C. J. Conti, *Oncogene* **9,** 3627 (1994).

[30] J.-J. Liu, J.-R. Chao, M.-C. Jiang, S.-Y. Ng, J. J. -Y. Yen, and H.-F. Yang-Yen, *Mol. Cell. Biol.* **15,** 3654 (1995).

[31] A. I. Robles, M. L. Rodriguez-Puebla, A. B. Glick, C. Trempus, L. Hansen, P. Sicinski, R. W. Tennant, R. A. Weinbergh, S. H. Yuspa, and C. J. Conti, *Genes Dev.* **12,** 2469 (1998).

[32] J. T. Winston, S. R. Coats, Y.-Z. Wang, and W. J. Pledger, *Oncogene* **12,** 127 (1996).

[33] A. W. Lin, M. Barradas, J. C. Stone, L. van Aelst, M. Serrano, and S. W. Lowe, *Genes Dev.* **12,** 3008 (1998).

[34] A. Sewing, B. Wiseman, A. C. Lloyd, and H. Landt, *Mol. Cell. Biol.* **17,** 5588 (1997).

[35] H. Aktas, H. Cai, and G. M. Cooper, *Mol. Cell. Biol.* **17,** 3850 (1997).

[36] M. Kawada, S. Yamagoe, Y. Murakami, K. Suzuki, S. Mizuno, and Y. Uehara, *Oncogene* **15,** 629 (1997).

[37] M. Serrano, H.-W. Lee, L. Chin, C. Cordon-Cardo, D. Beach, and R. A. DePinho, *Cell* **85,** 27 (1996).

[38] T. Kamijo, F. Zindy, M. F. Roussel, D. E. Quelle, J. R. Downing, R. A. Ashmun, G. Grosveld, and C. J. Sherr, *Cell* **91,** 649 (1997).

fibroblasts, however, inactivation of both the pRB/p16 and p53 pathways is required to overcome growth inhibition by Ras.[22] The Cdc25A-phosphatase is activated by Raf-1 kinase,[19] raising the possibility that Cdc25A may contribute to the induction of CDK activity by Ras. Collaborating oncogenesis may be mediated both through the accumulation of increased levels of CDK activity[39] and through the induction of distinct high molecular weight combinatorial interactions between the CDKs and as yet to be fully characterized protein complexes.[40]

The assays described herein are commonly used to assess cyclin D1–CDK4/6, cyclin E–CDK2, and CAK activity. Briefly, cyclin–CDK holoenzymes are immunoprecipitated from whole cell extracts onto agarose beads. The holoenzymes are activated and the kinase reaction is initiated with the appropriate purified substrate and [γ-^{32}P]ATP. Products of the kinase reaction are separated by sodium dodecyl sulfate–polyacrylamide gel electrophoresis (SDS–PAGE), and by the relative levels of substrate phosphorylation the kinase activity is assessed. There are several important caveats to the utility of CDK assays in assessing Ras transforming function. The substrates used in these assays (pRB, pRB-related pocket proteins, and histone H1) may reflect only one of several targets of CDK action perturbed by Ras. Thus, although cyclin D1 is not required for G_1-phase progression in mammalian cells that lack pRB, suggesting a critical role for pRB phosphorylation in G_1-phase progression,[41] pRB binding-independent domains of cyclin D1 that collaborate in cellular transformation have been identified.[42] As cyclin D1 is capable of binding several other proteins, including the estrogen receptor,[43] coactivator proteins,[44] and Myb-related proteins,[45] it is formally possible that these interactions may contribute to the transforming properties of cyclin D1 in a cell type-specific manner.[44,46] Second, the CDKs may directly phosphorylate and thereby regulate the activity of several different transcription factors such as B-Myb, E2Fs, estrogen receptor, and components of the basal transcription apparatus.[24] The role of these non-pRB substrates in CDK-regulated cellular transformation is poorly understood. Analysis of these additional protein–protein interactions may provide complementary insights into the transforming properties of Ras.[45]

[39] G. Leone, J. Degregori, R. Sears, L. Jakoi, and J. R. Nevins, *Nature* (*London*) **387,** 422 (1997).

[40] I. Perez-Roger, S.-H. Kim, B. Griffiths, A. Sewing, and H. Land, *EMBO J.* **18,** 5310 (1999).

[41] J. Lukas, J. Bartkova, M. Rohde, M. Strauss, and J. Bartek, *Mol. Cell. Biol.* **15,** 2600 (1995).

[42] J. Zwicker, S. Brusselbach, K. U. Jooss, A. Sewing, M. Behn, F. C. Lucibello, and R. Muller, *Oncogene* **18,** 19 (1999).

[43] E. Neuman, M. H. Ladha, N. Lin, T. M. Upton, S. J. Miller, J. DiRenzo, R. G. Pestell, P. W. Hinds, S. F. Dowdy, M. Brown, and M. E. Ewen, *Mol. Cell. Biol.* **17,** 5338 (1997).

[44] R. M. L. Zwijsen, R. S. Buckle, E. M. Hijmans, C. J. M. Loomans, and R. Bernards, *Genes Dev.* **12,** 3488 (1998).

[45] H. Hirai and C. Sherr, *Mol. Cell. Biol.* **16,** 6457 (1996).

[46] R. M. L. Zwijsen, E. Wientjens, R. Klompmaker, J. van der Sman, R. Bernards, and R. J. A. M. Michalides, *Cell* **88,** 405 (1997).

II. Materials

A. Preparation of Cell Extracts

Cultured cells

Phosphate-buffered saline (PBS), per liter: 8 g of NaCl, 0.2 g of KCl, 1.44 g of Na_2HPO_4, 0.24 g of KH_2PO_4. Adjust to pH 7.4 with HCl.[47] Store at room temperature. Prior to use, chill on ice

Lysis buffer: 50 m*M* HEPES [N-(2-hydroxyethyl) piperazine-*N*′-(2-ethanesulfonic acid), pH 7.2], 150 m*M* NaCl, 1 m*M* EGTA [ethylene glycol-bis(β-aminoethyl ether)-*N,N,N*′,*N*′-tetraacetic acid], 1 m*M* EDTA (ethylenediaminetetraacetic acid). Store at room temperature. Prior to use, aliquot the appropriate volume and chill on ice. Just prior to use, add polyoxythylene sorbitan monolaurate [Tween 20; to a final concentration of 0.1% (v/v)], dithiothreitol (DTT; 1 m*M* final), sodium orthovanadate (0.1 m*M* final), leupeptin (2.5 μg/ml final), and phenylmethylsulfonyl fluoride (PMSF; 0.1 m*M* final).[11,41,48–50,*]

Protein quantitation reagent: e.g., Bradford assay reagent [Bio-Rad (Hercules, CA) protein assay reagent]

Cell scraper

B. Preparation of Kinase Substrates

Vector encoding GST–pRB fusion substrate (pGEX-RB; for cyclin D1-dependent kinase assays[51,52,**] or GST–CDK2 (for CAK assays)

Competent bacteria (e.g., *Escherichia coli* DH5α)

Bacterial growth medium: Luria–Bertani medium [LB: 10 g of Bacto-Tryptone (Difco, Detroit, MI), 5 g of Bacto-Yeast extract, 10 g of NaCl in

[47] T. Sambrook, E. R. Fritsch, and T. Maniatis, "Molecular Cloning: A Laboratory Manual," 2nd Ed. Cold Spring Harbor Laboratory Press, Cold Spring Harber, New York, 1989.

[48] S. T. Eblen, M. P. Fautsch, R. A. Anders, and E. B. Leof, *Cell. Growth. Differ.* **6,** 915 (1995).

[49] E. A. Musgrove, A. Swarbrick, C. S. L. Lee, A. L. Cornish, and R. L. Sutherland, *Mol. Cell. Biol.* **18,** 1812 (1998).

[50] O. W. J. Prall, B. Sarcevic, E. A. Musgrove, C. K. W. Watts, and R. L. Sutherland, *J. Biol. Chem.* **272,** 10882 (1997).

[*] Other groups have reported the use of similar lysis buffer [50 m*M* HEPES (pH 7.5), 150 m*M* NaCl, 1 m*M* EDTA, 2.5 m*M* EGTA, 1 m*M* DTT, 0.1% (v/v) Tween 20, 10% (v/v) glycerol, 0.1 m*M* PMSF, leupeptin (10 μg/ml), aprotinin (10 μg/ml), 10 m*M* β-glycerophosphate, 1 m*M* NaF, and 0.1 m*M* sodium orthovanadate]. Note also that the inhibitory activity and $p27^{Kip1}$ binding *in vitro* depend on incubation of extracts at physiological temperature or in the presence of a reducing agent.[48] Reported kinase buffers also differ [e.g., 50 m*M* HEPES (pH 7.5), 1 m*M* DTT, 2.5 m*M* EGTA, 10 m*M* $MgCl_2$, 0.1 m*M* sodium orthovanadate, 1 m*M* NaF, 10 m*M* β-glycerophosphate].[11,41,49,50] In our hands, our buffer conditions for lysis and kinase assay performed similarly or better for cyclin D1- and cyclin E-dependent kinase assays.

1 liter, pH 7.0[47]] or 2 × YT (16 g of Bacto-Tryptone, 10 g of Bacto-Yeast extract, 5 g of NaCl in 1 liter, pH 7.0)
Ampicillin
Isopropyl-β-D-thiogalactopyranoside (IPTG)
NETN buffer: 120 m*M* NaCl, 1 mM EDTA, 50 m*M* Tris-HCl (pH 7.5), and 0.5% (v/v)
Nonidet P-40 (NP-40), with pepstatin (1 μg/ml), leupeptin (1 μg/ml), aprotinin (1 μg/ml), 1 m*M* PMSF
Glutathione–Sepharose 4B (Pharmacia Biotech, Piscataway, NJ)
Glutathione (Sigma, St. Louis, MO).

C. *Immunoprecipitation*

1. *Binding of Antibody to Agarose*

Protein A–agarose (Boehringer Mannheim, Indianapolis, IN)
Lysis buffer
Antibodies to cyclin D1 (clone DCS-11; NeoMarkers, Fremont, CA), cyclin E (clone M-20; Santa Cruz Biotechnologies, Santa Cruz, CA), and the CDK7 component of CAK (Upstate Biotechnology, Lake Placid, NY)[†]
Shaking/rotating platform

2. *Precipitation of Cyclin Complexes*

Lysis buffer
Cell extracts
Shaking/rotating platform

D. *Cyclin-Dependent Kinase Assay*

Lysis buffer

[51] M. E. Ewen, H. K. Sluss, C. J. Sherr, H. Matsushime, J.-Y. Kato, and D. M. Livingston, *Cell* **73,** 487 (1993).

[52] M. Kitagawa, H. Higashi, H.-K. Jung, I. Suzuki Takahashi, M. Ikeda, K. Tamai, J.-Y. Kato, K. Segawa, E. Yoshida, S. Nishimura, and Y. Taya, *EMBO J.* **15,** 7060 (1996).

** Our pRB C-terminal construct contains amino acids 773 to 928, encoded in the pGEX30X plasmid.[51] Serine residue 780 of pRB has been shown to be phosphorylated in a cell cycle-dependent manner specifically by cyclin D1–CDK4, and not by cyclin E–CDK2 or cyclin A–CDK2. Because cyclin–CDK holoenzymes have different consensus motifs for phosphorylation, it is therefore essential for the cyclin D1-dependent kinase assay that the pRB substrate include serine residue 780.[52]

† For these assays, in which immunoprecipitation of a holoenzyme complex is required, selection of precipitating antibody is important. Antibodies that may target cyclin–CDK interaction domains or that may otherwise disrupt the holoenzyme are therefore not useful. The antibodies described here have been shown to be effective for immunoprecipitation of active kinase complexes.

Kinase assay buffer: 50 m*M* HEPES (pH 7.0), 10 m*M* $MgCl_2$, 5 m*M* $MnCl_2$, 1 m*M* DTT.[53] Prepare fresh, before use; chill on ice

ATP

Substrates for cyclin D1-dependent kinase assay (GST–pRB) or cyclin E-dependent kinase assay (histone H1; Boehringer Mannheim)

[γ-^{32}P]ATP (6000 Ci/mmol, 370 MBq/ml, 10 mCi/ml)

E. Sodium Dodecyl Sulfate–Polyacrylamide Gel Electrophoresis

SDS dye/sample buffer (6×): 0.35 *M* Tris-HCl (pH 6.8), 10.28% (w/v) SDS, 36% (v/v) glycerol, 0.6 *M* DTT, and 0.012% (w/v) bromphenol blue

Molecular weight markers (Rainbow markers; Amersham, Arlington Heights, IL)

SDS–12% (w/v) polyacrylamide gel (prepared according to Ref. 47)

SDS running buffer (prepared according to Ref. 47)

Whatman paper or chromatography paper (Fisher, Fairlawn, NJ)

Plastic wrap

F. CDK-Activating Kinase Assays

Vector encoding GST–CDK2

CAK buffer: 50 m*M* HEPES (pH 7.5), 30 m*M* $MgCl_2$, 1 m*M* DTT, 90 μ*M* ATP

III. Methods

A. Preparation of Cell Extracts

To prepare extracts suitable for assays of cyclin-dependent kinase activity, it is critical that the cyclin/cyclin-dependent kinase complex be maintained as an intact holoenzyme; therefore, the lysis buffer used has only 0.1% (v/v) Tween 20. To ensure that at each time point there is enough protein for the kinase assay, it is recommended that at least one 100-mm plate at 50–70% confluence be used.[‡] Typically, 1×10^6 cells should suffice. Larger plates or greater numbers of plates may be used.

1. Aspirate culture medium from the cells.

2. Rinse each plate with 5 ml of cold phosphate-buffered saline (PBS), and aspirate the PBS wash (for loosely adherent cells it is important to avoid dislodging the cells while washing with PBS). Repeat.

[53] M. Meyerson and E. Harlow, *Mol. Cell. Biol.* **14,** 2077 (1994).

[‡] Cell confluency should be carefully monitored for these assays. The effect of a mitogenic stimulus can be optimally measured if the cells are allowed the maximal capacity for growth. Because cells may arrest on contact with neighboring cells, overconfluent cultures may affect the kinase activity of these G_1-phase holoenzymes.

3. Add 5 ml of PBS and scrape the plate until all cells have been detached. Transfer the cells to a 15-ml conical tube. Add an additional 5 ml of PBS to rinse the plate further; transfer to the 15-ml tube.

4. Pellet the cells by centrifugation at 1500 *g* for 5 min at 4°.

5. Remove all but 1 ml of the supernatant PBS.

6. Resuspend the cell pellet; transfer to 1.5-ml capped tubes.

7. Centrifuge at 15,000 rpm for 1 min at 4°.

8. Remove the supernatant.

9. Resuspend the pellet in an equivalent volume of cold lysis buffer. Once the cell pellet has been resuspended, pipette the suspension 10–15 times to ensure complete disruption of the cell pellet.

10. Freeze the suspension at −80° or on dry ice. The lysates may be kept frozen at −80° for long-term storage. The freeze–thaw process is essential for complete extraction.

11. Thaw the lysates on ice. Centrifuge the lysates at 10,000 *g* for 10 min at 4°.

12. Quantify the lysate protein concentration by the Bradford (or other) assay, as per the manufacturer protocol.

B. Preparation of Kinase Substrates

1. Transform the GST fusion substrate vector into appropriate competent host *E. coli* (e.g., DH5α), using standard techniques. Grow the bacteria at 37° in LB broth or 2× YT broth supplemented with ampicillin (50 μg/ml), until the optical density at 595 nm is 0.5.

2. Induce expression of the fusion protein by addition of IPTG at a final concentration of 0.2 m*M*. Incubate with shaking for 4 hr at 30°.

3. Recover the cells by centrifugation (5000 *g* for 10 min) at 4°.

4. Lyse the cells on ice by sonication in a 1/10 volume of NETN buffer. Centrifuge at 40,000 rpm for 15 min at 4° to clear the lysates.

5. Mix the lysates with glutathione–Sepharose 4B (Pharmacia Biotech), that has been equilibrated in NETN buffer and incubate for 2 hr at 4°.

6. Wash the beads three times with NETN buffer and twice with kinase buffer: centrifuge the beads to pellet, carefully remove supernatant, add buffer, and invert to mix.

7. The GST fusion protein is released at 4° by incubation in kinase buffer containing 2 m*M* reduced glutathione. Centrifuge to pellet the beads and save the supernatant. Protein concentration can be determined, as described above.

C. Immunoprecipitation

Purification of the cyclin–CDK holoenzyme requires two steps. First, the antibody must be bound to a solid support (agarose), via protein A. Second, the cyclin protein and its partner CDK are precipitated from the cell extracts, and cleared of other cellular proteins.

1. Binding of Antibody to Agarose

1. Completely suspend the protein A–agarose beads in the accompanying solution and remove a total amount of the suspension equivalent to 30 μl per sample (i.e., for five samples, remove 150 μl); place the entire volume of beads into one 1.5-ml capped microcentrifuge tube.
2. Centrifuge the beads at 15,000 rpm for 1 min at 4° to pellet. Carefully remove the supernatant.
3. Resuspend the beads in a total volume of 800 μl of lysis buffer: add buffer to 800 μl, recap the tubes, and invert to resuspend the beads. Centrifuge to pellet the beads, remove the supernatant, and repeat.
4. After twice washing the beads, add lysis buffer to 800 μl. Add 2 μl of antibody per sample to the supernatant. Resuspend the beads and incubate at 4° on a shaking or rotating platform for 1 hr.

2. Precipitation of Cyclin Complexes

1. Centrifuge the beads at 15,000 rpm for 2 min at 4°. Carefully remove the supernatant.
2. Divide the beads into the appropriate number (N) of 1.5-ml microcentrifuge tubes: add lysis buffer to the agarose beads to a total volume of $(N + 1) \times 100$ μl. For example, for seven samples, add enough lysis buffer to the beads so that the total volume is 800 μl. Resuspend the beads by inversion of the capped microcentrifuge tube. Aliquot 100 μl of beads to each of the N tubes.*
3. To each tube, add a separate cell lysate. For most assays, 100 μg of cell lysate is sufficient for detection of differences in cyclin-dependent kinase activity.
4. Normalize the total volume in each sample to 800 μl with lysis buffer.
5. Resuspend the bead pellet by inversion of the capped tubes, and incubate at 4° on a shaking or rotating platform for at least 6 hr to overnight.

* To ensure that samples are exposed to equivalent amounts of antibody, the volume of antibody-bound agarose beads must be equivalent. Therefore, proper resuspension and distribution of the agarose in step 2 of Section III,C,2 is essential. To check for even distribution of beads, the tubes may be centrifuged and examined for a gross estimation of equivalent volume of beads per tube. In all subsequent steps involving the agarose beads, including washing of immunoprecipitates, it is also critical that pipette tips do not disturb pelleted beads and that care is taken when supernatants are removed. For washes of immunoprecipitates in Section III,D, steps 1 and 2, disruption of the pellet can be avoided by careful removal of the supernatant, leaving approximately 40–50 μl above the pellet. After the second wash in step 2 of Section III,D, centrifuge the agarose again at 15,000 rpm for 1 min, and carefully remove as much of the supernatant as possible without disruption of the pellet.

D. Cyclin-Dependent Kinase Assay

1. Centrifuge the samples at 15,000 rpm for 1 min at 4°. Carefully remove the supernatant, leaving the agarose beads undisturbed. Wash the beads with 800 μl of lysis buffer. Repeat.

2. Centrifuge and wash the sample beads twice with kinase assay buffer. Carefully remove the supernatant from the beads again.

3. Resuspend the beads in 40 μl of kinase assay buffer containing 20 μ*M* ATP, 5 μCi of [γ-^{32}P]ATP, and the protein substrate for either cyclin D1-dependent kinase (2 μg of soluble glutathione *S*-transferase–pRB fusion protein) or for cyclin E-dependent kinase (2 μg of histone H1). Vortex briefly for 5 sec.

4. Incubate at 30° for 20 min. Resuspend the beads during the 20-min incubation by gentle vortexing, after 5 and 12 min of incubation at 30°.

E. Sodium Dodecyl Sulfate–Polyacrylamide Gel Electrophoresis

1. After brief centrifugation, add 10 μl of 6 × SDS dye to each sample. Mix/resuspend by gentle vortexing.

2. Denature the samples by boiling for 5 min.

3. Centrifuge the samples at 15,000 rpm for 3 min at room temperature.

4. Without disturbing the beads, carefully remove 45 μl of each supernatant and place into fresh microcentrifuge tubes.[**]

5. Load 25 μl of each sample into each well of the SDS-polyacrylamide gel. Include one lane of molecular weight markers (4 μl of Rainbow markers, 4 μl of 6 × SDS dye, 17 μl of kinase assay buffer).

6. Run the gel until the dye front nears the bottom of the gel.[†]

7. Once finished, use a razor blade to carefully cut off and dispose of the gel at the dye front. This region of the gel contains the free [γ-^{32}P]ATP, which will increase the background radioactivity of the gel. The phosphorylated substrates should be well above the level of the dye, and use of prestained molecular weight markers will aid in identification and assignment of phosphorylated proteins.

8. Place the gel on chromatography paper by adhering the gel to the paper and peeling the gel from the glass plate. Use three sheets of paper that are several centimeters larger in each dimension than the gel. Cover with plastic wrap. Dry on a gel drying device, at 80° for 1 hr.

[**] Transfer of the supernatant to fresh microcentrifuge tubes is important to ensure that equivalent volumes of supernatant are loaded in the SDS–polyacrylamide gel. If the supernatant is removed from the tubes with agarose, agarose may plug the pipette tip and prevent loading of the appropriate volume. Because this assay measures *relative* kinase activity, it is important to load equivalent volumes on the gel.

[†] For optimal resolution of the phosphorylated substrate, we use 16 × 16 cm large gels. After the samples have entered the stacking layer of the gel, carefully remove the top chamber buffer to reduce background radioactivity. Replace with fresh running buffer.

9. Phosphorylated proteins can be visualized by autoradiography of the dried gels and quantitation can be performed by densitometry. Phosphorylated proteins can also be visualized by phosphoimaging analysis.

F. CDK-Activating Kinase Assays

The protocol is derived from the method of Musgrove *et al.*[49] Preparation of GST–CDK2 substrate follows that for preparation of the GST–pRB substrate in Section III,B. Most steps are similar to those in the cyclin-dependent kinase assay, with modifications of buffers.

1. Follow steps through Section III,C for immunoprecipitation of CAK. Proceed with steps 1 and 2 of Section III,D, using the CAK buffer described in Section II,F instead of the kinase assay buffer in step 2 of Section III,D.
2. Incubate the immunoprecipitated CAK in 30 μl of CAK buffer with 10 μCi of [γ-^{32}P]ATP and 10 μg of GST–CDK2, at 30° for 20 min. Terminate the reactions with SDS–PAGE sample buffer and perform SDS–PAGE according to Section III,E.

Acknowledgments

This work was supported in part by grants from the NIH (R01CA70897, RO1CA75503), and by awards from the Susan G. Komen Breast Cancer Foundation, Irma T. Hirschl, and the Pfeiffer Foundation (R.G.P.). Work conducted at the Albert Einstein College of Medicine was supported by Cancer Center Core National Institute of Health grant 5-P30-CA13330-26.

[14] STAT Proteins: Signal Tranducers and Activators of Transcription

By JACQUELINE BROMBERG and XIAOMIN CHEN

STAT (Signal transducer and activator of transcription) proteins are latent transcription factors that become activated by phosphorylation on a single tyrosine (at about residue 700 in each protein), typically in response to extracellular ligands.[1,2] Virtually every cytokine and growth factor (polypeptide ligands) can cause STAT tyrosine phosphorylation through either cytokine receptors plus associated janus kinase (JAK) kinases or growth factors [e.g., epidermal growth factor (EGF),

[1] J. E. Darnell, Jr., *Science* **277,** 1630 (1997).

[2] G. R. Stark, I. M. Kerr, B. R. Williams, R. H. Silverman, and R. D. Schreiber, *Annu. Rev. Biochem.* **67,** 227 (1998).

0076-6879/00 $35.00

platelet-derived growth factor (PDGF), colony-stimulating factor 1 (CSF-1)] acting through intrinsic receptor tyrosine kinases, clearly implicating STAT activation in numerous biological events; and indeed knockout experiments confirm their importance in cytokine and growth factor signaling (Table I[3–15]).

An active STAT dimer is formed via the reciprocal interactions between the Src homology 2 (SH2) domain of one monomer and the phosphorylated tyrosine of the other.[16] The dimers accumulate in the nucleus, recognize specific DNA elements, and activate transcription. The STAT proteins are subsequently inactivated by tyrosine dephosphorylation and return to the cytoplasm.[17,18] In addition to tyrosine phosphorylation, STATs 1, 3, 4, and 5 are serine phosphorylated within their transcription-activating domain (1).[19,20] Serine phosphorylation is critical for the transcription-activating capacity of STATs 1 and 3.[21] In addition, STATs 1, 3, 4, and 5 exist in two forms (a long α form and a shorter β form), which occur as a result of alternative splicing in the cases of STATs 1, 3, and 4. Protein processing

[3] J. E. Durbin, R. Hackenmiller, M. C. Simon, and D. E. Levy, *Cell* **84,** 443 (1996).

[4] M. A. Meraz, J. M. White, K. C. Sheehan, E. A. Bach, S. J. Rodig, A. S. Dighe, D. H. Kaplan, J. K. Riley, A. C. Greenlund, D. Campbell, K. Carver-Moore, R. N. Du Bois, R. Clark, M. Aguer, and R. D. Schreiber, *Cell* **84,** 431 (1996).

[5] D. H. Kaplan, V. Shankaran, A. S. Dighe, E. Stockert, M. Aguer, L. J. Old, and R. D. Schreiber, *Proc. Natl. Acad. Sci. U.S.A.* **95,** 7556 (1998).

[6] C. Schindler, (1999).

[7] K. Takeda, K. Noguchi, W. Shi, T. Tanaka, M. Matsumoto, N. Yoshida, T. Kishimoto, and S. Akira, *Proc. Natl. Acad. Sci. U.S.A.* **94,** 3801 (1997).

[8] K. Takeda, T. Kaisho, N. Yoshida, J. Takeda, T. Kishimoto, and S. Akira, *J.Immunol.* **161,** 4652 (1998).

[9] R. S. Chapman, P. C. Lourenco, E. Tonner, D. J. Flint, S. Selbert, K. Takeda, S. Akira, A. R. Clark, and C. J. Watson, *Genes Dev.* **13,** 2604 (1999).

[10] S. Sano, S. Hani, K. Takeda, M. Tarutani, Y. Yamaguchi, H. Miura, K. Yoshikawa, S. Akira, and J. Takeda, *EMBO J.* 4657 (1999).

[11] W. E. Thierfelder, J. M. van Deursen, K. Yamamoto, R. A. Tripp, S. R. Sarawar, R. T. Carson, M. Y. Sangster, D. A. Vignali, P. C. Doherty, G. C. Grosveld, and J. N. Ihle, *Nature (London)* **382,** 171 (1996).

[12] X. Liu, G. W. Robinson, K. U. Wagner, L. Garrett, A. Wynshaw-Boris, and L. Hennighausen, *Genes Dev.* **11,** 179 (1997).

[13] S. Teglund, C. McKay, E. Schuetz, J. M. van Deursen, D. Stravopodis, D. Wang, M. Brown, S. Bodner, G. Grosveld, and J. N. Ihle, *Cell* **93,** 841 (1998).

[14] G. B. Udy, R. P. Towers, R. G. Snell, R. J. Wilkins, S. H. Park, P. A. Ram, D. J. Waxman, and H. W. Davey, *Proc. Natl. Acad. Sci. U.S.A.* **94,** 7239 (1997).

[15] M. H. Kaplan, U. Schindler, S. T. Smiley, and M. J. Grusby, *Immunity* **4,** 313 (1996).

[16] X. Chen, U. Vinkemeier, Y. Zhao, D. Jeruzalmi, J. E. Darnell, Jr., and J. Kuriyan, *Cell* **93,** 827 (1998).

[17] R. L. Haspel, M. Salditt-Georgieff, and J. E. Darnell, Jr., *EMBO J.* **15,** 6262 (1996).

[18] R. L. Haspel and J. E. Darnell, Jr., *Proc. Natl. Acad. Sci. U.S.A.* **96,** 10188 (1999).

[19] K. C. Goh, S. J. Haque, and B. R. Williams, *EMBO J.* **18,** 5601 (1999).

[20] J. Turkson, T. Bowman, J. Adnane, Y. Zhang, J. Y. Djeu, M. Sekharam, D. A. Frank, L. B. Holzman, J. Wu, S. Sebti, and R. Jove, *Mol. Cell. Biol.* **19,** 7519 (1999).

[21] Z. Wen, Z. Zhong, and J. E. Darnell, Jr., *Cell* **82,** 241 (1995).

TABLE I

STAT PROTEINS, PHENOTYPE OF STAT-DEFICIENT MICE, AND LIGANDS TNAT CAN LEAD TO STAT ACTIVATION

	STAT1	STAT2	STAT3	STAT4	STAT5	STAT6
Phenotype of STAT-deficient mice	No innate response to viral or bacterial infection[a,b]; increased propensity to tumor formation[c]	Defective in IFN-α signaling[d]	Embryonic Lethal[e]; conditional knockout leads to changes in regulation of apoptosis and cell migration[f–h]	Defective Th1 development[i]	Defective lobuloalveolar development in the breast[j,k]; sexual dimorphism of body growth rates[k,l]	Defective Th2 development[m]
Ligands leading to STAT activation[n]	IFN-γ (+++) IFN-α (+++) EGF (+++) PDGF (+++) FGF (+++) AngII (+++) v/c-Eyk (+) GH (+) CSF-1 (+) TPO (+) Abl (+) IL-6, -11, -10, LIF, OSM, CNTF (+)	IFN-α (+++) AngII (+)	EGF (+++) PDGF (+++) G-CSF (+++) IL-10 (+++) AngII (+++) Leptin (+++) Src (+++) Elk (+++) Eyk (++) Abl (++) IFN-α (++) IL-6, -11, LIF, OSM, CNTF (+++) TPO (++) IL-12, IL-7, IL-2, IL-9, Kit (+) Fps, Fes, Lck, Sis, Ros (++) HTLV-1 (+++)	IL-12 (+++) IFN-α (++)	IL-3 (+++) IL-5 (+++) GM-CSF (+++) PRO (+++) EPO (+++) Abl (+++) HTLV-1 (+++) c-Kit (++) EGF (++) ErbB4 (++) PDGF (++) IL-2, -7, -9 (+++)	IL-4 (+++) IL-13 (+++)

[a] Durbin *et al.* (1996).[3]
[b] Meraz *et al.* (1996).[4]
[c] Kaplan *et al.* (1998).[5]
[d] Schindler (1999).[6]
[e] Takeda *et al.* (1997).[7]
[f] Takeda *et al.* (1998).[8]
[g] Chapman *et al.* (1999).[9]
[h] Sano *et al.* (1999).[10]
[i] Thierfelder *et al.* (1996).[11]
[j] Liu *et al.* (1997).[12]
[k] Teglund *et al.* (1998).[13]
[l] Udy *et al.* (1997).[14]
[m] Kaplan *et al.* (1996).[15]
[n] Abbreviations: FGF, fibroblast growth factor; AngII, angiotensin II; GH, growth hormone; TPO, thrombopoietin; LIF, leukemia inhibitory factor; OSM, oncostatin M; CNTF, ciliary neurotrophic factor; PRO, prolactin.

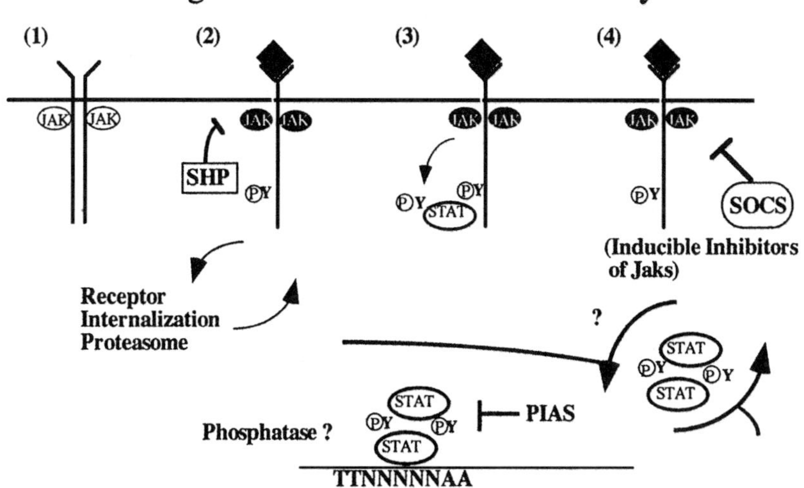

FIG. 1. An example of how STAT proteins are activated by nonreceptor tyrosine kinases, via JAK kinases; and potential molecules or pathways involved in downegulating this pathway.

may play a role in the formation of the α and β forms of STAT5.[22] The β forms of the STAT proteins lack transcription-activating domains and may be naturally occurring dominant-negative proteins.

The number of regulators of STAT activation are numerous and include the positive effectors of this pathway such as the tyrosine kinases (JAK kinases, intrinsic receptor tyrosine kinases, and nonreceptor tyrosine kinases including Src and Abl) and the more recently identified negative regulators. The tyrosine phosphatase directly responsible for dephosphorylating STAT has not been identified but is expected to be a crucial enzyme in regulating STAT activity.[18] Internalization of an STAT-associated receptor/kinase complex via the ubiquitin/proteasome pathway has been suggested to be responsible for downregulating STAT activity.[17] A family of cytokine-inducible inhibitors of the JAK kinases, SOC proteins, have been identified as critical inhibitors of STAT activation.[23] Another group of proteins known as PIAS (protein inhibitor of activated STAT) proteins function as specific inhibitors of the activated or dimeric form of the STAT molecule[24] (Fig. 1).

[22] M. Azam, C. Lee, I. Strehlow, and C. Schindler, *Immunity* **6,** 691 (1997).

[23] T. Naka, M. Narazaki, M. Hirata, T. Matsumoto, S. Minamoto, A. Aono, N. Nishimoto, T. Kajita, T. Taga, K. Yoshizaki, S. Akira, and T. Kishimoto, *Nature* (*London*) **387,** 924 (1997).

[24] C. D. Chung, J. Liao, B. Liu, X. Rao, P. Jay, P. Berta, and K. Shuai, *Science* **278,** 1803 (1997).

Detection of Activated STAT Proteins

Activated STAT proteins are in a dimeric form that requires tyrosine phosphorylation.[21,25,26] STATs are transiently activated by ligands. The time it takes to observe activated STAT protein is on the order of minutes after exposure to a ligand and usually peaks within 30 min, at which point the protein becomes tyrosine dephosphorylated and is subsequently "inactivated." STATs are persistently activated under circumstances in which a tyrosine kinase capable of mediating STAT phosphorylation is constitutively active. For example, in a number of cancerous cell lines or primary tumors kinases are persistently activated, such as the EGF receptor (EGFR) in squamous cell carcinomas, JAKs in multiple myeloma, and Bcr-Abl in chronic myelogenous leukemia.[27–29] Nevertheless, if the aim is to determine whether a particular growth factor can lead to STAT activation, a time course following the addition of the ligand is encouraged.

The most sensitive means of detecting activated STAT protein is by electrophoretic mobility shift assay (EMSA). Briefly, an extract containing tyrosine-phosphorylated STAT protein is incubated with a radiolabeled double-stranded binding site (probe) for a STAT protein and resolved on a nondenaturing polyacrylamide gel. Free probe migrates quickly, while the mobility of a protein-bound probe is inhibited. The following protocol is a basic one applicable to all STAT proteins. Table II describes ligands, cell types, and optimal binding sites[30] to use as appropriate positive controls for all the STAT proteins. Of course, many other cell types, ligands, binding sites, and antisera can be used, which can lead to the activation of various STATs. Table II is meant only to aid those who need good positive controls.

The relative mobilities of STAT dimers differ from one another by EMSA. In order of slowest to fastest moving complex (excluded are the β forms of the STAT proteins because they are not readily observed): STAT2/STAT1/p48 (ISGF3), STAT3 homodimer, STAT3 : STAT1 heteodimer, STAT5 homodimer, STAT4 homodimer, STAT6 homodimer, STAT1 homodimer.[30,31] See Fig. 2 for examples of STAT3 homodimer, STAT3 : STAT1 heterodimer, and STAT1 homodimer.

[25] K. Shuai, C. M. Horvath, L. H. Huang, S. A. Qureshi, D. Cowburn, and J. E. Darnell, Jr., *Cell* **76,** 821 (1994).

[26] X. Zhu, Z. Wen, L. Z. Xu, and J. E. Darnell, Jr., *Mol. Cell. Biol.* **17,** 6618 (1997).

[27] J. R. Grandis, S. D. Drenning, A. Chakraborty, M. Y. Zhou, Q. Zeng, A. S. Pitt, and D. J. Tweardy, *J. Clin. Invest.* **102,** 1385 (1998).

[28] R. Catlett-Falcone, T. H. Landowski, M. M. Oshiro, J. Turkson, A. Levitzki, R. Savino, G. Ciliberto, L. Moscinski, J. L. Fernandez-Luna, G. Nunez, W. S. Dalton, and R. Jove, *Immunity* **10,** 105 (1999).

[29] K. Shuai, J. Halpern, J. ten Hoeve, X. Rao, and C. L. Sawyers, *Oncogene* **13,** 247 (1996).

[30] H. M. Seidel, L. H. Milocco, P. Lamb, J. E. Darnell, Jr., R. B. Stein, and J. Rosen, *Proc. Natl. Acad. Sci. U.S.A.* **92,** 3041 (1995).

[31] N. G. Jacobson, S. J. Szabo, R. M. Weber-Nordt, Z. Zhong, R. D. Schreiber, J. E. Darnell, Jr., and K. M. Murphy, *J. Exp. Med.* **181,** 1755 (1995).

TABLE II
POSITIVE CONTROLS FOR STAT ACTIVATION

Parameter	STAT1	STAT2	STAT3	STAT4	STAT5	STAT6
Cell lines	NIH 3T3 Bud8[a]	Hela[a]	HepG2[a]	Human T cells (Percoll gradient centrifugation)	U-937[a]	U-937[a]
Ligands	IFN-γ (100 U/ml; murine and human[b])	IFN-α (200 U/ml; murine α[c], human α[d]	IL-6 (500 U/ml[b])	IL-12 (100 U/ml[d])	GM-CSF (5 ng/ml[d])	IL-4 (30 ng/ml[d])
Optimal binding sites	5′-GATCTGC TTCCCGTA ACGT-3′	5′-GATCCTC GGGAAAG GGAAACC GAAACTG AAGCC-3′	5′-GATCTG CTTCCCGT AACGT-3′	5′-GATCTGC TTTCCCAG AAACGT-3′	5′-GATCTGC TTCCTGGA ACGT-3′	5′-GATCTG CTTCCTGG AACGT-3′
Antisera	Westerns[e] IPs, and supershift[f]; phosphotyrosine for western[g]	Westerns, IP, and supershift[h];	Westerns[h] IPs, and Supershift[f]; Phosphotyrosine for Western[g]	Westerns[f,h] IPs, and supershift[f,h]	Westerns,[a] IPs, and supershift[f] Phosphotyrosine for Western[h]	Westerns,[f,h] IPs, and supershift[f,h]

[a] ATCC, American Type Culture Collection (Manassas, VA).
[b] Boehringer Mannheim (Indianapolis, IN).
[c] Sigma (St. Louis, MO).
[d] R&D Systems (Minneapolis, MN).
[e] Transduction Laboratories (Lexington, KY).
[f] Santa Cruz Biotechnology (Santa Cruz, CA).
[g] New England BioLabs (Beverly, MA).
[h] Zymed Laboratories (South San Francisco, CA).

Detailed Protocol for STAT EMSAs

1. Pour 20 × 20 cm (1-mm Teflon spacer) 5% (w/v) acrylamide (19 : 1, acrylamide : bisacrylamide) gels, using 0.25 × TBE (Tris–borate–EDTA); insert a 22-well comb and allow the gel to polymerize for 1 hr. The bottom spacer and comb are removed and the gel is prerun at 4° in 0.25 × TBE at 400 V (constant volts) until the current is approximately 10 mA. Fresh TBE should be used. TBE that has some precipitate will result in "fuzzy"-looking bands, high background, and retention of protein/probe in the wells (Fig. 2). Combs with smaller wells will decrease the band resolution, sharp clean wells will improve the resolution, and shorter minigels give rise to "smiling" bands that are poorly resolved. Running the gel in the cold will improve resolution and prevent overheating. Acrylamide at 4% (w/v) should be used when examining ISGF3 (IFN-α-induced trimeric complex).

2. Preparing the probe: The best STAT-binding probes contain a single high-affinity site flanked by overlapping bases that can be "filled in" to high specific activity with radiolabeled nucleotides (see Table II). Equal molar amounts of

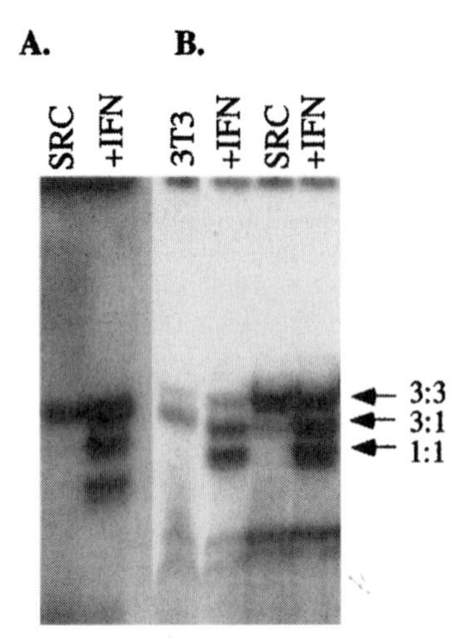

FIG. 2. Gel-shift analysis (EMSA) of nuclear extracts from NIH 3T3 cells (3T3) and v-*src*-transformed NIH 3T3 cells (SRC), using a ^{32}P-labeled double-stranded binding site (5′-GATCCATTTCCCGTAAATC-3′); either untreated or treated for 30 min with murine IFN-γ (IFN). STAT 3 homodimers, 3 : 3; STAT 1 homodimers, 1 : 1; STAT 3 : STAT 1 heterodimer, 3 : 1. (A) TBE containing a precipitate was used. Note the high background. (B) Fresh TBE used.

single-stranded complementary DNA are mixed together in TE with 100 m*M* NaCl. The DNA is heated to 100° for several minutes and then allowed to reanneal slowly to room temperature. The DNA is diluted to a concentration of 100 ng/μl in the same buffer. Unlabeled, double-stranded DNA (100 ng) is mixed with 4 μl of each radioactive [α-^{32}P] dNTP; Klenow (1000 U) and Klenow buffer are added. The reaction is stopped after 15 min at 37° with EDTA and an excess of cold nucleotides. Free nucleotides are removed by running the reaction through a Sepharose G-50 spin column in TE. The DNA is then precipitated with ethanol, ammonium acetate and glycogen as carrier for several hours at −70°. The DNA is resuspended in 100 μl of TE. One should expect $\sim 1 \times 10^5$ dpm/ng of DNA.

3. Reaction mixture: Compose a premix of 1 ng of probe, 2 μl of 5 × EMSA [100 m*M* HEPES (pH 7.9), 20% (v/v) Ficoll, 5 m*M* $MgCl_2$, 200 m*M* KCl, 0.5m*M* EGTA, 5 m*M* dithiothreitol (DTT)], 4 μg of poly(dI-dC) (Amersham, Arlington Heights, IL), and H_2O to 9 μl. To this add 1 μl of protein extract (approximately 5–10 μg by Bradford), spin the sample down in a microcentrifuge, and load immediately. Antisera (usually a 1 : 100 dilution) used against specific STATs for "supershifting" may be added to the premix or added to the protein extract. One can also wait to load and leave the sample at room temperature for 10 min or on ice for more than 30 min. Carefully load the sample, using thin, capillary sample loading pipette tips. Use DNA loading buffer containing bromphenol blue and xylene cyanole at one end of the gel as a marker. Run the gel for ~2–3 hr at 400 V

at 4° or until the bromphenol blue has just run off the gel. Running the gel more slowly (200 V) leads to slightly better resolution.

4. Isolation of protein extracts containing tyrosine phosphorylated STAT proteins: Tyrosine phosphorylation of STAT proteins takes place in the cytoplasm, but shortly thereafter STAT dimers are translocated to the nucleus. Thus, the highest concentration of phosphorylated protein is in the nucleus. Although whole cell extracts will contain phosphorylated STATs, nuclear extracts are enriched for them and constitute a cleaner preparation.

From tissue culture cells: Adherent cells. A total of $1–5 \times 10^6$ cells grown in a six-well dish (3.5 cm) is sufficient to detect activated STATs. In general, higher cell density (~80%) will ensure better detection of activated STATs. Wash the cells once with phosphate-buffered saline (PBS) and scrape the cells off the dish, using a cell scraper. One can also trypsinize the cells but the quality of the extracts is not as reproducible. For spinner cultures, the same number of cells is resuspended in PBS and treated as follows. The cells are collected in an Eppendorf tube and spun briefly at 3000 rpm. All the following steps should be done at 4°. *Nuclear extracts:* The pellet is resuspended in . . . (50 : 1; v/v) or in ~200 μl of hypotonic lysis buffer [10 m*M* KCl, 20 m*M* HEPES (pH 7.9), 1 m*M* EDTA, 1 m*M* DTT, 1 m*M* phenylmethyl sulfonyl fluoride (PMSF), 1× complete protease inhibitors, 10% (v/v) glycerol] and left on ice for 5–10 min. The cell suspension is then homogenized, using a Teflon Dounce (Fisher Scientific, Pittsburgh, PA), which can fit into an Eppendorf tube, for 5–10 strokes or twists. Hypotonic lysis buffer containing 0.5% (v/v) NP-40 obviates the need for homogenization. However, this amount of NP-40 can inhibit the binding of STAT3 to DNA and therefore the use of nonionic detergents is not recommended. The sample is spun at 13,000 rpm at 4° in a microcentrifuge for 5 min. The supernatant is the "cytoplasmic" extract. The pellet (usually 10–20 μl in volume) is resuspended in 1–2 pellet volumes of nuclear extract buffer [420 m*M* NaCl, 10 m*M* KCl, 20 m*M* HEPES (pH 7.9), 20% (v/v) glycerol, 1 m*M* EDTA, 1 m*M* DTT, 1 m*M* PMSF, 1× complete protease inhibitors (Boehringer Mannheim, Indianapolis, IN)] and left on ice for 20–30 min. The sample is centrifuged for 5 min at 4°, 13,000 rpm in a microcentrifuge. The supernatant is "nuclear" extract. A Bradford assay should be performed on 1 μl of extract and concentrations of 5–20 μg/μl should be obtained. Alternatively, whole cell extracts (WCEs) can be obtained by resuspending the original pellet in WCE buffer [50 m*M* Tris (pH 8), 280 m*M* NaCl, 0.5% (v/v) NP-40, 0.2 m*M* EDTA, 0.2 m*M* EGTA, 10% (v/v) glycerol, 1 m*M* DTT, 1 m*M* PMSF, 1× complete protease inhibitors].

From tissue: Fresh tissue should be coarsely dissected and homogenized 20 times in hypotonic lysis buffer (100 : 1, v/v; see above) at 4°. After centrifugation in a microcentrifuge at 13,000 rpm for 5 min, the pellet is resuspended in 4 volumes of hypotonic lysis buffer, homogenized (using a Dounce) 10 times, and recentrifuged for 5 min at 13,000 rpm. The pellet containing nuclei is resuspended in 2 volumes of nuclear extract buffer and extracted for 30 min at 4°. After centrifugation at 13,000 rpm the supernatant is the "nuclear" extract. When small samples are used it is recommended to dilute the nuclear extract 10-fold in hypotonic buffer and to concentrated it, using a Microcon-50 ultrafiltration column (Amicon Darvers, MA), by centrifuging for 15 min at 4000 *g*.[32] For clean preparations, nuclei can be purified from Dounce-homogenized tissue by centrifugation through 2.2 *M* sucrose and then extracted with 0.2 *M* NaCl.[33,34] If fresh tissue is unavailable then frozen tissue can be used by grinding the tissue, using a mortar (in crushed dry ice) and pestle. The pulverized sample is then treated identically to fresh tissue (see above).

Interestingly, the truncated tyrosine-phosphorylated versions or β forms of STAT3 and STAT5 are more stable than the α forms.[35,36] One can observe better binding per microgram of protein analyzed. Their mobilities differ on EMSA gels, and, surprisingly, STAT3β (the shorter form) migrates more slowly than STAT3α, presumably because of differences in charge. Because all STAT (homodimer) protein can bind to a canonical STAT-binding site (TTNNNNNAA) fairly well, one must utilize antisera specific for a STAT protein in order to help differentiate one STAT dimer from another (Table II). In addition, positive controls (see Table II) can also be helpful because the mobilities of the various STAT dimers differ significantly by gel shift analysis. One must be careful which antiserum is used, especially if one is trying to differentiate between the α and β forms and an antiserum has been generated against the COOH terminus (generally specific for the α form). The affinities of various STAT dimers for DNA differ. For example, the off rate of a STAT1 homodimer from DNA is ~3 min, while that of STAT3 is 17 min.[37,38] The STAT-DNA complexes nonetheless are relatively stable and once the protein extract has been incubated with DNA one can load it on a gel immediately.

STAT-binding sites in nature are usually relatively low in affinity and occur in tandem on natural promoters.[37] In fact, STAT dimers can form tetramers with one another and enhance, in a synergistic manner, DNA-binding affinity for weak sites. This increased affinity can translate into enhanced transcriptional activation.[37] Thus, if one is screening for STAT-binding sites on a "favorite" promoter, a high-affinity site is not necessary, and if one attempts attempts to show binding by gel

[32] C. Vaisse, J. L. Halaas, C. M. Horvath, J. E. Darnell, Jr., M. Stoffel, and J. M. Friedman, *Nat. Genet.* **14,** 95 (1996).

[33] S. Ruff-Jamison, K. Chen, and S. Cohen, *Science* **261,** 1733 (1993).

[34] S. Ruff-Jamison, Z. Zhong, Z. Wen, K. Chen, J. E. Darnell, Jr., and S. Cohen, *J. Biol. Chem.* **269,** 21933 (1994).

[35] O. K. Park, T. S. Schaefer, and D. Nathans, *Proc. Natl. Acad. Sci. U.S.A.* **93,** 13704 (1996).

[36] G. Luo and L. Yu-Lee, *J. Biol. Chem.* **272,** 26841 (1997).

[37] U. Vinkemeier, S. L. Cohen, I. Moarefi, B. T. Chait, J. Kuriyan, and J. E. Darnell, Jr., *EMBO J.* **15,** 5616 (1996).

[38] J. Bromberg, M. H. Wrzeszczynska, G. Devgan, Y. Zhao, R. G. Pestell, C. Albanese, and J. E. Darnell, Jr., *Cell* **98,** 295 (1999).

shift of a STAT dimer to the weak site then it is unlikely that binding will be observed.

Western Blot Analysis

STAT proteins are usually abundantly produced in many cell types. It is the activated or tyrosine-phosphorylated form that is of interest. In addition to gel shift analysis one can also determine the abundance of activated STAT protein by Western blot analysis, using antiserum that is specific to the tyrosine-phosphorylated form of a STAT protein (Table II). Alternatively, immunoprecipitation (IP) with antisera to both the active and inactive form of a STAT protein, followed by Western blot analysis with anti-phosphotyrosine antisera is also appropriate. As mentioned earlier, the sensitivity of Western blot analysis is far less than that of gel shift analysis for the tyrosine-phosphorylated form of STAT proteins. The molecular mass of STAT proteins varies between 80 and 110 kDa. A 7–8% (w/v) polyacrylamide gel is appropriate to use to differentiate between α and β forms (the β forms are usually 30–50 amino acids shorter) as well as various phosphorylated forms. It is recommended to resolve the proteins until 60- to 70-kDa prestained molecular mass marker is at the bottom of a 6-cm minigel. Only in the case of STAT1α can one differentiate between the tyrosine- and non-tyrosine-phosphorylated form by Western blot analysis, the tyrosine-phosphorylated form migrating somewhat more slowly.[39,40] The serine-phosphorylated forms of STAT3 and STAT4 can be differentiated by Western blot because the serine-phosphorylated forms migrate more slowly than the non-serine-phosphorylated forms. Both serine- and non-serine-phosphorylated forms of STAT3 and STAT4 can be tyrosine phosphorylated.[41,42] Serine- and tyrosine-phosphorylated STAT5A do not reveal any molecular weight differences, while STAT5B does, but the latter cannot be attributed to any of the known phosphorylated residues.[43]

Immunohistochemistry

Immunohistochemical methods are only just being developed to detect activated STAT proteins in tissue. Using antiserum that recognizes both tyrosine- and non-tyrosine-phosphorylated forms of a STAT protein is not appropriate to detect

[39] K. Shuai, C. Schindler, V. R. Prezioso, and J. E. Darnell, Jr., *Science* **259,** 1808 (1992).

[40] H. B. Sadowski, K. Shuai, J. E. Darnell, Jr., and M. Z. Gilman, *Science* **261,** 1739 (1993).

[41] T. G. Boulton, Z. Zhong, Z. Wen, J. E. Darnell, Jr., and G. D. Yancopoulos, *Proc. Natl. Acad. Sci. U.S.A.* **92,** 6915 (1995).

[42] S. S. Cho, C. M. Bacon, C. Sudarshan, R. C. Rees, D. Finbloom, R. Pine, and J. J. O'Shea, *J. Immun.* **157,** 4781 (1996).

[43] H. Yamashita, J. Xu, R. A. Erwin, W. L. Farrar, R. A. Kirken, and H. Rui, *J. Biol. Chem.* **273,** 30218 (1998).

the active form of the protein even if there is apparent strong nuclear localization of the protein.[44] Unfortunately, the tyrosine phospho-specific STAT antibodies are difficult to use reproducibly for immunohistochemistry. These antisera are quite sensitive to tissue fixation techniques and antigen retrieval.[45]

Measuring Transcriptional Competence of Activated STAT

To determine whether a STAT protein complex is transcriptionally active it is important to examine the ability of these transcription complexes to mediate transcriptional activation. If one is evaluating cell lines in tissue culture, then transfection with a STAT reporter construct (usually an array of STAT-binding sites followed by a luciferase gene) is appropriate. A single STAT-binding site followed by a luciferase gene does not lead to significant transcriptional activation. Multiple sites are needed: three or more in parallel, spaced by 10 base pairs, is probably optimal.[30,37,46] Using such reporter constructs will allow one to determine whether a STAT protein is indeed transcriptionally active. In general there is a concordance between the ability to bind DNA and the ability to activate transcription.[30] However, serine phosphorylation is required for optimal transcriptional activation of STAT1 and STAT3 and serine phosphorylation does not influence the ability of the STAT dimers to bind DNA.[21,47] In general, when performing transient transfections of STAT reporter luciferase constructs into cell lines one should titrate the amount of reporter being used in order to optimize the amount of induction. One should observe at least a 5- to 10-fold induction of luciferase activity after the addition of a STAT-activating ligand [6 hr after the addition of a ligand such as interferon γ (IFN-γ) or interleukin 6 (IL-6)]; and it is not uncommon to observe a 100-fold induction.

If, however, one is using a natural promoter containing a single STAT-binding site, this may be sufficient because STAT proteins do not function in isolation but in concert with other transcription factors.[1] Of course, natural promoter–reporter constructs will vary in efficacy as a function of the cell line used, presumably in relation to the abundance of the other cofactors and transcription factors present. Similarly, Northern blot or reverse transcriptase-polymerase chain reaction (RT-PCR) analysis of "known" STAT target genes may differ in response to a STAT-activating ligand when different cell lines are used. This may be due to differences in the level or duration of STAT activation as well as to the amounts of other transcription factors necessary to drive transcription of a particular gene.[48] In addition,

[44] F. E. Jones, T. Welte, X. Y. Fu, and D. F. Stern, *J. Cell Biol.* **147,** 77 (1999).
[45] J. Bromberg, (1999).
[46] X. A. Xu, Y. L. Sun, and T. Hoey, *Science* **273,** 794 (1996).
[47] Z. Wen and J. E. Darnell, Jr., *Nucleic Acids Res.* **25,** 2062 (1997).
[48] T. Nosaka, T. Kawashima, K. Misawa, K. Ikuta, A. L. Mui, and T. Kitamura, *EMBO J.* **18,** 4754 (1999).

different STAT protein dimers can lead to the activation of the same genes, perhaps to varying levels. For example, both STAT3 and STAT5 can lead to the activation of cyclin D1 and Bcl-XL.[28,38,48,49] Understanding STAT-mediated gene-specific transcription is still an open area of research.[1]

If one observes STAT activation in a particular cell type, it is important to determine whether there are any biological consequences to the activation. The best way to study the role of an individual STAT protein is by the analysis of cells that are deficient for a particular STAT. STAT-deficient animals [with the exception of STAT2 (C. Schindler, personal communication) and STAT3, which results in early embryonic lethality] have been generated and should be available to study.[3,4,7,11–15,50] Another alternative is to express dominant-negative STAT proteins in the cell line(s). A number of these constructs have been demonstrated to inhibit the activity of the wild-type protein.[27, 28,51–57] The most commonly available are those that contain a phenylalanine in place of the critical tyrosine and the COOH end of the molecule. It is recommended that the dominant-negative construct be epitope tagged (FLAG, histidine, Myc) to distinguish it from the endogenous protein. Once stable lines have been established, one should determine whether the "dominant-negative" protein is indeed inhibiting activation of the endogenous STAT.

Purification of STAT Proteins

Various forms of full-length and truncated STAT proteins can be overexpressed in insect cells or in *Escherichia coli*. A number of vectors can be used including pET20b(+) (Novagen, Madison, WI) for bacterial expression of truncated STATs and pAcSG2 (PharMingen, San Diego, CA) for baculovirus expression of full-length STATs in Sf9 cells. These constructs will result in the purification of reasonable to high yields (4–10 mg/liter of culture) of STAT proteins by adhering to

[49] I. Matsumura, T. Kitamura, H. Wakao, H. Tanaka, K. Hashimoto, C. Albanese, J. Downward, R. G. Pestell, and Y. Kanakura, *EMBO J.* **18,** 1367 (1999).

[50] K. Takeda, T. Tanaka, W. Shi, M. Matsumoto, M. Minami, S. Kashiwamura, K. Nakanishi, N. Yoshida, T. Kishimoto, and S. Akira, *Nature* (*London*) **380,** 627 (1996).

[51] J. Turkson, T. Bowman, R. Garcia, E. Caldenhoven, R. P. De Groot, and R. Jove, *Mol. Cell. Biol.* **18,** 2545 (1998).

[52] M. Minami, M. Inoue, S. Wei, K. Takeda, M. Matsumoto, T. Kishimoto, and S. Akira, *Proc. Natl. Acad. Sci. U.S.A.* **93,** 3963 (1996).

[53] J. F. Bromberg, C. M. Horvath, D. Besser, W. W. Lathem, and J. E. Darnell, Jr., *Mol. Cell. Biol.* **5,** 2553 (1998).

[54] M. Kortylewski, P. C. Heinrich, A. Mackiewicz, U. Schniertshauer, U. Klingmuller, K. Nakajima, T. Hirano, F. Horn, and I. Behrmann, *Oncogene* **18,** 3742 (1999).

[55] K. Nakajima, Y. Yamanaka, K. Nakae, H. Kojima, M. Ichiba, N. Kiuchi, T. Kitaoka, T. Fukada, M. Hibi, and T. Hirano, *EMBO J.* **15,** 3651 (1996).

[56] R. M. Moriggl, V. Gouilleux-Gruart, R. Jahne, S. Berchtold, C. Gartmann, X. Liu, L. Hennighausen, A. Sotiropoulos, B. Groner, and F. Gouilleux, *Mol. Cell. Biol.* **16,** 5691 (1996).

[57] J. F. Bromberg, Z. Fan, C. Brown, J. Mendelsohn, and J. E. Darnell, Jr., *Cell Growth Differ.* **9,** 505 (1998).

the following published protocol with some minor modifications.[37] For example, the following procedure is used to purify human STAT1 core(132–712) from *E. coli* strain BL21 (DE3). Briefly, cells are lysed by sonication in lysis buffer [20 m*M* HEPES (pH 7.5), 100 m*M* KCl, 10% (v/v) glycerol, 1 m*M* EDTA, 20 m*M* DTT, complete protease inhibitors [1 tablet/50 ml; Boehringer Mannheim]. Nucleic acid contamination is removed by polyethyleneimine (PEI) (0.1%, w/v) precipitation. STAT protein is separated from others by two subsequent ammonium sulfate precipitation (35 and 55%, respectively) steps. An alkylation reaction is carried out [with 20 m*M* *N*-ethylmaleimide (NEM) in 50 m*M* sodium phosphate (pH 7.0), 1 m*M* EDTA, complete protease inhibitors (1 tablet/50 ml)] to prevent aggregation caused by cysteine cross-linking. The reaction is quenched with 50 m*M* 2-mercaptoethanol. The STAT protein is then purified on a fast protein liquid chromatography (FPLC) system (Pharmacia, Uppsala, Sweden). One-eighth volume of saturated $(NH_4)_2SO_4$ solution is added to the sample after alkylation. The mixture is loaded onto a high-performance phenyl-Sepharose column (Pharmacia), preequilibrated with 50 m*M* Tris-HCl (pH 7.4), 1 m*M* EDTA, 2 m*M* DTT, and 900 m*M* $(NH_4)_2SO_4$. The proteins are eluted with a linear gradient of 900 to 50 m*M* $(NH_4)_2SO_4$ in the same buffer. STAT-containing fractions are pooled and concentrated to 10–20 ml and diluted at least 10-fold, using the equilibration buffer [40 m*M* morpholine ethane sulfonic acid (MES, pH 6.5), 10% (v/v) glycerol, 0.5 m*M* EDTA, 0–500 m*M* KCl, 4 m*M* DTT] for Sepharose S column purification. The diluted sample is then loaded onto a Sepharose S column (Pharmacia). Elution is achieved with a linear pH/salt gradient of 40 m*M* MES (pH 6.5), 50 m*M* KCl to 40 m*M* HEPES (pH 7.2), 500 m*M* KCl in the same buffer. STAT-containing fractions are pooled and concentrated to 10–20 ml and diluted at least 10-fold, using the equilibration buffer [20 m*M* Tris-HCl (pH 8.0), 1 m*M* EDTA, 2 m*M* DTT, 50 m*M* KCl] for heparin column purification. The diluted sample is then loaded onto a HiTrap heparin cartridge (Pharmacia) and eluted with a linear gradient of 50 to 500 m*M* KCl in the same buffer. Fractions containing STAT1 core are pooled and concentrated to a minimal volume and loaded onto a Superdex 200 column (Pharmacia), and purified in a buffer containing 10 m*M* HEPES (pH 7.2), 100 m*M* KCl, 2 m*M* DTT, 0.5 m*M* EDTA. Purified STAT protein is concentrated and aliquots are flash frozen in liquid nitrogen and stored at −80°.

STAT1 protein can be phosphorylated *in vitro* by the EGF receptor,[58] which can be prepared from human carcinoma A431 cells.[59] A large-scale phosphorylation reaction and the subsequent purification of the phosphorylated protein can be done according a published procedure[37] with minor modifications. Briefly, STAT-containing fractions after Sepharose S column chromatography are pooled and

[58] F. W. Quelle, W. Thierfelder, B. A. Witthuhn, B. Tang, S. Cohen, and J. N. Ihle, *J. Biol. Chem.* **270,** 20775 (1995).

[59] Y. Yarden, I. Harari, and J. Schlessinger, *J. Biol. Chem.* **260,** 315 (1985).

concentrated by ultrafiltration instead of ammonium sulfate precipation and resuspension. Addition of EGF and additional ATP is not necessary for the reaction. Catalytic metal ion $MnCl_2$ is added to 5 mM instead of 10 mM. The reaction is allowed to proceed for more than 20 hr at 4°. The phosphorylation efficiency is typically about 80% and the unphosphorylated protein can be separated from the phosphorylated protein by a HiTrap heparin cartridge. Finally, the phosphorylated STAT is further purified on a Superdex 200 column.

A different procedure has been published for the production and purification of phosphorylated STAT3 core fragment.[60] Briefly, the protein is overexpressed in bacterial strain TKB1 (Stratagene La Jolla, CA), which coexpresses the Elk tyrosine kinase. The protein is quantitatively phosphorylated at the correct tyrosine residue and can be purified by ammonium sulfate precipitation followed by gel filtration.

Early efforts to dissect the STATs into separable domains with distinct functions such as DNA binding have met with limited success. The crystal structures[16,61] show that the core fragment of a STAT protein contains four tandem structural domains. Each of the four domains is fused to the adjacent domains by the formation of a contiguous hydrophobic core. The presence of extensive interdomain interfaces explains why previous efforts at constructing smaller units encompassing the distinct functions of STATs have met with failure. On the other hand, the loop connecting the N-domain and the rest of the molecule is susceptible to proteolysis, making it difficult to obtain diffraction-quality crystals of full-length STATs.

[60] S. Becker, G. L. Corthals, R. Aebersold, B. Groner, and C. W. Muller, *FEBS Lett.* **441,** 141 (1998).
[61] S. Becker, B. Groner, and C. W. Muller, *Nature* (*London*) **394,** 145 (1998).

[15] Integrin Regulation of Receptor Tyrosine Kinase and G Protein-Coupled Receptor Signaling to Mitogen-Activated Protein Kinases

By RUDOLPH L. JULIANO, ANDREW E. APLIN, ALAN K. HOWE, SARAH SHORT, JUNG WEON LEE, and SURESH ALAHARI

Overview

It has become clear that integrins are not only involved in cell adhesion and the determination of cytoskeletal organization and cell shape, but are also intimately involved in signal transduction cascades. There are really two types of integrin-related signal transduction. The first is direct signaling, where binding of integrins

0076-6879/00 $35.00

to extracellular matrix proteins or other ligands activates intracellular signaling events. The second type concerns integrin modulation of signaling pathways activated by soluble growth or differentiation factors such as those that impinge on receptor tyrosine kinases (RTKs) or G protein-coupled receptors (GPCRs). In many cases integrin signaling or signal modulation is associated with activation of elements of the Ras–mitogen-activated protein (MAP) kinase cascade, but other signaling cascades can be affected as well.[1–4]

In many respects assays for integrin effects on signal transduction are similar to assays used to evaluate the effects of other stimuli, such as polypeptide growth factors, on signaling cascades. The key difference is that the assays need to be appropriately adapted to the situation in which the presence or absence of integrin engagement is the parameter of interest. Often this involves depriving cells of their anchorage to the tissue culture plate, and then allowing them to reengage with a solid substratum via their integrins. It should be recognized that the event of cell detachment per se could have important impacts on signaling processes. Thus it is important to be aware of the extent and kinetics of signals triggered by the manipulations involved in the assays. Another key consideration is the use of appropriate controls; it is important to discriminate between integrin-specific effects and effects due to nonspecific binding of cells to "sticky" surfaces. In the sections below we describe protocols that we have found useful in studying the effects of integrin-mediated cell adhesion on various signaling events. Because there is an abundance of information on signaling pathways in this volume, we start with the assumption that the reader knows the basic elements of the Ras–MAP kinase pathway and other common pathways, and we discuss protocols that are modified to focus on the role of integrin-mediated adhesion and cytoskeletal organization.

Our laboratory has worked primarily on several aspects of the linkage between integrins and the Ras–MAP kinase cascade, working mainly with mouse fibroblasts (3T3 cells) and human endothelial cells (HUVECs). Some of the assays that have worked well for us in this context are described below. We have tried to provide detailed step-by-step protocols with indications for specific reagents. Naturally some of these protocols may need modification if the assays are done with other types of cells or tissue samples.

Specific Attachment of Cells to Substratum, Using Anti-Integrin Antibodies

Cells can be anchored via specific integrins, using anti-integrin subunit monoclonal antibodies. Such integrin-mediated attachment will directly, but transiently,

[1] A. E. Aplin, S. M. Short, and R. L. Juliano, *J. Biol. Chem.* **274,** 31223 (1999).
[2] A. K. Howe and R. L. Juliano, *J. Biol. Chem.* **273,** 27268 (1998).
[3] F. G. Giancotti and E. Ruoslahti, *Science* **285,** 1028 (1999).
[4] M. A. Schwartz, *J. Cell Biol.* **139,** 575 (1997).

stimulate the MAP kinase cascade. In addition, subsequent to the initial activation by adhesion, the cells then remain in a state that is permissive for efficient signaling to MAP kinase by RTKs or GPCRs.[1,2,5] In the antibody approach only a single integrin α/β heterodimer is involved in anchoring the integrin to the plate or dish. This contrasts with the situation in which dishes are coated with extracellular matrix proteins such as fibronectin, collagen, or laminin (the natural ligands for many integrins), because each of these matrix proteins can interact with several different integrin heterodimers. Good results have been obtained with specially treated anti-mouse IgG-precoated MicroCellector flasks (Applied Immune Science, Santa Clara, CA). However, ordinary petri dishes can also be used by adsorbing anti-mouse IgG.[1] These dishes are then incubated with anti-integrin monoclonal antibodies at 2 μl/ml at 4° overnight and excess reagent is removed by rinsing with buffer. The dishes are then blocked with a solution of 1% (w/v) bovine serum albumin (BSA) in buffer prior to use. A variety of antibodies that recognize human integrins are available from GIBCO-BRL Gaithersburg, MD or Chemicon (Temecula, CA). Similar reagents to analyze the role of mouse integrins are not so widely available.[1] For example, P1E6, P1D6, and P1B6 (GIBCO-BRL), which recognize human integrin α_2, α_5, and α_3 subunits, respectively, are used to prepare dishes that adhere endothelial cells in a subunit-specific manner.[1] Similarly, P4C10 can be used to attach cell specifically via β_1-integrins (see Fig. 1). As positive and negative controls, cells can be allowed to attach to fibronectin (10–20 μg/ml)-coated dishes (which will engage several integrins as well as syndecans in many cells), or can be allowed to attach to polylysine (5–100 μg/ml, depending on cell type)-coated dishes, where cells bind by simple charge–charge interactions that do not specifically involve integrins.

Cell Adherence and Preparation of Cell Lysates

For experiments comparing attached versus nonattached (suspended) cells, confluent cells are serum starved for 4–6 hr before detachment with trypsin–EDTA; trypsin activity is subsequently neutralized with soybean trypsin inhibitor (1 mg/ml; GIBCO-BRL). Cells are suspended in Dulbecco's modified Eagle's medium (DMEM) with 2% (w/v) BSA (NIH 3T3) or Earle's balanced medium (EBM) with 2% (w/v) BSA (HUVECs) and incubated in suspension at 37° for 45 min in a rotator to allow kinases to become quiescent. Cells are then plated onto antibody- or matrix protein-coated dishes or maintained in suspension and incubated at 37° for the indicated times. After incubation, cells are washed twice with cold phosphate-buffered saline (PBS) and then lysed in a modified radioimmunoprecipitation assay (RIPA) buffer.[6] Total cell lysates are cleared by centrifugation

[5] S. Short, G. Talbot, and R. L. Juliano, *Mol. Biol. Cell.* **9,** 169 (1998).

[6] Q. Chen, M. S. Kinch, T. H. Lin, K. Burridge, and R. L. Juliano, *J. Biol. Chem.* **269,** 26602 (1994).

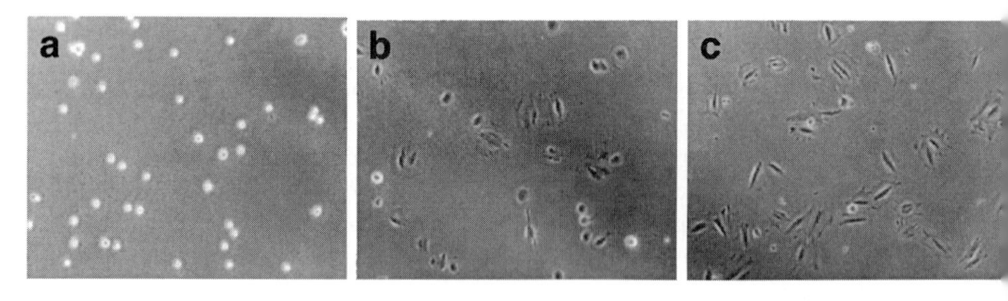

FIG. 1. Cell adhesion to antibody-coupled surfaces. Quiescent WI38 cells were detached and replated on (a) poly-D-lysine (10 μg/ml); (b) anti-β_1(2 μl/ml); (c) fibronectin (10 μg/ml) in DMEM–1% (w/v) BSA. After 90 min, the ability of cells to spread on the different surfaces was analyzed by phase-contrast microscopy. Cells spread on fibronectin- and anti-β_1-coated surfaces to a much greater degree than did cells on polylysine.

at 16,000 *g* for 5 min at 4°. Protein concentration in the lysates is determined by the bicinchoninic acid assay (Pierce, Rockford, IL).

Measurement of Mitogen-Activated Protein Kinase by Immunoprecipitation and *in Vitro* Kinase Assay

One of the major issues in integrin-mediated signaling has been activation of the MAP kinase cascade. This is the basic assay for evaluating the activity of the extracellular signal-regulated kinase 1/2 (ERK 1,2) forms of MAP kinase. Although there are now commercial antibodies available that detect activated forms of MAP kinases (see below), it is still desirable to use a quantitative biochemical measurement of this type. The assay involves phosphorylation of myelin basic protein (MBP), a commonly used substrate for ERKs. This protocol is similar to ones we have reported previously.[6,7] This assay can be used to examine direct effects of integrin-mediated anchorage on MAP kinase, or can be used to evaluate effects of growth factors on attached or suspended cells.

1. Wash the cells twice with ice-cold PBS.
2. Add modified RIPA lysis buffer [0.5% (w/v) deoxycholate, 1.0% (v/v) Nonidet P-40 (NP-40), 50 m*M* Tris (pH 7.4), 150 m*M* NaCl, 1 m*M* EDTA] containing phosphatase and protease inhibitors [e.g., use phenylmethylsulfonyl fluoride (PMSF), aprotinin, pepstatin, NaF, and sodium orthovanadate]. Use 0.5–1.0 ml of lysis buffer for a 10-cm plate (depending on cell density), and scale up or down accordingly.
3. Scrape the lysate and transfer to a microcentrifuge tube.

[7] Q. Chen, T. Lin, C. Der, and R. Juliano, *J. Biol. Chem.* **271,** 18122 (1996).

4. Vortex (10 sec) and incubate on ice for 10 min.

5. Spin in a microcentrifuge for 10 min at 4° at 14,000 rpm.

6. Transfer the supernatant lysate to a new microcentrifuge tube. Freeze at –80° or proceed directly to immunoprecipitation (IP). At this point preclear the lysate by incubating it with 30 μl of protein G–Sepharose for 30 min at 4° with rocking, centrifuge the lysate in the cold for 1 min at 14,000 rpm, and transfer the supernatant lysate to a fresh tube. Depending on the cell type being used this may or may not be necessary; however, it is always better to do it.

7. Determine the protein concentration by whatever method is customary and familiar.

8. Typically, 150–200 μg of protein is used per immunoprecipitation. Less can be used (as little as 50 μg) but the higher amount gives a good, easily detectable amount of activity.

9. To the lysate, add 1.5 μg of anti-ERK antibody [SC154 (ERK2) or SC93 (ERK1); Santa Cruz Biotechnology, Santa Cruz, CA] and incubate, with rocking, at 4° for 1 hr.

10. Add 20–30 μl of protein G–Sepharose (50% slurry) and incubate at 4° for 30 min to 1 hr.

11. Prepare kinase buffer [50 m*M* HEPES or Tris (pH 7.4), 10 m*M* $MgCl_2$, 10 m*M* $MnCl_2$, 1 m*M* dithiothreitol (DTT)] and reaction mixture, which is 15 μ*M* ATP and MBP (500 μg/ml) in kinase buffer.

12. Wash the immunoprecipitate three times with RIPA buffer and once with kinase buffer.

13. Add 40 μl of the reaction mixture, along with 15 μCi of [γ-^{32}P] ATP (3000 Ci/mmol; New England Nuclear, Boston, MA), to the washed beads and mix gently by tapping the tube (try to avoid allowing the beads to stick to the side of the tube, above the level of the reaction mixture).

14. Incubate at 25° (room temperature) for 25 min, mixing occasionally.

15. Add sample buffer to 1× (usually done by adding three 20-μl volumes), and boil the samples for 3–5 min.

16. Run samples on a 12 or 15% (w/v) polyacrylamide gel. Stain the gel, dry it, and expose it to film or a PhosphorImager screen (Molecular Dynamics, Sunnyvale, CA). Usually between 5 and 10 μl is loaded, providing a strong signal overnight. The sample can also be stored, after boiling, at –20°.

Coupled Two-Stage Raf Immunoprecipitation and *in Vitro* Kinase Assay

In cell adhesion studies it is often important to examine intermediate steps in the MAP kinase cascade. This assay involves immunoprecipitation of the Raf kinase and then testing its activity by its ability to phosphorylate and activate MAP/ERK kinase (MEK), which in turn phosphorylates and activates MAP kinase (ERK). The final readout is the phosphorylation of myelin basic protein by the ERKs. This method is adapted from that developed by Marshall and colleagues,[8,9] and has been used extensively by us.[2,7,10] This protocol uses commercially available recombinant MEK and MAP kinase (MAPK). However, we have noticed that reagents from some suppliers often give high background activity. Thus, we recommend using laboratory-prepared reagents. Expressing and purifying one's own batches of MEK and MAPK is highly recommended, as it is far more economical, the results are generally much more reproducible, and quality control is in the hands of the investigator. Expression plasmids for glutathione *S*-transferase (GST)- or hexahistidine (His6)-tagged MEK and MAPK have been generously provided to the research community by several investigators.[8,11] The amount of protein required for this assay is generally 500 μg of lysate per IP.

1. Cells are lysed with modified RIPA buffer and Raf-1 is immunoprecipitated with anti-Raf-1 (c12; Santa Cruz Biotechnology) followed by incubation with protein G–Sepharose.
2. Spin down the immunoconjugates after immunoprecipitation and aspirate the excess.
3. Wash three times with 0.5 ml of buffer A and twice with 0.5 ml of buffer B (recipes below). These are termed the "IPs."
4. Add 30 μl of first reaction mix (recipe below) to each IP, to an empty tube labeled "Reaction Only." and to one tube containing 30 μl of 2× sample buffer [sodium dodecyl sulfate (SDS) containing sample buffer for polyacrylanide gel electrophoresis (PAGE)], labeled "Reaction Stop"; these last two samples will serve as controls.
5. Mix gently and incubate at room temperature for 30 min or at 30° for 15 min, with occasional, gentle mixing.
6. During incubation, prepare MBP reaction mix (recipe below) and aliquot 35 μl into the appropriate number of microcentrifuge tubes, and then add 1 μl of [^{32}P]ATP to each tube; Also, if desired, add 10 μl of 2× sample buffer to another set of microcentrifuge tubes.
7. Briefly spin all first reaction tubes.
8. Remove 10 μl and add to MBP reaction mix tubes; remove another 10 μl and add to 2× SDS sample buffer tubes (FS) (this is to monitor background in the absence of substrate phsophorylation); incubate the MBP reaction mix tubes at room temperature for 15 min.

[8] D. R. Alessi, P. Cohen, A. Ashworth, S. Cowley, S. J. Leevers, and C. J. Marshall, *Methods Enzymol.* **255,** 279 (1995).

[9] M. A. Bogoyevitch, C. J. Marshall, and P. H. Sugden, *J. Biol. Chem.* **270,** 26303 (1995).

[10] T. H. Lin, A. E. Aplin, Y. Shen, Q. Chen, M. Schaller, L. Romer, I. Aukhil, and R. L. Juliano, *J. Cell Biol.* **136,** 1385 (1997).

[11] C. A. Lange-Carter and G. L. Johnson, *Methods Enzymol.* **255,** 290 (1995).

9. Remove and discard the remaining supernatant (~10 μl) from the IP tubes and add 30–50 μl of 1× SDS sample buffer (IP blot).

10. Add 45 μl of 2× SDS sample buffer to the MBP reaction mix tubes ("MBP").

11. Boil all the tubes for 5 min.

12. Run 10 μl of FS on 10% (w/v) polyacrylamide gels and blot with anti-MAPK (Sc094; Santa Cruz Biotechnology) and active MAPK antibodies (Promega, Madison, WI).

13. Run 15–25 μl of IP blot on a 7.5% (w/v) polyacrylamide gel and blot with anti-Raf antibody.

14. Run 20–30 μl of MBP on a 15% (w/v) polyacrylamide gel, stain, dry, and expose to film or a PhosphorImager plate.

Buffer A: 10 m*M* Tris (pH 7.4), 5 m*M* EDTA, 50 m*M* NaF, 50 m*M* NaCl, 1% (v/v) Triton X-100, 0.1% (w/v) BSA

Buffer B: 50 m*M* Tris (pH 7.4), 0.1 m*M* EGTA, 0.5 m*M* sodium vanadate, 0.1% (v/v) 2-mercaptoethanol

Buffer C: 60 m*M* Tris (pH 7.4), 0.2 m*M* EGTA, 0.2% (v/v) 2-mercaptoethanol, 0.6% (v/v) Brij 35, 20 m*M* $MgCl_2$

First reaction mix (per 100 μl): 50 μl of 2× buffer C, 5 μl of 1 m*M* ATP (final, 50 μ*M*), 1 μl of GST–MEK (Upstate Biotechnology, Lake Placid, NY) (final, 2.5 μg/ml), 1.25 μl of MAPK (Santa Cruz Biotechnology) (final, ~20 μg/ml), 42.75 μl of H_2O

MBP reaction mix (1 ml): 50 μl of Tris, pH 7.4 (final, 50 μ*M*), 0.5 μl of EGTA (final, 0.1 m*M*), 12.5 μl of $MgCl_2$ (final, 12.5 μ*M*), 66.7 μl of MBP (stock, 2.99 mg/ml; final, ~0.2 mg/ml), 870 μl of H_2O

Use of Phospho-Specific Antibodies to Analyze Activation of Signal Transduction Pathways

Activation of individual protein kinase cascades can now be analyzed rapidly and without the necessity for radioactivity by use of phospho-specific antibodies. These reagents recognize phosphorylated and, hence, theoretically either active or inactive versions of kinases. Phospho-specific antibodies are available for many of the signal proteins involved in growth control, development, and apoptotic pathways. This approach is particularly useful in the study of the kinetics of signal pathway activation. For example, it might be desirable to measure the kinetics of activation of MAP kinase or c-Jun kinase in response to cell attachment mediated by a particular integrin. We have often utilized[2,5,12] an anti-active ERK antibody (Promega) that preferentially recognizes the dually phosphorylated pT-E-pY motif within the MAP kinases, ERK1 and ERK2. This antibody can be used at relatively

[12] A. E. Aplin and R. L. Juliano, *J. Cell Sci.* **112,** 695 (1999).

low dilutions (1 : 3000–1 : 5000) for Western blot analysis. Total levels (phosphorylated and nonphosphorylated) of ERK1 and ERK2 in samples should also be determined by Western blotting (anti-ERK, K-23; Santa Cruz Biotechnology).

Another useful approach has been to use phospho-specific antibodies to evaluate levels of active protein kinase B (PKB)/Akt kinase, which is also activated by receptor tyrosine kinases acting through Ras; this assay is described in detail as an example of use of such antibodies. Proteins (15 μg/sample) in cell lysates are resolved by 8% (w/v) SDS–PAGE, electroblotted to polyvinylidene difluoride (PVDF) membranes, and blocked with 5% (w/v) nonfat dried milk in PBS–0.05% (v/v) Tween 20. The membrane is probed with rabbit polyclonal anti-phospho-S473-PKB/Akt antibody (New England Biolabs, Beverly, MA) at a volume dilution of 1 : 1000 in buffer overnight at 4°. Blots are also probed for total PKB by using sheep polyclonal anti-rat PKB/Akt antibody (Upstate Biotechnology) overnight at 4° at 0.45 μg/ml. After washes with PBS–0.05% (v/v) Tween 20 for 1 hr, membranes are incubated with fluorescein-conjugated anti-immunoglobulins for 1 hr at room temperature. Fluorescein-conjugated anti-rabbit IgG [supplied in an enhanced chemifluorescence (ECF) kit from Amersham, Arlington Heights, IL] is used at 1 : 1000 dilution and fluorescein-conjugated anti-sheep IgG (Molecular Probes, Eugene, OR) is used at 1 : 5000. After washing as described above, membranes are further incubated with alkaline phosphatase (AP)-conjugated anti-fluorescein antibody (included in the ECF kit) at 1 : 5000 for 1 hr at room temperature. After washing as described above, the signals from membranes are detected by using a chemifluorescence-based kit (ECF kit; Amersham), and quantitation of the band intensities is done with a chemifluorescence scanner on a PhosphorImager with Image-QuaNT software (Storm 840; Molecular Dynamics).

Simultaneous Examination of Morphology and Signaling by Transient Cotransfections

One of the major issues in integrin signaling is the role of the cytoskeleton. The assays below allow simultaneous visualization of effects on the cytoskeleton and detection of effects on the MAP kinase cascade.[12] For example, one could test whether Rho family GTPases, which are known to affect actin filament organization, also affect the activity of MAP kinase. Active or dominant-negative forms of other GTPases, or kinases, or other proteins, can just as easily be substituted for the Rho GTPases. Variations on this procedure could be used to look at the relationship between cytoskeletal effects and other signaling pathways.

To analyze the effects of Rho GTPases in NIH 3T3 fibroblasts, we have transiently cotransfected cells with constitutively active forms of these constructs along with cDNA encoding the green fluorescent protein (GFP) (pGreen Lantern-1; GIBCO-BRL). After allowing transfected cells to readhere to fibronectin for a given time the changes in the actin cytoskeleton and focal adhesion structures are

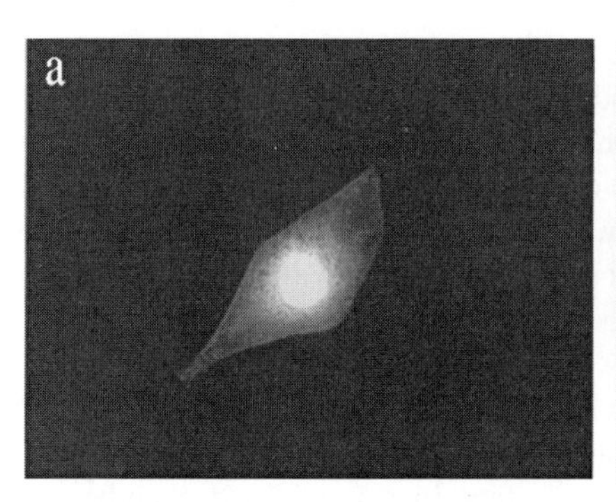

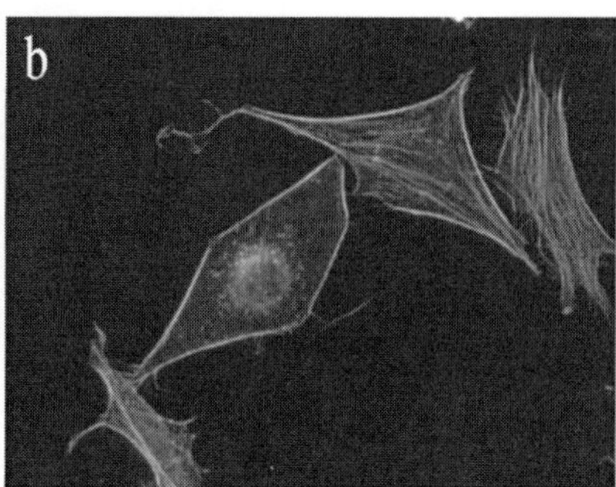

FIG. 2. Analysis of modulation of the actin cytoskeleton by transient transfection in NIH 3T3 cells. NIH 3T3 cells were transiently cotransfected with a plasmid encoding the Rho inhibitor C3, and with GFP. Serum-starved cells were replated on fibronectin-coated coverslips in serum-free medium for 2 hr. Cells were then fixed and the actin cytoskeleton was stained with rhodamine-conjugated phalloidin. Transfected cells were identified by the fluorescence emitted from GFP (a). Staining of the actin cytoskeleton showed a disruption of actin stress fibers in C3-transfected cells compared with neighboring untransfected cells (b).

analyzed with tetramethyl rhodamine isothiocyanate (TRITC)–phalloidin (Sigma, St. Louis, MO) or anti-phosphotyrosine (Upstate Biotechnology), anti-vinculin (Sigma), or anti-paxillin (Transdction Laboratories, Lexington, KY). Transfected cells can be readily identified in the fluorescein channel by their expression of GFP (see Fig. 2). Effects on MAP kinase can be evaluated by cotransfection of an epitope-tagged MAPK construct[10] (and see below).

The entire population of cells can also be manipulated by using various drugs to affect the cytoskeleton. For example, cytochalasin D and latrunculin promote disassembly of actin filaments, while jasplakinolide promotes filament stabilization.[13] Thus, one can change the degree of actin filament network assembly, evaluate this effect by immunofluorescence, and examine the effects on signaling to MAP kinase through RTKs or other receptors by using activation-specific antibodies or *in vitro* kinase assays as described above.

Immunofluorescence Protocol

1. Prepare coated coverslips by flaming the coverslips (cover glass No. 1, 22 mm^2; Corning, Corning, NY) and inserting them in the wells of a six-well dish. Coat overnight with fibronectin (10 μg/ml) at 4°.
2. Wash the coverslips three times with PBS, and block for 1 hr at 37° with DMEM–2% (w/v) BSA.
3. Contransfect cells with the plasmid of interest with pGreen Lantern-1 (GIBCO-BRL) encoding GFP at a plasmid ratio of 1 : 4.

[13] A. Sotiropoulos, D. Gineitis, J. Copeland, and R. Treisman, *Cell* **98,** 159 (1999).

4. After the recovery period (40 hr), seed cotransfected cells to give approximately 50% density; allow the cells to adhere and spread for 2–3 hr.

5. After the adherence period wash the cells twice with cold PBS, and fix them for 10 min at room temperature in 3.7% (v/v) formaldehyde–PBS in wells.

6. Wash three times in PBS (5-min wash each).

7. Permeabilize the cells for 5 min at room temperature with 0.5% (v/v) Triton X-100 in PBS.

8. Wash three times in PBS.

9. Block in 2% (w/v) BSA–PBS for 1 hr at room temperature.

10. Remove each coverslip and incubate upside down on a drop of primary antibody (50–75 μl) on a level Parafilm surface in a humidified chamber either overnight at 4° or for 1 hr at room temperature. Antibody is diluted in 2% (w/v) BSA–PBS [for anti-PY 4G10 (Upstate Biotechnology), use a 1 : 100 dilution].

11. Remove and place each coverslip face up in a six-well dish. Wash three times, 10 min each with PBS.

12. Perform the secondary antibody incubation [1 : 100 dilution in 2% (w/v) BSA–PBS] in a humidified chamber for 1 hr at room temperature. Use a TRITC-labeled secondary antibody, either goat anti-mouse conjugated to TRITC (Sigma) or goat anti-rabbit conjugated to TRITC (Sigma).

13. Wash twice with PBS (10 min each) and once with water.

14. Drain off the excess water, mount in Permafluor (Thomas Scientific, Swedesboro, NJ) and visualize with a fluorescence microscope.

The alterations in the actin cytoskeleton and focal adhesion complexes induced by Rho GTPases, as well as the ability of cells to signal, can be analyzed under the same conditions. NIH 3T3 cells are cotransfected with the construct of interest and cDNA encoding a hemagglutinin (HA)-tagged version of ERK1 (pcDNAI-ERK1-HA) (originally obtained from J. Pouysseguer, Nice, France).

Epitope-Tagged Kinase Assays

1. Transfect cells with GFP and pcDNAI-ERK1-HA plasmids.

2. Transfected cells are allowed to readhere to fibronectin-coated dishes for a given time and then stimulated with growth factor.

3. Wash the cells twice with ice-cold PBS and then lyse in modified RIPA buffer [50 m*M* HEPES (pH 7.5), 1% (v/v) Nonidet P-40 (NP-40), 0.5% (w/v) sodium deoxycholate, 150 m*M* NaCl, 50 m*M* NaF, 1 m*M* sodium vanadate, 1 m*M* nitrophenylphosphate, 5 m*M* benzamidine, 0.2 μ*M* calyculin A, 2 m*M* PMSF, and aprotinin (10 μg/ml)].

4. Lyse for 20 min on ice and clarify the lysates by centrifugation at 16,000 *g* for 10 min at 4°.

5. Preclear the lysates with 30 μl of a 1 : 1 slurry of protein G–Sepharose Fast Flow (Pharmacia Biotech, Piscataway, NJ) on a rotator at 4° for 30 min.
6. Incubate precleared lysates with anti-HA antibody on ice for 2 hr.
7. Add 30 μl of protein G beads and incubate on a rotator at 4° for 2 hr.
8. Wash the immunocomplex once with cold lysis buffer.
9. Wash twice with cold wash buffer (0.1 *M* NaCl, 0.25 *M* Tris-HCl, pH 7.5).
10. To the washed immunocomplex, add 40 μl of reaction mixture [10 m*M* Tris-HCl (pH 7.5), 10 m*M* $MgCl_2$, 1 m*M* dithiothreitol (DTT), 25 μ*M* ATP, 5 μCi of [^{32}P] ATP (Du Pont NEN), and 10 μg of myelin basic protein (Upstate Biotechnology or GIBCO)] per assay and incubate on a shaking platform at room temperature for 30 min.
11. Add 13 μl of 4× SDS sample buffer and boil for 5 min to stop the reaction.
12. Use SDS–PAGE [15% (w/v) polyacrylamide gel] to analyze the incorporation of ^{32}P into myelin basic protein, and Western blotting [10% (w/v) polyacrylamide gel] to obtain levels of expressed HA–ERK1.

Assays for G Protein-Coupled Receptor Responses in Integrin-Anchored and Nonanchored Cells

In addition to effects on the "classic" RTK–MAP kinase pathway, integrin-mediated cell anchorage can also affect signaling in other pathways, particularly pathways regulated by G protein-coupled receptors (GPCRs). For example, in endothelial cells signaling from P2Y receptors through the G_q family of heterotrimeric G proteins and thence to MAP kinase is dramatically affected by integrin-mediated cell adhesion. The effects on MAP kinase or Raf-1 can be monitored by the protocols described above. Upstream events in this pathway include activation of phopholipase Cβ, generation of inositol phosphates, and triggering of intracellular calcium transients.[14] these can also be monitored in the context of integrin-mediated adhesion. For example, in the next section we provide a protocol to compare calcium transients in integrin-anchored cells and in cells maintained in suspension or plated on a nonspecific substratum.

Ca^{2+} Measurement

After serum deprivation for 6 hr, HUVECs are replated on glass coverslips coated with either fibronectin (20, μg/ml) or poly-L-lysine (100 μg/ml) at a cell density of approximately 25% confluence. Potentially, cells anchored via specific anti-integrin antibodies could be used instead. Cells are loaded with Fura-2AM (1–2 μ*M*) for 30–60 min, rinsed with Hanks' balanced salt solution (HBSS) containing 1 m*M* $CaCl_2$ and 1 m*M* $MgCl_2$, and mounted in a flowthrough chamber (0.2-ml

[14] S. Short, J. Boyer, R. Juliano, *J. Biol. Chem.* **275,** 12970, 2000.

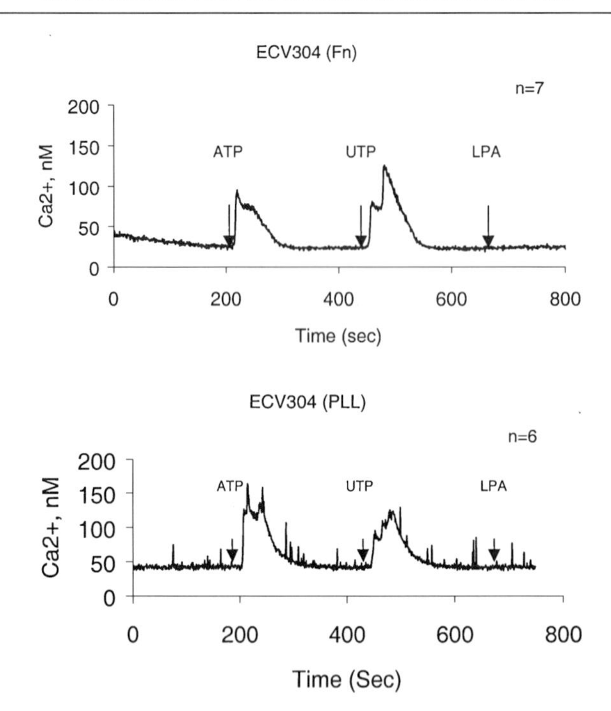

FIG. 3. Upstream events in GPCR signaling. ATP and UTP show anchorage-independent release of intracellular calcium in ECV304 cells. Cells were plated on fibronectin (Fn) or held in suspension for 1–3 hr. Suspension samples were replated on poly-L-lysine (PLL) for brief periods for calcium analysis. Cells were loaded with Fura-2 prior to challenge with 30-sec pulses of agonist.

volume) continuously perfused with HBSS. Addition of GPCR agonists such as ATP (for P2Y receptors) or lysophosphatidic acid (LPA) (for G_i-coupled receptors) is done for 30 sec at various concentrations, and is controlled by a valve attached to a gravity-driven six-well reservoir. The flowthrough chamber is secured on the stage of a Nikon (Garden City, NY) Diaphot inverted fluorescence microscope. Cells are exposed to alternating excitation wavelengths of 340 and 380 nm while fluorescence emission at 510 nm is monitored by a silicon-intensified tube (SIT) camera. The ratio of emission fluorescence determined at 340 and 380 nm is converted to $[Ca^{2+}]_i$, using the equation of Grynkiewicz *et al.*[15] Data are recorded and processed with an image I/FL digital imaging system (Universal Imaging, West Chester, PA). Figure 3 illustrates data on calcium transients in cells that are adherent to fibronectin, via integrins, or nonspecifically attached via polylysine. It demonstrates that ATP or UTP triggers calcium transients independently of cell adhesion.

[15] G. Grynkiewicz, M. Poenie, and R. Tsien, *J. Biol. Chem.* **260,** 3440 (1985).

Summary

Assays for use in integrin-mediated or -modulated signaling are essentially the same as those used in signaling studies involving growth factors or hormones. The major differences are the manipulations of the cells to compare the effects of gain or loss of anchorage, or the role of specific adhesion receptors. This needs to be done with some care, and with thought given to the overall biology of the particular cell type under investigation. Harsh treatment of cells, for example, prolonged suspension culture, may result in irreversible nonphysiological effects in some types of cells.

[16] R-Ras Regulation of Integrin Function

By PAUL E. HUGHES, BEAT OERTLI, JAEWON HAN, and MARK H. GINSBERG

Introduction

Integrins are heterodimeric cell–cell and cell–matrix adhesion receptors that play a central part in controlling cell growth, survival, migration, and tumor metastasis. A characteristic feature of certain integrins is their ability to dynamically modulate their affinity for ligand in response to extracellular cues, a process termed inside-out signaling or activation.[1] Integrin activation is central to such diverse processes as platelet aggregation, the organization of a fibronectin matrix, and the control of cell migration, which is dependent on rapid, controlled changes in integrin-dependent cell adhesion, suggesting that the coordinated activation and deactivation of integrins play a role in this process.

Presently, the signal transduction pathways controlling integrin activation are poorly understood. However, studies have demonstrated that the members of the Ras family of small GTP-binding proteins and their downstream effectors are key players in regulating integrin activation.[2–6] Zhang *et al.* originally reported that

[1] P. E. Hughes and M. Pfaff, *Trends Cell Biol.* **8,** 359 (1998).

[2] T. Sethi, M. H. Ginsberg, J. Downward, and P. E. Hughes, *Mol. Biol. Cell* **10,** 1799 (1999).

[3] P. E. Hughes, M. W. Renshaw, M. Pfaff, J. Forsyth, V. M. Keivens, M. A. Schwartz, and M. H. Ginsberg, *Cell* **88,** 521 (1997).

[4] Z. Zhang, K. Vuori, H. Wang, J. C. Reed, and E. Ruoslahti, *Cell* **85,** 61 (1996).

[5] M. Osada, T. Tolkacheva, W. Li, T. O. Chan, P. N. Tsichlis, R. Saez, A. C. Kimmelman, and A. M. Chan, *Mol. Cell. Biol.* **19,** 6333 (1999).

[6] J. X. Zou, B. Wang, M. S. Kalo, A. H. Zisch, E. B. Pasquale, and E. Ruoslahti, *Proc. Natl. Acad. Sci. U.S.A.* **96,** 13813 (1999).

0076-6879/00 $35.00

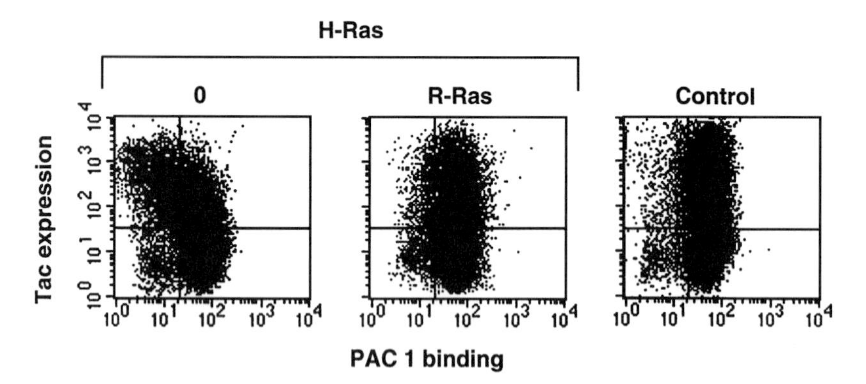

FIG. 1. Activated R-Ras rescues the suppression of integrin activation by H-Ras (G12V). αβ-py cells were transiently transfected with an expression vector encoding Tac-α_5 alone, or Tac-α_5 plus H-Ras(G12V). In a separate transfection Tac-α_5 plus H-Ras(G12V) was cotransfected with a plasmid encoding R-Ras(G38V). The cells were harvested and stained for Tac expression (ordinate) and PAC1 binding (abscissa). In the Ras(G12V)-transfected cells there is a leftward shift of the dot plot in the upper quadrants representing a reduction in PAC1 binding. This shift is reversed by the cotransfection of activated R-Ras(G38V). In the empty vector control transfection there was no suppression of PAC1 binding in the Tac-α_5-expressing cells. [Reprinted from *Molecular Biology of the Cell* (1999, volume 10, 1799–1809) with permission by the American Society for Cell Biology.]

expression of an activated variant of R-Ras promoted the adhesion of mouse 32D.3 and human U937 cells to fibronectin and vitronectin, most likely by stimulating integrin activation.[4] Studies by Osada *et al.* suggest that phosphatidylinositol (PI) 3-kinase may be the critical R-Ras effector responsible for promoting the adhesion of 32D.3 cells.[5]

In contrast to R-Ras, activated H-Ras and its downstream effector kinase Raf-1 were found to suppress the activation of integrins in chinese hamster ovary (CHO) cells. Whereas R-Ras(G38V) does not directly activate integrins in CHO cells, it can via the activation of an unknown R-Ras effector reverse the suppression of integrins induced by the Ras/Raf-initiated integrin suppressor pathway.[2,3] The effect of H-Ras and R-Ras on integrin activation in CHO cells is illustrated in Fig.1. These observations suggests that R-Ras could act in concert with Ras to regulate integrin affinity, and thus the adhesive properties of a cell.

In this chapter, we describe protocols designed to analyze the role of both Ras and R-Ras in regulating integrin function. Specifically, we describe in detail a fluorescence activated cell sorting (FACS)-based assay to determine the effect of activated H-Ras and R-Ras on integrin ligand-binding affinity. In addition, we also describe methods designed to assay the effect of activated R-Ras on integrin-mediated cell adhesion.

Considerations

When designing a method to examine the role of Ras proteins in regulating integrin affinity, several factors need to be taken into consideration. First, it is necessary to use a cell type expressing an integrin whose affinity state can be routinely measured. Second, given the well-documented affects of Ras and R-Ras on cell growth and morphology it is highly advantageous if the chosen cell line can efficiently express recombinant H-Ras or R-Ras. In addition, the routine transfection of the chosen cell line will allow the rapid testing of multiple combinations of potential regulators of integrin function.

To address these problems we have generated CHO cell lines stably expressing the platelet-specific integrin $\alpha_{IIb}\beta_3$ (A5 cells) and $\alpha_{IIb}\beta_3$ chimeras such as $\alpha_{IIb}\alpha_{6A}\beta_3\beta_1$ ($\alpha\beta$-py cells).[7,8] This chimeric integrin is composed of the extracellular domains of the platelet integrin $\alpha_{IIb}\beta_3$ and the cytoplasmic domains of $\alpha_{6A}\beta_1$. This integrin has the ligand-binding properties of $\alpha_{IIb}\beta_3$, while its activation state is controlled by the $\alpha_{6A}\beta_1$ cytoplasmic domains. In contrast to wild-type $\alpha_{IIb}\beta_3$ this chimera is constitutively activated in CHO cells and binds activation-dependent ligands with high affinity.[9]

The advantage of using the $\alpha_{IIb}\beta_3$ extracellular domain as the reporter of activation is the availability of natural ligands and several monoclonal antibodies that bind with high affinity only to the activated conformation of $\alpha_{IIb}\beta_3$, the most notable of these being a ligand-mimetic antibody called PAC-1.[10]

We have used the ligand-mimetic antibody PAC-1 to develop a two-color flow cytometry-based assay to determine the activation of $\alpha_{IIb}\beta_3$ and $\alpha_{IIb}\beta_3$ chimeras stably expressed in CHO cells. A flow diagram illustrating the various steps in this procedure is illustrated in Fig. 2. To specifically measure PAC-1 binding in cells transiently transfected with the plasmid of interest we contransfect an expression vector, pCMV-Tac-α_5, to facilitate the identification of transfected cells. Tac-α_5 is composed of the Tac subunit of the interleukin 2 (IL-2) receptor fused to the cytoplasmic domain of the integrin α_5 subunit cytoplasmic domain.[11] This construct was chosen to provide a marker for transfected cells, as it is well expressed in CHO cells and the hybridoma producing the anti-Tac monoclonal antibody is readily available from the American Type Culture Collection (ATCC, Manassas, VA).

[7] E. K. Baker, E. C. Tozer, M. Pfaff, S. J. Shattil, J. C. Loftus, and M. H. Ginsberg, *Proc. Natl. Acad. Sci. U.S.A.* **94,** 1973 (1997).

[8] T. E. O'Toole, J. C. Loftus, X. P. Du, A. A. Glass, Z. M. Ruggeri, S. J. Shattil, E. F. Plow, and M. H. Ginsberg, *Cell Regul.* **1,** 883 (1990).

[9] T. E. O'Toole, Y. Katagiri, R. J. Faull, K. Peter, R. Tamura, V. Quaranta, J. C. Loftus, S. J. Shattil, and M. H. Ginsberg, *J. Cell Biol.* **124,** 1047 (1994).

[10] S. J. Shattil, J. A. Hoxie, M. Cunningham, and L. F. Brass, *J. Biol. Chem.* **260,** 11107 (1985).

[11] S. E. LaFlamme, S. K. Akiyama, and K. M. Yamada, *J. Cell Biol.* **117,** 437 (1992).

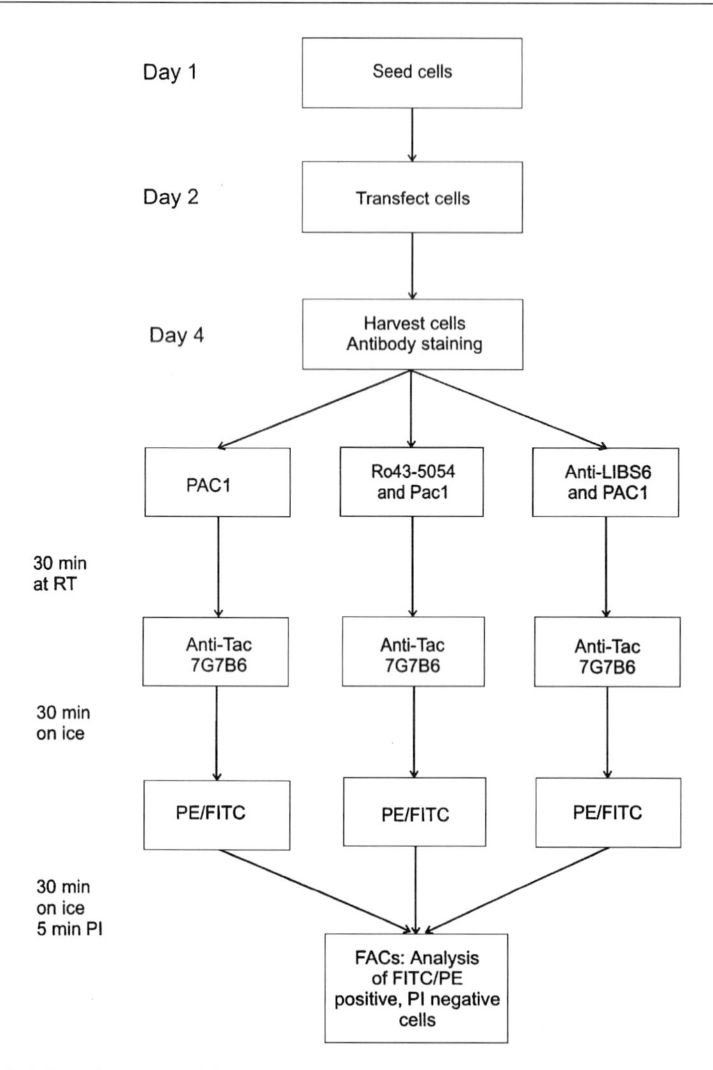

FIG. 2. A flow chart summarizing the PAC1-binding assay used to measure integrin activation.

PAC-1 Binding Assay

Transfection of CHO Cells

The following protocol outlines the various steps involved in analyzing effect of R-Ras and H-Ras on integrin activation in αβ-py cells. Twenty-four hours prior to transfection cells are plated in 100-cm dishes at a density of 2×10^6 cells per plate. The αβ-py cells are then transfected with LipofectAMINE (GIBCO-BRL, Gaithersburg, MD), which in our hands gives high levels of transfection;

typically between 30 and 60% of cells express Tac-α_5. For each experiment it is necessary to include a control plate in which $\alpha\beta$-py cells are transfected with Tac-α_5 alone. In the other dishes the cells are cotransfected with 2 μg of DNA encoding Tac-α_5 and expression vectors encoding cDNAs of interest. To achieve optimum suppression of integrin activation in $\alpha\beta$-py cells we transfect each 100-cm dish with 4 μg of pcDR-H-Ras(G12V). Suppression can resuced by the cotransfection of 2 μg of pSG5-R-Ras(G38V). For each transfection we standardize the total amount of DNA presented to each plate to a total of 10 μg by the addition of plasmid vector DNA.

For each transfection mix up to 10 μg of plasmid DNA with 180 μl of Dulbecco's modified Eagle's medium (DMEM) and 20 μl of LipofectAMINE in 5-ml polystyrene round-bottom tubes (Falcon; Becton Dickinson Labware, Lincoln Park, NJ). The tubes are then incubated at room temperature for 10 min, before adding 3.8 ml of DMEM to bring the final volume to 4 ml. At this time remove the medium from each of the plates and wash the cells with prewarmed phosphate-buffered saline (PBS). Immediately after the wash gently add DMEM–DNA–LipofectAMINE mix to the cells. After 6 hr remove the DNA–LipofectAMINE mix, wash the cells once with prewarmed complete DMEM, and leave them to recover in fresh complete DMEM.

Harvesting and Staining of Transfected Cells

For each plate to be analyzed add diluted PAC-1 ascites to three 5-ml polystyrene round-bottom tubes (Falcon). Prior to the start of the experiment the PAC-1 ascites should be titered and the appropriate dilution for FACS analysis determined. The first of the tubes measures basal PAC-1 binding. The second and third of these tubes function as controls to measure background, that is, non-integrin-mediated, PAC-1 binding, and maximal PAC-1 binding. To the second tube in the set of three we add an $\alpha_{IIb}\beta_3$ antagonist such as Ro43-5054 (a kind gift of B. Steiner, Hoffman-LaRoche, Basel, Switzerland) to block PAC-1 binding. Any $\alpha_{IIb}\beta_3$ antagonist can be substituted here, including eptifibatide (Integrilin; COR Therapeutics, South San Francisco, CA), tirofiban (Aggrestat; Merck, Rahway, NJ), or lamifiban (Hoffman-LaRoche). To achieve maximal PAC-1 binding we add an activating anti-β_3 antibody, anti-LIBS6, to a third tube in each set. It should be noted that there are less expensive alternatives to both Ro43-5054 and anti-LIBS6. In general, ligand binding to integrins can be blocked by 5 mM EDTA and integrins can be activated by the addition of 0.25 mM $MnCl_2$. Finally, for each experiment it is also necessary to set up the following three compensation controls for the FACS: (1) a no-antibody control; (2) a fluorescein isothiocyanate (FITC)-positive only control; and (3) a phycoerythrin (PE)-positive only control.

Forty-eight hours after the start of the transfection remove the medium from the cells and wash once with PBS. Then add 3 ml of trypsin–EDTA solution (Irvine Scientific, Santa Ana, CA), and incubate the cells for 3 min at room temperature.

When the cells begin to detach after mild agitation add 3 ml of complete medium to neutralize the trypsin and transfer the cells into a 15-ml centrifuge tube. Spin the cells for 5 min at 1000 rpm in tabletop centrifuge at room temperature and aspirate off the supernatant. Wash the cells once in 10 ml of DMEM, pellet again by centrifugation, and resuspend the cells in 100 μl of DMEM. Add 25 μl of the cell suspension to the tubes containing the diluted PAC-1 and incubate at room temperature for 30 min.

Add 0.5 ml of DMEM to each tube and centrifuge at 1000 rpm for 5 min at 4°. Decant the liquid and add 50 μl of DMEM–2% (v/v) biotinylated 7G7B6 and incubate the cells on ice for 30 min. Wash the cells by adding 0.5 ml of DMEM to each tube and centrifuging them at 1000 rpm for 5 min at 4°, remove the DMEM, and resuspend the cells in 50 μl of DMEM containing 4% (w/v) streptavidin–phycoerythrin (Molecular Probes, Eugene, OR) and 4% (w/v) FITC-labeled anti-mouse IgM μ chain antibodies (Biosource, Howell, NJ). The cells are then incubated on ice in the dark for 30 min. In the last 5 min of the incubation add propidium iodide (Sigma, St. Louis, MO) at 2 μg/ml to allow for the identification of dead cells. Wash the cells once with 0.5 ml of cold PBS, pH 7.4, spin the cells down at 1000 rpm for 5 min at 4°, and resuspend them in 0.5 ml of cold PBS. Analyze the cells by fluorescence, using a FACSCaliber flow cytometer (Becton Dickinson, Mountain View, CA) and CellQuest software.

Data Analysis

PAC-1 binding (FL1 channel) is analyzed on the gated subset of live cells (propidium iodide negative, FL3) that is strongly positive for Tac-α_5 expression (FL2 channel). To obtain numerical estimates of PAC-1 binding we calculate a numeric activation index (AI) defined as $100(F_0 - F_r)/(F_0\text{LIBS6} - F_r)$, where F_0 is the median fluorescence intensity of PAC-1 binding; F_r, is the median fluorescence intensity of PAC-1 binding in the presence of competitive inhibitor (Ro43-5054, 1 μM) and F_0LIBS6 is the median fluorescence intensity of PAC-1 binding in the presence of 2 μM anti-LIBS6. A percent inhibition of PAC-1 binding can also be calculated as $100(\text{AI}_0 - \text{AI})/\text{AI}_0$, where AI_0 is the activation index in the absence of a cotransfected suppressor of integrin activation and AI is the activation index in its presence.

Alternative Assays to Measure Effect of R-Ras on Integrin Activation

FACS-based assays similar to the assay described above have been devised to examine soluble-ligand binding to $\alpha_v\beta_3$ and $\alpha_4\beta_1$. In the case of $\alpha_v\beta_3$ a novel anti-$\alpha_v\beta_3$ ligand-mimetic antibody (WOW-1) was used to probe activation, whereas a vascular cell adhesion molecule (VCAM)–Cκ fusion protein composed of the seven immunoglobulin domains of human VCAM-1 coupled to the murine Fc heavy chain was used to measure the affinity state of $\alpha_4\beta_1$.[12,13] The development of these reagents opens the possibility to the design of experiments analyzing the role of R-Ras and other Ras family members in regulating the activation of these integrins in a wide variety of cell types. Indeed, Rose *et al.* have already demonstrated that H-Ras(G12V) can supress the activation of $\alpha_4\beta_1$ expressed in CHO cells.[13]

The flow cytometry-based assays described above have many advantages, the greatest being the possibility to test rapidly whether R-Ras and its downstream effectors are influencing integrin affinity modulation. However, one drawback of these types of FACS-based assays involving the transient expression of R-Ras and other regulators of integrin function is the comparative difficulty in obtaining exact measurements of changes in ligand-binding affinity. However, as demonstrated by Pampori *et al.*, it is possible to analyze ligand binding by flow cytometry and obtain an apparent K_d of ligand binding.[12]

Cell Adhesion Assays

In addition to stimulating soluble ligand binding to integrins, activated R-Ras can also promote integrin-mediated adhesion to collagen, vitronectin, and fibronectin.[4–6] Prior to assaying the effect of R-Ras on cell adhesion it is desirable to establish stable lines in the cell type of interest, in which the expression of R-Ras is placed under the control of an inducible promoter. Placing the expression of the R-Ras variants under the influence of an inducible promoter reduces the possibility of changes in adhesive properties being the result of long-term overexpression of R-Ras. In addition, prior to comparing the adhesive properties of different cell lines it is necessary to confirm that integrin expression is comparable in all the cell lines being examined, to ensure that any observed differences in adhesion are not a product of variations in the number of integrin receptors on the cell surface.

Fluorescent Cell Adhesion Assay

The following flourescent cell adhesion assay has been successfully used for a wide variety of different cell types, for example, k562, Jurkat, and Hut78. Before measuring changes in adhesion due to the activation of R-Ras or other genes of interest, initial dose–response experiments must be performed for each individual substrate and cell type to determine the optimum coating concentration. For most substrates (e.g., fibronectin and fibrinogen) maximal adhesion can normally be obtained with coating concentrations of substrate in the range of 10 μg/ml.

Precoat Immunolon II 96-well plates with a range of substrate concentrations, for example, serial dilutions from 10 μg/ml, making sure that some wells are left

[12] N. Pampori, T. Hato, D. G. Stupack, S. Aidoudi, D. A. Cheresh, G. R. Nemerow, and S. J. Shattil, *J. Biol. Chem.* **274,** 21609 (1999).

[13] D. M. Rose, P. M. Cardarelli, R. R. Cobb, and M. H. Ginsberg, *Blood* **95,** 602 (2000).

empty for nonspecific binding controls and background fluorescence measurements. To allow for pipetting errors coat the substrate in triplicate at each coating concentration. Incubate the substrate at the desired concentrations in 0.1 *M* $NaHCO_3$ (pH 8.0) for 2 hr at 37°, or overnight at 4°. The plates are then blocked by adding a final concentration of 2% (w/v) heat-inactivated (12 min at 85°) BSA for 2 hr at room temperature. Coat one row of empty wells with BSA also. These will be control wells for nonspecific adhesion. After blocking with BSA, wash the plates four times in modified Tyrode's buffer [12 m*M* $NaHCO_3$, 150 m*M* NaCl, 2.5 m*M* KCl, BSA (1 mg/ml), dextrose (1 mg/ml), pH 7.4].

Wash the cell suspension twice in serum-free medium (this is necessary as serum can contain esterases that can cleave the fluorescent dye and make it impermeable to the cell membrane), and resuspend the cells at a concentration of 1×10^6 cells/ml. For the majority of cell lines incubate the fluorescent dye (CellTracker Green CMFDA, excitation wavelength of 492 nm, emission wavelength of 516 nm; Molecular Probes) at 2 μ*M* for 30 min at 37° in serum-free medium. It should be noted that different cell types may not retain the fluorescent dye with high efficiency, possibly because of the activity of nonspecific ion channels. If this is the case, the ion channel inhibitor probenecid can be added to facilitate dye retention.

Wash the cells twice in modified Tyrode's buffer and resuspend them at a final concentration of 1×10^6 cells/ml. Plate 50 μl of cells in each well of a 96-well plate, leaving some wells empty for the measurement of background fluorescence.

To allow the cells to adhere to the matrix place the plates in a 37° incubator for at least 1 to 1.5 hr, and then read the fluorescence on a Cytofluor II plate reader. With prewarmed modified Tyrode's buffer, gently wash the plates four times to remove nonadherent cells. This process can be modified according to how efficiently each cell line adheres to the substrate. Generally, wash the plates sufficiently well that the cell number in the BSA-coated control wells is less than 10% of the number of cells in the wells in which maximal adhesion is supposed to occur. This can be monitored on an inverted microscope.

After washing, read the plate on a Cytofluor II plate reader. The efficiency of cell adhesion at each substrate coating concentration can be expressed as the percentage of adherent cells, using the following formula: $(\text{Fluor}_{\text{af}} - \text{Fluor}_{\text{bkgd}})/(\text{Fluor}_{\text{bf}} - \text{Fluor}_{\text{bkgd}}) \times 100$, where Fluor_{bf} is the fluorescence before washing, Fluor_{af} is the fluorescence after washing, and $\text{Fluor}_{\text{bkgd}}$ is the fluorescence in wells without cells. To obtain a graph illustrating the efficiency of cell adhesion, plot the percent cell adhesion, plot the percent cell adhesion against substrate coating concentration.

Alternative Cell Adhesion Assays

The method described has the advantage of being able to recover the cells after the adhesion assay for further analysis. However, this method does depend on the availability of a fluorescence plate reader. Should one not be available it is possible to measure cell adhesion by the assay described above, with the following modifications. Do not label the cells with the fluorescent dye; instead, add unlabeled cells to each well and proceed as described above. Prior to washing the plate to remove unattached cells add an identical number of cells to three empty wells and allow the cells to settle. Do not wash these cells. To fix these cells, add 8 μl of 50% (v/v) glutaraldehyde and incubate at room temperature for 20 min. The signal from these wells will be used to defined 100% cell adhesion. After the final wash fix the cells in 100 μl of 2% (v/v) glutaraldehyde in PBS for 10 min. Remove the PBS–glutaraldehyde and allow the plates to air dry for approximately 5 min. Stain the cells carefully by adding 100 μl of 0.1% (w/v) crystal violet in 20% (v/v) methanol. When adding the stain take care not to leave droplets on the side of the well. After 45 min remove the dye and wash the cells three times in PBS and allow the plate to air dry after the final wash. Solubilize the cells in 100 μl of 100 m*M* sodium citrate and allow sufficient time for the signal to develop before measuring the absorbance of the plate at 560 nm, using a microplate reader.

The efficiency of cell adhesion at each substrate coating concentration can be expressed as the percentage of adherent cells, using the following formula: ($Abs_{560\ nm}$ adherent cells/$Abs_{560\ nm}$ of total number of cells) × 100. To obtain a graph illustrating the efficiency of cell adhesion, plot percent cell adhesion against substrate coating concentration.

Analysis of Cell Adhesion Assays

Integrin-dependent cell adhesion assays can be influenced by several factors; changes in cell shape, variations in integrin receptor number, changes in integrin affinity, and local density can all affect cell adhesion. Therefore, caution must be used when interpreting the results of cell adhesion assays in 96-well plates, with each of these factors taken into consideration. Variations in integrin expression can be assayed by FACS, and changes in integrin affinity can be assayed by the soluble ligand-binding assays described earlier in this chapter. If necessary, cell shape can be examined by inverted phase microscopy.

Acknowledgments

We thank Dr. Darren Woodside for insightful comments on cell adhesion assays. P. E. H. is grateful for the support of the Leukemia Society of America. B. O. is supported by grants from the Swiss National Science Foundation and Novartis, J. H. is supported by a fellowship from the Arthritis Foundation, and M. H. G. is supported by grants HL57900, HL48728, and AR27214 from the National Institutes of Health.

[17] Caveolin and Ras Function

By ROBERT G. PARTON and JOHN F. HANCOCK

Introduction

Ras proteins must be localized to the inner surface of the plasma membrane to be biologically active. This reflects the requirement for Ras effectors to be recruited to the plasma membrane in order to signal. For example, the plasma membrane recruitment of Raf allows access to specific kinases and phospholipids that result in Raf activation. Interestingly, an ever-increasing body of data indicates that although the three ubiquitously expressed Ras isoforms H-, N-, and K-Ras all interact with the same set of effectors *in vitro,* they activate these effectors with different potency *in vivo.*[1] This is important because such quantitative biochemical differences can offer a rational explanation for biological differences between the Ras proteins, including the essential role for K-Ras but not H- or N-Ras in mouse development and the selective activation of specific Ras isoforms in different human tumors.

It is now clear that these *in vivo* differences can be attributed in large part to the different membrane anchors that are posttranslationally attached to the Ras C termini. The C-terminal C*aa*X motif (where *a* indicates any aliphatic amino acid) of all three Ras proteins is farnesylated, *aa*X proteolyzed, and methyl esterified. The H- and N-Ras anchor is then completed by palmitoylation of adjacent cysteine residues, but with different stoichiometries, whereas the second motif of the K-Ras anchor comprises an adjacent polybasic domain of six contiguous lysines.[2] These anchors direct N- and H-Ras proteins from the endoplasmic reticulum (ER) to the plasma membrane through the exocytic pathway, but traffic K-Ras through an alternate route to the cell surface.[3,4] Once at the plasma membrane H- and K-Ras may function in discrete domains of the plasma membrane, this segregation being facilitated by the membrane association mechanisms of the two isoforms. We have postulated that direct interaction of the palmitoyl group of H-Ras with lateral arrays of specific lipids, so-called lipid rafts, might provide a means of concentrating H-Ras in discrete regions of the plasma membrane.[5,6] Lipid rafts rely on the lateral association of lipids (particularly cholesterol and glycosphingolipids) within

[1] J. Yan, S. Roy, A. Apolloni, A. Lane, and J. F. Hancock, *J. Biol. Chem.* **273,** 24052 (1998).
[2] J. F. Hancock, H. Paterson, and C. J. Marshall, *Cell* **63,** 133 (1990).
[3] A. Apolloni, I. Prior, M. Lindsay, R. G. Parton, and J. F. Hancock, *Mol. Cell Biol.* **20,** 2475 (2000).
[4] E. Choy, V. K. Chiu, J. Silletti, M. Feoktisitov, T. Morimoto, D. Michaelson, I. E. Ivanov, and M. R. Philips, *Cell* **98,** 69 (1999).
[5] S. Roy, R. Luetterforst, A. Harding, A. Apolloni, M. Etheridge, E. Stang, B. Rolls, J. F. Hancock, and R. G. Parton, *Nat. Cell Biol.* **1,** 98 (1999).
[6] T. V. Kurzchalia and R. G. Parton, *Curr. Opin. Cell Biol.* **11,** 424 (1999).

METHODS IN ENZYMOLOGY, VOL. 333
0076-6879/00 $35.00

the plane of the membrane, based on the biophysical properties of those lipids, and provide a simple mechanism for segregating specific lipids or lipid-anchored proteins from the bulk of the plasma membrane.[7,8] Although palmitoylated H- and N-Ras may associate with raft domains, we have proposed that because a polybasic domain anchors K-Ras to the membrane through a charge interaction, it would not be functionally associated with lipid rafts.

Evidence in support of this general model was provided by studies of a particular type of lipid raft domain, the caveola. Caveolae represent a specialized form of lipid microdomain with a characteristic morphology (55- to 80-nm uncoated flask-shaped pit) and by the presence of caveolins (caveolins-1, -2, and -3).[6,9,10] Caveolin-1, the best-characterized member of the family, binds cholesterol[11] and can generate caveolae *de novo* when expressed in cells normally lacking caveolae.[12] As well as being involved in caveola formation, caveolin-1 has been implicated in regulating and directly transporting free cholesterol.[13] Caveolae have been proposed to represent sites of free cholesterol efflux.[14] At the same time, caveolins have been shown to have direct interactions with signaling molecules including Raf, protein kinase C isoforms, and trimeric G protein α subunits, and to play a modulatory role in growth factor-stimulated pathways leading to mitogen-activated protein (MAP) kinase activation.[9,15,16] Consistent with this role, it has been shown that ablation of caveolin expression can trigger uncontrolled cell proliferation whereas heterologous expression of caveolin reverses the transformed phenotype.[17] Cell transformation *in vitro* has also been shown to cause downregulation of caveolae and caveolin.[18] Moreover, caveolin-1 was also independently identified as a candidate tumor suppressor gene by screening for genes with expression that was downregulated in human mammary carcinomas.[19] The antiproliferative activities of caveolin-1 expression have also been suggested to be related to the role of

[7] K. Simons and E. Ikonen, *Nature (London)* **387,** 569 (1997).

[8] R. G. Parton and K. Simons, *Science* **269,** 1398 (1995).

[9] T. Okamoto, A. Schlegel, P. E. Scherer, and M. P. Lisanti, *J.Biol. Chem.* **273,** 5419 (1998).

[10] R. G. Parton, *Curr. Opin. Cell Biol.* **8,** 542 (1996).

[11] M. Murata, J. Peranen, R. Schreiner, F. Wieland, T. V. Kurzchalia, and K. Simons, *Proc. Natl. Acad. Sci. U.S.A.* **92,** 10339 (1995).

[12] A. M. Fra, E. Williamson, K. Simons, and R. G. Parton, *Proc. Natl. Acad. Sci. U.S.A.* **92,** 8655 (1995).

[13] C. J. Fielding and P. E. Fielding, *J. Lipid Res.* **38,** 1503 (1997).

[14] P. E. Fielding and C. J. Fielding, *Biochemistry* **34,** 14288 (1995).

[15] R. G. Anderson, *Annu. Rev. Biochem.* **67,** 199 (1998).

[16] J. A. Engelman, X. Zhang, F. Galbiati, D. Volonte, F. Sotgia, R. G. Pestell, C. Minetti, P. E. Scherer, T. Okamoto, and M. P. Lisanti, *Am. J. Hum. Genet.* **63,** 1578 (1998).

[17] F. Galbiati, D. Volonte, J. A. Engelman, G. Watanabe, R. Burk, R. G. Pestell, and M. P. Lisanti, *EMBO J.* **17,** 6633 (1998).

[18] A. J. Koleske, D. Baltimore, and M. P. Lisanti, *Proc. Natl. Acad. Sci. U.S.A.* **92,** 1381 (1995).

[19] R. Sager, S. Sheng, A. Anisowicz, G. Sotiropoulu, Z. Zou, G. Stenman, and K. Swisshelm, *Cold Spring Harbor Symp. Quant. Biol.* **59,** 537 (1994).

caveolin in regulating intracellular cholesterol.[20] Disruption of cellular cholesterol is known to have drastic effects on signaling pathways, including those involving p42/44 MAP kinase.[21]

To address these issues and, in particular, to address the question of how caveolins can play a role in both signal transduction and cholesterol homeostasis, a novel screening approach for caveolin dominant-negative mutants was adopted.[5] This resulted in the identification of two mutant caveolin proteins that acted as inhibitors of simian virus 40 (SV40) infection, a caveola-mediated process.[22] These mutants were then screened for their effects on Ras signaling in a Raf activation assay.[5] One of these, termed Cav^{DGV}, was shown to be a potent inhibitor of H-Ras-mediated Raf activation but had minimal effects on K-Ras signaling. The Cav^{DGV} mutant was not targeted to the plasma membrane but accumulated in lipid-rich membrane-bound vesicles. This suggested that the effects of this inhibitory mutant were not mediated through direct interaction with Ras or Raf, which were recruited normally to the plasma membrane as in control (wild-type caveolin transfected or untransfected) cells. A clue to the selective effect of the Cav^{DGV} mutant on H-Ras signaling was provided by the finding that a cholesterol probe accumulated in the Cav^{DGV}-containing cytoplasmic bodies. In addition, the density of H-Ras-, but not K-Ras-containing membranes, was significantly affected by the caveolin mutant. These data indicated that an effect on cellular cholesterol might underlie the selective effect on H-Ras. This was directly tested by adding cholesterol back to the Cav^{DGV}-expressing cells and showing that cholesterol addition rescued the inhibition, whereas cholesterol removal from the plasma membrane replicated the inhibition. Together, these results suggest that caveolin may regulate cholesterol transport to plasma membrane signaling domains and it is this transport that is disrupted by the Cav^{DGV} mutant. Moreover, they show that H-Ras and K-Ras signaling domains are affected differently by cholesterol depletion, and are therefore presumably localized in distinct surface domains.[5] Alternatively, access of H-Ras, but not K-Ras, to a common signaling domain might depend on an appropriate cholestrol-containing environment.

In this chapter we describe two experimental techniques that allow selective manipulation of Ras/plasma membrane interactions without affecting Ras posttranslational processing. Although the two approaches of expressing dominant-negative caveolin or treating cells with cholesterol-binding drugs appear fundamentally different, they have a common mechanism of action: N-terminally truncated caveolin proteins interrupt the flow of cholesterol to the cell surface, while cholesterol-binding drugs directly extract cholesterol from the plasma membrane. The result is a disordering and loss of integrity of lipid rafts.

[20] C. J. Fielding, A. Bist, and P. E. Fielding, *Biochemistry* **38,** 2506 (1999).

[21] T. Furuchi and R. G. Anderson, *J. Biol. Chem.* **273,** 21099 (1998).

[22] E. Stang, J. Kartenbeck, and R. G. Parton, *Mol. Biol. Cell* **8,** 47 (1997).

Monitoring the effect of cholesterol depletion on cell surface domains has classically involved the use of sucrose or other gradients to separate membrane microdomains on the basis of their buoyant density. This approach is usually coupled with a Triton solubilization step to further purify lipid rafts on the basis of their detergent insolubility. When studying Ras, however, such a step must be excluded because, although palmitoylated Ras is apparently functionally associated with cholesterol-dependent lipid "rafts," it is nevertheless completely solubilized out of cell membranes by detergents. We describe two gradient protocols that can be used to resolve Ras- and caveolin-containing microdomains from tissue culture cell membranes.

The conclusions that can be drawn from analyzing sucrose gradients are necessarily limited. Fractions are characterized by the presence of specific membrane markers identified by immunoblotting. A common mistake is overinterpretation of results: thus while the comigration of a signaling protein with caveolin over a sucrose gradient is consistent with that protein being found in caveolae, it is not conclusive proof, because other membranes are invariably found to be present when gradient fractions are analyzed by electron microscopy. There is no substitute for directly visualizing the localization of the protein of interest at the cell surface. We describe here a novel protocol for visualizing the inner surface of the plasma membrane. The technique preserves the morphology of the plasma membrane and thus allows for quantitative immunoelectron microscopic localization of Ras and other signaling proteins.

Methods

Caveolin Mutants and Transfection

A number of caveolin constructs have been generated from wild-type caveolin-3 by standard polymerase chain reaction (PCR) methodology and confirmed by sequencing.[23] Cav^{DGV} corresponds to residues 54–151 of caveolin-3 and therefore lacks most of the N-terminal cytoplasmic domain. In all our experiments we have used Cav^{DGV} appended with a C-terminal hemagglutinin (HA) epitope tag, or an N-terminal green fluorescent protein (GFP) tag. Constructs are expressed under the control of a cytomegalovirus (CMV) promoter in the CB6 vector.[24] For transfection purposes DNA should be purified by an alkaline lysis method.[25]

1. Split baby hamster kidney (BHK) cells onto dishes/coverslips the day before the transfection, ensuring that cells are well dispersed. The confluence of the cells

[23] R. Luetterforst, E. Stang, N. Zorzi, A. Carozzi, M. Way, and R. G. Parton, *J. Cell Biol.* **145,** 1443 (1999).

[24] M. Way and R. G. Parton, *FEBS Lett.* **376,** 108 (1995).

[25] I. Feliciello and G. Chinali, *Anal. Biochem.* **212,** 394 (1993).

at the time of transfection depends on the particular application for which the cells are intended (e.g., 40% for immunofluorescence to 80% for biochemistry).

2. To transfect cells in a 3.5-cm-diameter dish, either seeded on coverslips, or grown on the dish: mix 1 μg of DNA with 5 μl of LipofectAMINE in 100 μl of Opti-MEM (supplemented with L-glutamine) (GIBCO-BRL, Gaithersburg, MD) and incubate for 15 min at room temperature. Add 5 μl of LipofectAMINE Plus (GIBCO-BRL) reagent to another 100 μl of Opti-MEM, mix with the DNA–LipofectAMINE solution, and incubate for 30 min.

3. Wash the cells with 1 ml of Opti-MEM. Add 800 μl of the same medium and add the DNA–LipofectAMINE mix.

4. Leave the mix on the cells for 5–7 hr (routine method for immunofluorescence), and then add an equal volume of normal growth medium. Alternatively, to increase transfection efficiency (e.g., for biochemistry) leave the cells in transfection mixture for up to 24 hr.

5. Assay transfection efficiency on coverslips by immunofluorescence, using standard methods.

Cholesterol Depletion with Cyclodextrin

The selective effect of Cav^{DGV} on H-Ras function is faithfully mimicked by acute depletion of cholesterol from the plasma membrane, using cyclodextrin, and can be fully reversed by cholesterol repletion, using a mixture of cyclodextrin and cholesterol. The following protocols for manipulating the cholesterol content of cell surface cholesterol-rich microdomains are based on those first published by Furuchi and Anderson.[21]

1. Make up stock solutions of: 10% (w/v) β-methylcyclodextrin (Sigma, St. Louis, MO) in Dulbecco's modified Eagle's medium (DMEM) and filter sterilize; and cholesterol (Sigma), 20 mg/ml in 100% ethanol.

2. For cholesterol depletion experiments, transfect BHK cells with LipofectAMINE as described above and incubate for 18 hr in DMEM without added serum. Replace the medium of the serum-starved cells with DMEM containing 2% (w/v) cyclodextrin, made by diluting the 10% (w/v) stock solution 1 : 5 with prewarmed DMEM. Incubate the cells for 1 hr and then harvest.

3. Cholesterol replenishment is carried out for 1 hr on cells that have not been serum starved. Make up a solution of DMEM supplemented with cholesterol (16 μg/ml) and 0.4% (w/v) cyclodextrin exactly as follows: Mix 5 ml of 10% (w/v) cyclodextrin with 100 μl of cholesterol (20 mg/ml). Sonicate for 1 min and incubate at 40° until clear, and then filter sterilize. Add 2 ml of the cyclodextrin–cholesterol mix to 50 ml of DMEM.

4. Harvest the cells for cell fractionation and/or biochemical assays.

Sucrose Gradient A

Cell lysates are prepared in alkaline sodium carbonate because this does not strip Ras proteins off membranes. After a sonication step to disrupt cell membranes the whole-cell lysate is adjusted to 45% (w/v) sucrose and centrifuged under lower concentration sucrose cushions. In this bottom-loaded gradient light membranes float through the 35% (w/v) sucrose cushion to an interface with the 5% (w/v) cushion. Caveolin- and Ras-containing microdomains are enriched at the 35%/5% interface. However, it is clear from electron microscopy and Western analysis of these fractions that they do not represent a pure caveolar fraction because other membranes cofractionate. The following protocol is based on that described by Song *et al.*,[26] but has been adapted for use with smaller volumes and includes a second centrifugation step to isolate membranes from the soluble cytosolic proteins that remain at the bottom of the gradient.

1. Prepare stock solutions of 90% (w/v) sucrose, 70% (w/v) sucrose, and 10% (w/v) sucrose in MBS [25 m*M* morpholineethanesulfonic acid (MES, pH 6.5), 150 m*M* NaCl]. On the day of the experiment prepare fresh 0.5 *M* Na_2CO_3, pH 11.0, and make up working solutions of 35% (w/v) sucrose–0.25 *M* Na_2CO_3 and 5% (w/v) sucrose–0.25 *M* Na_2CO_3 by mixing equal volumes of 0.5 *M* Na_2CO_3 with the 70 and 10% (w/v) sucrose stock solutions, respectively.
2. Harvest three 10-cm plates of confluent cells in 1.5 ml of phosphate-buffered saline (PBS), and pellet gently at 1500 rpm for 5 min at 4°. Remove all the PBS and resuspend the cells in 0.5 ml of ice-cold 0.5 *M* Na_2CO_3, passing the lysate through a 23-gauge needle 15 times. Then sonicate for 5 min at 4° in a sonicating water bath, or on ice, using a microprobe sonicator (three 20-sec bursts) (both procedures give comparable results).
3. Weigh out 0.333 g of 90% (w/v) sucrose (250 μl) into an Eppendorf tube and add 250 μl of cell lysate. (Snap freeze the remaining lysate and store at –70°.) Vortex and microcentrifuge for about 2 sec to collect the mixture.
4. Pipette the lysate–sucrose mix into the bottom of a thin-walled centrifuge tube (2.4 ml) (Beckman, Fullerton, CA). Layer on 1.4 ml of 35% (w/v) sucrose–0.25 *M* Na_2CO_3 and then 0.3 ml of 5% (w/v) sucrose–0.25 *M* Na_2CO_3. Apply each new layer carefully onto the previous, making sure there is no mixing.
5. Centrifuge for 6 hr at 55,000 rpm in a TLS-55 rotor at 10°. Remove 0.22-ml fractions from the top of the gradient. Transfer a 10-μl aliquot from each fraction into 0.8 ml of PBS for a Bradford protein assay. To the remainder of each fraction add 1.2 ml of MBS, vortex hard to fully dissolve all the sucrose, and spin in a fixed-angle TL100.3 rotor at 60,000 rpm (100,000 *g*) for 30 min at 4°. Remove

[26] S. K. Song, S. Li, T. Okamoto, L. A. Quilliam, M. Sargiacomo, and M. P. Lisanti, *J. Biol. Chem.* **271,** 9690 (1996).

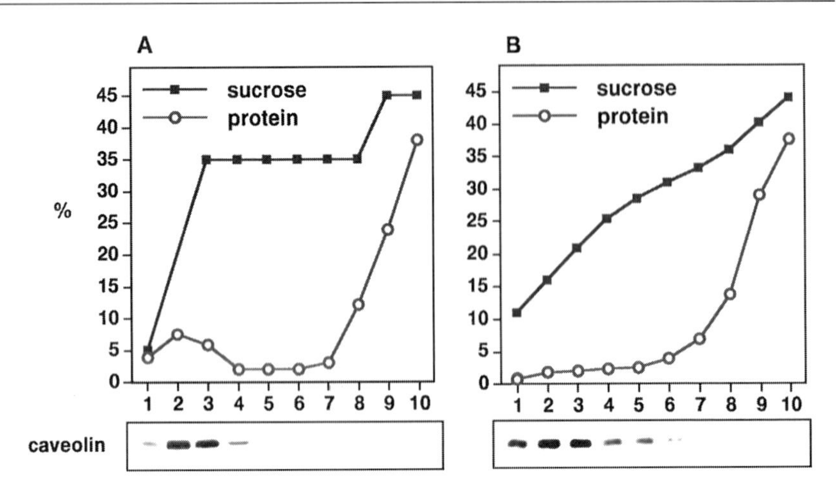

FIG. 1. Comparison of the resolving power of two sucrose gradient protocols. BHK cells were fractionated according to gradient protocol A or B. The graphs show the sucrose concentration (%) measured in each fraction after centrifugation for 6 hr (A) or 16 hr (B). Fraction 1 is the top fraction of each gradient. The total protein content of each fraction is also shown (expressed as a percentage of the total protein in the cell lysate). The high protein content of the lower fractions is due to the cytosolic proteins in the lysate. Each fraction was then diluted with MBS and the membranes present isolated by centrifugation. An equivalent aliquot of each fraction was immunoblotted for caveolin 1, using a mouse monoclonal antibody (Transduction Laboratories). Note that the membrane proteins that collect at the 5%/35% sucrose interface in gradient A are resolved across seven fractions in gradient B. In consequence, specific membrane microdomain markers that all colocalize with caveolin over gradient A can be resolved from caveolin over gradient B (I. A. Prior, A. Harding, J. Yan, J. Sluimer, R. G. Parton, J. F. Hancock, *Nature Cell Biol.*, in press, 2001).

all the MBS, add 60 μl of 2× sodium dodecyl sulfate (SDS) sample buffer to the membrane pellet, and heat at 95° for 4 min with vortexing. Analyze 20 μl of each fraction by immunoblotting.

Figure 1A shows a typical fractionation of BHK cells over this type of gradient, with endogenous caveolin detected by immunoblotting.

Sucrose Gradient B

We have devised a protocol that by virtue of a much shallower sucrose gradient yields better resolution of caveolin- and Ras-containing microdomains. After an overnight centrifugation a linear gradient develops, ranging from 12 to 45% over 10 fractions.

1. On the day of the experiment prepare fresh 0.5 M Na_2CO_3, pH 11.0, and make up working solutions of 35% (w/v) sucrose–0.25 M Na_2CO_3, 30% (w/v) sucrose–

0.25 *M* Na_2CO_3, 25% (w/v) sucrose–0.25 *M* Na_2CO_3, and 5% (w/v) sucrose–0.25 *M* Na_2CO_3 by mixing equal volumes of 0.5 *M* Na_2CO_3 with appropriate 2× sucrose stock solutions prepared in MBS.

2. Harvest three 10-cm plates of cells in 0.5 ml of 0.5 *M* Na_2CO_3 and sonicate as described in step 2 of the previous section.

3. Mix 0.5 ml of the lysate with 0.5 ml of 90% (w/v) sucrose. Vortex and microcentrifuge for about 2 sec to collect the mixture.

4. Pipette the lysate–sucrose mix into the bottom of a thin-walled centrifuge tube (5 ml) (Beckman). Layer on sequentially: 1 ml of 35% (w/v) sucrose–0.25 *M* Na_2CO_3, 1 ml of 30% (w/v) sucrose–0.25 *M* Na_2CO_3, 1 ml of 25% (w/v) sucrose–0.25 *M* Na_2CO_3, 1 ml of 5% (w/v) sucrose–0.25 *M* Na_2CO_3. Apply each new layer carefully onto the previous, making sure there is no mixing.

5. Centrifuge for 16 hr at 48,000 rpm in an SW-55 rotor at 10°. Remove 0.4-ml fractions from the top of the gradient. Transfer a 10-μl aliquot from each fraction into 0.8 ml of PBS for a Bradford protein assay. To the remainder of each fraction add 1 ml of MBS, vortex to fully dissolve all of the sucrose, and spin in a fixed-angle TL100.3 rotor at 60,000 rpm (100,000 *g*) for 30 min at 4°. Remove all the MBS, add 60 μl of 2× SDS sample buffer to the membrane pellet, and heat at 95° for 4 min with vortexing. Analyze 15 μl of each fraction by immunoblotting.

Figure 1B shows a typical fractionation of BHK cells over this type of gradient, with endogenous caveolin detected by immunoblotting. Note that all the proteins that were concentrated at the 5%/35% sucrose interface in gradient A (fractions 2 and 3 in Fig. 1A) are now arrayed over fractions 1–8 in gradient B (Fig. 1B).

Plasma Membrane "Rip Off" and Electron Microscopy

To visualize the localization of protein complexes associated with the cytoplasmic face of the plasma membrane we have optimized conditions to obtain sheets of plasma membrane with the cytoplasmic surface exposed. This technique, a variation on that of Sanan and Anderson,[27] allows a two-dimensional view of the plasma membrane to be obtained and maximizes visualization of the spatial relationship between different gold-labeled antigens and defined organelles, such as caveolae and coated pits. The method is simple and avoids sectioning, and thus is possible in any laboratory with access to an electron microscope. Moreover, the complete accessibility of antigens without the need to section the material or to use detergents optimizes the efficiency of labeling. Most important, the technique offers a chance to determine the distribution of antigens quantitatively.

In this method, polylysine-coated grids are allowed to adhere to the upper surface of cultured cells. The grids, together with attached plasma membrane sheets,

[27] D. Sanan and R. Anderson, *J. Histochem. Cytochem.* **39,** 1017 (1991).

are then removed. The grids with adherent material are then fixed and labeled according to standard immunogold labeling techniques. Importantly, the final step of this method uses a methylcellulose–uranyl acetate mixture to produce a supportive layer around the membrane. As in the Tokuyasu cryosectioning procedure,[28] the use of the methylcellulose supporting mixture avoids the collapse of cellular material that would otherwise occur on air drying.[29] While the plasma membrane itself is a thin layer and so is not greatly affected by air drying, this modification has optimized visualization of the associated membranous organelles such as clathrin-coated pits and caveolae (Fig. 2).

1. Prepare polylysine-coated Formvar-coated grids (e.g., 200 mesh hexagonal; ProSciTech, Thuringowa Queensland, Australia) by incubating the grids on drops of polylysine (1 mg/ml; Sigma) for 15–30 min. Rinse by transferring to drops of water on Parafilm (four times, 1 min each). Allow to air dry in a dust-free environment.
2. Culture cells (e.g., BHK, A431, or 3T3-L1 adipocytes) on 1-cm glass coverslips and transfect if necessary.
3. Place the grids onto filter paper in a group (up to 10 in area smaller than the coverslips area) with the polylysine-coated Formvar side facing up.
4. Immediately before use, wash the cells with warm (37°) PBS to remove any debris. Then rinse the cells with intracellular KOAc buffer [25 m*M* HEPES (pH 7.4), 115 m*M* potassium acetate, 25 m*M* $MgCl_2$]. Immediately remove from KOAc buffer and touch the edge of the coverslip to filter paper to remove excess solution. Place the coverslip cell side down onto the grids. Place filter paper on top of the coverslip and exert slight pressure, using a rubber bung, for 1 sec (3T3-L1) to 5 sec (BHK, A431).
5. Remove the filter paper and turn the coverslip over so that grids are uppermost. Carefully add KOAc buffer to the grids so that they float up onto the top of the drop.
6. Wash the grids by transferring them to another drop of KOAc buffer on Parafilm and then fix at room temperature with either 8% (w/v) paraformaldehyde, or 4% (w/v) paraformaldehyde containing 0.01–1% (v/v) glutaraldehyde.
7. After 15 min of fixation, wash the grids four times with drops of PBS and then quench free aldehyde groups with 25 m*M* glycine in PBS for 10 min.
8. Immunolabel according to standard techniques, using primary antibodies followed by either protein A–gold or species-specific anti-IgG gold conjugates.
9. Wash the labeled grids with five changes of water over 10 min and then incubate on ice for 10 min in a mixture of 9 parts 2% (w/v) methylcellulose in distilled water to 1 part saturated aqueous uranyl acetate (filtered through a 0.2-μm

[28] K. T. Tokuyasu, *J. Cell Biol.* **57,** 551 (1973).
[29] G. Griffiths, "Fine Structure Immunocytochemistry." Springer-Verlag, Berlin, 1993.

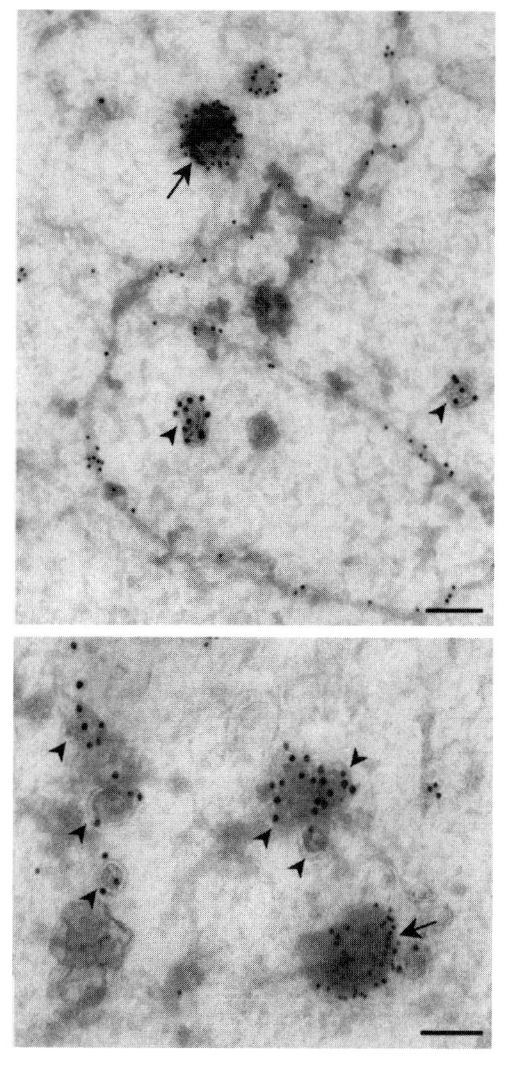

FIG. 2. Ripped-off plasma membrane technique for immunolocalization of plasma membrane antigens. Plasma membrane sheets were immunogold labeled for dynamin with HUDY1 antibodies followed by 10-nm anti-mouse IgG gold and for caveolin 1 followed by 15-nm goat anti-rabbit IgG. The plasma membrane fragments were processed by the methylcellulose/uranyl acetate method to yield good preservation of the plasma membrane architecture, including coated pits (labeled with dynamin, arrows) and caveolae (labeled with caveolin 1, arrowheads). Note the high labeling efficiency, especially with the dynamin antibody. Bars: 100 nm.

pore size filter). Finally, remove excess mixture and allow to dry. The grids are then ready for viewing.

After immunogold labeling, the distribution of gold can be easily assessed with respect to defined morphological features and this can be readily quantified. To give a theoretical example, random sampling might show that an area of 1×1 μm contains on average 50 gold particles. This indicates a labeling density of 50 gold particles/μm^2. The plasma membrane area of a BHK cell is approximately 2200 μm^2 [30] and so this would indicate that the minimum number of antigenic sites is 1.1×10^5 per cell, assuming 100% labeling efficiency (i.e., a 1 : 1 gold-to-antigen ratio). In fact, the actual number of antigenic sites is likely to be underestimated and to depend on variables such as the antibody and the accessibility of the antigen; labeling efficiency of between 1 and 10% is common for immunogold labeling of sectioned material.[29] In a similar way we can estimate the density of caveolae as approximately 1/μm^2 (see Fig. 2) and threfore in one BHK cell there are approximately 2000 caveolae (assuming a fairly uniform distribution over the plasma membrane). These quantitative estimates can be extremely useful for understanding the dynamics of caveola association and the relative concentration of signaling proteins in such domains.

The ability to accurately and quantitatively determine the localization of plasma membrane components with respect to caveolae should help resolve controversies regarding caveola localization on the basis of biochemical methods.[31,32] This technique is a powerful approach to examine the distribution of different Ras isoforms, Ras-related proteins, and other signaling complexes after transfection or after growth factor stimulation, and may therefore lead to the ultrastructural identification of plasma membrane signaling domains.

Summary

Experimental protocols that allow confident assignment of signaling proteins to specific subdomains of the plasma membrane are essential for a full understanding of the complexities of signal transduction. This is especially relevant for Ras proteins, where the different membrane anchors of the Ras isoforms target them to functionally distinct microdomains that in turn allow quantitatively different signal outputs from otherwise highly homologous proteins. The methods outlined in this chapter, in addition to being invaluable in addressing Ras function, should also have wide utility in the study of many mammalian signal transduction pathways.

[30] G. Griffiths, R. Back, and M. Marsh, *J. Cell Biol.* **109,** 2703 (1989).

[31] J. E. Schnitzer, D. P. McIntosh, A. M. Dvorak, J. Liu, and P. Oh, *Science* **269,** 1435 (1995).

[32] R. V. Stan, W. G. Roberts, D. Predescu, K. Ihida, L. Saucan, L. Ghitescu, and G. E. Palade, *Mol. Biol. Cell.* **8,** 595 (1997).

Acknowledgments

We thank Stuart Kornfeld for helpful advice; Angus Harding and Ian Prior for providing the images for Figs. 1A and 2; Margaret Lindsay, Rob Luetterforst, and Sandrine Roy for technical assistance; and Sandra Schmid for providing the antibody to dynamin. This work was supported by grants from the National Health and Medical Research Council of Australia to R. G. P. and J. F. H. and the Australian Research Council to R. G. P. J. F. H. is also supported by the Royal Children's Hospital Foundation of Queensland.

Section II

Biological Analyses

[18] Analyses of M-Ras/R-Ras3 Signaling and Biology

By LAWRENCE A. QUILLIAM, JOHN F. REBHUN, HUI ZONG,
and ARIEL F. CASTRO

Introduction

The Ras subfamily of GTPases consists of about 19 members that share 40–98% identity with H-, K-, and N-Ras or each other.[1] M-Ras (also referred to as R-Ras3[2]) was originally cloned by rapid amplification of cDNA ends (RACE) as a GTPase expressed in a C2 myoblast cDNA library.[3] It was also identified as a novel Ras-like sequence in the National Center for Biotechnology Information (NCBI) expressed sequence tag (EST) database.[2,4,5] Finally, M-Ras was isolated by representational difference analysis as a gene induced by interleukin 9(IL-9) stimulation of helper T cell lines.[6] It shares closest homology with TC21/R-Ras2 and K-Ras4B and has a 10-amino acid extension at its N terminus compared with the classic H-, K-, and N-Ras proteins.[3] M-Ras has moderate transforming ability; a Q71L mutant has ~10% of the focus-forming activity of the equivalent H-Ras(61L), promotes weak extracellular signal-regulated kinase (ERK) activation that is required for its transforming ability, and can interact weakly with a number of other Ras effectors.[2,4] Interestingly, M-Ras protein is broadly expressed, can be coimmunoprecipitated by the Ras monoclonal antibody Y13-259, and is activated by known Ras guanine nucleotide exchange factors (GEFs) (Sos and guanine nucleotide release factor [GRF]).[4,5] Therefore many studies using the Y13-259 monoclonal antibody to measure the activity state (GTP/GDP ratio) of Ras *in vivo* will also have immunoprecipitated M-Ras. In addition, any studies using the dominant inhibitory mutant H-Ras (17N), which acts by titrating out GEFs, thus preventing their activation of endogenous Ras proteins, will also block M-Ras activation. Therefore, although the physiological function of M-Ras or its contribution to human cancer is still not clear, its activity may have mistakenly been attributed to Ras in many previous studies. This chapter describes several

[1] S. L. Campbell, R. Khosravi-Far, K. L. Rossman, G. J. Clark, and C. J. Der, *Oncogene* **17,** 1395 (1998).

[2] A. Kimmelman, T. Tolkacheva, M. V. Lorenzi, M. Osada, and A. M. Chan, *Oncogene* **15,** 2675 (1997).

[3] K. Matsumoto, T. Asano, and T. Endo, *Oncogene* **15,** 2409 (1997).

[4] L. A. Quilliam, K. R. Graham, A. F. Castro, C. B. Martin, C. J. Der, and C. Bi, *J. Biol. Chem.* **274,** 23850 (1999).

[5] G. R. Ehrhardt, K. B. Leslie, F. Lee, J. S. Wieler, and J. W. Schrader, *Blood* **94,** 2433 (1999).

[6] J. Louahed, L. Grasso, C. De Smet, E. Van Roost, C. Wildmann, N. C. Nicolaides, R. C. Levitt, and J.-C. Renauld, *Blood* **94,** 1701 (1999).

METHODS IN ENZYMOLOGY, VOL. 333 0076-6879/00 $35.00

assays used to test the transforming ability and signaling properties of M-Ras and to identify which Ras family GEFs lead to M-Ras activation.

NIH 3T3 Focus-Forming Assay

Transformation can result in many cellular changes *in vitro,* including increased growth rate, reduced serum dependence, altered cell morphology, loss of contact inhibition, and anchorage-independent growth. The NIH 3T3 cell focus-forming assay is probably the most sensitive and widely used procedure to test for the ability of a gene to induce cellular transformation.[7] This assay measures the ability of a gene to promote a combination of density-independent growth and morphological transformation. The morphology of the transforming foci varies with different oncogenes. Typically Ras-induced foci form swirling patterns of elongated, refractile cells (see Fig. 1) with occasional balls or stellate patterns, particularly after prolonged incubation.

NIH 3T3 Cell Culture

NIH 3T3 cells are purchased from the American Type Culture Collection (Manassas, VA; *ATCC.org*) and cultured in Dulbecco's modified Eagle's medium (DMEM) supplemented with 10% calf serum (Colorado Serum Company, Denver, Co), 2 m*M* glutamine, penicillin (100 U/ml), and streptomycin (100 μg/ml) and maintained at 37° in a humidified 10% CO_2 atmosphere. We currently use donor calf serum from Life Technologies (Rockville, MD). This serum is less expensive but requires prior testing of sample lots for background/spontaneous transformation. Cell stocks are generated by plating 10^3 cells/100-mm-diameter culture dish. Cells are fed every 48 hr for 7–10 days (or until cells within each colony appeared ~50% confluent). Cells are then trypsinized [medium is removed, excess serum is diluted by washing with 10 ml of phosphate-buffered saline (PBS), and 1 ml of 0.05% (w/v) trypsin–0.53 m*M* EDTA is added for 3–5 min or until the cells can be dislodged). Cells are then pooled (to provide uniformity between assays), pelleted at 800 rpm in a tabletop centrifuge for 5 min at 4°, and then resuspended and frozen in 1 ml of ice-cold DMEM–20% (v/v) calf serum–10% dimethyl sulfoxide (DMSO) per dish. Freezing can be accomplished overnight at −80° by placing the vials in a polystyrene box or a cardboard Revco (Huntington Beach, CA) sample box insulated by placing four "C-fold" paper towels in the lid and in the base. Long-term stocks are maintained under liquid nitrogen. For reconstitution, cells are rapidly thawed, diluted in warm culture medium, and divided into 3 × 100 mm dishes; it will take ~4 days for the cells to reach 80% confluence (the time will vary, depending on the initial freezing density). Cells are then passaged, seeding at

[7] G. J. Clark, A. D. Cox, S. M. Graham, and C. J. Der, *Methods Enzymol.* **255,** 395 (1995).

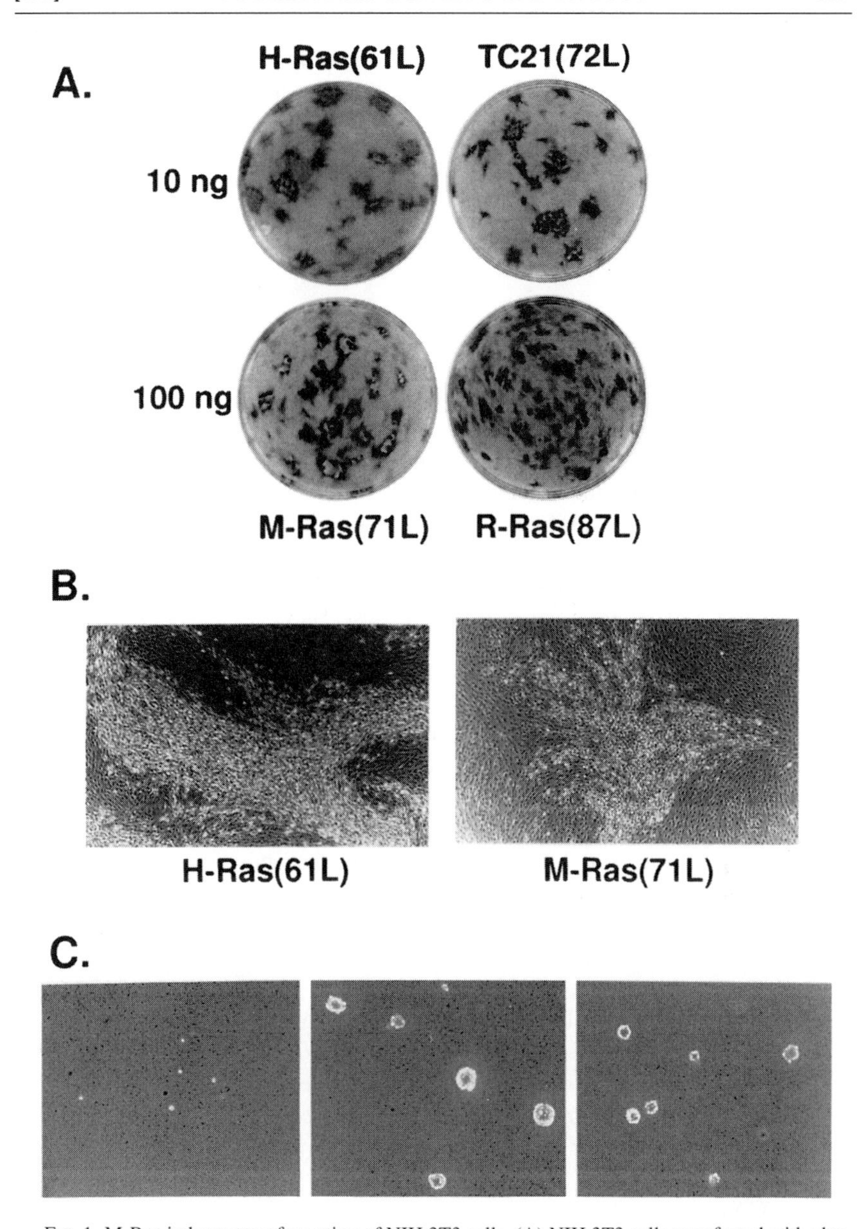

FIG. 1. M-Ras induces transformation of NIH 3T3 cells. (A) NIH 3T3 cells transfected with plasmids encoding the indicated proteins were grown in regular culture medium for 12–18 days prior to staining foci with crystal violet. (B) Individual foci from (A) were photographed at ×4 magnification. (C) NIH 3T3 cells stably transfected with empty vector (left panel), H-Ras(61L) (middle panel), or M-Ras(71L) (right panel) were grown in soft agar for 2 weeks prior to photographing colonies at ×4 magnification. [Reproduced with permission from the American Society for Biochemistry and Molecular Biology.[4]]

1×10^5/100 mm dish. These cells will be ready to divide again (80% confluent) in 3–4 days. Cell stocks are used for just four passages for focus-forming assays, or establishing stable cell lines, to reduce the possibility of spontaneous focus formation. Typically we see fewer than one background focus per 10 plates. For other procedures such as luciferase reporter assays, cell stocks can be maintained much longer. For experimentation, cells are plated at 2.5×10^5/60-mm dish and used the following day for transfection. Typically we perform focus assays in quadruplicate.

Transfection Protocol

There are many variations of the calcium phosphate DNA precipitation transfection method[7,8] and commercial kits, for example, from Life Technologies, are available. It is also possible to use liposomal reagents, yielding much higher transfection and transformation efficiencies while using less plasmid DNA. However, this can become expensive if large focus assays are frequently performed.

Stock Solutions for Transfection

HEPES-buffered saline (HBS): Add the following to 900 ml of deionized water: 8 g of NaCl, 0.37 g of KCl, 1 g of glucose, 5 g of cell culture-tested HEPES, and 10 ml of 100× dibasic sodium phosphate (1.9 g of $Na_2HPO_4 \cdot 7H_2O$ in 100 ml of H_2O). Adjust to exactly pH 7.05 with 5 *M* NaOH, adjust the volume to 1 liter, and sterilize by filtration or autoclaving. The correct pH is critical for good DNA precipitation. Store at room temperature

HBS–glycerol: Add 15% (v/v) glycerol to the above-described HBS and sterilize by autoclaving

Calcium chloride (1.25 *M*): 18.38 g of $CaCl_2 \cdot 2H_2O$ is added to 90 ml of H_2O; when dissolved add

H_2O to a final volume of 100 ml. Autoclave

Calf thymus DNA: The best quality high molecular weight carrier DNA is obtained from Roche Molecular Biochemicals (Indianapolis, IN). The stock is diluted to 1 mg/ml in sterile deionized water and stored at −20°

Plasmid DNA: Plasmid DNA can be prepared by CsCl density gradient centrifugation[9] or by using one of various commercially available ion-exchange columns. We currently use Maxi-preps from Qiagen (Chatsworth, CA)

[8] C. A. Hauser, J. K. Westwick, and L. A. Quilliam, *Methods Enzymol.* **255,** 412 (1995).

[9] J. Sambrook, E. F. Fritsch, and T. Maniatis, "Molecular Cloning: A Laboratory Manual." Cold Spring Harbor Laboratory Press, Cold Spring Harbor, New York, 1989.

All subsequent manipulations are performed in a sterile tissue culture hood. Sufficient HBS is aliquoted into a sterile tube to provide >0.5 ml/plate and supplemented with carrier DNA (40 μg/ml). DNA is placed in sterile 14-ml snap-cap polystyrene tubes and diluted with 0.5 ml of HBS–carrier DNA per plate. After mixing, $CaCl_2$ solution at 50 μl/plate is added dropwise with vortexing. A light milky precipitate should begin to form almost immediately because of the presence of fine fibers. It can be best visualized by holding a tube to the light or against a dark background. Precipitation should be complete by 20 min, at which point the precipitate is added to the cell culture medium at 0.5 ml/60 mm dish. It will be necessary to visually test new lots of HBS for their ability to produce a good precipitate before proceeding with experiments. Too much precipitation, resulting in large aggregates, is as undesirable as little or no precipitation. Typically, using the pZIP-Neo SV(X)1 plasmid,[10] 10–20 ng of H-Ras (61L) or 1–200 ng of the equivalent M-Ras (71L) will produce 40–60 transforming foci per dish (see Fig. 1).[4] However, the amount of DNA required for other plasmids and genes of interest will have to be determined empirically.

On inspection of the cells after 3–4 hr, there should be a fuzzy coating of DNA precipitate attached to the cells. At this point the transfection efficiency can be greatly enhanced by glycerol shock to increase DNA uptake. The medium is removed from the cells by aspiration and the monolayers are gently washed with 3 ml of PBS or serum-free medium. One milliliter of HBS–15% (v/v) glycerol is then added to each plate, with rocking to coat the entire plate. The HBS–glycerol is then aspirated and 3 ml of PBS added to the plates, with swirling to dilute the residual glycerol. The time of exposure to glycerol is critical, as prolonged exposure can be toxic to the cells. Three to 4 min should elapse between the addition of glycerol and the subsequent addition of PBS, and this time should be kept the same for all plates within a transfection. Finally, the PBS is replaced with 4 ml of fresh growth medium and the cells are returned to the incubator. Medium is changed every 48 hr and Ras-induced foci are detectable in 8–10 days. They are usually scored after 12–16 days. This rate will vary depending on the strength of the plasmid promoter and on the transforming ability of the gene of interest. For quantitative analysis foci are counted with a phase-contrast microscope equipped with a ×4 objective.

To obtain a permanent record, the medium is removed and the monolayer is washed twice (5 ml each) with PBS. Five milliliters of fixative [10% (v/v) acetic acid, 10% (v/v) methanol] is added for 10 min; this is then replaced with 1 ml of stain (made by dissolving 0.4 g of crystal violet in 10 ml of methanol, and then bringing the solution to 100 ml by adding 80 ml of H_2O and 10 ml of acetic acid) for an additional 10 min. Dye can be recovered for reuse and plates gently but

[10] C. L. Cepko, B. E. Roberts, and R. C. Mulligan, *Cell* **37,** 1053 (1984).

extensively washed with H_2O and allowed to dry. It is possible to count foci on these stained plates, but it is easier to do so with live cells.

Secondary Focus-Forming Assay

Many weakly transforming genes do not promote focus formation in the above-described assay but foci may be observed in a secondary focus assay, using stably transfected cell lines.[11] This assay has also proved useful for weakly transforming M-Ras effector-binding domain mutants. Here, after transfection with a plasmid (50–500 ng) that confers antibiotic resistance, the cells are cultured for 3 days in regular growth medium as outlined above. Cells are then removed by trypsinization and one-third are plated on a 100-mm dish in 10 ml of growth medium supplemented with selection drug. The amount of drug required may vary with different cell types and plasmid promoter, but we typically use G418 (active ingredient, 400 μg/ml; Life Technologies) for Neo^r plasmids, hygromycin B (100–200 μg/ml; Roche), or puromycin (4 μg/ml; Roche). Antibiotic-containing medium is replaced every 3–4 days and selection is complete (individual colonies of cells should be visible) in 10–14 days. Individual colonies may be isolated with cloning cylinders or the population pooled by trypsinization, and cells are plated at 5×10^5/dish and fed every 48 hr as described above. Transforming foci may appear in 1–2 weeks. This procedure is not desirable for strongly transforming genes as, because of loss of contact inhibition, cells will pile up into a thick monolayer without distinguishable foci.

Soft Agar Assay

Monolayer cultures require attachment to a solid matrix in order to grow and survive. However, after transformation many cell types, including NIH 3T3 cells, will proliferate in an anchorage-independent manner. The growth of individual cells can be readily scored with a semisolid medium such as soft agar. Results from such assays do not correlate perfectly with the ability of cells to promote tumors in animals, but this assay probably provides the best *in vitro* correlate with tumorigenic potential.[7]

Reagents/Materials Required

A 42° water bath is required, preferably adjacent to the culture hood. Bacto-agar (Difco, Detroit, MI) 1.8 g/100 ml in H_2O, is autoclaved and cooled to 65°. Solid stock can be melted in a microwave and similarly cooled to 65°. A 2× concentrate

[11] S. J. Mansour, W. T. Matten, A. S. Hermann, J. M. Candia, S. Rong, K. Fukasawa, G. F. Vande Woude, and N. G. Ahn, *Science* **265,** 966 (1994).

of DMEM can be made from powdered medium by using half the recommended volume, as per the manufacturer instructions, and filter sterilized.

Assay Procedure

Typically we perform these assays in triplicate. The recipe below is sufficient for 15 dishes. Mix 50 ml of 2× DMEM, 20 ml of sterile H_2O plus 10 ml of calf serum and warm to 37°. Add 33 ml of molten agar to the above-described medium and rapidly transfer 5 ml of the mix into each 60-mm petri dish (using a 25-ml pipette). The remaining agar–DMEM mix should be immediately placed in the 42° water bath to prevent it from setting.

Once the agar plates have set, place them in the incubator to keep warm. Trypsinize the cells and dilute to 15,000/1.5 ml of warm growth medium per warm 14-ml sterile tube. Taking the plates for each set out as required (so they do not cool down), mix 1.5 ml of cells with 3 ml of molten agar/medium and rapidly pipette 1.5 ml onto each dish and spread. Once set, return the plates to the 37° CO_2 incubator. Cells may be fed weekly with 0.5 ml of regular growth medium. Transforming colonies can be counted or photographed under the microscope (as in Fig. 1) after ~2 weeks.

C2 Cell Differentiation Assay

In response to serum deprivation myoblasts differentiate and fuse to form myotubes. This process is inhibited by the expression of activated H- or M-Ras.[4] C2 myoblast cells are grown in DMEM supplemented with 15% (v/v) fetal bovine serum. One microgram of pZIP plasmid encoding wild-type or activated M-Ras(Q71L) or Ha-Ras(Q61L) protein is transfected into C2 cells by calcium phosphate precipitation. After 24 hr cells are selected in growth medium supplemented with G418 (400 μg/ml) to establish mass populations of cells expressing the indicated Ras proteins. Individual clones can also be selected with cloning cylinders and may give more uniform differentiation than a pooled population. To induce differentiation, C2 cells are grown to 70% confluence, then fed with DMEM containing 2% (v/v) horse serum and insulin (10 μg/ml; Life Technologies). After approximately 7 days, cells are scored for acquisition of differentiated characteristics (see Fig. 2).

Elk-1 Transcription Assay

On activation by the Ras–Raf–MEK (MAPK/ERK kinase) cascade ERKs/MAPKs (mitogen-activated protein kinases) translocate to the nucleus, where they phosphorylate the Ets family transcription factor, Elk-1, resulting in gene induction. Eukaryotic transcription factors typically consist of functionally separable

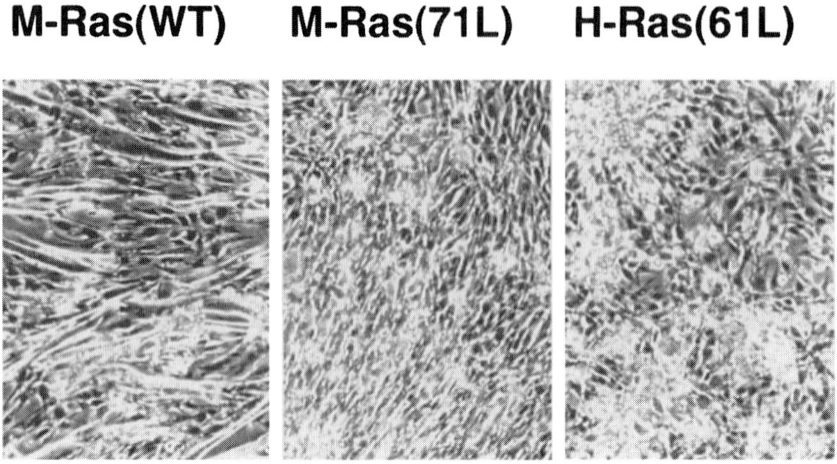

M-Ras(WT) M-Ras(71L) H-Ras(61L)

FIG. 2. Activated M-Ras attenuated the fusion of C2 myoblasts. Stable transfectants were shifted from growth medium containing 15% (v/v) FBS to 2% (v/v) horse serum–insulin (10 μg/ml) to induce muscle differentiation. *Left:* Similar to control cells, cells expressing wild-type M-Ras differentiate and fuse to form myotubes. In contrast, cells expressing mutationally activated M-Ras (Q71L) (*middle*) or Ha-Ras (Q61L) (*right*) fail to form myotubes. Instead, these cells became transformed and overcame density arrest. [Reproduced with permission from the American Society for Biochemistry and Molecular Biology.[4]]

domains for directing DNA binding and transcriptional activation. Indeed, this is the basis of the yeast two-hybrid assay. Using an Elk-1 *trans*-activation domain (that contains the ERK phosphorylation sites) fused to the DNA-binding domain of the yeast *GAL4* promoter it is possible to induce the expression of luciferase from a *GAL4*-luciferase reporter plasmid.

Transfection Procedure

NIH 3T3 cells are plated 1×10^5/35-mm dish the day prior to experimentation. Typically experiments are performed in duplicate. For each set of two plates 1 ml of HBS (see NIH 3T3 Focus-Forming Assay, above) 40 μg of calf thymus carrier DNA, 0.2 μg of GAL4-Elk, and 2 μg of 5× GAL4-Luc plasmids[12] are mixed. DNA of interest, for example, 0.5–1 μg of pZIP or pZIP-M-ras (71L),[4] is diluted in sterile water and placed in 4-ml sterile polystyrene snap-cap tubes to which 1 ml of the above-described mixture is added. A volume of 100 μl of 1.25 *M* $CaCl_2$ is then added with vortexing and a DNA precipitate is formed as described above. This is added to the culture medium and cells are glycerol shocked as outlined above. After 24 hr, cells are serum starved overnight (~18 hr) prior to harvest.

[12] R. Marais, J. Wynne, and R. Treisman, *Cell* **73,** 381 (1993).

Luciferase Assay

Cell monolayers are washed twice with ice-cold PBS and lysed in 250 μl of 0.1 *M* sodium phosphate (pH 7.8), 0.5% (v/v) Triton X-100, 1 m*M* dithiothreitol (DTT) or lysis buffer from a commercially available kit. Assays tend to be more reproducible if buffer is left on the cells for 15–30 min on ice before scraping into a microcentrifuge tube. Insoluble material is pelleted by centrifugation (12,000 rpm, 10 min, 4°) and 100 μl of the supernatant is used to measure luciferase activity. Lysates can be kept on ice for 1–2 hr or frozen at −80° if not assayed immediately. We obtain lysis buffer, ATP, and luminol solutions from PharMingen (San Diego, CA; enhanced luciferase assay kit). Using a luminometer with automatic injectors, 100 μl each of ATP and luminol is mixed with the sample and light production is counted for 20 sec. It is typical to use a second *Renilla* luciferase (Promega, Madison, WI) or β-galactosidase reporter plasmid[8] as an internal control for transfection/recovery efficiency. However, we have found that Ras proteins strongly induce the promoters of many standard vectors and prefer to confirm results with independent preparations of test plasmids.

Mitogen-Activated Protein Kinase Assay

NIH 3T3 cells (plated the previous day at 6.5×10^5/100-mm dish) are shifted to 5 ml of serum and antibiotic-free medium and contransfected with 1 μg of pcDNA3-HA-ERK2 [encoding hemagglutinin (HA) epitope-tagged ERK2] plus 2 μg of pcDNA3 alone or encoding Ha-Ras (Q61L) or M-Ras (Q71L), using 20 μl of NovaFECTOR (Venn Nova, Pompano Beach, FL). DNAs are diluted in 0.5 ml of serum and antibiotic-free DMEM and 20 μl of NovaFECTOR is diluted in a similar volume. The two solutions are mixed, incubated for 15 min at room temperature, diluted to 5 ml with serum/antibiotic-free DMEM (DMEMϕ), and placed on the cell monolayer. After 3 hr, 5 ml of DMEMϕ–20% (v/v) calf serum is added. The next morning the medium is replaced with regular growth medium and in the evening cells are serum starved (0.1%, v/v) overnight. Monolayers are washed twice with ice-cold PBS before addition of 1 ml of 20 m*M* Tris-HCl (pH 7.4), 1 m*M* EDTA, 1 m*M* EGTA, 0.27 *M* sucrose, 0.1% (w/v) sodium dodecyl sulfate (SDS), 1% (v/v) Triton X-100, 0.5% (w/v) deoxycholate, 1 m*M* Na_3VO_4,* 5m*M* $NaPP_i$, 50 m*M* NaF, 0.1% (v/v) 2-mercaptoethanol, 10 m*M* β-glycerophosphate,

* A 100 m*M* Na_3VO_4 (sodium orthovanadate) stock is made by dissolving 1.84 g in 90 ml of H_2O and adjusting the to pH 10.0 with HCl. This results in the generation of a bright orange color. The solution is then heated (boiled) on a conventional hot plate until colorless, cooled, and readjusted to pH 10.0 (if needed). The cycle of heating–cooling and pH adjustment should be repeated if the pH is still high. Care should be taken to avoid allowing the pH to drop below pH 10.0. The solution can then be stored at 4° or aliquoted at −20°

aprotinin (19 μg/ml), and 1 m*M* phenylmethylsulfonyl fluoride (PMSF, from a 100 m*M* stock in 2-propanol).

After 5 min on ice, cells are scraped into ice-cold microcentrifuge tubes and clarified by centrifugation (10 min, 14,000 rpm, 4°). At this point 50-μl aliquots are removed for subsequent Western blotting of total lysates. The remaining supernatant is transferred to a fresh tube with 20 μl of protein A/G–agarose (Santa Cruz Biotechnology, Santa Cruz, CA); tumble for 10 min at 4° to preclear. Beads are pelleted (microcentrifuge, 10-sec pulse) and the supernatant is transferred to a tube containing 1 μl of HA monoclonal (BAbCo, Berkeley, CA) and 30 μl of protein A/G–agarose. After tumbling for 60 min at 4°, beads are pelleted (microcentrifuge, 10-sec pulse) and washed twice with 1 ml of lysis buffer supplemented with 500 m*M* NaCl and twice with wash buffer [20 m*M* Tris, 1 m*M* EGTA, 0.05% (v/v) Triton X-100, and 0.1% (v/v) 2-mercaptoethanol].

Kinase activity is determined with myelin basic protein [Sigma (St. Louis, MO) or Life Technologies] as substrate, essentially as described.[13] Beads are pelleted and all remaining liquid is carefully removed with a 20-μl pipette. Fifty microliters of kinase buffer [40 m*M* Tris-HCl (pH 7.4), 1 m*M* EGTA, 0.2 m*M* DTT, 10 m*M* $MgCl_2$, 100 μ*M* ATP, [γ-^{32}P]ATP (3000 Ci/mmol, 5 μl/ml), 200 μ*M* Na_3VO_4, 0.5 μ*M* okadaic acid (Life Technologies), myelin basic protein (0.4 mg/ml), 50 m*M* NaF] is added and incubated for 30 min at 30°; vortex every few minutes to mix the beads. The reaction is terminated by pelleting the beads and adding 40 μl of the supernatant to 25-mm-diameter Whatman (Clifton, NJ) P81 paper disks (prenumbered in pencil) and placing them in 1% (w/v) phosphoric acid. The filters are washed four times (5 min each), plus a final rinse with acetone, air dried, place in scintillation vials with 2 ml of fluid, and counted. (To wash filters, a hot glass rod was used to introduce multiple holes in the sides of a 400-ml plastic beaker. This is then suspended within a 1-liter beaker containing ~200 ml of phosphoric acid plus a stir bar. This enables efficient washing of the filters without them becoming damaged.)

Yeast Two-Hybrid Analysis of M-Ras Effector Interactions

The ability of M-Ras to interact with candidate effectors can be evaluated with the yeast two-hybrid system, essentially as described[14] (also see [17] in this volume[14a]). Briefly, one hybrid is M-Ras (Q71L) fused to the LexA DNA-binding domain. The second hybrid is a candidate effector [full-length RalGDS (Ral guanine nucleotide dissociation stimulator), cRaf-1, A-Raf, B-Raf, or phosphatidylisositol 3-kinase (PI3K) p110δ, truncated AF6, Rin1, or cRaf-1] fused

[13] D. R. Alessi, P. Cohen, A. Ashworth, S. Cowley, S. J. Leevers, and C. J. Marshall, *Methods Enzymol.* **255,** 279 (1995).

[14] A. B. Vojtek and S. M. Hollenberg, *Methods Enzymol.* **255,** 331 (1995).

[14a] R. G. Parton and J. F. Hancock, *Methods Enzymol.* **333,** [17] 2001 (this volume).

to nuclear-localized VP16 acidic activation domain.[4] The two hybrids are introduced into *Saccharomyces cerevisiae* strain L40 by simultaneous or sequential yeast small-scale transformation.[14] The specific interaction of two hybrids activates the two reporter genes in the L40 strain; thus cells grow on medium lacking histidine and express β-galactosidase, the activity of which can be measured by a simple color reaction.

Controls

To confirm the specificity of the interactions, it is always critical to include a control in which the LexA–bait fusion plasmid is introduced together with the VP16 plasmid (empty vector). The H-Ras(G12V) hybrid is used to verify positive interactions, as all the candidates evaluated are known Ras effectors. M-Ras(S27N) is used as a negative control. It is known that effectors interact with GTP- but not GDP-bound Ras. Therefore they should fail to interact with the GDP-bound M-Ras(S27N) dominant inhibitory mutant. This control is especially useful when a cDNA library is being screened to identify new effector proteins. We noticed that the recommended use of LexA–lamin to eliminate false positives in these screenings could lead to discarding real effectors. Indeed, we observed that the M-Ras effector AF6 interacts with LexA–lamin, but fails to interact with M-Ras(S27N).

β-Galactosidase Assay

β-Galactosidase activity can be assessed by a filter assay when only a qualitative result is desired (refer to [17] in this volume[14a]). Alternatively, a quantitative β-galactosidase assay can be performed on liquid cultures. For the latter assay the L40 strain cotransformed with the two hybrids is grown overnight in 5 ml of YC medium −Leu −Trp to mid- or late-log phase (OD_{600} of 0.5 to 1.0). Cells are centrifuged at 2500 rpm for 5 min at 4°, resuspended in 1 ml of Z buffer (60 m*M* $Na_2HPO_4 \cdot 7H_2O$, 40 m*M* $Na_2H_2PO_4 \cdot H_2O$, 10 m*M* KCl, 1 m*M* $MgSO_4 \cdot 7H_2O$, 50 m*M* 2-mercaptoethanol, pH 7.0), and the OD_{600} is determined. The cells are permeabilized by adding 1 drop of 0.1% (w/v) SDS and 2 drops of chloroform, and cleared by centrifugation. The supernatant is mixed with 0.2 ml of *o*-nitrophenyl-β-D-galactopyranoside (4 mg/ml), incubated at 30°, and timed until a medium-yellow color develop. Adding 0.5 ml of 1 *M* Na_2CO_3 terminates the reaction, and the OD_{420} is determined. The β-galactosidase activity is expressed as $(OD_{420} \times 1000)/(OD_{600} \times \text{time})$.

Association of M-Ras with Guanine Nucleotide Exchange Factors

Guanine nucleotide exchange factors (GEFs) act by promoting the release of GDP and stabilizing the nucleotide-free Ras protein. GEFs bind much tighter to the nucleotide-free from of Ras and this is the basis of the action of dominant inhibitory Ras 17N and 15A mutants that block the ability of GEFs to activate endogenous Ras proteins.[15] We have taken advantage of this high-affinity binding to determine which GEFs bind to M-Ras and which Ras family GTPases bind to various novel putative GEFs. Identifying these GEF–Ras interactions *in vitro* enables accurate prediction of regulator/substrate preferences and reduces the need to perform extensive exchange assays.[16] Briefly, epitope-tagged GEFs expressed in 293T cells can be precipitated by nucleotide-free GST–Ras fusion proteins and this association detected by Western blotting.

Production and Purification of Glutathione S-Transferase Fusion Proteins

Escherichia coli strain BL21-DE3(lysE) (Novagen, Madison, WI) is transformed with the various pGEX-Ras plasmids. These bacteria are grown overnight to saturation and then diluted 1 : 10 and grown to the start of log phase (~90 min). They are then induced with 0.2 m*M* isopropyl-β-D-thiogalactopyranoside (IPTG) for 2 hr at 37°. Bacteria are isolated and resuspended in lysis buffer containing 20 m*M* Tris-HCl (pH 8.0), 500 m*M* NaCl, 1% (v/v) Triton X-100, 10 μ*M* GDP, 1 m*M* $MgCl_2$, 1 m*M* PMSF, and aprotinin (0.05 trypsin inhibitory units/ml). The cells are lysed by pressure [12,000 psig (pounds per square inch gauge)] generated with a French press. Alternatively, sonication may be used or bacteria can by lysed by freeze–thawing. In the latter situation, cells are resuspended in lysis buffer without detergent. The suspension is frozen in liquid nitrogen or a dry ice–95% (v/v) ethanol bath and rapidly thawed at 37° three times. Detergent is then added and cells are lysed (DNA sheared) by passing the lysate 15 times through an 18-gauge syringe needle. The lysate is cleared by centrifugation (10,000*g*, 10 min, 4°) and tumbled with 0.5 ml of preswollen glutathione–agarose beads (Sigma) for ~4 hr. The beads are washed four times with 10 ml of wash buffer containing 20 m*M* Tris-HCl (pH 8.0), 20% (v/v) glycerol, 1 m*M* dithiothreitol, 50 m*M* NaCl, 5 m*M* $MgCl_2$, 10 μ*M* GDP, and 1 m*M* PMSF. The beads are resuspended to 50% in wash buffer and NaN_3 is added to 0.02% (w/v). Proteins are kept at 4° until needed and are stable for months under these conditions.

GEF–Ras Binding Assay

To remove bound GDP, glutathione *S*-transferase (GST)–Ras-bound beads containing ~20 μg of protein are mixed with empty glutathione–agarose beads (50% slurry) to give a final volume of 35 μl. The beads are rinsed twice with 1 ml of Ras buffer [20 m*M* Tris-HCl (pH 7.6), 5% (v/v) glycerol, 50 m*M* NaCl, 0.1% (v/v) Triton X-100, and 1 m*M* dithiothreitol] and then incubated with the Ras buffer spiked with 10 m*M* EDTA for 10 min at room temperature. This action chelates

[15] L. A. Feig, *Nat. Cell Biol.* **1,** E25 (1999).

[16] J. F. Rebhun, H. Chen, and L. A. Quilliam, *J. Biol. Chem.* **275,** 13406 (2000).

Mg^{2+} ions dissociating GDP from the Ras proteins, allowing tight association of the specific GEFs. Cells expressing FLAG epitope-tagged GEFs are lysed with lysis buffer (1 ml/100-mm dish) containing 20 m*M* Tris-HCl (pH 7.4), 1% (v/v) Igepal (Sigma) [or 1% (v/v) Nonidet P-40], 1 m*M* EDTA, 100 m*M* NaCl, aprotinin (19 μg/ml), and 1 m*M* PMSF. Depending on the expression level of the GEF we typically use 40% of the cell lysate from a 100-mm dish for interaction with 20 μg of Ras protein.[16] After incubation on ice for 5 min, the cellular debris is removed by microcentrifugation (13,000*g*, 5 min, 4°) and 400 μl of lysate is added to the rinsed Ras-bound beads. This slurry is tumbled for 2 hr at 4° and then washed four times with the above-described Ras buffer supplemented with EDTA to 5 m*M*. Fifty microliters of 2× sample buffer [20 m*M* sodium phosphate (pH 7.0), 4% (w/v) SDS, 20% (v/v) glycerol, 0.02% (w/v) bromphenol blue, and 10% (v/v) 2-mercaptoethanol] is added to the beads. A 20-μl sample is separated by SDS–polyacrylamide gel electrophoresis (PAGE) and transferred to polyvinylidene difluoride (PVDF) membrane (Immobilon-P, 25 mA, 4 hr; Millipore, Bed ford, MA), blocked in Tris-buffered saline (TBS)–0.05% (v/v) Tween 20–5% (w/v) milk powder (60 min), and probed with M2 anti-FLAG antibody (60 min, 1 : 2000 dilution; Sigma). Horseradish peroxidase-conjugated secondary antibody, diluted 1 : 5000, is applied for 30 min, followed by ECL (enhanced chemiluminescence; Amersham, Arlington Heights, IL) as per the manufacturer recommendations.

In Vivo Exchange Assay Using Metabolically ^{32}P-Labeled Cells

In vivo guanine nucleotide exchange assays are performed essentially as described.[4,17] Sixty-millimeter dishes of 293T cells are transiently transfected (using 1 μg of each DNA plus 10 μl of NovaFECTOR as outlined above) with plasmids encoding the Ras and GEF proteins. After 36 hr, cells are incubated in serum and phosphate-free medium for 30 min, followed by similar medium supplemented with 150 μCi of $^{32}P_i$ (orthophosphate) for an additional 4 hr. Cells are washed with ice-cold PBS, lysed in 50 m*M* Tris-HCl, (pH 7.4), 500 m*M* NaCl, 6 m*M* $MgCl_2$, 1 m*M* EDTA, 1% (v/v) Triton X-100, 0.5% (w/v) sodium deoxycholate, 0.05% (w/v) SDS, aprotinin (19 μg/ml), and 1 m*M* PMSF, and clarified by microcentrifugation for 10 min, and Ras proteins are immunoprecipitated with anti-HA or FLAG antibody. Beads are washed five times with 1 ml lysis buffer. The buffer is removed and Ras proteins are denatured at 68° in 0.2% (w/v) SDS–2 m*M* EDTA (20 μl), and guanine nucleotides are separated by thin-layer chromatography after spotting 10 μl on 20-cm polyethyleneimine (PEI) cellulose plates, using 0.75 *M* KH_2PO_4/HCl, pH 3.4, as solvent. Spots are placed 20 mm from the bottom of the plate, making sure not to dip the origin into the solvent. The dye front is allowed to migrate 70–100% up the plate. Chromatograms are air dried and spots

[17] T. Gotoh, Y. Niino, M. Tokuda, O. Hatase, S. Nakamura, M. Matsuda, and S. Hattori, *J. Biol. Chem.* **272,** 18602 (1997).

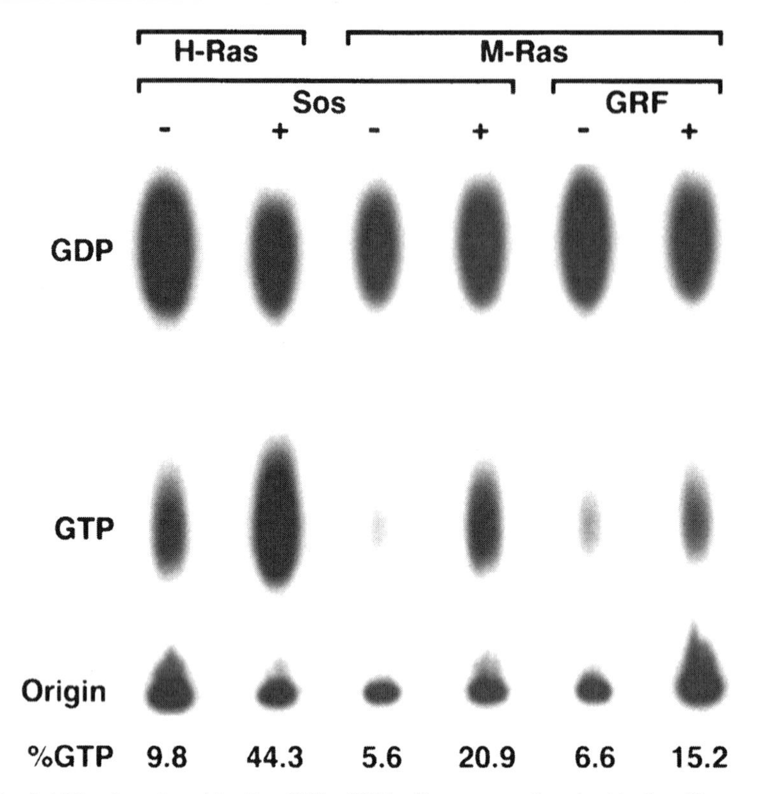

FIG. 3. M-Ras is activated by Ras GEFs. 293T cells were transfected with plasmids encoding epitope-tagged wild-type M-Ras or Ha-Ras along with pRCbac, pRCbac-(5′-SosF),[18] pcDNA3, or pcDNA3 c-CDC25-C*aa*X as indicated. After 36 hr the guanine nucleotide pool was metabolically labeled in the absence of serum, using $^{32}P_i$. Cells were then lysed and Ras-bound GTP and GDP were determined by immunoprecipitation and thin-layer chromatography. Sos1 and GRF1/CDC25 elevated the GTP levels of both Ha-Ras and M-Ras. %GTP, Average [GTP cpm/(1.5× GDP)cpm + GTP cpm] ×100 ratio from two (Sos) or three (GRF) independent experiments. [Reproduced with permission from the American Society for Biochemistry and Molecular Biology.[4]]

are quantified with an AMBIS (San Diego, CA) β-scanner. A PhosphorImager can similarly be used or, after overnight autoradiography (Fig. 3), spots can be scraped off and quantified by scintillation counting.

Nonradioactive Determination of Ras–GTP Levels

Nonradioactive determination of Ras–GTP levels takes advantage of the observation that the Ras-interacting domains of Ras effectors (Raf, Rlf, RalBP1, AF6,

[18] A. Aronheim, D. Engelberg, N. Li, N. al-Alawi, J. Schlessinger, and M. Karin, *Cell* **78,** 949 (1994).

etc.) can be isolated and preferentially bind to activated (GTP-bound) versus inactive (GDP-bound) Ras proteins. Thus, these Ras activation (RA) domains can be used to "fish out" activated Ras from cell lysates and the amount of bound Ras is directly proportional to the level of Ras activation in the cell.[19] In this discussion we focus on the assay for M-Ras, but further derivations of this concept can be used with other Ras-related GTPases. Currently, we use the following isolated RA domains for the aforementioned Ras proteins: Rlf for M-Ras and Rap2B; RalBP1 for RalA; and Raf(1-140) for Rap1A, Ha-Ras, and R-Ras.

Preparation of GST–Rlf Protein

The GST–Rlf RA domain fusion is made essentially as described above for Ras proteins, except that GDP and $MgCl_2$ are replaced with 1 m*M* EDTA in the lysis and wash/storage buffers. Purified protein levels are estimated by SDS–PAGE with a known standard/size marker, for example, glycerol-3-phosphate dehydrogenase (Sigma). Proteins are stored on beads at 4°. Proteins have a half-life of ~2 weeks, so we prepare fresh proteins every 1–2 weeks from frozen bacterial cell pastes.

In Vivo Exchange Assay

293T cells are plated at 50% confluency (2.7×10^5 cells) in 60-mm dishes 24 hr prior to transfection. HA-tagged M-Ras (750 ng) and 750 ng of the guanine nucleotide exchange factor are diluted in 250 μl of serum and antibiotic-free DMEM (DMEMϕ) in one tube and 8 μl of NovaFECTOR in 250 μl of DMEMϕ in another. The solutions are then mixed and allowed to stand for 15 min at room temperature. The cells are rinsed once with 5 ml of DMEMϕ. After the 15-min period the DNA–lipid solution is diluted to 2.5 ml with DMEMϕ and added to the cells. After 4 hr, an equal volume (2.5 ml) of DMEM supplemented with 20% (v/v) fetal bovine serum (FBS) is added to the cells. Twenty-four hours later, the medium is changed to DMEM containing the antibiotics and 0% FBS. After 18 hr in serum-free medium the cells are rinsed twice in ice-cold PBS. If the 293T cells lift off the plate with PBS during the wash step, then dislodge the remaining cells from the plate by trituration and pellet them in 1.5-ml microcentrifuge tubes. It is necessary to keep the cells on ice to minimize the intrinsic GTPase and exchange activity of the Ras proteins. The cells are lysed with, per dish, 0.5 ml of Rlf buffer [50 m*M* Tris-HCl (pH 7.4), 10% (v/v) glycerol, 200 m*M* NaCl, 2.5 m*M* $MgCl_2$, 1% (v/v) Igepal (Sigma), aprotinin (19 μg/ml), 1 m*M* PMSF] and gentle trituration. After 5 min on ice the lysate is cleared of cell debris by a 3-min spin at 13,000*g*. Approximately 10 μg of GST-Rlf bound to 35 μl of glutathione–agarose beads (50% slurry) is tumbled with 400 μl of cell lysate for 1 hr at 4°. The beads are washed four times with 1 ml of Rlf buffer, omitting the protease inhibitors. Fifty

[19] J. de Rooij and J. L. Bos, *Oncogene* **14,** 623 (1997).

microliters of SDS–PAGE sample buffer is added to the beads and the activated M-Ras protein that bound to the GST-Rlf is visualized by immunoblotting of the HA-tagged M-Ras protein.

GTPase-Activating Protein Assay

To measure GTPase-activating protein (GAP) activity GST–M-Ras and H-Ras fusion proteins are expressed from pGEX vectors (Pharmacia, Piscataway, NJ) in *E. coli* BL21-DE3(LysE) (Novagen) as outlined above. Purified p120 RasGAP was obtained from G. Bollag (Onyx Pharmaceuticals, Richmond, CA).[4] However, the p120 or NF1 GAP catalytic domains[20] may also be generated in *E. coli* as GST–fusion proteins.

Bead-bound proteins are loaded with GTP by incubation (10 min at 30°) in 50 m*M* Tris-HCl (pH 7.4), 150 m*M* NaCl, 5 m*M* EDTA, bovine serum albumin (BSA, 1 mg/ml), 6 μM GTP, 5 μCi of [γ-^{32}P]GTP, and 1 m*M* dithiothreitol. $MgCl_2$ is added to 10 m*M* and free nucleotide is removed by washing the beads three times with the loading buffer minus GTP. Proteins are then eluted by incubating the beads with 100 μl of 100 m*M* Tris-HCl (pH 8.0), 10 m*M* $MgCl_2$, and 20 m*M* glutathione on ice for 20 min. GTP-bound Ras (0.2 pmol, determined by scintillation counting) is then incubated for the indicated times with or without 5 ng of p120 GST–Ras GAP in a 50-μl final reaction volume containing 25 m*M* Tris-HCl, BSA (1.5 mg/ml), 7.5 m*M* $MgCl_2$, and 2 m*M* dithiothreitol at room temperature for the indicated times. The reaction is stopped by addition of 4 μl of 0.5 *M* EDTA and 1 μl of 10% (w/v) SDS, Ras proteins are denatured by incubation at 68° for 5 min, and guanine nucleotides are separated by thin-layer chromatography on PEI cellulose plates using 0.75 *M* KH_2PO_4/HCl, pH 3.4, as solvent. Chromatograms are dried and quantified with an AMBIS β-scanner. It is also possible to use [α-^{32}P]ATP and to pass the sample over nitrocellulose filters to measure loss of hydrolyzed phosphate as in Ref. 21. However, the thin-layer chromatography method specifically measures GDP generated and avoids any complications due to loss of label as a result of GTP exchange/release.

[20] G. F. Xu, B. Lin, K. Tanaka, D. Dunn, D. Wood, R. Gesteland, R. White, R. Weiss, and F. Tamanoi, *Cell* **63,** 835 (1990).

[21] L. A. Quilliam, C. J. Der, R. Clark, E. C. O'Rourke, K. Zhang, F. McCormick, and G. M. Bokoch, *Mol. Cell. Biol.* **10,** 2901 (1990).

[19] Analyses of TC21/R-Ras2 Signaling and Biological Activity

By SUZANNE M. GRAHAM, KELLEY ROGERS-GRAHAM, CLAUDIA FIGUEROA, CHANNING J. DER, and ANNE B. VOJTEK

TC21/R-Ras2 is a member of the Ras superfamily of proteins.[1] Within this family, TC21 shares the strongest amino acid identity with R-Ras and R-Ras3/M-Ras. TC21 functions as a GDP/GTP-regulated switch. Ras GTPase-activating proteins [p120 GTPase-activating protein (GAP) and, to a lesser extent, NF1-GAP] can serve as negative regulators of TC21, whereas Ras guanine nucleotide exchange factors (SOS1 and RasGRF) can serve as positive regulators of TC21.[2] Like Ras proteins, TC21 also terminates in a carboxyl-terminal C*aa*X tetrapeptide sequence (C, cysteine; *a,* aliphatic amino acid; X, terminal amino acid) that signals for posttranslational modifications that promote membrane association and are critical for biological activity.[3] However, whereas Ras proteins are modified by the C_{15}-farnesyl isoprenoid, TC21 is modified by the more hydrophobic C_{20}-geranylgeranyl isoprenoid.[4] Thus, TC21 is not inhibited by the farnesyltransferase inhibitors that can block Ras isoprenylation and function.

TC21 shares strong sequence identity with Ras residues important for interaction with downstream effectors (Ras residues 25–45). Consequently, it is not surprising that they exhibit biological properties similar to those of Ras. GTPase-deficient mutants of TC21 (with mutations analogous to the G12V or Q61L mutations of Ras) exhibit the transforming activities comparable to those of oncogenic mutants of Ras when assayed in NIH 3T3 and other cells.[3,5,6] Mutational activation, as well as overexpression, of TC21 has been observed in human tumor cells.[6–9]

[1] S. L. Campbell, R. Khosravi-Far, K. L. Rossman, G. J. Clark, and C. J. Der, *Oncogene* **17,** 1395 (1998).

[2] S. M. Graham, A. B. Vojtek, S. Y. Huff, A. D. Cox, G. J. Clark, J. A. Cooper, and C. J. Der, *Mol. Cell. Biol.* **16,** 6132 (1996).

[3] S. M. Graham, A. D. Cox, G. Drivas, M. G. Rush, P. D'Eustachio, and C. J. Der, *Mol. Cell. Biol.* **14,** 4108 (1994).

[4] J. M. Carboni, N. Yan, A. D. Cox, X. Bustelo, S. M. Graham, M. J. Lynch, R. Weinmann, B. R. Seizinger, C. J. Der, and M. Barbacid, *Oncogene* **10,** 1905 (1995).

[5] S. M. Graham, S. M. Oldham, C. B. Martin, J. K. Drugan, I. E. Zohn, S. Campbell, and C. J. Der, *Oncogene* **18,** 2107 (1999).

[6] G. J. Clark, M. S. Kinch, T. M. Gilmer, K. Burridge, and C. J. Der, *Oncogene* **12,** 169 (1996).

[7] A. M. Chan, T. Miki, K. A. Meyers, and S. A. Aaronson, *Proc. Natl. Acad. Sci. U.S.A.* **91,** 7558 (1994).

[8] Y. Huang, R. Saez, L. Chao, E. Santos, S. A. Aaronson, and A. M. Chan, *Oncogene* **11,** 1255 (1995).

[9] K. T. Barker and M. R. Crompton, *Br. J. Cancer* **78,** 296 (1998).

0076-6879/00 $35.00

Finally, like Ras, TC21 activity can also induce the differentiation of PC12 rat pheochromocytoma cells and block C2 myoblast differentiation.[5,10]

Presently, there is disagreement regarding whether TC21, like Ras, can activate Raf and the extracellular signal-regulated kinase (ERK) mitogen-activated protein kinase cascade. In our analyses, we found that TC21 could not interact with and cause the activation of full-length Raf-1.[2] Although we did find that ERK activity was constitutively upregulated in TC21-transformed NIH 3T3 cells,[3] the inability of TC21 to cause activation of ERKs when assessed in transient expression assays argued that this ERK activation was not a direct consequence of TC21 activity.[2] In contrast to our results, two other groups have reported that TC21 can activate the Raf/ERK pathway.[10,11] The basis for these different observations is not clear. However, the inability of activated TC21 to cause senescence of primary rodent fibroblasts is consistent with an inability of TC21 to activate Raf.[12] Oncogenic Ras induction of senescence has been shown to be mediated by activation of the Raf/ERK pathway alone.[12,13] This chapter summarizes experimental approaches for evaluating TC21 function. The first part of this chapter discusses approaches for evaluating TC21 signaling and biological activity in mammalian cells. The second part describes yeast two-hybrid binding analyses for characterizing the interaction of TC21 with Ras effectors and for the isolation of novel interacting proteins.

Mammalian Expression Vectors for Wild-Type and Mutant TC21 Proteins

The properties of wild-type and mutant TC21 proteins are summarized in Table I. The original TC21 sequence[14] omitted nucleotides at positions 11 (C), 20 (G), and 33 (C).[4,7] Hence, our earlier designations of the 71L mutations have since been revised to 72L. Full-length cDNA sequences encoding wild-type, constitutively activated mutants of human TC21 (23V and/or 72L), the 46A effector domain mutant,[3] and the 26A and 28N dominant-negative mutants[2] were introduced into the retroviral vector pZIP-NeoSV(x)1 (Fig. 1)[15]. Expression of TC21 in this vector is under the control of the strong Moloney long terminal repeat (LTR) retrovirus promoter. These vectors also encode a neomycin resistance gene, under the control of the simian virus 40 (SV40) promoter, to allow selection in growth medium containing geneticin (G418). These constructs can be used as

[10] N. Movilla, P. Crespo, and X. R. Bustelo, *Oncogene* **18,** 5860 (1999).

[11] M. Rosario, H. F. Paterson, and C. J. Marshall, *EMBO J.* **18,** 1270 (1999).

[12] M. Serrano, A. W. Lin, M. E. McCurrach, D. Beach, and S. W. Lowe, *Cell* **88,** 593 (1997).

[13] J. Zhu, D. Woods, M. McMahon, and J. M. Bishop, *Genes Dev.* **12,** 2997 (1998).

[14] G. T. Drivas, A Shih, E. Coutavas, M. G. Rush, and P. D'Eustachio, *Mol. Cell. Biol.* **10,** 1793 (1990).

[15] C. L. Cepko, B. E. Roberts, and R. C. Mulligan, *Cell* **37,** 1053 (1984).

TABLE I
MUTANTS OF TC21/R-Ras2

Protein[a]	Amino acid substitution (codon)[b]	Properties
TC21		Wild-type human protein
TC21(23V)	Gly (GGC) → Val (GTC)	Analogous to Ras G12V; GTPase deficient, transforming
TC21(72L)	Gln (CAA) → Leu (CTA)	Analogous to Ras Q61L; GTPase deficient, transforming
TC21(23V,72L)	Gly (GGC) → Val (GTC) Gln (CAA) → Leu (CTA)	Analogous to Ras G12V/G61L; GTPase deficient, transforming
TC21(46A,72L)	Thr (ACC) → Ala (GCC) Gln (CAA) → Leu (CTA)	Analogous to Ras T35A effector domain mutant; nontransforming
TC21(26A)	Gly (GGC) → Ala (GCC)	Analogous to Ras G15A dominant negative; inhibitor of Ras GEFs
TC21 (28N)	Ser (TCG) → Asn (AAT)	Analogous to Ras S17N dominant negative; inhibitor of Ras GEFs

[a] The mutated residue in TC21 is indicated in parentheses.
[b] The nucleotide substitutions are underlined.

plasmid DNAs for transfection into mammalian cells. Alternatively, they may be transfected into ecotropic or amphotropic retrovirus packaging cell lines, such as the Phoenix cell lines, for generation of infectious retrovirus for introduction into rodent or human cells, respectively. The pZIP-NeoSV(x)1 retroviral vector

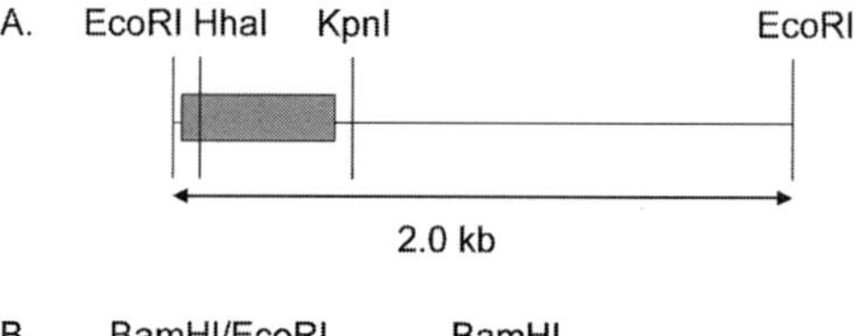

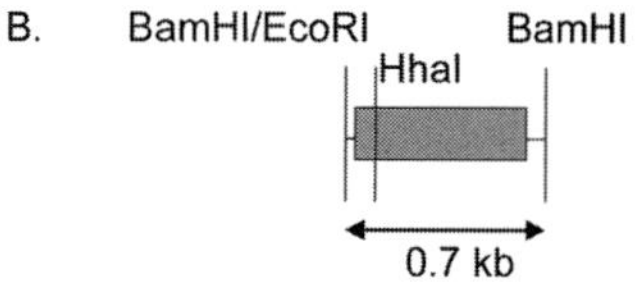

FIG. 1. Mammalian expression vectors encoding wild-type and mutant human TC21. A 2.0-kb *Eco*RI fragment encoding wild-type, 23V, 72L, 23V/72L, and 46A/72L mutants of TC21 was inserted into the *Eco*RI site of a variant of pZIP-NeoSV(x)1. An *Hha*I restriction enzyme site was used to construct cDNA sequences encoding the 23V and 72L mutants from the double 23V/72L mutant. Digestion with *Hha*I creates an NH_2-terminal 0.1-kb fragment containing codons for TC21 residues 1 to 27 and a 1.9-kb fragment encoding the remaining TC21 sequences and 3′ noncoding sequences. A 0.7-kb *Bam*HI fragment was used to construct pZIP-NeoSV(x)1 expression vectors encoding the 26A and 28N dominant-negative mutants of TC21.

contains a low copy number pBR322 bacterial origin of replication. Therefore, when preparing these retroviral DNAs for transfection of cells, the bacterial cultures should be treated with chloramphenicol to amplify the plasmid DNA. In addition, the cultures are commonly grown in the presence of ampicillin/carbenicillin and kanamycin to select for the plasmid and to select against those vectors that have undergone recombination between the LTR direct repeats. We have used the pZIP-NeoSV(x)1 expression vectors of TC21 for both stable and transient expression analyses.

pCGN-hyg mammalian expression vectors[16] encoding wild-type and activated (72L) TC21 proteins were also generated.[3] *Bam*HI fragments derived from the pGEX-2T TC21 bacterial expression vectors (described below) were ligated into the unique *Bam*HI site of pCGN-hyg. This results in the addition of NH_2-terminal sequences encoding a hemagglutinin (HA) protein epitope tag onto the NH_2 terminus of the encoded TC21 protein. Fusions to the amino terminus of TC21 do not affect its biological activity. Expression of the chimeric HA–TC21 proteins in transfected mammalian cells can then be detected by Western blot analyses using the anti-HA monoclonal antibody (BAbCo, Berkeley, CA). Expression of TC21 in pCGN-hyg is regulated by the human cytomegalovirus (CMV) promoter. These constructs also encode resistance for hygromycin, which allows for selection of transfected or infected mammalian cells using growth medium supplemented with hygromycin at 400 μg/ml. We have used the pCGN-hyg expression vectors of TC21 for both stable and transient expression analyses.

Transformation Analyses

The transforming activity of TC21 is assayed by essentially the same approaches that are used to assay for Ras transforming activity. GTPase-deficient, constitutively activated mutants of TC21 (G23V or Q72L) can cause growth transformation of a variety of fibroblast and epithelial cells, including NIH 3T3 mouse fibroblasts, Rat-1 rat fibroblasts, RIE-1 rat intestinal epithelial cells, and MCF-10A human breast epithelial cells.

The most extensively employed cell type for assessing TC21 transforming activity is the mouse fibroblast NIH 3T3 cell line. Because NIH 3T3 cells exhibit a high frequency of spontaneous transformation, only cultures of NIH 3T3 cells that have been appropriately maintained to avoid the appearance of spontaneously transformed cells are useful for TC21 transformation analyses. The proper selection of NIH 3T3 cell lines, and their maintenance, has been described previously.[17] Furthermore, it is worth noting that NIH 3T3 cell lines from different laboratories

[16] B. S. Yang, C. A. Hauser, G. Henkel, M. S. Colman, C. Van Beveren, K. J. Stacey, D. A. Hume, R. A. Maki, and M. C. Ostrowski, *Mol. Cell. Biol.* **16,** 538 (1996).

[17] G. J. Clark, A. D. Cox, S. M. Graham, and C. J. Der, *Methods Enzymol.* **255,** 395 (1995).

are not identical in their transforming properties and that the methods and results from transformation assays may vary. Our methods have been developed to provide optimal reproducibility for the transformation assays done in the NIH 3T3 cell line that we have employed in our studies, sometimes referred to as the NIH 3T3(UNC) cell line.

Details for focus formation analyses using our NIH 3T3 cell line have been described previously.[17] Therefore, we provide only a brief summary of these analyses. For focus formation transformation assays, we use calcium phosphate precipitation to introduce plasmid DNA into NIH 3T3 cells. TC21(23V), TC21(72L), and TC21(23V,72L) exhibit potent focus-forming activities that are comparable to those seen with pZIP-NeoSV(x)1 vectors encoding 12V or 61L mutants of H-Ras ($4–7 \times 10^3$ foci/μg plasmid DNA). Typically, we transfect 10 to 50 ng of purified plasmid DNA per 60-mm dish (plated with 2×10^5 cells per dish). For focus formation assays using Rat-1 cells, 2–5 μg of plasmid DNA is transfected per 60-mm dish. The appearance of transformed foci of cells is visualized by staining with 0.4% (w/v) crystal violet. In contrast to focus formation analyses using NIH 3T3 cells, Rat-1 focus formation assays revealed a much less potent focus-forming activity for TC21(23V), whereas H-Ras(61L) and TC21(72L) displayed comparable activities. To ensure that differences in generation of foci among tested constructs are not due simply to variations in transformation efficiency, the transformation assays are repeated twice in triplicate with two different DNA preparations. Alternatively, 12 to 16 hr after removing the calcium phosphate precipitate, one-third or one-fourth of each transfected culture is replated in selection medium [medium supplemented with geneticin (G418), when using pZIP-NeoSV(x)1 expression vectors] to determine the number of transfectants, and the other half is plated in medium without drug and then subsequently stained with crystal violet after approximately 14 days to reveal the number of foci.

For analyses of other transformed phenotypes, NIH 3T3 cells stably transfected with plasmid DNAs encoding wild-type or mutant TC21 are established. Typically, 50 to 100 ng of plasmid DNA is transfected per 60-mm dish of NIH 3T3 cells. Three days after transfection, one-third of each transfected culture is replated into a 100-mm dish containing growth medium supplemented with G418 (Geneticin, 400 μg/ml; GIBCO, Grand Island, NY). After approximately 2 weeks, the several hundred drug resistance colonies that have arisen are then trypsinized and pooled together, with one-tenth of the culture replated into another 100-mm dish containing growth medium supplemented with G418. These mass cell populations are then used to assess the ability of TC21 expression to cause a reduced dependence on serum growth factors, to allow proliferation in suspension, and to form tumors in experimental animals. Our assays for low serum growth (0.5 to 2%), colony formation in soft (0.3%, w/v) agar, and tumorigenicity in athymic nude mice (5×10^5 cells per site) have been described in detail previously.[17] Untransformed NIH 3T3 cells do not proliferate on plastic in growth medium supplemented with low serum, do not form colonies in soft agar, and are negative for tumor formation for up to 4 weeks. However, NIH 3T3 cells expressing activated forms of TC21 or H-Ras can proliferate in low serum, grow in soft agar, and form large and progressively growing tumors in nude mice.

Other Biological Assays for TC21 Function

Activated TC21 and Ras share comparable biological activities when assayed in a variety of cell types. Activated TC21 can cause focus formation when transfected into cultures of Rat-1 rat fibroblasts or RIE-1 rat intestinal epithelial cells.[3,5] Activated TC21 also causes neurite induction and cessation of proliferation when stably overexpressed in PC12 rat pheochromocytoma cells.[5,10] Finally, activated TC21 can also block C2 mouse myoblast differentiation or MyoD-induced myoblast differentiation of C3H10T1/2 cells.[5] These assays are done essentially as we have described for Ras.[5]

Signaling Analyses

We have utilized retrovirus expression vectors encoding wild-type or mutant TC21 proteins for various assays to delineate the signaling pathways that are regulated by TC21. Ectopic expression of constitutively activated mutants has been used to define downstream signaling events activated by TC21. Dominant-negative mutants have been used to determine signaling pathways that may require TC21 function.

The signaling activities of TC21 can be assessed using either transiently transfected or stably transfected cells. Transient transfection expression analyses are used to determine the signaling activities that are stimulated by TC21. The analyses of cells stably overexpressing TC21 allow delineation of alterations in specific signaling molecules due to sustained TC21 activity. These may involve changes caused by activated TC21 or changes due to the transformed phenotype caused by TC21 transformation.

Activation of TC21 and Ras leads to alterations in gene expression as a result of activation of various nuclear transcription factors, including Jun, Elk-1, serum response factor, Fos, Myc, and NF-κB. Two transient expression analyses to assess Ras- or TC21-induced alterations in gene expression have been described. One assay assesses the potency of TC21-stimulated *trans*-activation of nuclear transcription factors, such as Elk-1. The second assay assesses the ability of TC21 or Ras to activate a minimal promoter element.

To evaluate the ability of TC21 to stimulate the *trans*-activation domain of nuclear transcription factors, including Elk-1, Jun, Fos, and RelB, an expression plasmid encoding TC21 is introduced into cells along with three additional plasmid DNAs. The first plasmid encodes a chimeric transcription factor where the

DNA-binding domain of Gal4 is fused to the *trans*-activation domain (TA) of Elk-1 or other transcription factor. The second plasmid encodes the luciferase gene under the control of a minimal promoter that contains tandem copies of the Gal4 DNA-binding sequence. The third plasmid expresses β-galactosidase (β-Gal), an internal control to normalize for variation in transfection efficiency. Briefly, 1 μg of either pCGN-hyg empty vector control, or pCGN-*tc21*(72L), is cotransfected with 1 μg of the desired Gal-TA expression plasmid and 1 μg of the β-Gal expression vector into COS or NIH 3T3 cells (plated at 1.25×10^5 cells/well in six-well plates). The plasmid DNAs are added to 500 μl of Dulbecco's modified Eagle's medium (DMEM) together with 3 μg of carrier (calf thymus) DNA (Boehringer Mannheim, Indianapolis, IN). The DNA mixture is then incubated for 15 min with a premixed solution of 500 μl of DMEM and 20 μl of LipofectAMINE (GIBCO). This reaction mixture is then added to washed cells containing 4 ml of DMEM (without serum) and allowed to incubate for 5–8 hr. Five milliliters of DMEM supplemented with 20% (v/v) fetal calf serum is then added for an overnight incubation. The following day, the cultures are rinsed with DMEM and then incubated with DMEM supplemented with 0.5% (v/v) fetal calf serum for 48 hr. Cells are then lysed in lysis buffer [20 m*M* Tris (pH 7.4), 1% (v/v) Triton X-100, 10% (v/v) glycerol, 137 m*M* NaCl, and 2 m*M* EDTA, with fresh additions of 25 m*M* BGP, 2 m*M* *p*-nitrophenylene phosphate (PNPP), 1 m*M* Na_3VO_4, 0.5 m*M* phenylmethylsulfonyl fluoride (PMSF), and leupeptin (10 μg/ml)]. Protein concentration is then determined by the bicinchoninic acid (BCA) protein assay (Pierce, Rockford, IL).

TC21 activation of the *trans*-activation domain of a particular Gal4–TA fusion protein results in stimulation of its ability to activate transcription and expression of luciferase. Thus, production of luciferase protein expression can provide a measure of TC21 activation of a particular transcription factor. TC21 can activate Elk-1, Jun, Fos, Myc, and RelB to varying levels. The chapter by Hauser and colleagues in this volume[17a] provides a detailed description of Gal4–*trans*-activation domain fusion reporter plasmids.

To evaluate the ability of TC21, or Ras, to activate a minimal promoter element, plasmid DNAs in which expression of the chloramphenicol acetyltransferase (CAT) or luciferase gene is under the control of promoter elements that contain various Ras-responsive promoter elements are introduced into cells. These include minimal promoter elements harboring tandem NF-κB or ets/AP-1 DNA-binding sites. Briefly, 2 to 5 μg of pZIP-NeoSV(x)1 plasmid DNA encoding activated TC21 is cotransfected together with 1 μg of the desired reporter plasmid into NIH 3T3 cultures. Forty-eight hours after transfection, cell lysates are prepared and the CAT or luciferase activity is then assayed as described previously.[2] A detailed

[17a] G. Foos, C. K. Galang, C.-F. Zhang, and C. A. Hauser, *Methods Enzymol.* **333,** Chap. 7, 2001 (this volume).

description of the Ras-responsive promoter reporter plasmids can be found in Galang *et al.*[18]

We have also used retrovirus vector constructs encoding the different TC21 proteins for either transient or stable transfection of NIH 3T3 cells for analyses of activation of the ERK mitogen-activated protein kinase (MAPK) cascade.[2] Briefly, total cell extracts (equal protein concentrations) are resolved by sodium dodecyl sulfate–polyacrylamide gel electrophoresis (SDS–PAGE) and then transferred to Immobilon membranes (Millipore, Bedford, MA) for analysis by western immunoblotting. The phospho-ERK antibody [rabbit anti-ERK antiserum 691 from Santa Cruz Biotechnology (Santa Cruz, CA) or PhosphoPlus p44/42 MAPK antibody from New England BioLabs (Beverly, MA)] is used to detect phosphorylated, activated forms of both p42 ERK2 and p44 ERK1; and the anti-ERK antiserum (p44/42 MAP kinase antibody; New England BioLabs) is used to detect total ERK (phosphorylated and unphosphorylated forms). An increase in phospho-ERK relative to total ERK indicates that the introduced cDNA can activate the ERK kinase. Transient cotransfection with a plasmid vector encoding HA-tagged ERK2, followed by an *in vitro* kinase assay using myelin basic protein as substrate, is also used to determine ERK activation by TC21. No activation of ERKs is seen in transient expression assays. However, ERK activation is seen in TC21-transformed cells. TC21 does activate both the Jun N-terminal kinase (JNK) and p38 MAPKs when assayed in transient expression assays in COS-7 cells. Here, cotransfection with plasmid DNAs encoding glutathione *S*-transferase (GST)–Jun or GST–ATF-2 is done to assay for JNK and p38 activation, respectively, using myelin basic protein as a substrate in the *in vitro* kinase assays.

In Vitro Biochemical Analyses

We have utilized GST fusions to TC21 for various *in vitro* biochemical analyses.[2,4] Full-length coding sequences of wild-type and mutant (72L or 46A/72L) TC21 are introduced into the pGEX-2T bacterial expression vector (Pharmacia-LKB, Uppsala, Sweden). An *Nco*I–*Bam*HI fragment containing the TC21-coding sequences and flanking 3′ noncoding sequences is ligated into the *Bam*HI site of pGEX-2T, using a linker containing a 5′ *Bam*HI site and 3′ *Nco*I-compatible ends.

Bacterially expressed GST–TC21 proteins are isolated by conventional methods for the expression and purification of GST fusion proteins. Briefly, XL1-Blue bacteria transformed with each pGEX plasmid are grown overnight at 37° in 5 ml of Luria broth (LB) supplemented with carbenicillin, to maintain selection for the plasmid, and then diluted in 250 ml of fresh LB and incubated for 2 hr. After the 2-hr incubation, 0.1 m*M* isopropylthio-β-D-galactopyranoside (IPTG) is added and the

[18] C. K. Galang, C. J. Der, and C. A. Hauser, *Oncogene* **9,** 2913 (1994).

culture is incubated for an additional 3 hr. Bacteria are then collected by centrifugation, resuspended in 5 ml of ice-cold lysis buffer [20 m*M* EDTA, 50 μ*M* GTP, lysozyme (0.2 mg/ml, added fresh), and protease inhibitors in phosphate-buffered saline (PBS)], and then incubated for 30 min on ice. This 30-min incubation is followed by an additional 20-min incubation, after the addition of 5 m*M* $MgCl_2$ and DNase I. Lysates are then freeze–thawed three times in a dry ice–ethanol bath, after which NaCl is added to 0.5 *M*, Triton X-100 to 1.0% (v/v), and dithiothreitol (DTT) to 5 m*M*. After multiple passages through a 21-gauge needle to shear bacterial DNA, the lysate is incubated with a 50% solution of glutathione beads in PBS overnight at 4°. We have used recombinant GST–TC21 proteins for *in vitro* intrinsic and GAP-stimulated GTP hydrolysis assays.[2] We have also used these proteins for *in vitro* protein prenylation studies.[4]

tc21 cDNA sequences are also introduced into the pGEX2T-XL bacterial expression vector (a pGEX-2T derivative encoding the A kinase-phosphorylatable RRASV sequence 5′ of the *Bam*HI site) to express GST–TC21 fusion proteins that can be radiolabeled with ^{32}P for use as a probe to screen an expression library to identify TC21-interacting proteins. GST beads with bound GST–TC21 proteins are washed three times with PBS, and then three times with kinase buffer [20 m*M* Tris (pH 7.5), 100 m*M* NaCl, 12 m*M* $MgCl_2$, and 50 μ*M* GTP]. This is followed by a kinase reaction using [γ-^{32}P]ATP and 125 units of protein kinase A (Sigma, St. Louis, MO) for 30 min at 4°, and then for 10 min at 30°. The kinase reaction is stopped by the addition of 10 m*M* $NaPO_4$ (pH 8.0), 10 m*M* sodium pyrophosphate, 10 m*M* EDTA, bovine serum albumin (BSA, 1 mg/ml), and 50 μ*M* GTP. Beads are washed five times with TENN [0.5% (v/v) Nonidet P-40 (NP-40), 20 m*M* Tris (pH 8.0), 1 m*M* EDTA, 100 m*M* NaCl, and 50 μ*M* GTP]. The GST–TC21 fusion protein is eluted off the beads by rocking the beads for 20 min at 4° in 20 m*M* glutathione, 100 m*M* Tris (pH 8.0), 100 m*M* NaCl, and 50 μ*M* GTP. Beads are then centrifuged and labeled proteins in the supernatant are counted in a scintillation counter.

To identify TC21-interacting proteins, we have screened a 16-day-old mouse embryo cDNA expression library cloned into the λEXlox expression vector (Novagen, Madison, WI). BL21 (DE3) pLysE bacteria are infected with the TC21 expressing phage for 30 min at 37°, and then layered over 150-mm plates containing solidified 2× YT medium. After 8 hr of growth at 37°, the infected bacterial plates are overlaid with IPTG-impregnated filters for overnight incubation to induce protein expression. The following day, purified GST–TC21(72L) protein is preloaded with [γ-^{32}P]GTP (as described above) and incubated for 2.5 hr with filters that have been preblocked for 1 hr at room temperature in 50 m*M* Tris (pH 7.5), 100 m*M* NaCl, 5% (w/v) nonfat milk, 0.1% (v/v) Tween 20, 1 m*M* EDTA, 1 m*M* DTT, 30 m*M* $MgCl_2$. Filters are washed four times at 4° with wash solution [block solution with 0.2% (w/v) BSA substituting for the 5% (w/v) nonfat milk and with excess cold GTP] and exposed to film overnight at −80°. All positive clones are amplified and purified to tertiary clones. All positive tertiary clones are then analyzed in the same fashion for binding to GST, GST–H-Ras(12V), and GST–TC21(46A, 72L).

Interaction with Effectors

We have employed yeast two-hybrid binding analyses to evaluate the ability of TC21 to interact with known and candidate effectors of Ras. Briefly, two hybrid proteins are expressed in yeast. One is a fusion of a protein of interest to a DNA-binding domain. The second is a fusion of a protein of interest to an activation domain. By themselves, the DNA-binding domain and the activation domain cannot interact in yeast. If, however, the DNA-binding and activation domains are fused to proteins that can interact, then this interaction can be readily detected in yeast. We describe here a two-hybrid system that uses the LexA DNA-binding domain and a nuclear-localized VP16 activation domain.[19] Plasmids expressing fusions to LexA or VP16 are introduced into the yeast L40 reporter strain. The association of the two fusion proteins results in the specific and potent *trans*-activation of a *His3* and *lacZ* reporter gene. The *trans*-activation of the *His3* reporter enables yeast to grow in the absence of histidine, whereas the *trans*-activation of the *lacZ* reporter results in blue yeast, when yeast are lysed with liquid nitrogen and then incubated with X-Gal. The materials and methods to determine whether two proteins interact in the yeast two-hybrid system are outlined below. We have described previously the use of the two-hybrid system to identify novel partners for proteins of interest.[19,20]

Materials

Plasmids

pBTM116-ADE2. By cloning genes of interest into the polylinker of vector pBTM116-ADE2, in-frame fusions to the LexA DNA-binding domain are generated. This vector contains the complete coding sequence of LexA, expressed from the *ADH1* promoter, a yeast 2μ origin of replication, a bacterial origin of replication and β-lactamase gene, the yeast *TRP1* gene, and the yeast *ADE2* gene. The β-lactamase gene is a selectable marker for *Escherichia coli* (ampicillin or carbenicillin). The *TRP1* gene is a selectable marker for yeast. The *ADE2* gene can be used to score plasmid loss during two-hybrid library screens. The polylinker

[19] A. B. Vojtek, S. M. Hollenberg, and J. A. Cooper, *Cell* **74,** 205 (1993).

[20] A. B. Vojtek, J. A. Cooper, and S. M. Hollenberg, *in* "The Yeast Two-Hybrid System" (S. Fields and P. L. Bartel, eds.), p. 29. Oxford University Press, New York, 1997.

and frame of pBTM116-ADE2 are shown below:

GAA TTC CCG GGG ATC CGT CGA CCT GCA G

*Eco*RI *Sma*I *Bam*HI *Sal*I *Pst*I

pVP16. By cloning genes of interest into the polylinker of vector pVP16, in-frame fusions to a nuclear-localized VP16 activation domain are generated. This vector contains the SV40 large T antigen nuclear localization sequence fused to the VP16 acidic activation domain, expressed from the *ADH1* promoter, a yeast 2μ origin of replication, a bacterial origin of replication and β-lactamase gene, and the yeast *LEU2* gene. The *LEU2* gene is a selectable marker for yeast. The polylinker and frame of pVP16 are shown below:

TGG ATC CCC GGG TAC CGA GCT CAA TGG CGG CCG CTA GAT AGA TAG AAG

*Bam*HI *Not*I Stop

CTA GCT CGA ATT C

*Eco*RI

pVP16 can be replaced with other activation domain plasmids, such as those encoding the Gal4 activation domain, as long as the Gal4 plasmids have a yeast *LEU2* selectable marker.

Strains and Media

Strain

The plasmids expressing the LexA and VP16 fusions of interest are introduced into *Saccharomyces cerevisiae* strain L40 (*Mat*a *his3Δ200 trpl-901 leu2-3,112 ade2 LYS2*::(lexAop)4-*HIS3 URA3*::(lexAop)8-*lacZ GAL4*).

Composition of Media

YPAD, per liter: 10 g of yeast extract, 20 g of Bacto-peptone, 0.1 g of adenine, 2% (w/v) glucose (added after autoclaving), 20 g of agar, as required

YC–WHULK medium, per liter: 1.2 g of yeast nitrogen base, without amino acids and ammonium sulfate; 5 g of ammonium sulfate; 10 g of succinic acid; 6 g of sodium hydroxide; 0.75 g of amino acid mix lacking tryptophan (W), histidine (H), uracil (U), leucine (L), and lysine (K); 2% (w/v) glucose (added after autoclaving); and 20 g of agar, as required

YC –L –W medium, per liter: YC –WHULK with 0.05 g of histidine, 0.1 g of uracil, and 0.1 g of lysine

Amino acid mix –WHULK: Mix together 1 g each of adenine sulfate, arginine, cysteine, and threonine; and 0.5 g each of aspartic acid, isoleucine, methionine, phenylalanine, proline, serine, and tyrosine. With a pestle or the back of a spoon, pulverize the mixture until it is of an even consistency

Small-Scale Transformation of Yeast

1. Inoculate a single colony of the L40 yeast strain in 5 ml of YPAD. Grow overnight with shaking at 30°.
2. Dilute the overnight culture in YPAD to an OD_{600} of 0.5 (a 100-ml culture is enough for 10 transformations).
3. Incubate for 2 to 4 hr at 30° with shaking.
4. Harvest the cells by centrifugation at 4000 rpm for 5 min at room temperature.
5. Resuspend the cells in 1× lithium acetate–TE [100 m*M* lithium acetate in 1× TE (10 m*M* Tris, 1 m*M* EDTA, pH 7.5)]. Centrifuge as in step 4.
6. Resuspend the cells in 1 ml of lithium acetate–TE. Incubate the cell suspension for 10 min at room temperature.
7. To an Eppendorf tube, add 0.2 to 0.4 μg of each plasmid and 8 μl of carrier DNA (10-mg/ml stock). Then, add 100 μl of the cell suspension from step 6.
8. Add 700 μl of PEG–Lithium acetate [40% (w/v) polyethylene glycol (PEG) 3350, 1× lithium acetate 1× TE). Mix by pipetting.
9. Incubate at 30° for 30 min.
10. Add 88 μl of dimethyl sulfoxide (DMSO), mix, and heat shock for 7 min at 42°.
11. Centrifuge the cells briefly (1 min, 7000 rpm). Wash the cell pellet with 200 μl of TE.
12. Centrifuge the cells briefly. Resuspend the pellet in 150 μl of TE. Plate 50 μl to YC –L –W plates. Incubate the plates at 30° for 2 to 3 days.

Assessing *Trans*-Activation of *His3* Reporter Gene

1. Inoculate a single yeast transformant in YC –W –L. Grow overnight at 30° with shaking. Yeast plasmids containing a 2μ origin of replication are readily lost when selection for the plasmids is not maintained during growth. Omitting tryptophan maintains selection for the bait pBTM116 vector; omitting leucine maintains selection for the pVP16 vector.
2. Dilute each culture to an OD_{600} of 0.1 in YC –WHULK or TE (dilution A). Further dilute A 1 : 10 (dilution B), 1 : 100 (dilution C), and 1 : 1000 (dilution D).
3. Spot 1 μl of each dilution (A to D) on YC –WHULK and YC –L –W plates.
4. Incubate at 30° for 2 to 3 days.

5. Compare growth on YC –WHULK plates with growth on YC –L –W plates. Growth on YC –WHULK plates indicates a positive interaction in the yeast two-hybrid system.

The plasmid encoding the LexA–bait fusion should always be introduced into the L40 strain together with the VP16 plasmid. Transformants resulting from this plasmid combination should not grow on YC –WHULK. If growth is observed, this suggests that the bait is nonspecifically *trans*-activating the *His3* reporter gene. To reduce the level of nonspecific *trans*-activation of the *His3* reporter, 3-aminotriazole (3-AT), a competitive inhibitor of the *His3* gene, can be spread into the YC –WHULK plates prior to spotting the dilutions onto the plate. In addition, a positive control [e.g., L40 transformed with pLexA-Ras and Rip51 (an amino-terminal domain of the Ras effector Raf)] should be included in each experiment.

In Fig. 2, we show the interaction of oncogenic TC21 and Ras with various Ras effectors. TC21 was introduced into the yeast L40 strain together with the expression vectors encoding the Ras effectors c-Raf, Ral guanine nucleotide dissociation stimulator (RalGDS), and phosphatidylinositol 3-kinase (PI3K). Yeast expressing LexA–TC21 fusion proteins and the amino-terminal domain of c-Raf-1, full-length RalGDS, or PI3K grow in the absence of histidine (Fig. 2) and turn blue in the presence of X-Gal (not shown). The interaction between LexA–TC21 and the Ras effectors requires an intact TC21 effector domain.

Assessing *Trans*-Activation of *lacZ* Reporter Gene

1. Using a sterile toothpick or yellow tip for a P20 pipettor, transfer (patch) yeast transformants to a YC –L –W plate. Transfer four colonies from each transformation. Incubate the plate at 30° for 2 days.
2. Transfer the yeast patches from step 1 to a supported 0.45-μm pore size, 82-mm nitrocellulose filter.
3. Lyse the yeast by immersing the filter in liquid nitrogen. Place the filter, yeast colony side up, on an aluminum foil boat and immerse the filter/boat in liquid nitrogen for 5 sec. Remove the filter from the boat and thaw, colony side up.
4. Place the thawed nitrocellulose filter on a Whatman No.1 filter circle soaked in X-Gal (10 to 30 μl of X-Gal, 50 mg/ml in DMSO) diluted in Z buffer [60 m*M* Na_2HPO_4, 40 m*M* NaH_2PO_4, 10 m*M* KCl, 1 m*M* $MgSO_5$ (pH 7.0)]. The X-Gal–Z buffer solution should be placed in the lid of a petri dish prior to adding the Whatman filter circle. Cover the petri dish with the bottom and incubate at 30° for 15 min to overnight. Score the blue color development over time. Blue color indicates a positive interaction in the two-hybrid system.

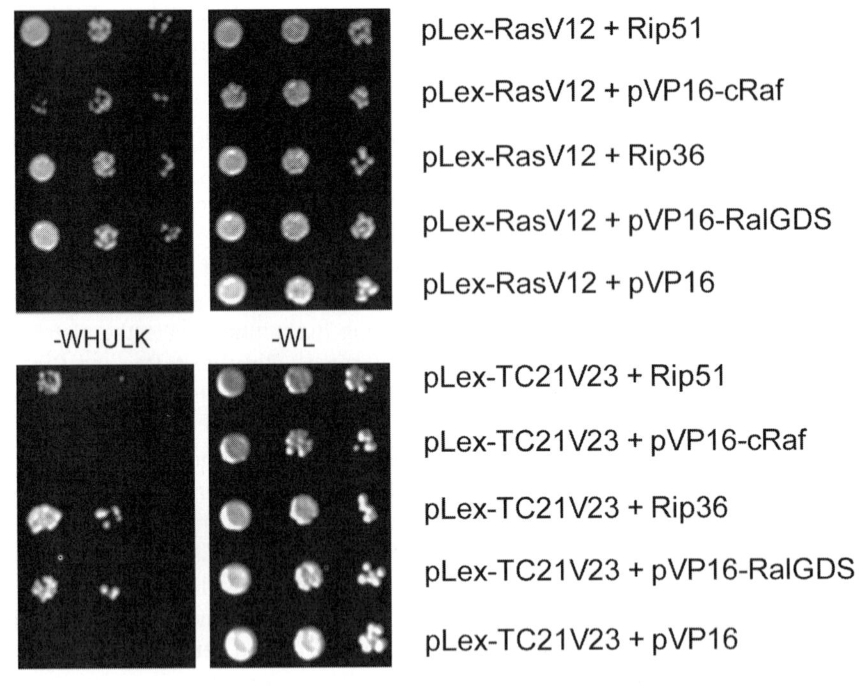

FIG. 2. Association of Ras and TC21 with downstream effector proteins. The indicated plasmid combinations were introduced into the L40 yeast strain, templates were prepared as described in this chapter, and growth was scored after 2 days. pLex-RasV12 expresses activated full-length H-Ras as a fusion protein to the LexA DNA-binding domain. pLex-TC21V23 expresses an activated TC21 as a fusion protein to the LexA DNA-binding domain. Rip51 expresses an amino-terminal domain of c-Raf as a fusion protein to the nuclear-localized VP16 acidic activation domain. pVP16-cRaf expresses VP16 fused to the entire c-Raf1 protein. Rip36 expresses an amino-terminal domain of phosphatidylinositol-3-kinase. pVP16-Ral GDS expresses VP16 fused to RalGDS. As has been previously observed, TC21 interacts with an isolated amino-terminal domain of c-Raf but not with the intact, full-length protein.

Acknowledgments

We thank Andrea Kusa for preparation of the plasmid maps. Our research studies were supported by Public Health Service grants CA42978, CA55008, and CA63071 (C.J.D.) from the National Cancer Institute and by grants from the DOD Breast Cancer Research Program (DAMD17-98-1-8319, A.B.V.) and the American Diabetes Association (A.B.V.).

[20] Characterization of Rheb Functions Using Yeast and Mammalian Systems

By JUN URANO, CHAD ELLIS, GEOFFREY J. CLARK, and FUYUHIKO TAMANOI

Introduction

Rheb (Ras homolog enriched in brain) is a new member of the Ras superfamily of G proteins that is highly conserved in a wide range of organisms. Homologs have been identified in human, rat, *Saccharomyces cerevisiae, Schizosaccharomyces pombe,* fruitfly, zebrafish, sea squirt, *Botrytis cinerea,* and *Candida albicans.*[1–5] These Rheb homologs share several key features. First, the third position of the G1-box, which is usually a glycine in other Ras-like G proteins, is perfectly conserved as an arginine. It is speculated that the presence of an arginine at this position results in the Rheb proteins being in a constitutively active form.[1,4] Second, the effector domain is well conserved between the Rheb proteins, with a consensus sequence of F-V-E/D-S-Y-Y/D-P-T-I-E-N-E/Q/T-F-T/S/N-R/K-x-x. Last, the Rheb proteins maintain a C-terminal C*aa*X farnesylation motif (C is for cysteine, *a* is for aliphatic amino acids, and X is usually a serine, cysteine, alanine, methionine, or glutamine). Rheb shares greatest homology with the Rap subfamily; however, close sequence analysis suggests that it occupies its own separate branch.[4,6]

Rheb shares some of the biological properties of Rap proteins, such as binding nonproductively to Raf-1 and antagonizing Ras transformation[7]; however, Rheb also has unique attributes. Rheb shows immediate-early gene characteristics. The Rheb transcript is increased in response to maximal electroconvulsive seizures as well as to *N*-methyl-D-aspartate (NMDA)-mediated synaptic activity and growth factors.[1] In addition, the protein is farnesylated and the protein is localized to the plasma membrane.[7] Moreover, instead of activating B-Raf, like Ras and Rap, Rheb inhibits B-Raf.[8] Thus, the physiological role of Rheb is likely to be unique.

[1] K. Yamagata, L. K. Sanders, W. E. Kaufmann, W. Yee, C. A. Barnes, D. Nathans, and P. F. Worley, *J. Biol. Chem.* **269,** 16333 (1994).

[2] P. S. Gromov, P. Madsen, N. Tomerup, and J. E. Celis, *FEBS Lett.* **377,** 221 (1995).

[3] N. Mizuki, M. Kimura, S. Ohno, S. Miyata, M. Sato, H. Ando, M. Ishihara, K. Goto, S. Watanabe, M. Yamazaki, A. Ono, S. Taguchi, K. Okumura, M. Nogami, T. Taguchi, A. Ando, and H. Inoko, *Genomics* **34,** 114 (1996).

[4] J. Urano, A. P. Tabancay, W. Yang, and F. Tamanoi, *J. Biol. Chem.* **275,** 11198 (2000).

[5] J. Urano and F. Tamanoi, unpublished data (2000).

[6] S. L. Campbell, R. Khosravi-Far, K. L. Rossman, G. J. Clark, and C. J. Der, *Oncogene* **17,** 1395 (1998).

[7] G. J. Clark, M. S. Kinch, K. Rogers-Graham, S. M. Sebti, A. D. Hamilton, and C. J. Der, *J. Biol. Chem.* **272,** 10608 (1997).

[8] C. Ellis and C. J. Clark, unpublished data (2000).

0076-6879/00 $35.00

Function of the Rheb homolog in *S. cerevisiae, ScRHEB,* was examined by creating a disruption strain and analyzing it for phenotypes.[4] This analysis revealed that ScRheb is critical for resistance to canavanine and thialysine, which are analogs of the basic amino acids arginine and lysine, respectively. Further investigations showed that ScRheb is involved in the regulation of arginine and lysine uptake by the yeast. In addition, experiments using various mutant forms of ScRheb showed that all the sequence features that define the Rheb proteins (the conserved arginine in the G1-box, effector domain, and C*aa*X) are important for full activity. Genetic studies using the yeast will likely uncover additional proteins involved in the Rheb pathway as well as the mechanism of action of this novel G protein.

The presence of a C-terminal C*aa*X motif indicates that the Rheb proteins are likely to be modified by isoprenoid lipids. Analysis has now shown that Rheb, like Ras, is modified by farnesyl.[4,7] Farnesylation is thought to play a role in proper subcellular localization as well as in mediating protein–protein interactions.[9] In addition, farnesylation has been shown to be essential for the normal physiological functions of proteins such as Ras, RhoB, and yeast mating factor.[10–12] Farnesylation has been shown to be critical for Rheb functions in *S. cerevisiae* and *Sch. pombe.*[4,13]

In this chapter we present the methods used in the study of ScRheb in the yeast, *S. cerevisiae,* as well as those used to study mammalian Rheb. Furthermore, we present methods used to address the C-terminal farnesylation of Rheb proteins and the requirement of this modification in Rheb function.

Functional Analysis of ScRheb in *Saccharomyces cerevisiae*

Phenotypic analysis of a yeast strain disrupted for the *ScRHEB* gene identified that this yeast strain is defective in resistance to canavanine and thialysine, which are analogs of the amino acids arginine and lysine, respectively. The disruptants also show increased uptake of arginine.[4] In this section we describe methods for generating *scrheb*-disrupted yeast strains as well as for assaying canavanine sensitivity and arginine uptake.

Yeast Strains and Media

Yeast strains used are TD1 (*MAT*α *gal2 his4 ura3 trp1*),[14] JU29-2 (*MAT*α *gal2 his4 ura3 trp1 scrheb*::*kanMX;* described below), and JU41-1 (*MAT*α *gal2*

[9] C. J. Marshall, *Science* **259,** 1865 (1993).
[10] F. L. Zhang and P. J. Casey, *Annu. Rev. Biochem.* **65,** 241 (1996).
[11] I. Sattler and F. Tamanoi, *in* "Regulation of the RAS Signaling Network" (H. Maruta and A. W. Burgess, eds.), pp. 95–137. R. G. Landes Company, New York, 1996.
[12] A. D. Cox and C. J. Der, *Biochim. Biophys. Acta* **1333,** F51 (1997).
[13] W. Yang, J. Urano, and F. Tamanoi, *J. Biol. Chem.* **275,** 429 (2000).
[14] D. I. Johnson, J. M. O'Brien, and C. W. Jacobs, *Gene* **90,** 93 (1990).

TABLE I
PRIMERS USED FOR CREATING AND CONFIRMING *scrheb*::*kanMX* DISRUPTION

Primer	Name	Sequence[a]
Disruption primers	ScRheb-5′	CGAAACAGTTTTATTGGGGCACGGCGAGAAAAGAT ATAACCTGTGACATAcagctgaagcttcgtacgc
	ScRheb-3′	ACCGCTTTCCTACTAATAAAACCAATGATATCCTAGG GACGCTCGAGTTTataggccactagtggatctg
Primer set A	ScRheb-Up	CTGGTCCTATAGTGTAGTGG
	kan-Up	GACAATTCAACGCGTCTGTG
Primer set B	ScRheb-Dn	ACACTCGCTGTTTGGTTCTG
	kan-Dn	TTTCGCCTCGACATCATCTG

[a] Sequences of the disruption primers that anneal to the ends of the *kanMX* cassette are written in lower-case letters.

his4 ura3 trp1 dpr1::*kanMX*). All yeast cultures are grown at 30° unless otherwise stated. Liquid cultures are grown in an orbital shaker at 300 rpm. Media used are YPD [1% (w/v) yeast extract, 2% (w/v) peptone, and 2% (w/v) dextrose], YPD+G418 (YPD supplemented with G418 at 200 mg/liter), and SDHUW [0.67% (w/v) yeast nitrogen base without amino acids (YNB), 2% (w/v) dextrose, histidine (20 mg/liter), uracil (20 mg/liter), tryptophan (20 mg/liter)]. Agar (2%, w/v) is added for solid media. YNB, G418, canavanine, thialysine, and amino acids are added after autoclaving and cooling to roughly 50°.

Disruption of ScRHEB Gene, using kanMX Cassette

Disruption of the *ScRHEB* gene in yeast provides a useful genetic approach by which the function of this gene can be studied. The *ScRHEB* gene is disrupted by a single-step polymerase chain reaction (PCR) method that utilizes the *kanMX* disruption cassette.[15] This cassette encodes the kanamycin resistance gene, which confers resistance to kanamycin or G418. Fifty nucleotides of sequence that flank the *ScRHEB* gene are fused to the ends of the *kanMX* disruption cassette by PCR. The primers used are composed of the *ScRHEB* flanking sequence at the 5′ end and approximately 20 nucleotides that anneal to the ends of the *kanMX* cassette at the 3′ end (Table I, disruption primers ScRheb-5′ and ScRheb-3′). One hundred nanograms of the pUG6 plasmid[15] is used as template per 100-μl reaction and the high-fidelity Vent polymerase (New England BioLabs, Beverly, MA) is used as the source of DNA polymerase. The PCR conditions are as follows: (1) an initial denaturation at 95° for 2 min, (2) 35 cycles of 30 sec at 95°, 30 sec at 50°, and 5 min at 72° followed by (3) a final extension of 10 min at 72°.

[15] U. Guldener, S. Heck, T. Fielder, J. Beinhauer, and J. H. Hegemann, *Nucleic Acids Res.* **24,** 2519 (1996).

Two reactions of 100 μl are pooled and then subjected to phenol–chloroform extraction with 200 μl of a phenol–chloroform–isoamyl alcohol (25 : 24 : 1, v/v/v) solution. The *scrheb*::*kanMX* disruption cassette is then ethanol precipitated by the addition of 20 μl of 3 *M* sodium acetate (pH 5.2) and 500 μl of ethanol. The precipitated DNA is resuspended in 30 μl of water and 5 μl is used per transformation. Generally, two or three transformations are performed per parent strain. The lithium acetate method[16] is used for transformation with the modification that prior to plating the cells on selective media, the cells are allowed to recover in YPD for 2–3 hr in a 30° orbital shaker at 200 rpm. The transformed cells are plated onto YPD containing G418 (200 mg/liter). This plate is incubated at 30° for 1–2 days and then replica plated onto a fresh YPD+G418 plate and incubated until colonies are evident (approximately two more days).

The transformants are streaked out to isolate single colonies and then confirmed for disruption by whole cell PCR.[17] Approximately 0.2 A_{600} unit equivalent of cells (roughly 5×10^6 cells) is resuspended in 250 μl of 0.04 *N* NaOH and incubated at 37° for 30 min. Part of this suspension (0.5 μl) is used as template in a 20-μl PCR. *Taq* (Qiagen, Valencia, CA) is utilized as the source of DNA polymerase. The primer sets are designed such that one primer is directed to *kanMX* and the second primer is directed to a region flanking the *ScRHEB* gene (primer set A or primer set B in Table I). If the disruption is present in the transformants, a product (718 and 788 bp for primer set A and B, respectively) is expected, but no product would be seen in the wild-type parents. In addition, a primer set with the primers directed to genomic sequence flanking the disruption cassette is used (primers ScRheb-Up and ScRheb-Dn; Table I). This PCR would produce a fragment that is roughly 1 kb greater in the disruptant compared with the wild type (2.5 vs. 1.5 kb) due to the size difference between the *kanMX* cassette and the *ScRHEB* coding region.

Assessing Sensitivities to Canavanine and Thialysine

Sensitivities to the amino acid analogs canavanine and thialysine are performed with solid SDHUW medium, which is supplemented with either canavanine (0.5 mg/liter) or thialysine (2.0 mg/liter). Minimal medium is used because the presence of arginine or lysine in the medium would eliminate the effects of their respective analogs. Cultures of the wild-type control yeast strain, TD1, and the *scrheb*-disrupted yeast strain, JU29-2, are grown overnight in 2 ml of YPD. These cultures are pelleted (centrifugation at 3000 *g* for 2 min at room temperature) and washed twice. Each wash involves resuspending the pellet in 5 ml of water followed by centrifugation. After the second wash, the pellet is resuspended in 5 ml of water. These cell suspensions are then diluted to an A_{600} of 1.0 and then serially

[16] D. Gietz, A. St. Jean, R. A. Woods, and R. H. Schiestl, *Nucleic Acids Res.* **20,** 1425 (1992).
[17] H. Wang, S. E. Kohalmi, and A. J. Cutler, *Anal. Biochem.* **237,** 145 (1996).

diluted five times by a factor of three. Five microliters of each dilution is then spotted onto the plates (no drug and with drug) and allowed to dry. The plates are then incubated at 30° for 3 days. A typical set of results is shown in Fig. 1A.

Assessing Uptake Rates of Arginine and Lysine

Uptake rates of these amino acids are followed by the use of ^{3}H-labeled arginine (40 Ci/mmol; Sigma, St. Louis, MO) or lysine (60 Ci/mmol; Sigma). Cultures of the wild-type control yeast strain, TD1, and the *scrheb*-disrupted yeast strain, JU29-2, are grown overnight in YPD. These cultures are then diluted into 20 ml of YPD and allowed to grow for at least three generations to an A_{600} of 0.2–0.5. These cultures are then collected and washed twice with 5 ml of SDHUW medium (prewarmed to 30°). The cells are resuspended in 1 ml of SDHUW medium and the A_{600} is measured. The volume is increased with additional prewarmed SDHUW medium so that the final A_{600} is 2.0. Part of this suspension (1.2 ml) is used for each time course (typically, five time points at $t = 0, 0.5, 1, 2$, and 5 min are used). The uptake assay is initiated by adding 17 μl of a mix of unlabeled and labeled amino acid (12 : 5 ratio of 10 m*M* unlabeled amino acid to ^{3}H-labeled amino acid) followed by incubation in a 30° water bath. At each time point, 200 μl of the cell suspension is removed and immediately diluted into 5 ml water and filtered with a vacuum manifold [we use a Millipore (Bedford, MA) 1225 sampling vacuum manifold]. This process is accomplished by injecting the 200-μl sample into 5 ml of water in a reservoir above the filter and then immediately applying vacuum. Glass microfiber filters [Fisherbrand G4 (Fisher Scientific, Pittsburgh, PA) or Whatman GF/C (Whatman, Maidstone, Kent, UK] are used to collect the cells. The filters are then rinsed two more times with 5 ml of water, dried under a heat lamp, and submerged in scintillation cocktail. The amounts of radioactive amino acid that have been taken up by the cells are then measured in a scintillation counter. A typical time course using arginine is shown in Fig. 1B.

Characterization of Mammalian Rheb: Evaluation of Rheb Effects on Ras Signaling and Transformation

Functional assays in mammalian cells include the ability of Rheb, like Rap, to antagonize transformation and signaling mediated by activated Ras. An *in vitro* assay is used to examine the ability of Rheb to bind Raf-1. Herein, we describe experimental protocols designed to evaluate mammalian Rheb protein biologically and biochemically.

Propagation and Transfection of NIH 3T3 Cells

The ease of transfection and transformation by activated Ras makes the NIH 3T3 cell line one of the most commonly used systems for investigating the function of Ras superfamily proteins.

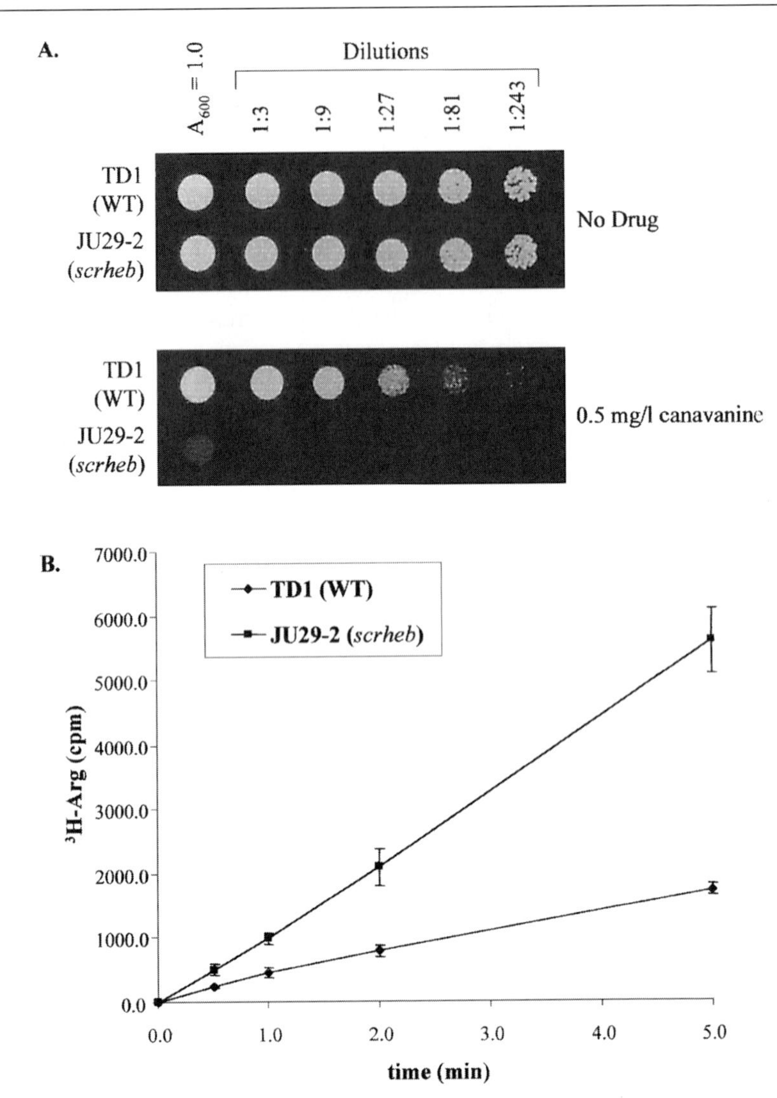

FIG. 1. *scrheb* disruptant strain phenotypes. (A) Overnight cultures are diluted to an A_{600} of 1 and then serially diluted by a factor of three. Five microliters of each dilution is spotted onto the indicated plates and incubated at 30° for 3 days. (B) Overnight cultures are diluted and grown to midlog phase (A_{600} of 0.2–0.5). Cells are collected and assayed for arginine uptake, using [^{3}H]arginine as described in text. The amounts of arginine taken up into the cells are reported as the amount of radioactivity. The results shown are the average of an experiment that was performed in triplicate. The error bars indicate 1 SD from the average. [Adapted from Urano *et al.* (2000).[4]]

NIH 3T3 cells are maintained in Dulbecco's modified Eagle's medium (DMEM) supplemented with 10% (v/v) calf serum and penicillin (100 U/ml)–streptomycin (100 μg/ml) at 37° in a 10% CO_2 incubator. For subculturing of NIH 3T3 cells, 100-mm dishes are seeded at 5×10^5 cells/plate. Growth medium should be changed every 3 days until the cells reach approximately 70% confluence, at which point they should be passed by trypsinization and reseeding. For transfection, 1×10^5 cells/60-mm plate should be set out. Seeding densities may be scaled accordingly to accommodate different culture vessels. In transfection experiments, cells are seeded 18–24 hr prior to use.

Calcium phosphate-mediated DNA transfection is a reproducible, efficient, and inexpensive means of introducing large quantities of DNA into NIH 3T3 cells.[18] The transfection protocol requires the following reagents.

$CaCl_2$ (1.25 *M*): Makes a 10× concentrated stock solution; autoclave to sterilize. Store at 25°

N-2-Hydroxyethylpiperizine-*N*′-2-ethanesulfonic acid (HEPES)-buffered saline (HBS), pH 7.05: To 0.9 liters of distilled H_2O, add 8 g of NaCl, 0.37 g of KCl, 0.19g of $Na_2HPO_4 \cdot 7H_2O$, 1 g of glucose, and 5 g of HEPES. Adjust the pH to exactly 7.05 with 5 *N* NaOH to ensure good DNA precipitation. Adjust the volume to 1 liter with H_2O and autoclave to sterilize

HBS–15% (v/v) glycerol: Mix glycerol (autoclaved) and HBS at a 15 : 85 ratio. Store at 25°

High molecular weight carrier DNA (HMW DNA; e.g., calf thymus DNA): High molecular weight DNA promotes the formation of calcium phosphate-plasmid DNA precipitates. Calf thymus DNA is available from many commercial vendors. We have found Roche Diagnostics (Indianapolis, IN) to be a reliable source

Plasmid DNA (empirically determined amount) is added to a solution of HBS–HMW DNA (40 μg/ml) and mixed thoroughly by vortexing. A 1 : 10 volume (final concentration, 0.125 *M*) of $CaCl_2$ is added followed by vortexing and 25° incubation. The calcium phosphate–DNA precipitate forms over 10–30 min and confers a faint milky appearance to the HBS solution. The precipitated DNA solution can be added in dropwise fashion directly to the growth medium. Cell culture dishes are rocked front to back, side to side, and swirled to ensure equal distribution of the DNA and placed at 37° in a 10% CO_2 incubator for 3–4 hr. A good precipitate is visible by phase-contrast microscopy and has the appearance of dark, fine grains of sand. Transfection ends with a brief (4-min) glycerol shock at 25° with 15% (v/v) glycerol–HBS followed by rinsing of the cells in fresh growth

[18] G. J. Clark, A. D. Cox, S. M. Graham, and C. J. Der, *Methods Enzymol.* **255,** 395 (1995).

medium. Glycerol shocking is quite stressful to the cells, and exceeding the 4-min incubation is not recommended. Refresh the growth medium every 2–3 days or as indicated by the experimental protocol.

Some considerations regarding volume, quantity of plasmid DNA, and number of replicates are worth noting. The volume of calcium phosphate–DNA precipitate should be approximately one-tenth the total volume of cell culture medium (i.e., 300–400 μl for 3–4 ml in a 60-mm dish). As the total amount of all plasmid DNAs (in grams) can influence transfection efficiency, this amount needs to be held constant and can be accomplished by supplementing experimental groups with varying amounts of empty vector to make up the difference. Finally, it is advisable to design transfection experiments such that each experimental group is performed in duplicate or triplicate.

Focus Assays

A common method to assess transformation potential of a protein is the focus formation assay in NIH 3T3 fibroblasts. This type of experiment measures the ability of an introduced gene to promote gross alterations in cellular morphology, growth control, and the loss of contact inhibition. In NIH 3T3 fibroblasts, *ras*-induced foci appear as swirling clusters of multilayered, spindle-shaped cells when viewed by phase-contrast microscopy (Fig. 2). The rate at which foci appear is a function of the nature of the protein, the strength of the promoter from which it is expressed, and the amount of DNA transfected. *ras*-induced NIH 3T3 foci typically appear after 10–14 days in culture. Accurate quantitation of foci is best performed on live cultures under a phase-contrast microscope at low magnification (×40).

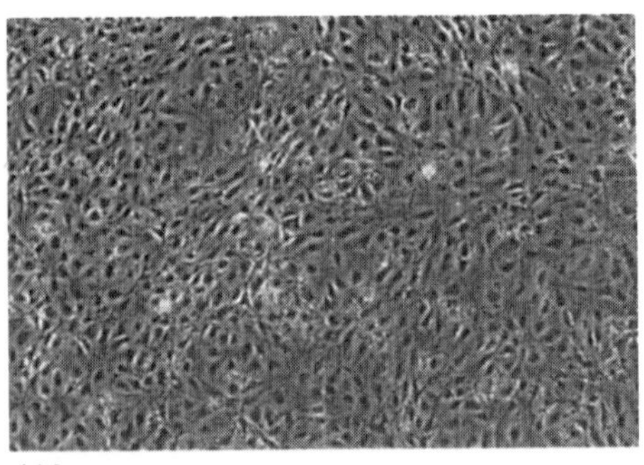

(A) vector

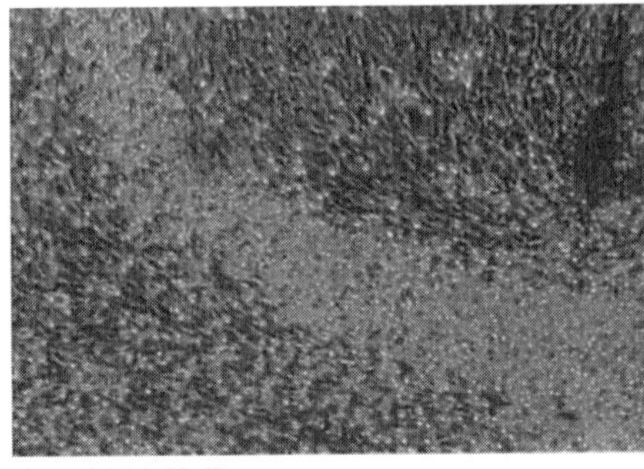

(B) 12V H-Ras

FIG. 2. Ras-induced focus formation in NIH 3T3 cells. Fifty nanograms of H-Ras 12V was transfected into NIH 3T3 cells as described in text. After 16 days foci were observed growing over the flat monolayer (B) as compared with control cells transfected with vector alone (A).

Focus assays are performed on NIH 3T3 cells plated at 1×10^5 cells in triplicate 60-mm dishes 1 day prior to transfection. A stock solution of HBS–calf thymus DNA (0.4 ml/dish) with a volume sufficient for all dishes is prepared. After setting up the vector alone control sample tube with 1.2 ml of HBS and 3 μg of vector, 600 ng of activated *ras* DNA [pZip-NEO SV(X)1 H-Ras 12V] is added to the remaining stock (2.4 ml). The stock is then vortexed and split in half. One half is then supplemented with 3 μg of vector, the other half with 3 μg of a Rheb expression construct.[7] DNA precipitation and transfection are carried out as described above. The cell culture medium is refreshed every 2–3 days for ~16 days.

After about 16 days the dishes may be scored for foci. Typically, the presence of the *rheb* gene will have reduced the number of Ras foci by 50%.

Establishing Stable Cell Lines

To study further the effects of exogenous Rheb on NIH 3T3 cell properties, stable cell lines may be established. Most expression vectors contain a selectable marker that confers resistance to a drug. Commonly used vectors that work well for Rheb expression are pZipNEO SV(X)1 and G418 (geneticin) or pCGN and hygromycin B.[7] Three days after transfection, cells are expanded approximately 1 to 10 in 100-mm dishes in the presence of the appropriate drug: geneticin (G418) at 500 μg/ml or hygromycin B at 500 μg/ml. Parallel dishes of untransfected cells are included in the drug selection as negative controls. Growth medium supplemented with drug should be refreshed every 1–2 days depending on the cell death rate. Selection of drug-resistant cells can take 7–14 days. Resistant cells may be pooled or individual clones produced by limiting dilution. Stable cell lines should be subjected to Western analysis for confirmation of protein expression prior to use.

Signal Transduction Assays in NIH 3T3 Cells

Quantifying the cellular effects of protein expression requires the ability to measure changes in biological end points (i.e., second-messenger generation, altered gene expression). Reporter assays are ideally suited for the study of signal transduction because they measure transcriptional activity. Provided the regulatory elements of responsive genes are known, reporter genes (i.e., firefly luciferase, β-galacosidase, chloramphenicol transferase) can be joined to the promoter region of these genes and cotransfected with the experimental plasmid of interest. Cells are then assayed for the presence of reporter mRNA, protein, or reporter protein activity. Suitable reporter genes are not endogenously expressed in the cell type of interest. The most convenient and sensitive assays currently rely on the luciferase reporters.

Here we describe an *in trans* reporter system to measure the effects of Rheb on Ras-mediated Elk-1 activation in NIH 3T3 cells.[19] The system is composed of a Gal–Elk fusion protein expression construct and a Gal-controlled promoter linked to the luciferase gene. Phosphorylation of the Elk domain of the Gal–Elk fusion protein serves to activate the Gal domain, which then promotes transcription of the luciferase gene. The Elk-1 transcription factor is a substrate for activated mitogen-activated protein kinase (MAPK) and thus Elk phosphorylation serves as a measure of Raf/MEK [MAPK/ERK (extracellular signal-regulated kinase) kinase]/MAPK pathway signaling.

Dishes (60 mm) are plated at 1×10^5 cells/dish and transfected the following day. Each dish is transfected with 100 ng of Gal–Elk, 200 ng of 5×Gal–Luc. To ensure equal addition of DNA, a stock solution of enough HBS–reporter DNA to transfect all the dishes is made up and aliquoted. To examine the effects of Rheb on Ras action, 100 ng of activated Ras in pZip-Neo is mixed with 1 μg of pZip-Rheb. After transfection, the cells are incubated for 48 hr before shifting to 1% serum overnight. The following day, the quiescent cells are lysed in the dish with luciferase lysis buffer (several commercial brands are available, e.g., PharMingen, San Diego, CA) with gentle rocking for 20 min. Twenty microliters of the lysate is then read in a luminometer, using commercially supplied solutions of luciferin and ATP (e.g., from PharMingen). Such experiments must be performed in duplicate at least three times.

These experiments are sufficiently sensitive to discriminate between the effects of Rap and Rheb. Cotransfection experiments with a B-Raf expression construct show that although activated Rap will enhance B-Raf-mediated Elk signaling, activated Rheb inhibits it.[8]

In Vitro Raf-Binding Assays

To examine the ability of Rheb to interact with known Ras effector molecules *in vitro,* recombinant Rheb may be produced in bacteria as a glutathione *S*-transferase (GST) fusion protein. Rheb cDNA was cloned into the GST fusion vector pGEX-2T (Amersham Pharmacia, Piscataway, NJ) and transfected into XL1-blue bacteria (Stratagene, La Jolla, CA). Five milliliters of an overnight culture is used to seed 500 ml of LB, which is grown at 37°, 250 rpm until an A_{600} of 1 is reached. At this point isopropyl-β-D-thiogalactopyranoside (IPTG, 0.5 m*M*) is added and the incubation allowed to proceed for an additional 4 hr. Bacteria are then pelleted and resuspended in 10 ml of phosphate-buffered saline (PBS) supplemented with 20 m*M* EDTA, lysozyme (2 mg/ml), 100 μ*M* GDP, benzamidine (100 mg/ml), aprotinin (2 μg/ml), leupeptin (1 μg/ml), and incubated for 10 min at room

[19] J. K. Westwick, A. D. Cox, C. J. Der, M. H. Cobb, M. Hibi, M. Karin, and D. A. Brenner, *Proc. Natl. Acad. Sci. U.S.A.* **91,** 6030 (1994).

temperature. The bacteria are then lysed by three cycles of freeze–thawing in a dry ice–ethanol bath. After the addition of 25 m*M* $MgCl_2$, 500 m*M* NaCl, 1% (v/v) Triton X-100, and 5 m*M* dithiothreitol (DTT) (final concentrations), the bacteria are subjected to sonication in order to fragment the viscous chromosomal DNA. The bacterial lysate is then centrifuged at 10,000 *g* for 20 min at 4°. The supernatant is rotated at 4° with glutathione-conjugated agarose beads for 4 hr. Beads are washed four times in wash buffer [20 m*M* Tris (pH 8.0), 20% (v/v) glycerol, 1 m*M* DTT, 5 m*M* $MgCl_2$, 50 m*M* NaCl, and 100 μ*M* GDP]. The purified protein is stored in wash buffer supplemented with gycerol to 50% (v/v) at −20°.

Raf protein is purified from Sf9 (*spodoxtera frugiperda* ovary) cells infected with baculovirus expressing Raf. The cells are grown at 27° in T-70 flasks. At 50% confluence (10^7 cells), the cells are infected at a multiplicity of infection of five. The infection is allowed to proceed for 48 hr before the cells are lysed in 20 m*M* Tris-HCl (pH 7.4), 100 m*M* KCl, 1 m*M* $MgCl_2$, 10% (v/v) glycerol, 1% (v/v) Triton X-100, 1 μ*M* leupeptin, 0.1 μ*M* aprotinin. The lysate is clarified by centrifugation at 10,000 rpm for 10 min at 4°. The supernatant serves as a satisfactory source of crude Raf protein.

GST–Rheb is then loaded with GTP or a GTP analog by incubation in 50 m*M* Tris-HCl (pH 7.5), 50 m*M* NaCl, 1 m*M* DTT, 5 m*M* EDTA, and 6 μ*M* GTP or GTP analog for 10 min at 37°. After loading, Rheb is rinsed in wash buffer lacking GDP. One microgram of GST–Rheb [estimated from Coomassie staining of a sodium dodecyl sulfate (SDS)–polyacrylamide gel against standards] is then incubated with one-twentieth of the total Raf protein preparation (equivalent to 1×10^5 cells) for 1 hr at 4°. The beads are then precipitated by centrifugation and washed four times in wash buffer. Raf binding is then measured by SDS–polyacrylamide gel electrophoresis (PAGE) and Western blotting with anti-Raf antiserum (C-12; Santa Cruz Biotechnology, Santa Cruz, CA).

Farnesylation of Rheb Proteins

Farnesylation appears to be critical for the function of Rheb proteins. Furthermore, farnesyltransferase inhibitor (FTI), which belongs to a novel class of anticancer drugs, is able to block the farnesylation of Rheb in mammalian systems.[7] Therefore, understanding the role that farnesylation plays in Rheb function is critical for predicting possible effects of FTI treatments.

Farnesylation is the transfer of a 15-carbon farnesyl group onto the cysteine of the C-terminal C*aa*X motif by protein farnesyltransferase (FTase). FTase is a heterodimeric enzyme composed of an α subunit and a β subunit. In mammalian cells, they are referred to as α_{FT} and β_{FT}, respectively, while in the yeast, they are Ram2p and Dpr1p (Ram1p), respectively. The α subunit is also utilized by another prenyltransferase, protein geranylgeranyltransferase type I (GGTase I).

GGTase I recognizes a C-terminal C*aa*L motif that is similar to C*aa*X, except that the terminal amino acid is usually a leucine or phenylalanine.

Farnesylation of proteins can be studied by a number of methods. Here we describe two methods using the yeast system and one method using mammalian cells. In the first method, a mutant form of ScRheb that is no longer able to be farnesylated is utilized. The second method utilizes a strain of yeast that is deficient in farnesylation. The third method describes a protocol for examining farnesylation of mammalian Rheb *in vitro*. Finally, we describe a method to examine subcellular localization of Rheb in mammalian cells.

Method 1: Mutation of CaaX Motif of ScRheb

The requirement of farnesylation for ScRheb function can be addressed by assessing if a farnesylation-deficient mutant form of ScRheb can complement the phenotypes of a *scrheb* disruptant yeast strain. A farnesylation-deficient mutant is created by mutating the cysteine of the C*aa*X motif to a serine. This mutant form of ScRheb, ScRhebSSIM, is hemagglutinin (HA) epitope tagged and is placed under the control of its own promoter in the yeast centromeric plasmid, pRS314.[20] The resulting plasmid is named pCWHAScRhebSSIM. The wild-type version of this plasmid, pCWHAScRheb, is used as the positive control and the vector pRS314 is used as the negative control. These plasmids are transformed into JU29-2 and assessed for sensitivity to canavanine as well as for uptake of arginine as described earlier in this chapter. Typical results for the canavanine sensitivities are shown in Fig. 3 (top). The expressions of these constructs are confirmed by Western analysis usingan anti-HA monoclonal antibody (BAbCo, Richmond, CA).

Method 2: ScRheb Function in Farnesyltransferase Mutant Yeast

A second method to address the requirement for farnesylation takes advantage of the availability of yeast strains that are completely deficient for FTase activity. If farnesylation is important for ScRheb function, a strain that is deficient for FTase activity would be predicted to exhibit the same phenotypes observed with the *scrheb* disruptant strains. Once this prediction is confirmed, it is possible to demonstrate that the major farnesylated protein involved in these phenotypes is ScRheb by using a mutant form of ScRheb that can be geranylgeranylated. Such a mutant can bypass the farnesylation requirement and can complement these phenotypes. Geranylgeranylation can be achieved by mutating the terminal amino acid to a leucine, which will convert the C*aa*X farnesylation motif into a C*aa*L geranylgeranylation motif. This geranylgeranylation is carried out in the cells by GGTase I. The geranylgeranyl group is a 20-carbon isoprenyl chain that is similar

[20] R. S. Sikorski and P. Hieter, *Genetics* **122,** 19 (1989).

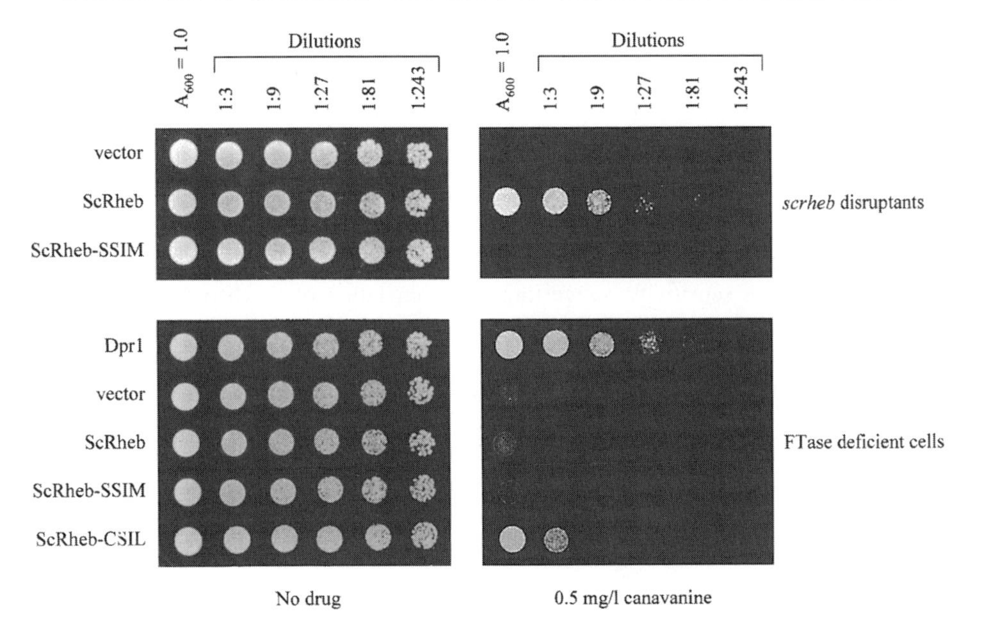

FIG. 3. ScRheb requirement for farnesylation. Overnight cultures are diluted to an A_{600} of 1.0 and then serially diluted by a factor of three. Five microliters of each dilution is spotted onto the indicated plates. *Top:* JU29-2 (*scrheb* disruptant) is transformed with vector (pR314), wild-type ScRheb (pCWHAScRheb), or the prenylation-defective ScRheb-SSIM (pCWHAScRhebSSIM). The plates are incubated at 30° for 3 days. *Bottom:* JU41-1 (FTase deficient) is transformed with vector (pRS31), Dpr1 (YCpDPR2.4), wild-type ScRheb (pCWHAScRheb), the prenylation-defective ScRheb-SSIM (pCWHAScRhebSSIM), or the geranylgeranylatable ScRheb-CSIL (pCWHAScRhebCSIL). The plates are incubated at 22° for 5 days. [*Top:* Adapted from Urano *et al.* (2000).[4]]

to the 15-carbon farnesyl moiety and has been shown to be able to functionally replace farnesylation in a number of cases.[13,21,22]

An FTase-deficient strain is created by disrupting the *DPR1* gene, which encodes the β subunit of FTase, using the *kanMX* cassette. JU41-1 (*MATα gal2 his4 ura3 trp1 dpr1::kanMX*) is an FTase-deficient strain created from the wild-type strain, TD1, through this process. Initially, this strain is transformed with either vector (pRS314) or a plasmid carrying the *DPR1* gene (YCpDPR2.4[23]) and assessed for canavanine sensitivity and arginine uptake. The results shown in Fig. 3 (first two rows of lower panels) demonstrate that the FTase deficiency does result in hypersensitivity to canavanine (compare Dpr1 and vector). The rate of arginine uptake is

[21] G. A. Caldwell, S. H. Wang, F. Naider, and J. M. Becker, *Proc. Natl. Acad. Sci. U.S.A.* **91,** 1275 (1994).

[22] M. S. Whiteway, C. Wu, T. Leeuw, K. Clark, A. Fourest-Lieuvin, D. Y. Thomas, and E. Leberer, *Science* **269,** 1572 (1995).

[23] L. E. Goodman, C. M. Perou, A. Fujiyama, and F. Tamanoi, *Yeast* **4,** 271 (1988).

also increased.[24] With the observation that both FTase-deficient yeast and *scrheb*-disrupted yeast exhibit the same phenotypes, the FTase-deficient strain is then also transformed with a plasmid carrying wild-type ScRheb (pCWHAScRheb), prenylation-deficient ScRheb (pCWHAScRhebSSIM), and the geranylgeranylatable mutant ScRheb (pCWHAScRhebCSIL). These transformants are then assessed for canavanine sensitivity and arginine uptake. A typical set of results for the canavanine sensitivity is shown in Fig. 3 (bottom). The results show that the geranylgeranylatable form of ScRheb (ScRhebCSIL) is able to complement the canavanine sensitivity, suggesting that prenylation of ScRheb is necessary. The prenylation-deficient (ScRhebSSIM) form of ScRheb is unable to complement the phenotype. In addition, the ScRheb wild type is unable to complement because of the lack of farnesylation. Consistent with these results, the arginine uptake phenotype of the FTase-deficient cells is complemented by the geranylgeranylated form of ScRheb and not by wild-type or prenylation-deficient ScRheb.[24] These results demonstrate that farnesylation is important for ScRheb function and that ScRheb is likely to be the major farnesylated protein responsible for the canavanine and arginine uptake phenotypes observed in FTase-deficient yeast cells. This type of approach can be used to assess the farnesylation requirements of other proteins whose phenotypes are also exhibited by FTase-deficient cells.

Method 3: Labeling of Recombinant Mammalian Rheb

GST fusion proteins of Rheb, Ras, and Rap may be prepared as described earlier (*in vitro* Raf-Binding Assays). Ras is used as a control that is preferentially farnesylated, while Rap is used as a control that is preferentially geranylgeranylated. The fusion proteins are incubated with 20 μl of rabbit reticulocyte (Promega, Madison, WI) and 2.5 μCi of [^{3}H]farnesyl pyrophosphate (New England Nuclear, Boston, MA) for 30 minutes at 37° C. The beads are then washed in PBS and resolved by SDS–PAGE. The gel is exposed to an autoradiographic image intensifier such as Autofluor (National Diagnostics, Atlanta, GA), dried, and exposed to film overnight. Two controls are required to correctly interpret the experiment. The process must be performed in parallel with [^{3}H]mevalonate (25 μCi) and also with [^{3}H]geranylgeranyl pyrophosphate (2.5 μCi). Mevalonate is a precursor for both farnesyl and geranylgeranyl and will therefore label all proteins that are capable of being processed by either lipid. Geranylgeranyl pyrophosphate will label only geranylgeranyl-specific proteins such as Rap.

Method 4: Cellular Localization of Mammalian Rheb

Cellular localization of Rheb proteins may be readily determined by transient transfection of cells with epitope-tagged expression constructs. The exogenous

[24] J. Urano, A. P. Tabancay, and F. Tamanoi, unpublished data (2000).

Rheb protein may then be detected by indirect immunoflouresence microscopy. A commonly used tag–antibody system is the HA tag. The anti-HA monoclonal works well for Western blotting, immunoprecipitation, and immunofluorescence (BAbCo).

To detect Rheb protein, NIH 3T3 cells seeded onto glass coverslips are transfected with pCGN-HA expression vector[25] encoding HA–Rheb and incubated at 37° for 48–72 hr. Cells are fixed with 3.7% (w/v) paraformaldehyde in PBS for 15 min at 25°, and then permeabilized with 0.5% (v/v) Triton X-100 in Tris-buffered saline [TBS: 150 m*M* NaCl, 50 m*M* Tris (pH 7.6), 0.1% (w/v) NaN_3] for 10 min at 25°. Permeabilized NIH 3T3 cells are blocked in 5% (w/v) bovine serum albumin (BSA) in PBS for 1 hr at 25° and then incubated with anti-HA monoclonal antibody (1–5 μg/ml; BAbCo) for 1 hr at 25°. Cells are gently washed three or four times with PBS and incubated with rhodamine-conjugated goat anti-mouse antibody (5 μg/ml) for 1 hr at 25°. Cells are gently washed three or four times with PBS, and coverslips are mounted onto glass slides and viewed with a Zeiss (Thornwood, NY) Axiophot microscope. As controls, NIH 3T3 cells are transfected with pCGN-HA, H-Ras/pCGN-HA, or Rap1A/pCGN-HA.[7]

Concluding Remarks

We have presented here methods for the analysis of Rheb functions, using yeast and mammalian systems. Because Rheb is a highly conserved protein, results obtained with different systems may enrich our understanding of the function(s) of this protein. The genetic system of the yeast can be used for mutational studies of the Rheb protein itself as well as for the identification of other members of the Rheb pathway. Such mutational studies will also likely result in the production of mutants/strains that will be invaluable for the further understanding of Rheb function. The methods presented here will be useful in the detailed analysis of various mutations that may affect Rheb activity. Furthermore, once other members of the Rheb signaling pathways are identified, these assays will be useful in investigating the roles that these other members may play in Rheb function(s).

Acknowledgments

We acknowledge A. P. Tabancay for contributing to the study of ScRheb and arginine uptake. We thank Wenli Yang for discussions on SpRheb. The work by J. U. and F. T. is supported by the National Institutes of Health grant CA41996.

[25] J. K. Westwick, Q. T. Lambert, G. J. Clark, M. Symons, L. Van Aelst, R. G. Pestell, and C. J. Der, *Mol. Cell. Biol.* **17,** 1324 (1997).

[21] Ras Regulation of Skeletal Muscle Differentiation and Gene Expression

By NATALIA MITIN, MELISSA B. RAMOCKI, STEPHEN F. KONIECZNY, and ELIZABETH J. TAPAROWSKY

Ras proteins are essential components in the signal transduction pathways controlling cell proliferation, differentiation, and apoptosis. Constitutive activation of Ras and its downstream effectors is a molecular feature common to a large percentage of human tumors. Ras proteins inhibit the expression of the muscle regulatory factors [MRFs: members of the basic helix–loop–helix (bHLH) transcription factor family] in both primary myoblasts (reviewed in Ref. 1) and in immortalized myoblast cell lines.[2,3] Furthermore, in a manner that is not completely understood, activated Ras proteins block the ability of MRF transcription complexes to activate muscle-specific gene expression.[2–6] Given that bHLH transcription factors are expressed in many cell types and play important roles in lineage determination and differentiation decisions, understanding how Ras signaling inhibits MRF activity in muscle cells can serve as a model of how oncogene activation impacts the function of bHLH factors in other tissues.

Beginning with the initial observation that C3H10T1/2 mouse fibroblasts are multipotential and can be converted to myoblasts by treatment with the demethylating agent 5-azacytidine,[7,8] C3H10T1/2 cells have been the model system of choice to define the molecular events critical for establishing a myogenic lineage. C3H10T1/2 cells were instrumental in identifying and characterizing the activities of the MRFs (MyoD, myogenin, Myf-5, and MRF4) as transcriptional regulators of myogenic determination and differentiation (reviewed in Ref. 9). Expression of an MRF cDNA in C3H10T1/2 cells, coupled with the induction of MRF activation by treatment with differentiation medium, causes the cells to exit the cell cycle, express a battery of muscle-specific genes, and fuse to form

[1] S. Alemá and F. Tató, *Semin. Cancer Biol.* **5,** 147 (1994).

[2] S. F. Konieczny, B. L. Drobes, S. L. Menke, and E. J. Taparowsky, *Oncogene* **4,** 473 (1989).

[3] A. B. Lassar, M. J. Thayer, R. W. Overall, and H. Weintraub, *Cell* **58,** 659 (1989).

[4] Y. Kong, S. E. Johnson, E. J. Taparowsky, and S. F. Konieczny, *Mol. Cell. Biol.* **15,** 5205 (1995).

[5] M. B. Ramocki, S. E. Johnson, M. A. White, C. L. Ashendel, S. F. Konieczny, and E. J. Taparowsky, *Mol. Cell. Biol.* **17,** 3547 (1997).

[6] M. B. Ramocki, M. A. White, S. F. Konieczny, and E. J. Taparowsky, *J. Biol. Chem.* **273,** 17696 (1998).

[7] S. M. Taylor and P. A. Jones, *Cell* **17,** 771 (1979).

[8] S. F. Konieczny and C. P. Emerson, Jr., *Cell* **38,** 791 (1984).

0076-6879/00 $35.00

multinucleate myotubes (reviewed in Ref. 9). C3H10T1/2 cells are used to study both quantitative and qualitative aspects of Ras signal transduction *in vitro*[10,11] and the impact of these signals on MRF function during myogenesis.[2–6] In addition, myogenic cell lines derived from the 5-azacytidine treatment of C3H10T1/2 cells (23A2)[8] or from the stable transfection of C3H10T1/2 cells with MRFs (10T1/2-MyoD; 10T1/2-MRF4)[12,13] are as useful as immortalized myoblast cell lines (e.g., C_2C_{12}, L6) to study molecular aspects of the terminal differentiation program of skeletal muscle cells.

Here we describe the routine procedures for maintaining myogenic cell lines in undifferentiated and differentiated states. Methods for transient and stable transfection of these lines are presented, along with the standard procedure for muscle-specific luciferase reporter gene assays and for the immunostaining of muscle-specific proteins in differentiated muscle cells.

Maintenance and Choice of Cell Lines

A number of cell lines may be used to study Ras-regulated gene expression in a skeletal muscle differentiation system. The cell lines commonly used in our laboratory are discussed in this section. Each of the cell lines discussed below is cultured in the appropriate serum-supplemented growth medium (GM) containing penicillin (100 U/ml) and streptomycin (100 μg/ml) (P/S). The cell lines are passaged with IX trypsin–EDTA (GIBCO-BRL, Grand Island, NY) warmed to 37° and are maintained at 37° in a humidified, 5% CO_2 chamber. Most lots of fetal bovine serum obtained from reliable vendors will support the growth of fibroblast and myoblast cell lines; however, we have observed a profound difference in the ability of individual serum lots to prevent spontaneous differentiation in the cultures. Therefore, we suggest that serum be prescreened rigorously for the ability to support myoblast cell growth and the subsequent ability of the cells to differentiate into skeletal myotubes expressing muscle-specific gene products. This screening can be achieved by plating myogenic cells at a clonal density of 100 cells per 100-mm gelatin-coated dish (described in detail below) in GM supplemented with distinct serum lots. After 2 weeks, the colonies are fixed with methanol (10 min at room temperature) and stained with a 4% (v/v) aqueous solution of Wright–Giemsa stain (EM Diagnostics Systems, Gibbstown, NJ). The total number of colonies that develop and the percentage of colonies that contain differentiated muscle cells are scored. Good serum lots should produce a high

[9] D. C. Ludolph and S. F. Konieczny, *FASEB J.* **9,** 1595 (1995).

[10] E. J. Taparowsky, M. L. Heaney, and J. T. Parsons, *Cancer Res.* **47,** 4125 (1987).

[11] C. M. Weyman, E. J. Taparowsky, M. Wolfson, and C. L. Ashendel, *Cancer Res.* **48,** 6535 (1988).

[12] R. L. Davis, H. Weintraub, and A. B. Lassar, *Cell* **51,** 987 (1987).

[13] S. J. Rhodes and S. F. Konieczny, *Genes Dev.* **3,** 2050 (1989).

number (80%) of well-formed colonies with the central, contact-inhibited region of each colony containing visible myotubes as shown in Fig. 1.

We prefer to transiently transfect the murine fibroblast cell line C3H10T1/2 (10T1/2) with muscle regulatory factor (MRF) cDNA expression vectors in order to assess the effects of Ras and related signaling events on skeletal myogenesis and muscle-specific gene expression. 10T1/2 cells, unlike myogenic cell lines, do not express endogenous MRF proteins. As a result, background reporter gene activation is eliminated and smaller increases or decreases in differentiation and gene expression can be measured reliably. The quantitative aspect of the morphological differentiation assay also is facilitated in the 10T1/2 model system because fewer cells are differentiation competent. The details of each of these experiments are described below (Transient Transfections). C3H10T1/2 cells [American Type Culture Collection (ATCC) CCL 226 clone 8, Manassas, VA] are cultured in GM consisting of basal modified Eagle's medium (BME) (GIBCO-BRL) supplemented with 10% (v/v) fetal bovine serum. The cells should be maintained at subconfluent densities and passaged every 3–4 days. If the fibroblasts are to be transfected with MRFs, the cells should be plated on gelatin-coated dishes (see below). Well-maintained C3H10T1/2 fibroblasts may be passaged over long periods of time without a loss of viability; however, after passage 22 (ATCC stocks are provided at passage 6), we have noticed a dramatic decrease in both transfection efficiency and differentiation competence of the cells.

Myoblast cell lines express MRF proteins in GM and thus are committed to a myogenic lineage. However, they do not differentiate unless stimulated by induction conditions that result in the activation of the MRF proteins. In differentiation medium (DM), myoblasts exit the cell cycle, fuse to form multinucleate myotubes, and express muscle-specific genes in addition to the MRFs. Myoblast cell lines are useful for experiments designed to detect quantitative differences in protein levels or enzyme activity, as well as for stable transfection experiments. In general, all myoblast cell lines must be cultured at significantly lower cell densities (50–70% confluent) compared with fibroblast cell lines because high density results in the spontaneous differentiation of the cells. The GM is supplemented with a higher level of serum (15–20%) and is replaced every 2–3 days. Cell stocks should be discarded and new vials thawed if the myoblasts display a reduced ability to differentiate in DM or a substantial level of spontaneous differentiation in GM. Myoblasts are grown on tissue culture plates that have been previously coated with a solution of 0.01% (w/v) sterile gelatin. Five milliliters of the gelatin solution is used to cover the growth surface of a 100-mm tissue culture dish. The solution should rest in the dish for at least 4 hr prior to removal by aspiration. The dishes may be stored in a sterile area at room temperature for approximately 1 month.

The 23A2 myoblast cell line is a clonally selected cell line derived from 10T1/2 cells that had been exposed to 5-azacytidine for a single round of DNA

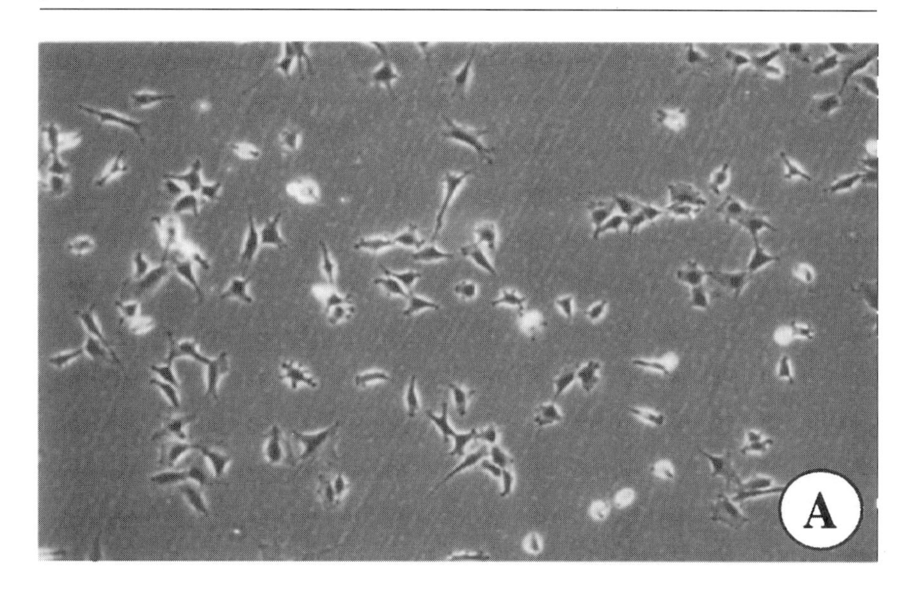

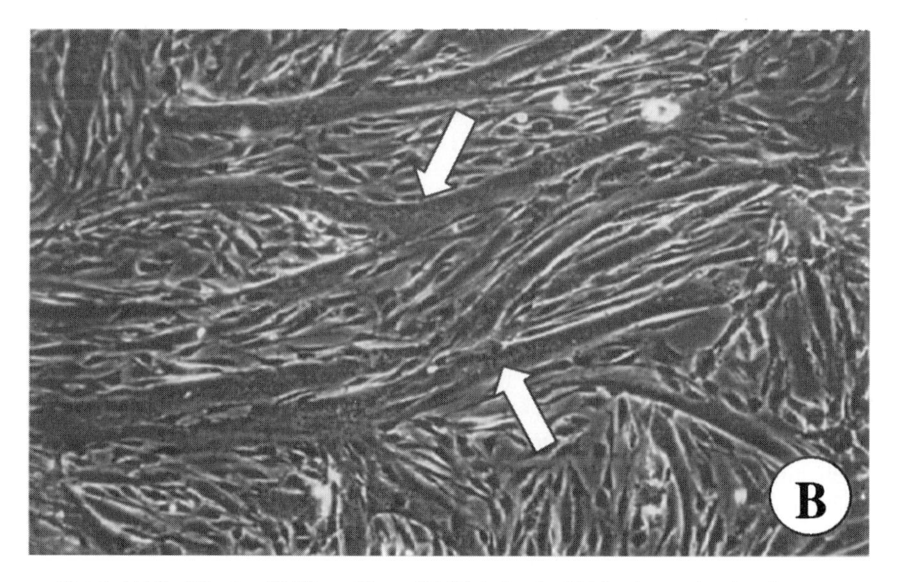

FIG. 1. (A) Proliferating 23A2 myoblasts. (B) High-density 23A2 cultures showing the presence of well-formed skeletal myotubes (arrows). Both cultures were viewed by phase-contrast microscopy and photographed at the same magnification.

replication.[8] These myoblasts are cultured in GM consisting of BME supplemented with 15% (v/v) fetal bovine serum. The 23A2-*ras* cell line[2,14] is a variant of 23A2 stably expressing a mutationally activated version of the human H-*ras* gene. These cells are differentiation defective in that they cannot fuse or express muscle-specific gene products. These cells are cultured in the same manner as the parental 23A2 line, using medium supplemented with a 400 μg/ml active concentration of geneticin (G-418; GIBCO-BRL).

The C_2C_{12} myoblast cell line (ATCC CRL 1772) is a fast-fusing subclone of the immortalized murine C_2 satellite cell line.[15] C_2C_{12} cells are cultured in GM consisting of high-glucose Dulbecco's modified Eagle's medium (hgDMEM) supplemented with 15% (v/v) fetal bovine serum. Like 23A2, C_2C_{12} cells must be maintained at subconfluent density. C_2C_{12} cells should not be passaged for more than 2–3 weeks as they tend to lose their differentiation competence.

The C3H10T1/2-MRF4 and C3H10T1/2-MyoD myogenic lines were generated through the stable cotransfection of 10T1/2 fibroblasts with the indicated MRFs and a neomycin selection gene.[12,13] As myogenic derivatives of 10T1/2, these cells are cultured in 10T1/2 GM supplemented with 15% (v/v) fetal bovine serum and geneticin (400 μg/ml).

To induce the differentiation of the cell lines described above, the GM is replaced with DM. Two versions of DM are recommended. When myofiber morphology (in addition to muscle-specific gene activation) will be a factor in the experimental analysis, low-glucose DMEM (lgDMEM) supplemented with 2% (v/v) horse serum and P/S should be used. For the analysis of reporter gene activation, a more rapid biochemical differentiation (sacrificing aspects of muscle syncytium formation) can be achieved by differentiating cells in lgDMEM containing 1× ITS [insulin (5 μg/ml), transferrin (5 μg/ml), selenium (5 ng/ml)] (GIBCO-BRL) and P/S.

Transient Transfections

Overview

Transient transfection analysis has proved to be a valuable tool in the study of gene expression regulated by Ras. The assays for gene expression are simple to perform and allow for the rapid testing of multiple constructs in parallel. This provides the opportunity to test Ras, Ras effectors, and signaling molecules well downstream of Ras for a role in the Ras-initiated phenotype. Transient transfection assays are particularly advantageous when using dominant-negative or

[14] T. B. Vaidya, C. M. Weyman, D. Teegarden, C. L. Ashendel, and E. J. Taparowsky, *J. Cell Biol.* **114,** 809 (1991).

[15] H. M. Blau, C. P. Chiu, and C. Webster, *Cell* **32,** 1171 (1983).

constitutively active signaling molecules to examine signaling pathways, because the prolonged exposure of cells to such proteins can cause secondary effects (i.e., cellular transformation or growth arrest) that complicate interpretation of the results.

Skeletal muscle differentiation has become a popular model for the study of oncogene-mediated effects because of several factors. Proliferating myoblasts are relatively easy to maintain, are excellent recipients for DNA transfection, and can be induced to biochemically and morphologically differentiate within 24 hr after experimental manipulation. Furthermore, the ability of 10T1/2 cells to be converted to myogenic precursors after expression of a single MRF provides a highly controlled experimental system in which the effects of signaling pathways can be measured against the differentiation-inducing potential of a single myogenic protein.

There are several reporter genes that are routinely used in myogenesis assays. Muscle-specific reporter genes consist of a cDNA, such as firefly luciferase (luc) or chloramphenicol acetyltransferase (CAT), cloned 3′ to transcriptional regulatory sequences that confer lineage-specific transcriptional activation during muscle differentiation. Some of the muscle-specific regulatory sequences (promoters) commonly used in these experiments have been characterized from the troponin I gene (TnI),[16] the muscle creatine kinase gene (MCK),[17,18] the myosin light chain gene (MLC1/3),[19] the α-cardiac actin gene (α-actin),[20] and the acetylcholine receptor gene (AChR).[21] Although there is no single regulatory element identified that alone accounts for the muscle specificity of these promoters, there is one feature that is common to most of the genes mentioned above: the presence of an "E-box" consensus site in the DNA of their regulatory regions. It is through these E boxes that heterodimers consisting of an MRF and a ubiquitously expressed E protein bind DNA to activate transcription in a muscle-specific fashion (reviewed in Ref. 22). Indeed, a simple reporter gene containing four tandem E-box sequences fused 5′ to a luciferase gene under the control of the minimal thymidine kinase promoter (4RTK-luc) demonstrates muscle-specific activation by the MRF MyoD in 10T1/2 cells.[23]

[16] K. E. Yutzey, R. L. Kline, and S. F. Konieczny, *Mol. Cell. Biol.* **9,** 1397 (1989).

[17] A. B. Lassar, J. N. Buskin, D. Lockshon, R. L. Davis, S. Apone, S. D. Hauschka, and H. Weintraub, *Cell* **58,** 823 (1989).

[18] T. J. Brennan and E. N. Olson, *Genes Dev.* **4,** 582 (1990).

[19] B. M. Wentworth, M. Donoghue, J. C. Engert, E. B. Berglund, and N. Rosenthal, *Proc. Natl. Acad. Sci. U.S.A.* **88,** 1242 (1991).

[20] B. A. French, K. L. Chow, E. N. Olson, and R. J. Schwartz, *Mol. Cell. Biol.* **11,** 2439 (1991).

[21] J. Piette, J. L. Bessereau, M. Huchet, and J. P. Changeux, *Nature (London)* **345,** 353 (1990).

[22] L. Li and E. N. Olson, *Adv. Cancer Res.* **58,** 95 (1992).

[23] H. Weintraub, V. J. Dwarki, I. Verma, R. Davis, S. Hollenberg, L. Snider, A. Lassar, and S. Tapscott, *Genes Dev.* **5,** 1377 (1991).

The relative strength of the muscle-specific promoter in the reporter gene should be taken into consideration when designing experiments, because the amount of reporter plasmid to be used depends on the predicted level of activation by the test constructs. If the strength of the regulatory sequence is unknown, it is advisable to titrate relative amounts of reporter DNA and activator (not to exceed 5 μg of each per well in a six-well plate format) until an acceptable level of activation is reached. After optimization of the activation, the amount of experimental test DNA also must be determined on the basis of a further linear activation or inhibition of the reporter gene. In all experiments, it is important to include a control group using the reporter gene and an equivalent amount of vector DNA (no cDNA insert) to determine endogenous reporter gene levels in the absence of activator or possible interference from the test DNA promoter elements in the presence of the activator.

Although a number of reporter genes and assays can be adapted for use in the skeletal muscle system, the luciferase reporter gene assay has several advantages. Luciferase assays are nonradioactive and can be completed within minutes of harvesting the cells. Most notably, the luciferase assay system is sensitive and remains linear within a wide range of activity. Therefore, transfections can be performed in a six-well format and a fairly large panel of proteins or chemical treatments can be analyzed in parallel.

Below is described a method for the analysis of oncogene-mediated effects on skeletal myogenesis, using the MyoD-induced activation of a muscle-specific luciferase reporter gene in transiently transfected 10T1/2 cells.

Reagents for Transient Transfection

HBS (2×): 280 m*M* NaCl, 50 m*M* HEPES, 1.5 m*M* Na_2HPO_4. Adjust to pH 7.1–7.15 with 1 *N* NaOH and filter sterilize. The pH of the HBS buffer is crucial, so we recommend checking and adjusting the pH (if necessary) on the following day. Store 2× HBS buffer at room temperature

$CaCl_2$, 2.5 *M*: Dissolve $CaCl_2$ in distilled water and filter sterilize. Store at room temperature.

TE (0.1×): Make from a stock of 1× TE [10 m*M* Tris, 1 m*M* EDTA (pH 8.0)]. Filter sterilize and store at room temperature

Plasmid DNA: DNA for transfections is prepared by the standard alkaline lysis procedure and purified by banding *twice* in CsCl density gradients (final density, 1.55 g/ml).[24] The concentration of each DNA is determined by the absorbance at 260 nm and confirmed by gel electrophoresis. Cells take up supercoiled plasmids more efficiently, so DNA conformation plays

[24] F. M. Ausubel, R. Brent, R. E. Kingston, D. D. Moore, J. G. Seidman, J. A. Smith, and K. Struhl (eds.), "Current Protocols in Molecular Biology." John Wiley & Sons, New York (1994).

an important role. DNA purity is another important factor, as RNA contamination significantly reduces transfection efficiency

Cell culture and media: 10T1/2 cells are maintained in GM as described in the previous section. Eighteen hours prior to transfection, cells are seeded at 1.2×10^5 cells per well in gelatin-coated, six-well plates

Transient Transfection Protocol

To test the effect of Ras or other signaling molecules on myogenesis, 10T1/2 cells seeded in six-well plates are transfected by the calcium phosphate DNA precipitation method. For each experimental group, a precipitate sufficient for two duplicate wells is prepared with 2 μg of the test cDNA (e.g., Ras G12V), 4 μg of the muscle-specific reporter plasmid (e.g., TnI-luc), and 1 μg of MyoD expression vector (e.g., pEM-MyoD) as an activator. To prepare the DNA precipitate, the plasmids are placed in a 1.5-ml microcentrifuge tube and the volume is adjusted to 270 μl with 0.1× TE. Thirty microliters of 2.5 *M* $CaCl_2$ is added to the DNA dropwise with mixing. Three hundred microliters 2× HBS is aliquoted to a round-bottom, 4-ml polystyrene tube (Falcon; Becton Dickinson Labware, Lincoln Park, NJ). The DNA–$CaCl_2$ solution is added dropwise to the HBS solution as the HBS is aerated by gently bubbling with a stream of sterile air (produced by a sterile micropipette attached to a Pipette Aide (AccuTek, San Diego, CA) with sterile tubing). Allow the solution to incubate undisturbed at room temperature for 30 min, during which time a precipitate should form and the solution should turn cloudy. Add 300 μl of the precipitate dropwise to the medium covering the cells in each of two wells. In our experience, a glycerol shock to enhance DNA uptake is not necessary for reporter gene assays. The six-well plates with precipitate are returned to the incubator for 5–6 hr. The precipitate is then removed and the cells fed fresh GM. Twenty-four hours after transfection, the GM is replaced with DM and the cells are incubated for an additional 24–48 hr before harvesting for luciferase assays. This procedure can be scaled up for use with larger plates or wells by adjusting the cell number and DNA amounts accordingly, using a factor representing the ratio of the surface area of the old to the new plating vehicles.

A large number of commercial transfection reagents have become available. Although the cost of these reagents is a consideration, especially if a large number of constructs is to be tested, they offer some advantages over the standard calcium phosphate precipitation method—-most notably, a higher transfection efficiency. Using the same amounts of DNA described above, we obtain 20-fold higher levels of TnI-luc activation (on average) when transfections are performed with the SuperFect transfection reagent (Qiagen, Valencia, CA).

Reagents for Luciferase Assays

Cell lysis buffer: 25 m*M* Tris, 2 m*M* EDTA, 10% (v/v) glycerol, 1% (v/v) Triton X-100, pH 7.8. Filter sterilize and store at 4°

Phosphate-buffered saline (PBS): 140 m*M*, NaCl, 2.7 m*M* KCl, 1.8 m*M* KH_2PO_4, 10 m*M* Na_2HPO_4. The pH of the buffer should be pH 7.3

Luciferase reagent: From Promega (Madison, WI)

Luciferase Assay Protocol

Cells are washed twice with cold PBS and lysed at room temperature for 15 min, using 200 μl of lysis buffer. Rock or tilt the plate during the lysis step to ensure complete coverage of the cells. The lysate from each well is transferred to a labeled 1.5-ml microcentrifuge tube and can be stored frozen at −80° for up to 2 weeks or assayed immediately for luciferase enzyme activity. Fifty microliters of luciferase reagent (premixed according to the manufacturer instructions) is aliquoted to the desired number of glass 12 × 75 mm tubes. Ten microliters of cell extract (prewarmed to room temperature) is added to the first tube, mixed by vortexing, and assayed in a luminometer (Lumat LB 9501/16; Berthold Systems, Bad Wildbad, Black Forest, Germany). Repeat this procedure with each of the remaining cell samples. It is critical that both the cell extracts and the luciferase reagent be at room temperature because luciferase activity is temperature dependent.

Several other factors should be kept in mind when performing reporter gene assays, one of which is the variation in transfection efficiency. Transfection efficiency is sensitive to changes in cell density, the amount of DNA used, and/or the DNA purity. To minimize fluctuations in transfection efficiency, care should be taken to faithfully duplicate assay conditions between experiments. To monitor transfection efficiency, an internal control also can be used. Most commonly, cells in a six-well format are cotransfected with 100 ng of a plasmid expressing β-galactosidase or a luciferase gene from a distinct species (i.e., sea pansy) and assayed in parallel with the test reporter. One potential complication with an internal control is that the constitutive promoters in these vectors can be influenced by the activities of the test DNA. In the case of the myogenesis assays described here, we have been unable to find an internal control gene whose expression is not modulated by the cotransfection of Ras or MyoD. If a suitable internal control is not available, transfections should be repeated in an identical manner several times (a minimim of three independent transfections in duplicate) and the luciferase values normalized to the total protein content in each cell extract. Protein content in cell extracts is most easily determined by the Bradford assay (Bio-Rad, Hercules, CA). Luciferase data are expressed as relative light units (RLU) per microgram of protein.

An example of a transient transfection assay is shown in Fig. 2. In this experiment, 10T1/2 cells were transfected with pEM-MyoD, TnI-luc, and either pDCR control vector or pDCR-H-Ras G12V. Cells receiving the H-Ras G12V vector display a low level of luciferase activity (10% of the pDCR control group), demonstrating that constitutive Ras signaling efficiently blocks the MyoD-induced activation of the muscle-specific troponin I reporter gene.

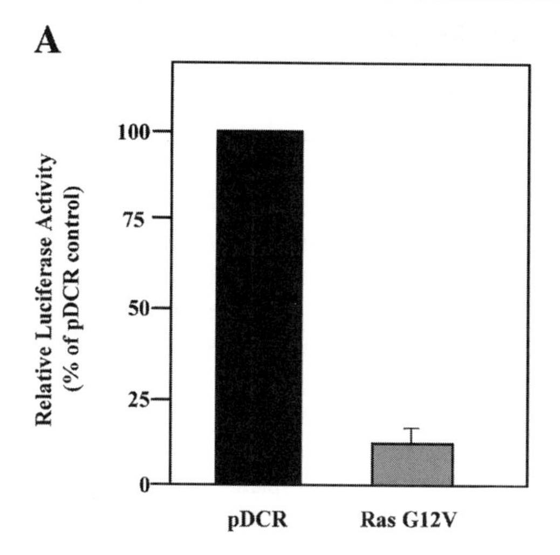

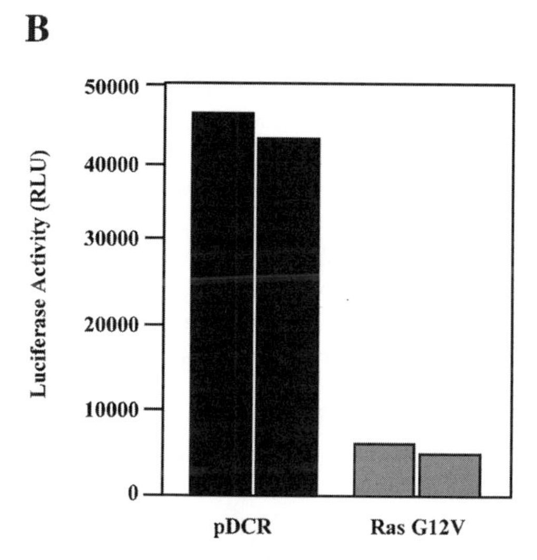

FIG. 2. (A) H-Ras G12V inhibits MyoD-specific *trans*-activation of the TnI-luc reporter gene. 10T1/2 cells were transiently transfected and induced to differentiate as described in text. Cell extracts were normalized for protein content and assayed for luciferase activity. Luciferase activity of the test group (Ras G12V) is expressed as light units per microgram of protein relative to the activity of the positive control (pDCR), which is set to 100. The standard error of the mean (SEM), derived from values obtained in three independent assays performed in duplicate, is indicated only for the test group. (B) Raw data from a representative transfection experiment performed in (A) are shown. Luciferase activity was measured for each duplicate sample. After normalization to protein content, the duplicates are averaged. The control average is set to 100 and the Ras G12V average is expressed as a percentage of the control. At least three independent sets of such duplicates are required to arrive at the final results shown in (A).

Stable Transfections

Overview

When the effects of oncogene expression need to be monitored for more than 72 hr, cell cultures are generated that have a gene or a cDNA expression vector stably integrated into the genome. Stable expression can be achieved by transfecting cells with DNA encoding both the protein of interest and a selectable drug resistance marker. If the plasmid selected cannot accommodate both cDNA expression cassettes, then a 10-fold molar excess of the test DNA is cotransfected with an expression vector for a drug resistance gene. The most commonly used selectable marker genes are bacterial genes rendering cells resistant to neomycin, puromycin, or hygromycin. Transfected cells are grown in the presence of the appropriate drug to produce resistant colonies that are subsequently isolated and analyzed for the presence and level of expression of the cointroduced gene. Stable transfection assays have a number of drawbacks. Because integration of plasmids is random, the number of gene copies integrated and the integration site can vary from cell to cell, resulting in a wide range of gene expression levels among individual clones. For certain assays, this variation can be overcome by combining ~500 individual clones as a "pooled" population. On the other hand, generation of individual clones that express different amounts of test DNA can be useful. For example, in the case of activated Ras, we have observed a dose-dependent effect on myogenesis after stable expression of H-Ras G12V in the 23A2 cell line.[2] Cell extracts prepared from stable Ras transfectants can be used for kinase assays, electrophoretic mobility shift assays, and/or coimmunoprecipitation experiments.

Here we describe a standard procedure for generating stable cell lines, using 23A2 myoblasts as recipients for the transfection of a vector driving expression of activated H-Ras and a gene conferring resistance to neomycin.

Reagents for Stable Transfections

$CaCl_2$, 0.5 *M:* Dissolve $CaCl_2$ in distilled water, filter sterilize, and store at room temperature

Carrier DNA: High molecular weight 23A2 genomic DNA is isolated as described[24] and used as a carrier DNA. We do not recommend the use of commercial DNA preparations of salmon sperm or calf thymus DNA for this purpose

Chloroquine, 100 m*M:* Chloroquine (Sigma, St. Louis, MO) is dissolved in BME and the pH is adjusted to 6.0–6.5. The stock is filter sterilized and stored in small aliquots at -20°

HBS (2×): See Reagents for Transient Transfection

Plasmid DNA: See Reagents for Transient Transfection

Cell culture and media: 23A2 cells are maintained in GM on gelatin-coated dishes. It is critical to passage cells at subconfluence to prevent them from spontaneously differentiating. Twenty-four hours prior to transfection, the cells are plated at 5×10^5 cells per 100-mm tissue culture dish in GM

Stable Transfection Protocol

Two hours before transfection, chloroquine is added directly to the medium in each dish to achieve a final concentration of 100 μM. A DNA–calcium phosphate precipitate is prepared as follows: 500 μl of 0.5 M $CaCl_2$ is added dropwise with mixing to a 4-ml tube (Falcon) containing 60 μg of carrier DNA and 100 ng of the pDCR-H-Ras G12V expression plasmid (also encoding neomycin resistance) in a final volume of 500 μl of 0.1$\times$ TE. The DNA–calcium chloride mix is then added dropwise to 1 ml of 2$\times$ HBS solution in a 14-ml tube (Falcon) with constant shaking. The final solution is incubated at room temperature for 30 min to allow a precipitate to form. During this incubation, the chloroquine-containing medium on the cells is replaced with GM. Each calcium phosphate–DNA precipitate is added to two 100-mm tissue culture dishes (1 ml/plate) and the plates are returned to the incubator for 6 hr, after which time the cells are fed GM. Twenty-four hours after transfection, each plate of cells is subcultured into six 100-mm dishes (12 plates per experimental group) and maintained on a 3- to 4-day feeding schedule with GM containing Geneticin (G-418) at 400 μg/ml. After 12 days in culture, individual drug-resistant colonies will be apparent. If colonies are to be pooled, the cells are trypsinized and replated. If individual clones are to be isolated, they can be removed from the plate with glass cloning rings. Each clone is replated on a gelatin-coated, 35-mm tissue culture dish. After expansion and freezing of the individual clones or pooled population, H-Ras expression is best assessed by making cell extracts from the stocks and performing Western blot analysis with Ras-specific antiserum.

Immunodetection of Muscle-Specific Proteins *in Situ*

Overview

The effects of Ras or other signaling molecules on skeletal muscle differentiation can be determined by immunostaining transfected cells with antibodies reactive with muscle-specific proteins. We routinely stain differentiated cells in culture with antibodies directed against the muscle-specific troponin T, muscle creatine kinase, or myosin heavy chain proteins. Several factors should be taken into account when performing a morphological differentiation assay that will be analyzed by immunostaining.

1. *Antibody specificity:* Most monoclonal antibodies are specific for their target proteins. However, when using any antibody for the first time, specificity should be confirmed. One approach is to perform a Western blot on extracts isolated from cells expressing the protein of interest. Unfortunately, some antibodies will

recognize only native proteins and therefore cannot be used for Western blots. In this case, nonmuscle cells (COS1 or 10T1/2) can be transiently transfected with a cDNA expression vector for the protein of interest (and the empty vector as a control), fixed, and immunostained. Transfected cells should exhibit enhanced immunoreactivity when compared with the nonexpressing cells on the same dish, or with the cells on the control dish.

2. *Fixation:* Differentiated cells can be fixed with 4% (v/v) formaldehyde, cold 100% methanol, or 70% (v/v) ethanol–formalin–acetic acid mix (20 : 2 : 1, v/v/v). The choice of fixative depends on the primary antibody and the epitope that it recognizes. For the optimal detection of the myosin heavy chain protein, we recommend the ethanol–formalin–acetic acid mix for fixation. Less harsh fixation conditions [e.g., 4% (v/v) formaldehyde] may compromise efficient MHC staining, yet may be used if the cells are to be costained for additional proteins (e.g., MyoD or Ras p21).

3. *Visualization of antibody–antigen complexes:* Immune complexes are visualized with secondary antibodies coupled to a fluorescent label or, if fluorescence microscopy is not available, to an enzyme such as horseradish peroxidase (HRP) or alkaline phosphatase (AP) that can be reacted with specific substrates to form a visible precipitate. Both methods of detection have been utilized in our laboratory to assess the differentiation of skeletal muscle cells in culture.

Below is described the standard procedure for fixing and immunostaining a 10T1/2 cell culture that has been transiently transfected with the MRF MyoD and induced to differentiate into skeletal muscle.

Reagents for Immunostaining

Fixative: Mix 70% (v/v) ethanol, 37% (v/v) formaldehyde, and glacial acetic acid at 20 : 2 : 1 (v/v/v). This solution can be stored at 4° for up to 1 year

Triton X-100 (0.1%, v/v): Dilute Triton X-100 (Sigma) in PBS to 0.1% (v/v). Store at room temperature

Horse serum: From GIBCO-BRL

Bovine serum albumin (BSA; Sigma): Dissolve in PBS to 0.1% (w/v) and store at 4° for up to 2 weeks

Primary antibody: MF-20 monoclonal antibody is panreactive for the myosin heavy chain protein and is isolated from the MF-20 mouse hybridoma (available from the Developmental Studies Hybridoma Bank, University of Iowa, Iowa City, IA)

Secondary antibodies: Anti-mouse HRP conjugate (Vector Laboratories, Burlingame, CA) is used in immunoperoxidase assays. Anti-mouse Texas Red-conjugated (Vector Laboratories) or fluorescein-conjugated (Vector Laboratories) secondary antibodies are used to detect myosin heavy chain by immunofluorescence

A

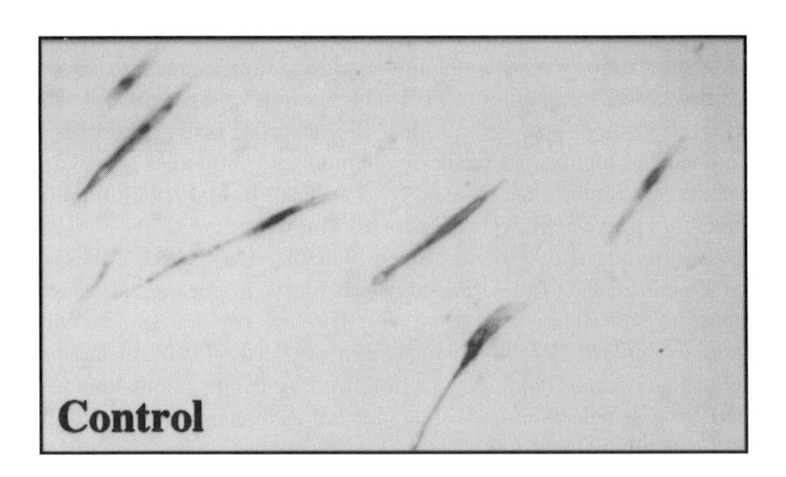

B

FIG. 3. The detection of myosin heavy chain protein in differentiated cultures. 10T1/2 cells were transiently transfected with the MyoD expression vector and either control DNA (A) or an expression vector for H-Ras G12V (B). Forty-eight hours after treatment with DM, the cells were fixed, permeabilized, and immunostained with MF-20 primary and HRP-conjugated secondary antibodies as described in text. The presence of skeletal myotubes with the distinct cytoplasmic staining of the MHC protein is observed in (A), but is absent in the culture coexpressing H-Ras G12V.

Immunostaining Protocol

All steps in the immunostaining procedure are performed at room temperature. After the fixation, permeabilization, and each immunoreaction step, the cells are washed two or three times in PBS. The solution volume for both the fixation and permeabilization steps is 5 ml for a 100-mm plate or 2 ml for cells in a six-well format. The solution volume for incubation with antibodies is kept at a minimum. One hundred microliters is used for a six-well format and 400 μl (immobilized under a strip of parafilm) is used for 100-mm dishes.

Differentiated cultures are fixed with the recommended fixative for 1 min [or 10 min if 4% (v/v) formaldehyde is used]. Permeabilization is achieved by incubation for 10 min with 0.1% (v/v) Triton X-100 in PBS. The MF-20 primary antibody [diluted 1 : 5 in PBS containing 0.1% (w/v) BSA] is incubated with the cultures for a minimum of 1 hr, followed by a 1-hr incubation with either the HRP-conjugated or fluorescein-conjugated secondary antibody [diluted 1 : 200 in PBS containing 0.1% (w/v) BSA]. Stained cultures can be overlayed with PBS and stored at 4° in a sealed box for up to 2 weeks. If a fluorescence-tagged secondary antibody is used, incubation should be performed in the dark (under a velvet blanket) and the stained plates stored in a light-tight box.

The number of myosin heavy chain-positive cells is determined by bright-field microscopy or fluorescence microscopy. Ten randomly chosen microscope fields per 100-mm dish, or 5 fields per well of a 6-well plate, are scored by counting the number of nuclei in myosin heavy chain-positive cells. Individual experiments should be repeated at least three times. The number obtained from the positive control group is set to 100% and the results from all test groups are expressed relative to the control.

An example of cell staining for morphological differentiation of 10T1/2 cells transiently transfected with pEM-MyoD (control) and pEM-MyoD plus activated H-Ras (Ras G12V) is shown in Fig. 3. The cultures were differentiated with lgDMEM and 2% (v/v) horse serum for 48 hr. Note the morphology of the myotubes in Fig. 3A and the absence of myosin heavy chain-positive cells in Fig. 3B.

Conclusions

The ability to measure both biochemical and morphological aspects of skeletal muscle differentiation using a controlled model system dependent upon the expression and activation of a single MRF is very attractive to scientists interested in examining the effects of oncogenes and growth factors on processes of mammalian cell determination and differentiation. Using this model system, our laboratories have reported unique effects of an activated Ras protein on the differentiation process. Our results implicate the existence of a new Ras-mediated signaling pathway(s) in the control of tissue-specific gene expression. We plan to exploit this system to learn more about the novel signaling intermediates downstream of Ras activation in muscle cells and, most importantly, to establish how the function of these intermediates exerts a negative effect on the expression and timely activation of MRFs during myogenesis.

Acknowledgments

We acknowledge the significant contributions of a number of former graduate students and post-doctoral fellows in the laboratories of E. J. Taparowsky and S. F. Konieczny toward the development of the protocols presented in this chapter. Special thanks to Dr. Arthur Kudla for comments on the manuscript.

[22] Induction of Senescence by Oncogenic Ras

By IGNACIO PALMERO and MANUEL SERRANO

Introduction

Oncogenic Ras is able to transform efficiently most immortal rodent cell lines.[1] In contrast, transformation of primary cells by Ras requires the cooperation of an immortalizing event.[2,3] Examples of this are overexpression of Myc, and inactivation of the tumor suppressors p53, p19^{ARF}, or p16^{INK4A}.[4–7] Also, several viral oncoproteins such as adenovirus E1A or herpesvirus E7 can cooperate with Ras.[4] Transformation of primary cells through the combined action of immortalizing alterations and oncogenic Ras is the basis for several models of multistep carcinogenesis both in culture cells and in animal models.[6–10] Normal, nonimmortal cells are therefore refractile to transformation by Ras, presumably because of the existence of built-in protection mechanisms against unlimited proliferation. Immortalizing

[1] M. Barbacid, *Annu. Rev. Biochem.* **56,** 779 (1987).
[2] H. Land, L. Parada, and R. A. Weinberg, *Nature* (*London*) **304,** 596 (1983).
[3] H. E. Ruley, *Nature* (*London*) **304,** 602 (1983).
[4] R. A. Weinberg, *Cancer Res.* **49,** 3713 (1989).
[5] N. Tanaka, M. Ishihara, M. Kitagawa, H. Harada, T. Kimura, T. Matsuyama, M. S. Lamphier, S. Aizawa, T. W. Mak, and T. Taniguchi, *Cell* **77,** 829 (1994).
[6] T. Kamijo, F. Zindy, M. F. Roussel, D. E. Quelle, J. R. Downing, R. A. Ashmun, G. Grosveld, and C. J. Sherr, *Cell* **91,** 649 (1997).
[7] M. Serrano, H.-W. Lee, L. Chin, C. Cordon-Cardo, D. Beach, and R. A. DePinho, *Cell* **85,** 27 (1996).
[8] T. Hunter, *Cell* **64,** 249 (1991).
[9] B. Vogelstein and K. W. Kinzler, *Trends Genet.* **9,** 138 (1993).
[10] L. Chin, J. Pomerantz, D. Polsky, M. Jacobson, C. Cohen, C. Cordon-Cardo, J. W. Horner, and R. A. DePinho, *Genes Dev.* **11,** 2822 (1997).

0076-6879/00 $35.00

alterations would have the effect of disabling some of these mechanisms, allowing transformation by Ras.

A detailed study of the effects of Ras in primary cells has not been achieved until recently, mainly because of the lack of efficient gene transfer techniques. Improvements in retrovirus-based gene transfer methods have made it possible to analyze the effect of the sustained expression of oncogenic Ras in primary cells.[11] Interestingly, it was observed that after an initial mitogenic response, prolonged action of Ras provokes a permanent cell cycle arrest. This is achieved through the simultaneous activation of the two main pathways for growth control in mammalian cells, the retinoblastoma protein (Rb) pathway and the p53 pathway, via the products of the *INK4A/ARF* locus, $p16^{INK4A}$ and $p19^{ARF}$.[11–14] The arrest induced by oncogenic Ras is reminiscent of cellular senescence in many respects. Senescence was originally defined as the cell cycle arrest that occurs in primary fibroblasts when they reach the end of their normal proliferative life span and it is typically accompanied by specific changes in morphology and gene expression.[15,16] The induction of a senescence-like arrest by Ras suggests that senescence is not only a passive consequence of the accumulation of cell divisions, but an intrinsic antioncogenic mechanism that can be activated in response to improper mitogenic stimuli.[11,17]

In this chapter we describe a methodological approach to analyze the effects of activated Ras on primary mouse and human fibroblasts, based on the use of retroviral transduction.

Preparation of Mouse Embryo Fibroblasts

Preparation of fibroblasts from total mouse embryos is a convenient and frequently used method to obtain large numbers of primary cells with few accumulated divisions.[18] Embryos at day 13.5 postconception offer a good balance between cell number and proliferative potential.

1. Kill a pregnant mouse.
2. Remove the uterus with embryos and keep them in a 50-ml Falcon tube with phosphate-buffered saline (PBS) containing antibiotics (1× antibiotic–antimycotic solution from GIBCO-BRL, Gaithersburg, MD), prewarmed to 37°.

[11] M. Serrano, A. W. Lin, M. E. McCurrach, D. Beach, and S. W. Lowe, *Cell* **88,** 593 (1997).

[12] I. Palmero, C. Pantoja, and M. Serrano, *Nature (London)* **395,** 125 (1998).

[13] J. Zhu, D. Woods, M. McMahon, and J. M. Bishop, *Genes Dev.* **12,** 2997 (1998).

[14] A. W. Lin, M. Barradas, J. C. Stone, L. van Aelst, M. Serrano, and S. W. Lowe, *Genes Dev.* **12,** 3008 (1998).

[15] D. Rohme, *Proc. Natl. Acad. Sci. U.S.A.* **78,** 5009 (1981).

[16] L. Hayflick, *Biochemistry* **62,** 1180 (1997).

[17] R. A. Weinberg, *Cell* **88,** 573 (1997).

[18] G. J. Todaro and H. Green, *J. Cell Biol.* **17,** 299 (1963).

3. Place the uterus in a clean tissue culture dish with PBS–antibiotics.

4. Using a pair of scissors and forceps, cut the uterus wall and remove individual embryos.

5. Transfer each embryo to a clean empty dish. Remove all red organs and head. If genotyping of individual embryos is required, heads can be used as a source of genomic DNA. For this, transfer each head to an Eppendorf tube and store at −70° until processed for DNA extraction.

6. Place each embryo in a clean 60-mm dish (with no PBS). Cut the embryo into pieces with a razor blade about 20–25 times, until no recognizable parts are left. Add 1 ml of a 2× trypsin–EDTA solution (diluted from a 10× stock solution; **GIBCO-BRL**). Leave the dish tilted in a CO_2 incubator at 37° for 20 min.

7. Disperse the cell suspension by pipetting up and down (about 15 times), using a plugged Pasteur pipette. Leave in the incubator for an additional 20 min.

8. Add 15 ml of prewarmed Dulbecco's modified Eagle's medium (DMEM; **GIBCO, Grand Island, NY**) containing 10% (v/v) fetal calf serum (FCS; Sigma, St. Louis, MO) and 1× antibiotic–antimycotic. Transfer to a 75-cm^2 flask.

9. Change the medium the following day.

10. Incubate the cells until confluent (usually 24–48 hr). At this stage, in addition to the monolayer formed by fibroblasts, a variable number of cell aggregates that have resisted trypsinization may be present, attached to the flask surface. These will be disaggregated in the following steps of the protocol and are a good source of cells.

11. Trypsinize. Add 25 ml of DMEM–10% (v/v) FCS and transfer to a 150-cm^2 flask. Incubate until confluent (48–72 hr).

12. At this point, the cells can be frozen and stored in liquid nitrogen. It is convenient to make aliquots with a reasonably high number of cells so that enough cells can be obtained after thawing without adding too many population doublings. We usually freeze 3–4 million cells per vial in 90% (v/v) FCS–10% (v/v) dimethyl sulfoxide (DMSO) in a 2-propanol chamber inside a −70° freezer. Once frozen, vials are transferred to liquid nitrogen for long-term storage.

13. For retroviral infection, frozen aliquots are thawed by quick immersion of the frozen vials in a water bath at 37°. When the contents of the vial are nearly thawed, pour them on a 100-mm plate containing DMEM–10% (v/v) FCS with antibiotic–antimycotic at 37°, swirl the plate, and transfer it to a CO_2 incubator (37°). Cultures should be confluent the day after thawing. Harvest the cells by trypsinization, count, and plate the cells for infection (see below).

Culture of Human Diploid Fibroblasts

Primary embryonic fibroblasts of human origin have a considerably longer life span than mouse embryo fibroblasts (MEFs) (about 70 population doublings in human vs. 20 in mouse).[15,18] Frequently used strains of human diploid fibroblasts

(HDFs) are IMR-90 and WI-38, both derived from fetal lung. Conveniently, both strains are available from the American Type Culture Collection (ATCC, Manassas, VA) at a population doubling level (PDL) of about 20. Cells at this passage level can still accumulate more than 50 population doublings before they enter senescence. This provides a considerable margin for experimental manipulation. Another common source of HDFs is foreskin biopsies.

HDFs divide at a much lower rate than MEFs, with a doubling time of 2–3 days for actively dividing early-passage cultures. HDFs should never be plated at a density lower than 40–50% because this causes a marked reduction in the division rate. For passage, HDFs should be split 1 : 2 every 3–4 days.

Retroviral Transduction

The efficiency of transfection of primary fibroblasts with standard methods (calcium phosphate, LipofectAMINE) is extremely low (about 1%). In turn, retroviral transduction allows efficiencies of 50% and higher.[19] Because of the limited life span of primary fibroblasts, particularly MEFs, selection and isolation of stably transfected clones are not feasible. Retroviral transduction provides the high efficiency required to obtain large amounts of dividing primary cells for analysis. Another relevant feature of the method is the fact that on average only one copy of the gene is integrated in the recipient cell and transmitted to its progeny. This results in moderate levels of overexpression through a long period of time. This is of importance because, in other experimental systems, the intensity and duration of the Ras signal may determine its biological effects (e.g., cell cycle arrest, proliferation).[11,13,14,20,21]

The first step in the retroviral transduction protocol is the transfection of so-called packaging cells with the gene of interest cloned into an appropriate retroviral vector. Packaging cells are stably transfected cell lines that express all the retroviral genes necessary for the production and release of viral particles. Most commonly used packaging systems are based on strains of the murine leukemia virus (MuLV). Viral particles exhibit a species tropism that depends on the particular type of MuLV Env protein expressed by the packaging cells. In this manner, ecotropic packaging cells produce viral particles that can enter only into rodent cells through their interaction with a specific murine transmembrane transporter called the ecotropic receptor. Similarly, amphotropic packaging cells produce viral particles able to enter into all mammalian cells, including human.

[19] C. Cepko and W. Pear, *in* "Current Protocols in Molecular Biology" (F. M. Ausubel, R. Brent, R. Kingston, D. D. Moore, J. G. Seidman, J. A. Smith, and K. Struhl, eds.), p. 9.9.1. John Wiley & Sons, New York, 1999.

[20] A. Sewing, B. Wiseman, A. Lloyd, and H. Land, *Mol. Cell. Biol.* **17,** 5588 (1997).

[21] D. Woods, D. Parry, H. Cerwinski, E. Bosch, E. Lees, and M. McMahon, *Mol. Cell. Biol.* **17,** 5598 (1997).

A variety of MuLV-derived retroviral vectors, such as pLXSN,[22] pBabe,[23] or their derivatives, are now available and a review can be found elsewhere.[22] These vectors do not encode any viral genes but contain *cis* elements that allow them to be substrates of viral replication and packaging. Therefore, when present in a packaging cell, these vectors are replicated and packaged into viral particles. Subsequently, the medium obtained from the packaging cells and containing the viral particles is placed in contact with the recipient cells. Because of the life cycle of retroviruses, integration of the viral DNA into the host genome requires passage through mitosis and, therefore, it is important that the recipient cells be actively dividing during infection.

In our protocol, the dish containing the recipient cells and the viral supernatant is centrifuged at low speed as a way to favor the entry of viruses into the cells. This centrifugation step causes a variable degree of cell death in some areas of the dish, but it significantly increases the efficiency of infection by poorly understood mechanisms.[19] Other alternative approaches have been described, including successive additions of fresh viral supernatants during long periods of time. Successfully infected cells are drug selected and used for the experiment.

Concerning biological safety, there are a number of considerations that can be made. It is preferable to use helper-free packaging cell lines and transient transfection with the retroviral constructs. In this way, the probability of recombination and generation of replication-competent retroviruses is significantly diminished. We use transient transfection of derivatives of 293 embryonic kidney cells, such as Bosc-23 or Phoenix.[19,24] It is also advisable to use ecotropic viral particles whenever possible. In this regard, it should be mentioned that human cells can be "murinized" by introduction of the ecotropic receptor, thus making them infectable by ecotropic retroviruses.[11] Safety guidelines must be obtained from the competent authority at each institution. The National Institutes of Health (NIH, Bethesda, MD) recommend standard tissue culture practices for both amphotropic and ecotropic murine retroviruses (see http://www4.od.nih.gov/oba/guidelines.html or a summary at http://www.collmed.psu.edu/ora/compliance/dna.htm).

A day-by-day description of the retroviral transduction protocol that we use is as follows.

Day 1. Plate 5×10^6 packaging cells in DMEM–10% (v/v) FCS per 100-mm dish for each transfection. One plate will produce enough viral supernatant for

[22] A. D. Miller, *in* "Current Protocols in Human Genetics" (N. Dracopoli, J. L. Haines, B. R. Korf, C. C. Morton, C. E. Seidman, J. G. Seidman, and D. R. Smith, eds.), p. 12.5.1, John Wiley & Sons, New York, 1996.

[23] J. P. Morgersten and H. Land, *Nucleic Acids Res.* **18,** 3857 (1993).

[24] W. S. Pear, G. P. Nolan, M. L. Scott, and D. Baltimore, *Proc. Natl. Acad. Sci. U.S.A.* **90,** 392 (1993).

two independent infections. It is important that cells be fresh and actively growing before splitting. Also, packaging cells tend to form clumps that should be well dispersed during trypsinization to ensure even distribution over the plate.

Day 2. Packaging cells should be 70% confluent at this point. Change the medium (10 ml) immediately before transfection. 293-derived packaging cells adhere loosely to plastic, and leaving cells without medium for a few minutes causes their detachment from the plates.

For transfection, we follow a standard calcium phosphate protocol, using 5 to 20 μg of the DNA of interest (adding carrier if necessary to complete 20 μg of total DNA). Incubate at 37° overnight (16–18 hr).

For a control of transfection efficiency, include a separate transfection with a convenient marker [such as LacZ or green fluorescent protein (GFP)] in a retroviral vector.

Day 3. Change the medium and put the plates in a CO_2 incubator at 32°. This increases the stability of viral particles. Incubate for 48 hr.

Optionally, the medium may be supplemented with dexamethasone to 1 μ*M* and sodium butyrate to 1 m*M*. This increases transcription from the long terminal repeat (LTR) present in the retroviral vectors and can represent an improvement, particularly for cells with low efficiency of infection, such as HDFs.

Day 4. Plate recipient cells (8×10^5 cells per 100-mm dish). It is critical that the cells be fresh and actively growing to ensure that there is a high proportion of dividing cells during infection. Incubate for 24 hr. This allows the cells to recover from trypsinization and to restore ecotropic receptor molecules lost during trypsinization.

Day 5. In the morning, collect medium from the transfected packaging cells. In the case of *lacZ*-transfected cells, keep the plates with fresh medium until further processing (see below, end of day 5). Filter the medium containing the retroviruses through a 0.45-μm pore size filter coupled to a syringe into a 50-ml Falcon tube. This step results in a decrease in viral titer (of about 50%), but is necessary to avoid contamination by packaging cells.

At this point, viral supernatants can be frozen at −80°. Once frozen, they can be thawed only once. The freezing step reduces the original titer by 50%.

Dilute the viral supernatant 1 : 2 (20-ml final volume) or 1 : 3 (30-ml final volume) with fresh medium. Viral supernatants are somewhat toxic to cells and cannot be used undiluted. Add Polybrene to a final concentration of 4 μg/ml. This polycation increases the efficiency of infection, presumably by facilitating aggregation of viral particles and attachment to cell membranes.

Replace medium of the recipient cells with the diluted viral supernatant (10 ml for a 100-mm dish). Spin the plates in a swinging rotor for 1 hr at 450*g* at 20° [1500 rpm in a tabletop centrifuge, Heraeus (Osterode, Germany) Megafuge 2.0 or similar]. Remove the plates immediately after the spin. Transfer to a CO_2 incubator set at 37° and leave overnight.

Stain control packaging cells transfected with the LacZ construct. Use the protocol described for senescence-associated ß-galactosidase (SA-ß-Gal) staining, but use PBS, pH 7, instead of a pH 6 buffer (see below). If GFP is used, inspect the cells under a fluorescence microscope or analyze by fluorescence-activated cell sorting (FACS). Efficiencies of about 50% should be obtained.

Day 6. In the morning, change the medium.

Day 7. In the afternoon, start the selection. Selection is usually more efficient in actively growing cells. In the case of MEFs, this requires that the infected cells be split 1 : 2, or 1 : 3, directly into selection medium to ensure that they are subconfluent during selection. As mentioned, HDFs divide more slowly and splitting is therefore not necessary.

As a selection control, add the appropriate drug-containing medium to a plate of noninfected cells. If using MEFs of different genotypes, use a control for each genotype because the rate of death may not be identical. We use the following selection conditions: puromycin (Sigma) 2 μg/ml for 2–3 days; hygromycin B (Boehringer Mannheim, Indianapolis, IN), 75 μg/ml for 4–5 days; G-418 (GIBCO), 300 μg/ml for 4–5 days.

Stain the nonselected LacZ-infected control, as described for the packaging cells. If GFP is used, cells can be observed under a fluorescence microscope or quantified by FACS analysis. At this point, we usually obtain about 30–50% LacZ-positive MEFs and 10–20% LacZ-positive HDFs, although this could be an underestimate of the actual number of infected cells. The efficiency of the whole procedure is variable and depends largely on the cell division rate of the recipient cells and on the particular retroviral construct used. As an indication, starting with 8×10^5 cells (see Day 4), it could be expected that at least $4–6 \times 10^6$ MEFs or $1–2 \times 10^6$ HDFs would be obtained after selection.

Once selected, it is not necessary to keep infected cells under selection for the duration of the experiment.

Analysis of Premature Senescence Induced by Oncogenic Ras

Premature senescence induced by activated Ras results in a number of characteristic phenotypical changes, namely, a block to cell division, accompanied by changes in cell morphology, and the appearance of senescence-specific markers.[11] It is worth mentioning that the different responses, particularly the cell cycle arrest, are usually more clear with HDFs than they are with MEFs. This most likely reflects the heterogeneous nature of early-passage MEF preparations and variability from one preparation to another. As opposed to this, HDF strains have been propagated for longer times and therefore are more homogeneous. In addition, the different responses appear with a characteristic timing. The first response of primary fibroblasts to Ras is in fact proliferative. This is usually obvious 2–3 days after the day of infection (Day 5 of the above protocol) and is accompanied by a change to a small, rounded, refractile morphology that is later lost. For HDFs, the cell cycle arrest is evident 4–5 days postinfection, while it is somewhat delayed in MEFs. Changes in morphology appear roughly at the same time as the cell cycle arrest, whereas induction of SA-ß-Gal activity is usually evident at day 6–7 postinfection (see below).

Many methods can be used to assess proliferation rates in these experiments, such as staining with crystal violet, metabolic dyes (MTT, XTT), incorporation of labeled thymidine, or FACS analysis.[11–14] There are different factors to be considered in the choice of method. We favor the use of [^{3}H] thymidine to analyze cell proliferation because of its simplicity and reliability. When a detailed analysis of cell cycle distribution is desired, a standard propidium iodide–bromodeoxyuridine (BrdU) FACS analysis can be used. Results so obtained are more informative but have the down side of requiring a much larger number of cells (10^4 for thymidine vs. 10^6 for PI–BrdU).

Thymidine Incorporation Assay

1. Plate cells in 96-well microtiter plates at a density of 2000–3000 cells per well in DMEM–10% (v/v) FCS. Plate enough wells for several measures at different time points. Use triplicates for each point.
2. For each time point, add 1 μCi of thymidine to each well and incubate for 24 hr. We use methyl[^{3}H]thymidine (2.0 mCi/mmol, 1 mCi/ml; Amersham, Arlington Heights, IL).
3. Trypsinize the cells with 1× trypsin–EDTA solution. Once the cells are detached, the plates can be frozen at −20° until use. Take care to dispose properly of all the solutions and plasticware.

It is advisable to repeat this measure several times during a 4- to 6-day period to obtain a more accurate result. Samples from the different time points should be stored frozen and counted at the same time. For counting, we use a Cell Harvester apparatus from Wallac (Gaithersburg, MD) that automatically transfers the contents of each well to a glass fiber filter, which is then counted in a scintillation counter. In a typical experiment, the thymidine incorporation of Ras-infected HDFs is always less than 10% of the control and can be as low as 1%.

Cell Morphology and Senescence-Associated ß-Galactosidase Activity

Primary cells expressing activated Ras suffer a characteristic change in morphology, which is particularly obvious for HDFs. As opposed to the thin elongated morphology of control cells, Ras-infected cells become flat, rounded, and enlarged. This morphology is reminiscent but not identical to that seen in HDFs that have reached senescence as a consequence of the accumulation of population doublings. In similar experiments using Ras effectors of the Raf–MEK [MAPK (mitogen-activated protein kinase)/ERK (extracellular signal-regulated kinase) kinase] pathway, cells adopt a small refractile morphology.[11,13,14] These differences in morphology between Ras and some of its effectors could be explained by a differential activation of Rho.

Cell morphology should be evaluated with cultures at low density. It is important that all cells to be compared are plated at the same density to avoid changes in cell morphology.

The observation that human senescent cells present a distinctive ß-galactosidase activity at neutral pH was originally made in the laboratory of J. Campisi and has become a widely used marker for senescence.[25] Most cells show a ß-galactosidase activity localized in lysosomes with an optimal acidic pH (pH 4.0). Apart from this activity, senescent cells show a characteristic ß-galactosidase activity detectable at neutral pH (pH 6.0). This activity is referred to as SA-ß-Gal (senescence-associated ß-galactosidase). The exact origin of this activity and its relevance to senescence is far from clear but it has been validated empirically as an excellent marker for senescence, because it is exclusively induced in senescent cells and not in cells arrested by serum deprivation or terminal differentiation. The validity of this marker is well characterized in human cells of different origins (fibroblasts, vein endothelial cells, mammary epithelial cells). Interestingly, in the case of mouse cells, the use of SA-ß-Gal as a marker for senescence seems to be restricted to MEFs from particular strains of mice, such as FVB.[25,26]

As mentioned, the distinct feature of SA-ß-Gal is its optimal neutral pH. Consequently, solutions should be made fresh before use and the pH should be measured as precisely as possible. Confluent cells may display a background SA-ß-Gal activity irrespective of other factors,[25] and therefore it is important that cells be at low density for staining.

1. Plate cells at low density (4×10^4 cells per well in six-well plates).
2. Incubate for 24 hr.
3. Remove the medium and wash the cells twice with PBS.
4. Fix with 2% (v/v) formaldehyde, 0.2% (v/v) glutaraldehyde in PBS, for 10–15 min at room temperature.
5. Wash twice with PBS.
6. Add staining solution (see preparation below; 1 ml per well).
7. Incubate at 37° until blue color appears. Staining can be detectable within 2 hr but usually requires 12–16 hr. Do not incubate longer than 16–18 hr. Long incubation times may result in increased background.

[25] G. P. Dimri, X. Lee, G. Basile, M. Acosta, G. Scott, C. Roskelley, E. E. Medrano, M. Liskens, I. Rubelj, O. Oliveria-Smith, M. Peacocke, and J. Campisi, *Proc. Natl. Acad. Sci. U.S.A.* **92,** 9363 (1995).

[26] J. Jacobs, K. Kieboom, S. Marino, R. A. DePinho, and M. van Lohuizen, *Nature* (*London*) **397,** 164 (1999).

8. Wash the plates with distilled water three times.
9. Inspect the stained cells under a phase-contrast microscope and estimate the percentage of blue cells. Usually, more than 60% of Ras-infected HDFs will be positive for SA-ß-Gal staining. Plates can be stored at 4° in the dark.

Staining Solution
For 20 ml of solution:

Citric acid (37 m*M*)–sodium phosphate (125 m*M*), pH 6.0	4 ml
Potassium ferrocyanide, 100 m*M*	1 ml
Potassium ferricyanide, 100 m*M*	1 ml
Sodium chloride, 5 *M*	0.6 ml
Magnesium chloride, 1 *M*	40 μl
X-Gal solution, 20 mg/ml in dimethylformamide	1 ml
Water	up to 20 ml

It is possible to replace the citric acid–sodium phosphate buffer with PBS–1 m*M* $MgCl_2$, pH 6.0.

Acknowledgments

We thank C. Pantoja and M. Barradas for critical review of the manuscript. Work in the authors laboratory is supported by the Spanish Ministry of Education, the Regional Government of Madrid, the Human Frontier Science Program, and by a core grant from the consortium between the Spanish Research Council and Pharmacia & Upjohn.

[23] Ras and Rho Protein Induction of Motility and Invasion in T47D Breast Adenocarcinoma Cells

By PATRICIA J. KEELY

Introduction

Cancer metastasis requires that cells invade through their extracellular matrix environment and migrate to distant sites *in vivo*. To understand the molecular mechanisms leading to metastasis and the role of Ras and Rho proteins in this event, it is important to have good models in which to study *in vitro* migration and invasion. Although much work has been done in fibroblastic cell lines, a large majority of cancers are epithelial carcinomas. Thus, epithelial models are needed in order to investigate the role Ras and Rho proteins play in carcinoma progression

0076-6879/00 $35.00

leading to invasion and metastasis. We describe here an *in vitro* model of breast epithelial migration and invasion.

Use of T47D Cells as Model System

We have found that T47D cells are responsive to activated Ras and Rho proteins, which promote migration and invasion of these cells. Because they are affected by Ras and Rho proteins, T47D cells represent a useful system with which to study signaling pathways related to the small GTPases, and how these pathways affect migration.

The T47D cell line was originally isolated as a well-differentiated epithelial substrain from an infiltrating ductal carcinoma. Infiltrating ductal carcinoma is the most prevalent single type of mammary carcinoma, accounting for 47–75% of all invasive breast carcinomas.[1,2] Importantly, early stages of infiltrating ductal carcinoma reiterate glandular/tubular characteristics despite having crossed the basement membrane into the collagen-rich stroma, and are graded as well to moderately differentiated partially on the basis of their tubule formation. Less differentiated, disorganized carcinomas with fewer tubule structures, more nuclear atypia, and more frequent mitotic figures are associated with poorer prognosis and increased metastasis.[1] Although T47D cells are transformed, they are well differentiated, can organize cell–cell junctions, and express estrogen receptors.[3] Consistent with this phenotype, we have found that these cells maintain the ability to polarize into tubule structures when cultured in three-dimensional collagen gels.[4] Because of their origin and phenotype, we view T47D cells as an ideal system in which to study molecules that mediate the transition of carcinomas from well differentiated to migratory and invasive.

A consideration in choosing a model system is the ease with which the cells can be cultured and manipulated. T47D cells are highly homogeneous, exhibit a classic epithelial phenotype, and are easily maintained in culture. These cells are reasonably transfectable in transient assays (20–25% efficiency is possible), and readily form stable cell lines upon selection. Importantly for this application, these cells are responsive to transfection with Ras and Rho proteins, which alter their phenotype in favor of migration and invasion.

Cell Culture

Culture medium: RPMI 1640, 10% (v/v) fetal bovine serum, insulin (0.8 μg/ml)

[1] F. A. Tavassoli, "Pathology of the Breast." Elsevier Science, New York, 1992.

[2] P. P. Rosen and D. Ernsberger, *Am. J. Surg. Pathol.* **11,** 351 (1987).

[3] H. Freake, C. Marcocci, J. Iwasaki, and I. MacIntyri, *Biochem. Biophys. Res. Commun.* **101,** 1131 (1981).

[4] P. Keely, A. Fong, M. Zutter, and S. Santoro, *J. Cell Sci.* **108,** 595 (1995).

Selection medium: Culture medium plus G418 (300 μg/ml) or hygromycin (150 μg/ml)

Cells are subcultivated by trypsinization and dilution into a fresh flask, 1 : 2–1 : 5. Typically, passage of cells diluted 1 : 3 or 1 : 4 twice per week is recommended. Dilutions greater than 1 : 6 are not well tolerated, and will result in slowed growth of the cells. We find that cell growth and morphology is best when the cells are cultured in Costar (Cambridge, MA) tissue culture flasks.

Stable Transfection. We have had success forming stable transfectants of T47D cells with wild-type or activated Ras and Rho proteins. Typically, pooled stable transfectants express the transfected protein at levels 2- to 4-fold above the endogenous protein. We have been unable to create stable cell lines expressing dominant-negative mutants of Rho proteins or of Ras, but have had success with dominant-negative R-Ras and TC21. To create stable cell lines, 3×10^6 T47D cells are seeded into a 75-cm^3 tissue culture flask and grown in complete medium overnight. The next day, cells are transfected with 20 μg of DNA, using 70 μl of LipofectAMINE (GIBCO-BRL, Gaithersburg, MD) for 5 hr, according to the manufacturer protocol. We have also had success with other cationic lipid products, including Superfect and Effectene (both from Qiagen, Valencia, CA), and LipofectAMINE Plus (GIBCO-BRL). Two days after transfection, the cells are passed into selection medium. Cells will continue to grow for several days, and then die off over a 7- to 10-day period. After this period, individual colonies of cells will begin to grow back over 3–6 weeks. Colonies can be scraped with a sterile plastic cloning loop into individual wells of a 12-well plate if clonal lines are desired. We typically pool the transfectants by trypsinizing the colonies and transferring the cells into a fresh flask. It is important to create control cell lines in parallel by transfecting with the expression plasmid that contains no insert.

Transient Transfections. Because it is difficult to create stable cell lines with growth-inhibitory proteins, such as dominant-negative Rho proteins, an alternative approach is to use transient transfection to assess the effect of these proteins on cell migration and invasion. Transient transfection also generally results in greater expression of the transfected protein, which may give a more striking cellular phenotype. The best efficiency of transient transfection we have been able to obtain with T47D cells is 20–25%. Because of this low efficiency, it is generally necessary to identify the cells that have been transfected. For this purpose, we usually cotransfect a plasmid that expresses green fluorescent protein (GFP) at a ratio of 1 : 4 with the plasmid encoding the Ras or Rho protein of interest. For transient transfections, T47D cells are seeded at a density of 3×10^6 cells in a 100-mm dish. On the next day, they are transfected with 12 μg of a DNA plasmid containing the Ras or Rho construct of interest and 4 μg of a DNA plasmid encoding GFP. This is mixed with 48 μl of FUGENE 6 (Boehringer Mannheim Biochemical, Indianapolis, IN) and incubated with the cells according to the manufacturer recommendations. The cells are used for migration assays, immunohistochemistry, or other assays 48 hr after transfection.

Migration Assays

Transwell Assays

Although there are several assays for cell migration, we most often use the Boyden chamber or transwell assay because it is easily quantified, readily adapted to a number of protocol adjustments, and of sufficient sensitivity to detect changes in cell migration of at least 1.5-fold.[4,5] In addition, several cells lines and conditions can be assayed at once. In the transwell or Boyden chamber assay, two chambers are separated by a polycarbonate membrane that contains pores of 8–12 μm in diameter. The membrane is coated with an adhesive ligand, and the cells are placed in the upper chamber and allowed to migrate to the lower chamber, where they are then fixed and counted. Because the pores are several times smaller than the cells, this process requires an active response by the cells to reach the lower surface of the membrane. The major limitation of this approach is that it is an end-point assessment of cell migration, and therefore will not give much information about membrane activity or cellular dynamics during the process of migration.

Several parameters of the assay can be altered, depending on the specific question being asked. We generally perform a haptotactic assay (immobilized gradient) in which the underside of the membrane is coated with an adhesive extracellular matrix protein, and the top of the membrane is left uncoated. This results in a step gradient in which the difference in adhesivity creates directionality. Alternatively, both sides of the membrane can be coated to assay random migration. Chemotaxis (diffusible gradient) can be assayed by adding a chemoattractant such as a growth factor to the lower chamber. Finally, chemokinesis, the effect of a factor on the overall migratory activity of a cell, can be assayed by adding a growth factor or other stimulus to both upper and lower chambers.

Migration of Stable Transfectants

Transwells (Costar 3403) (or a Boyden chamber and polycarbonate membrane with 8- to 12-μm pores)
Tissue culture dishes, 12 well
Collagen I for coating (Collaborative Biomedical Products, Bedford, MA), diluted into sterile water [other adhesive proteins may also be used, such as fibronectin or laminin diluted in sterile phosphate-buffered saline (PBS)]
RPMI plus bovine serum albumin (BSA, fraction V, fatty acid free) at 5 mg/ml
DiffQuick (Baxter, Morton Grove, IL)

[5] J. B. McCarthy, S. T. Hagen, and L. T. Furcht, *J. Cell Biol.* **102,** 179 (1986).

Permanent mounting solution (i.e., Cytoseal; Baxter)
Glass slides and coverslips for mounting

Transwells are seated into the wells of a 12-well tissue culture cluster, and the membrane coated on the underside by the addition of exactly 0.6 ml of sterile collagen I (3 μg/ml) to the lower chamber of the transwell. We have tested collagen at 0.1–300 μg/ml and found this to be the optimal concentration for migration. Alternatively, wells can be coated with other adhesive glycoproteins (i.e., fibronectin, laminin, or collagen IV at 3 μg/ml). T47D migration is greatest across collagen I or IV, and least across fibronectin. The transwells are then incubated at 37° for 4–24 hr. After this incubation, the collagen solution is removed by aspiration, and the lower chamber rinsed twice with sterile RPMI. Finally, 0.6 ml of RPMI is placed in the lower chamber, and the plate set aside until the cells are prepared. Smaller transwells can be used (Costar, 24 well). In this case, volumes are all decreased by half.

T47D cells or stable transfectants at ~70–95% confluency are detached with trypsin, washed once in RPMI–BSA, counted with a hemacytometer, and diluted to a concentration of 200,000/ml in RPMI–BSA, and 0.6 ml of this cell suspension is pipetted into the top chamber of the transwell. We perform these assays in the absence of serum, so that the only stimulus for migration is the extracellular matrix (ECM) substratum. The plate is then incubated at 37° for 5–20 hr. We typically incubate plates for 16 hr. To obtain consistent results, it is important to be accurate in counting, and careful and consistent in resuspending the cells at all times. In addition, when comparing different stable transfectants with control transfected cells, the results will be more reproducible if the cells are cultured in a consistent manner, all passed on the same day to the same confluency prior to being used for the migration assay.

The assay is terminated by wiping the top side of the membrane with a cotton swab to remove the cells that did not migrate. After a brief rinse in PBS, the migratory cells on the bottom of the membrane are fixed and stained with a Wright–Giemsa stain (DiffQuick; Baxter) for 30 sec per solution, and the top side swabbed once more. The transwells are then set aside until dry. Once completely dry, the membrane is excised with a scalpel and mounted on a glass slide with a permanent mounting solution. Cells are quantitated by counting a minimum of four microscope fields using a 10 × objective and bright-field illumination. To minimize bias, a consistent pattern of counting is established for each sample. We typically count every field across the horizontal and vertical diameters.

Because the counting is tedious, we have also tried other means of quantifying the migratory cells. To date, we have had the most success by radiolabeling the cells with $^{32}PO_4$ for 1 hr prior to the assay in phosphate-free RPMI. The assay is set up in the same manner, and terminated by swabbing off the nonmigratory cells in the top chamber and excising the membrane into a scintillation vial. We have had minimal success with colorimetric assays, because they have generally not been sensitive enough or consistent enough to detect even 3- to 4-fold differences in cell migration under these conditions.

Stable transfection of T47D cells with constitutively active versions of the Rho proteins Cdc42(12V), Rac(61L), or Rho(63L) induces a 3- to 5-fold increase in cell migration across collagen compared with cells transfected with vector alone[6] (Fig. 1A). Similarly, transfection with activated Ras proteins K-Ras(12V), N-Ras(12V), R-Ras(38V), and TC21(72L) induces a 2- to 3.5-fold increase in cell migration (Fig. 1A). We find various results when other ECM proteins are used.[6,7] For example, migration across fibronectin varies among cells transfected with Ras and Rho proteins (Fig. 1B).

Transient Migration Assay. The transwell migration assay can be used to investigate cells that have been transiently transfected, with some modification of the protocol. First, it is necessary to cotransfect the cells with a marker. We have found that GFP is ideal for this purpose. The transwells are coated and prepared as described above. Cells are transfected, and 48 hr later trypsinized and resuspended into RPMI–BSA. Cell number is determined by counting with a fluorescence microscope, and adjusted among samples to contain the same number of GFP-positive input cells for each sample, ignoring the nontransfected cells. The assay is then conducted as described above. To terminate the assay, the nonmigratory cells are swabbed off the top, and then the transwells are fixed in 4% (w/v) paraformaldehyde and mounted onto glass slides with glycerol or Vectashield (Vector Laboratories, Burlingame, CA). GFP-positive cells that have migrated are counted with a fluorescence microscope with a $10\times$ objective.

Addition of Inhibitors. To investigate signaling pathways downstream of Ras- and Rho-induced cell migration, we have performed several assays in the presence of pharmacological inhibitors. These assays are set up as described above, except that the cells are preincubated with the inhibitor at two times the final concentration in 0.6 ml of RPMI–BSA for 15 min (i.e., 50 μM LY294002 or PD98059). Cells are then plated into the top chamber of the transwell. Because the bottom chamber is 0.6 ml in volume, the final concentration of the inhibitor becomes $1\times$. When inhibitors are used, we incubate the assay for only 5–6 hr before termination.

We find that the Ras and Rho proteins show different sensitivities to pharmacological inhibitors. For example, LY294002, an inhibitor of phosphatidylinositol 3-kinase (PI3K), completely inhibits migration induced by Cdc42(12V) or Rac(61L), but only partially inhibits migration induced by R-Ras(38V) (Fig. 2). An inhibitor of MEK [mitogen-activated protein kinase (MAPK)/extracellular

[6] P. J. Keely, J. K. Westwick, I. P. Whitehead, C. J. Der, and L. V. Parise, *Nature (London)* **390,** 632 (1997).

[7] P. J. Keely, E. V. Rusyn, A. D. Cox, and L. V. Parise, *J. Cell Biol.* **145,** 1077 (1999).

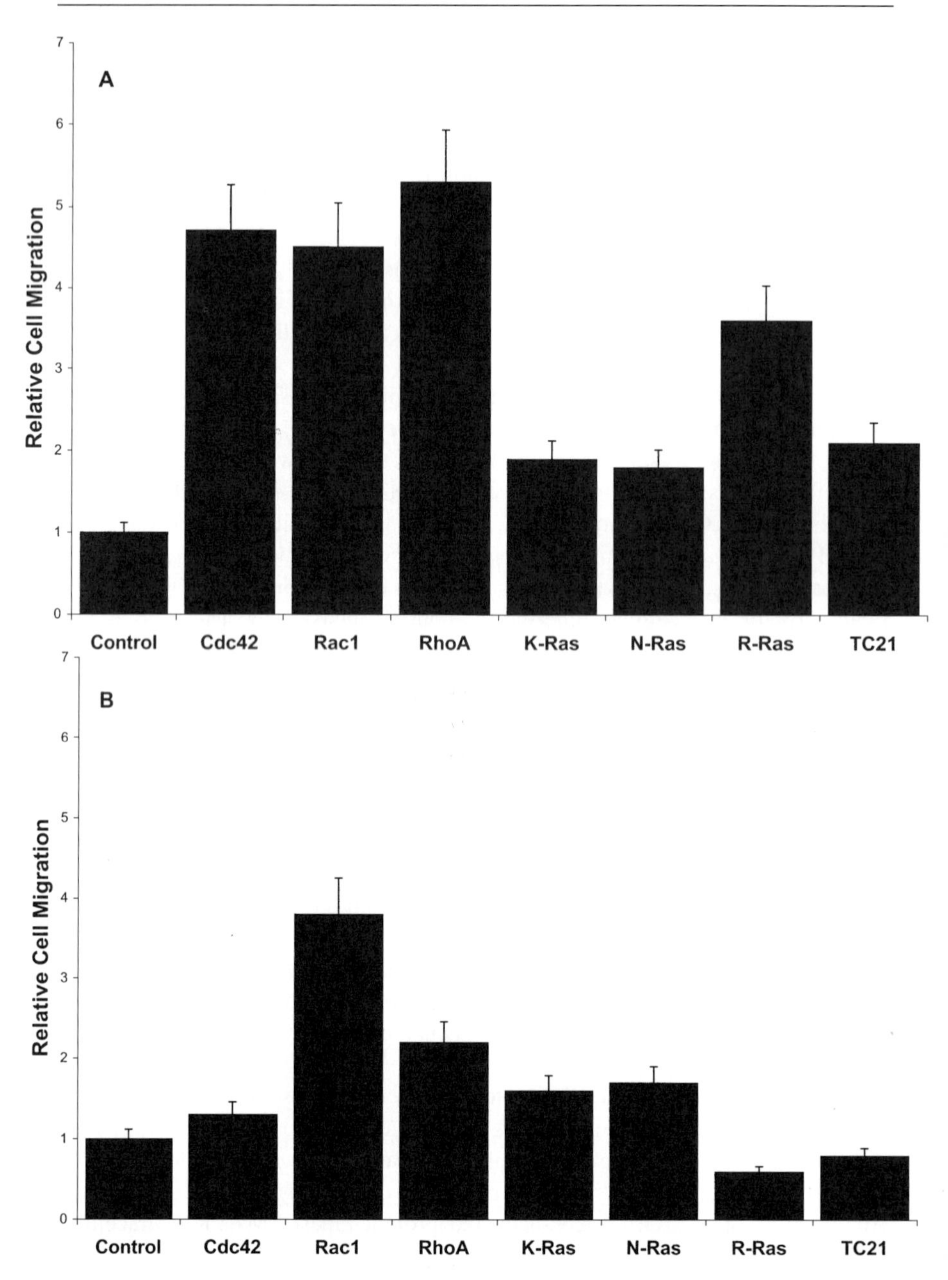

FIG. 1. Migration of T47D cells stably transfected with activated Rho and Ras proteins. Cells expressing empty vector (control) Cdc42(12V), Rac1(61L), RhoA(63L), K-Ras(12V), N-Ras(12V), R-Ras(38V), or TC21(72L) were assayed for their ability to migrate across transwells coated with collagen (A) or fibronectin (B). Cell migration is expressed relative to control migration ±SE.

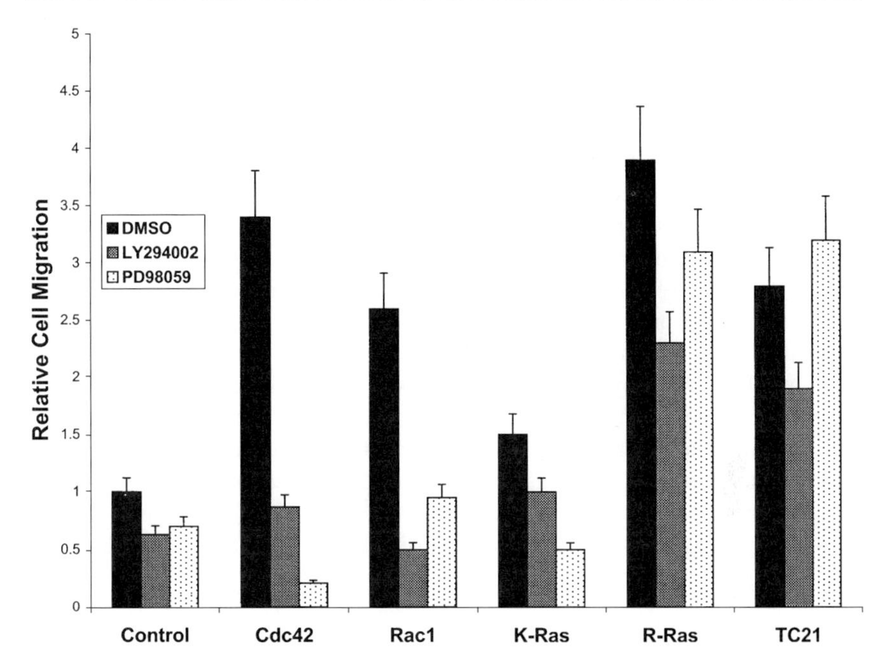

FIG. 2. Migration of T47D cells transfected with activated Rho and Ras proteins is differentially affected by pharmacological inhibitors. Cells were incubated with 25 μM LY294002, an inhibitor of PI3K, or 25 μM PD98059, an inhibitor of MEK. Migration was then assayed across collagen. Cell migration is expressed relative to control migration $\pm$SE.

signal-regulated kinase (ERK) kinase] PD98059,[8] inhibits migration induced by Cdc42(12V) and K-Ras(12V), but not by R-Ras(38V) (Fig. 2). Thus, we find that different Ras and Rho proteins induce migration via different downstream signaling pathways.

Other Assays of Cell Migration

Cell migration can be studied using other types of assays. "Wound" assays are performed by growing cells to a confluent monolayer, and then aspirating or scraping off a tract of cells to create an *in vitro* wound. Within a few hours, cells at the edge of the wound acquire a migratory phenotype, reorganize their actin cytoskeletons, and begin to migrate into the newly open space.[9] Wound assays are useful when performed on glass coverslips for the purpose of immunofluorescent

[8] D. T. Dudley, L. Pang, S. J. Decker, A. J. Bridges, and A. R. Saltiel, *Proc. Natl. Acad. Sci. U.S.A.* **92,** 7686 (1995).

[9] A. Huttenlocher, M. Lakonishok, M. Kinder, S. Wu, T. Truong, K. A. Knudsen, and A. F. Horwitz, *J. Cell Biol.* **141,** 515 (1998).

localization of proteins of interest. These assays are more difficult to quantitate than the transwell assay and are usually measured in terms of number of hours or days to "close" the wound. Although these assays are useful for a variety of cell types, we have found for T47D cells that wound assays are not quantitative enough to detect the 2- to 5-fold increased migration we see in cells expressing Ras and Rho proteins. Another consideration is that it is more difficult to control the properties of the substratum in these assays, because the cells are migrating onto substratum that may have been modified by cells previously attached. A modification of this protocol is to seed cells into the center of a cloning cylinder on a substratum coated with an adhesive protein, remove the cylinder, and let the cells migrate onto the new substratum that surrounds them.

Time-lapse video microscopy has been used by many investigators to study cell migration.[10] The advantage of video microscopy is that it is possible to obtain a kinetic and morphological assessment of the events occurring during cell migration. This approach is useful when assaying cells that have been microinjected or transiently transfected, because one can focus on only the cells of interest. However, video microscopy is tedious for studies in which many constructs or inhibitors are to be compared, and requires more extensive equipment to perform.

Invasion Assays

To invade, cells must migrate not only across a two-dimensional surface, but through a three-dimensional extracellular matrix. It is generally accepted that this requires cells not only to migrate, but also to proteolyze or alter their local ECM microenvironment.

Transwell Invasion Assay

To assay cell invasion *in vitro,* we use a modified transwell assay, in which a three-dimensional collagen gel has been placed in the upper chamber.[11] Transwells are coated from the underside with collagen, as described above. After 4–24 hr, the collagen solution is removed by aspiration, and the lower chamber is washed twice with sterile water. The transwell membrane is then allowed to dry completely. This is important so that surface tension will hold the unpolymerized collagen gel on the top; if the membrane is wet, the solution will leak through the pores of the membrane into the lower chamber before it has gelled. To make the collagen gel, a solution of collagen I (rat tail collagen; Collaborative) is neutralized by the addition of an equal volume of 2× PBS, 100 m*M* HEPES (pH 7.3), and a sufficient volume

[10] J. Lee, A. Ishihara, G. Oxford, B. Johnson, and K. Jacobson, *Nature* (*London*) **400,** 382 (1999).

[11] S. A. Santoro, M. M. Zutter, J. E. Wu, W. D. Staatz, E. U. M. Saelman, and P. J. Keely, *Methods Enzymol.* **245,** 147 (1994).

of RPMI to create a final collagen concentration of 1.3 mg/ml. This mixture is kept on ice until it is used. To cast the gels, 0.25 ml of the neutralized collagen is pipetted into the top chamber of the transwells and incubated at 37° for about 30 min until a firm gel has formed. An alternative to using collagen to form a three-dimensional matrix is to use Matrigel, a mixture of basement membrane proteins (collagen IV, laminin, proteoglycan, entactin, and other proteins). Matrigel should be gelled according to the manufacturer instructions and substituted for collagen in this assay if desired. One can also purchase transwell plates that are precoated with Matrigel, and need only be rehydrated for an invasion assay.

The cells for the transwell invasion assay are serum starved overnight by replacing the medium with RPMI. The next day, cells are released by trypsinization, washed in RPMI plus BSA (5 mg/ml), and diluted to a final concentration of 2×10^6/ml. Medium containing serum or another selected growth factor is placed into the lower chamber (0.6 ml), and 0.5 ml of cells is placed into the upper chamber. The cells are serum starved so that they will be more responsive to the presence of serum in the lower chamber, which is present as a chemoattractant to promote invasion directionally to the underside of the filter.

The invasion assay is incubated at 37° for 24 hr, and then terminated in the same manner as the migration assay. The collagen gel is removed, and the membranes are fixed, stained, and mounted. Parent T47D cells are minimally invasive, and it may be expected that less than 0.1% of the cells will invade. However, we find that expression of activated Cdc42, Rac, Rho, R-Ras, TC21, or K-Ras in these cells will significantly induce invasion through a collagen gel compared with control cells (Fig. 3).

Alternative Invasion Assays

Additional means to assay invasion *in vitro* also exist. Rather than monitor invasion across a transwell, it is possible to cast a gel into a tissue culture well, seed the cells on top, and optically monitor invasion into the gel. Cells invading into the gel are counted at successive 20 μm levels, using an inverted phase-contrast microscope.[12] Because this can be performed every day for several days, it is possible to obtain a kinetic assessment of invasion through the gel. However, this approach does not readily allow the addition of a chemoattractant on the other side of the gel. An alternative invasion assay is a Matrigel outgrowth assay. Cells are cultured in Matrigel for 2–10 days, and monitored for the presence of outwardly migrating cells.[13] Like the wound assay, this is a qualitative rather than quantitative approach.

[12] A. E. Faassen, J. A. Schrager, D. J. Klein, T. R. Oegema, J. R. Couchman, and J. B. McCarthy, *J. Cell Biol.* **116,** 521 (1992).

[13] S. N. Bae, G. Arand, H. Azzam, P. Pavasant, J. Torri, T. L. Frandsen, and E. W. Thompson, *Breast Cancer Res. Treat.* **24,** 241 (1993).

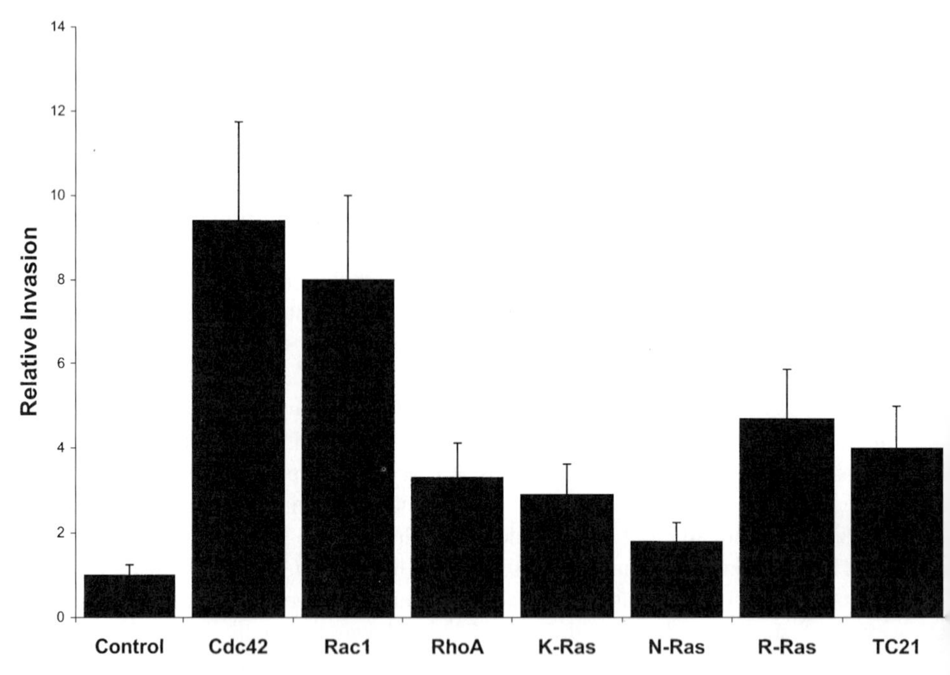

FIG. 3. Invasion of T47D cells through a collagen gel. Cells were stably transfected with empty vector (control), Cdc42(12V), Rac1(61L), RhoA(63L), K-Ras(12V), N-Ras(12V), R-Ras(38V), or TC21(72L), and invasion assayed as described. Cell invasion is expressed relative to invasion with control cells ±SE.

Summary

There has been much interest in how Ras and Rho proteins affect cell migration and invasion. Studies of the molecular mechanisms and signaling pathways that are involved will be aided by the development and use of relevant model systems and simple *in vitro* assays. While a complete understanding of metastasis will ultimately need to employ *in vivo* studies, the migration and invasion assays presented here are good initial assessments of events that are relevant to the metastatic cascade.

[24] Ras Regulation of Vascular Endothelial Growth Factor and Angiogenesis

By JANUSZ RAK and ROBERT S. KERBEL

Introduction: Linkage between Ras and Tumor Angiogenesis

There are several reasons why Ras oncoproteins represent an interesting case for studying the impact of cancer-associated genetic mutations and tumor angiogenesis. Thus, three existing *ras* genes (H-*ras,* N-*ras,* and K-*ras*) encoding four protein isoforms (respectively, H-, N-, and K-RasA and K-RasB) are frequent targets for activating mutations in human cancer.[1] In colorectal adenoma K-*ras* mutations are found at relatively early stages of tumor progression[2] and their frequency strongly correlates with the ability of adenomas to form large polypoid outgrowths.[3,4] Such unlimited, three-dimensional expansion of the tumor would be expected to depend strictly on formation of new blood vessels from a preexisting vasculature (angiogenesis).[5] Indeed, in many instances the presence of mutant Ras oncoproteins has been formally shown to be essential for continued tumor growth *in vivo.*[6–8] Furthermore, even in the absence of an overt *ras* mutation the biochemical activity of Ras protein is often chronically elevated in cancer cells, and thus it can contribute to induction and maintenance of their transformed phenotype.[9] Such a permanent increase in Ras activity (and GTP binding) is thought to result from upregulation, constitutive activation, or mutation of oncogenic upstream signaling molecules such as members of the epidermal growth factor receptor (EGFR) or platelet growth factor families.[9,10] Activated Ras is capable of triggering several

[1] R. Khosravi-Far and C. J. Der, *Cancer Metastasis Rev.* **13,** 67 (1994).

[2] E. R. Fearon and B. Vogelstein, *Cell* **61,** 759 (1990).

[3] T. Minamoto, K. Sawaguchi, M. Mai, N. Yamashita, T. Sugimura, and H. Esumi, *Cancer Res.* **54,** 2841 (1994).

[4] H. Hasegawa, M. Ueda, M. Watanabe, T. Teramoto, M. Mukai, and M. Kitajima, *Oncogene* **10,** 1413 (1995).

[5] J. Folkman, *J. Natl. Cancer Inst.* **82,** 4 (1990).

[6] S. Shirasawa, M. Furuse, N. Yokoyama, and T. Sasazuki, *Science* **260,** 85 (1993).

[7] L. Chin, A. Tam, J. Pomerantz, M. Wong, J. Holash, N. Bardeesy, Q. Shen, R. O'Hagan, J. Pantginis, H. Zhou, J. W. Horner, C. Cordon-Cardo, G. D. Yancopoulos, and R. A. DePinho, *Nature* (*London*) **400,** 468 (1999).

[8] N. E. Kohl, C. A. Omer, M. W. Conner, N. J. Anthony, J. P. Davide, S. J. deSolms, E. A. Giuliani, R. P. Gomez, S. L. Graham, K. Hamilton, L. K. Handt, G. D. Hartman, K. S. Koblan, A. M. Kral, P. J. Miller, S. D. Mosser, T. J. O'Neill, E. Rands, M. D. Schaber, J. B. Gibbs, and A. Oliff, *Nat. Med.* **1,** 792 (1995).

[9] A. Guha, M. M. Feldkamp, N. Lau, G. Boss, and A. Pawson, *Oncogene* **15,** 2755 (1997).

[10] T. Hunter, *Cell* **88,** 333 (1997).

METHODS IN ENZYMOLOGY, VOL. 333
0076-6879/00 $35.00

crucial signaling cascades[11] and in doing so can alter expression of a large numbers of Ras-responsive genes,[12,13] many of which could be relevant for triggering or contributing to tumor angiogenesis.

In addition to the induction of proangiogenic properties in affected tumor cells, activated Ras proteins may be involved in tumor–blood vessel interactions at several other levels. We discuss how, through a direct influence on tumor cell growth and survival requirements, *ras* mutations can decrease the relative vascular dependence of such cells.[14] Also, activation of wild-type Ras proteins in endothelial cells of the host stroma may play an important role in their responses to angiogenic stimuli.[15] In this context the possibility that inhibitors of Ras and/or other signal-transducing modules may act as antiangiogenic agents would appear to have a strong conceptual basis.

Materials and Methods

Experimental Models of ras Transformation in Epithelial Cells and Fibroblasts

Immortalized and nontumorigenic rat intestinal epithelial cell line (IEC-18) and NIH 3T3 fibroblasts are used as recipients of a mutated H-*ras* expression vector, as described previously.[16–18] This leads to the establishment of several model systems comprising stable *ras*-transformed clones (RAS-3, RAS-4, RAS-7) of IEC-18 cells, a pooled population of more than 100 clones derived from NIH 3T3 cells (3T3ras), transfectants of IEC-18 cells expressing mutant H-*ras* under the control of the Moloney murine leukemia virus (Mo-MuLV), metallothionein, or tetracycline-repressible promoters.[16,17,19] In addition, human colorectal cancer cell lines (DLD-1 and HCT116) harboring a mutant K-*ras* allele and their respective, poorly tumorigenic variants in which the oncogenic K-*ras* allele has been genetically disrupted[6] are included in the analysis. All cell lines are grown under conditions described previously.[18,19]

[11] R. Khosravi-Far, S. Campbell, K. L. Rossman, and C. J. Der, *Adv. Cancer Res.* **72,** 57 (1998).

[12] A. F. Chambers and A. B. Tuck, *Crit. Rev. Oncog.* **4,** 95 (1993).

[13] D. M. Bortner, S. J. Langer, and M. C. Ostrowski, *Crit. Rev. Oncog.* **4,** 137 (1993).

[14] J. Rak and R. S. Kerbel, *Cancer Metastasis Rev.* **15,** 231 (1996).

[15] M. Henkemeyer, D. J. Rossi, D. P. Holmyard, M. C. Puri, G. Mbamalu, K. Harpal, T. S. Shih, T. Jacks, and T. Pawson, *Nature (London)* **377,** 695 (1995).

[16] A. Quaroni and K. J. Isselbacher, *J. Natl. Cancer Inst.* **67,** 1353 (1981).

[17] R. N. Buick, J. Filmus, and A. Quaroni, *Exp. Cell Res.* **170,** 300 (1987).

[18] J. Rak, Y. Mitsuhashi, C. Sheehan, A. Tamir, A. M. Viloria-Petit, J. Filmus, S. J. Mansour, N. G. Ahn, and R. S. Kerbel, *Cancer Res.* **60,** 490 (2000).

[19] J. Rak, Y. Mitsuhashi, L. Bayko, J. Filmus, T. Sasazuki, and R. S. Kerbel, *Cancer Res.* **55,** 4575 (1995).

Analysis of Global Angiogenic Phenotype Induced by ras Transformation

Evaluation of angiogenic properties associated with *ras* transformation is based on several criteria, including (1) ability to stimulate growth and survival of growth factor-starved primary endothelial cells, (2) deregulated expression of selected angiogenesis inhibitors and stimulators (at the protein and/or mRNA level), (3) formation of vascularized tumors in mice, and (4) responsiveness of such tumors to angiogenesis inhibitors *in vivo.*

An endothelial cell growth/survival assay is conducted as follows: Nontransformed (IEC-18, NIH 3T3) cells, or their *ras*-transformed counterparts, are cultured to approximately 70% confluence as monolayers in their respective growth media. Fresh medium is incubated (conditioned) with such cultures for 24–72 hr, collected, spun at high speed (15,000 rpm), filtered through 0.22-μm pore size mesh, and added (1 : 1 dilution) to cultures (96-well plates) of human umbilical endothelial cells (HUVECs) or human dermal microvascular endothelial cells (HDMECs). Prior to addition of the conditioned medium, the endothelial growth medium [MCDB131 : basic fibroblast growth factor (bFGF, 5 ng/ml), heparin (10 U/ml), 10% (v/v) fetal bovine serum (FBS)] is removed and the cells are washed with growth factor-free assay medium [ExCell medium (J. R. H. Biosciences, Lenexa, KS) or Dulbecco's modified Eagle's medium (DMEM) with 5% (v/v) FBS]. Treatment is applied for 72 hr, at which point endothelial cell viability and mitotic activity in control groups are profoundly impaired, with morphological evidence of apoptosis. Against this background, the response of endothelial cells to conditioned medium is assayed by using MTS or [^{3}H]thymidine incorporation, as described previously.[20] Briefly, for the MTS assay, 3-(4,5-dimethylthiazol-2-yl)-5-(3-carboxymethoxyphenyl)-2-(4-sulfofenyl)-2*H*-tetrazolium substrate (MTS) and phenazine methanosulfate (PMS), purchased from Promega (Madison, WI), are mixed with Hanks' balanced salt solution (HBSS) at a ratio of 300 : 15 : 700 (v/v/v) and 100 μl/well of this mixture is added to endothelial cell cultures for the last 2–4 hr of the experiment. The color reaction reflecting cumulative metabolic activity (viability) of endothelial cells in the presence of conditioned or control medium is quantified as absorbance at 490 nm. Alternatively, the rate of DNA synthesis by conditioned medium-treated endothelial cells is assessed by incorporation of [^{3}H]thymidine (Amersham, Arlington Heights, IL). Briefly, 200 μCi of [^{3}H]thymidine is diluted in 5 ml of assay medium and added at 50 μl/well to cultures of endothelial cells for last 2–4 hr of the treatment. The incorporated radioactivity is quantified by using BetaPlate (Pharmacia, Piscataway, NJ) scintillation.[19]

[20] J. Rak, Y. Mitsuhashi, C. Sheehan, J. K. Krestow, V. A. Florenes, J. Filmus, and R. S. Kerbel, *Neoplasia* **1,** 23 (1999).

Analysis of Impact of Mutant Ras on Expression of Vascular Endothelial Growth Factor/Vascular Permeability Factor and Thrombospondin 1

Detection of either vascular endothelial growth factor (VEGF) or thrombospondin 1 (TSP-1) mRNA is conducted as described previously.[18] Testing for VEGF protein secreted into the conditioned medium is conducted by using a Quantikine enzyme-linked immunosorbent assay (ELISA) (R & D Systems, Minneapolis, MN) specific for either mouse or human VEGF. The mouse-specific assay detects rat but not human VEGF protein.[18]

Analysis of Molecular Mediators of Ras-Dependent Upregulation of Vascular Endothelial Growth Factor

Various *ras*-transformed epithelial or fibroblastic cell lines are treated with pharmacological signal transduction inhibitors at the concentrations indicated and in the presence of 1% (v/v) FBS, after which levels of VEGF mRNA or protein are assayed.[18]

Assessment of Tumor Growth, Metastasis, and Angiogenic Capacity

ras-transformed or control cell lines are injected subcutaneously into nude or severe combined immunodeficient (SCID) mice (Charles River, Wilmington, MA) as a single-cell suspension ($1–10 \times 10^6$ cells/0.2 ml per injection). Tumor growth is monitored by measurements of tumor size with a Vernier caliper. The presence and numbers of macroscopic metastases are determined by autopsy. Microscopic metastases in target organs (lungs) are detected by clonogenic colony formation assay (CFA) as described previously.[21,22] Vascularity of tumors derived from various *ras*-transformed cell lines is assessed[23] by staining of tumor tissue for endothelial cell-specific markers. Antibodies against von Willebrand factor (factor VM) (Dako, Carpinteria, CA), mouse platelet endothelial cell adhesion molecule 1 (PECAM-1)/CD31 (PharMingen, San Diego, CA), or *Griffonia simplicifolia* isolectin B4 (Serotec, Raleigh, NC) are employed for this purpose.

Assessment of Vascular Dependence of Various ras-Transformed Cell Lines

Mixtures of RAS-3 and RAS-7 cells are injected into nude mice, which results in formation of heterogeneous tumors. Mice are then injected with a fluorescent DNA-binding dye (Hoechst 33342) and killed within 15 min in order to create a

[21] C. J. Aslakson, J. W. Rak, B. E. Miller, and F. R. Miller, *Int. J. Cancer* **47,** 466 (1991).

[22] J. Rak, J. Filmus, and R. S. Kerbel, *Eur. J. Cancer* **32A,** 2438 (1996).

[23] A. M. Viloria-Petit, J. Rak, M.-C. Hung, P. Rockwell, N. Goldstein, and R. S. Kerbel, *Am. J. Pathol.* **151,** 1523 (1997).

perivascular gradient of staining.[24] Tumors are removed and enzymatically dissociated into a single-cell suspension, from which 10% of the "brightest" and 10% of the "dimmest" cells are sorted out by using a Coulter (Hialeah, FL) Epics V Elite fluorescence cell sorter. These cell populations represent the areas of the tumor that are either proximal or distal to perfused tumor blood vessels, respectively. The contribution of RAS-3 or RAS-7 cells to these "geographically" defined cell populations is then analyzed by using a quantitative colony formation assay in selective medium [for RAS-3 cells, –G418 plus hypoxanthine–aminopterin–thymidine (HAT); for RAS-7 cells, –thioguanine plus G418).[22]

Results and Discussion

Angiogenic "Switch" in Cells Harboring Mutant ras Oncogene: Upregulation of Vascular Endothelial Growth Factor/Vascular Permeability Factor

Constitutive expression of a mutant H-*ras* oncogene in rat intestinal epithelial cells (IEC-18) leads, apparently in one step, to the acquisition of an overt tumor-forming ability *in vivo*.[17,19] Because such tumors are able to grow beyond sizes achievable under avascular conditions,[5] we surmised that *ras* transformation itself must somehow trigger an angiogenic "switch."[19] Indeed, conditioned medium of *ras*-transformed cells contained an activity that was capable of sustaining the growth and survival of primary human endothelial cells (HUVECs).[19] An important component of this proangiogenic activity was subsequently identified as VEGF on the basis of the following criteria: (1) the activity could be neutralized by anti-VEGF antibodies[19]; (2) the conditioned medium of *ras*-transformed cells was able to activate VEGF receptor 2/flk-1[25]; (3) *ras* transformation strictly coincided with upregulation of immunodetectable VEGF protein and mRNA[18,19,25]; (4) transient expression of mutant H-*ras* in IEC-18 cells under the control of a metallothionein- or tetracycline-inducible promoter led to a corresponding increase in VEGF levels (Fig. 1[19]); and, in an independent set of experiments, (5) genetic disruption of the single mutant K-*ras* allele in two human colorectal cancer cell lines[6] led to a drastic reduction in tumor-forming capacity and to a severalfold downregulation of VEGF.[19] In addition, pharmacological inhibition of Ras protein activity by treatment with farnesyl protein transferase inhibitors (FTIs) was able to attenuate the oncogenic *ras*-dependent increase in VEGF production in various experimental systems (see Table IV). In one such study, human astrocytoma cells were found to express constitutively activated Ras protein even in the absence of the gene mutations.[9] Transfection of a dominant-negative mutant of *ras* (N17 *ras*) or treatment with Ras FTIs led in this case to a marked downregulation of

[24] P. L. Olive, D. J. Chaplin, and R. E. Durand, *Br. J. Cancer* **52,** 739 (1985).

[25] S. Grugel, G. Finkenzeller, K. Weindel, B. Barleon, and D. Marme, *J. Biol. Chem.* **270,** 25915 (1995).

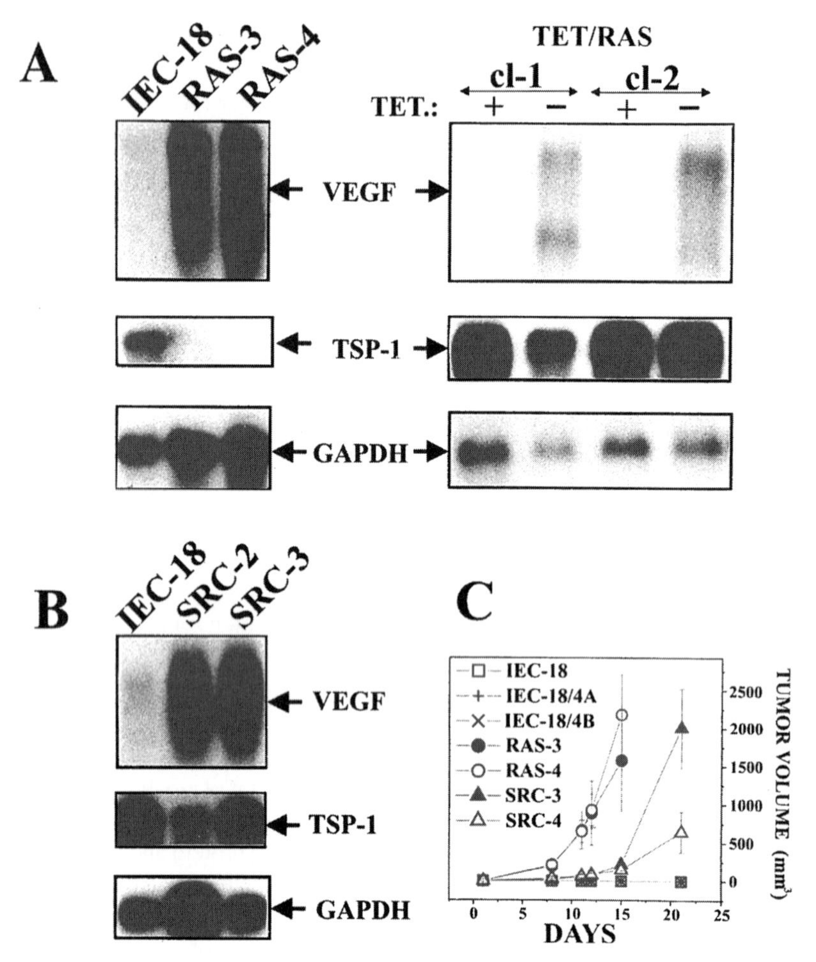

FIG. 1. Impact of mutant oncogenes on expression of VEGF and TSP-1 in epithelial cells. (A) Upregulation of VEGF and downregulation of TSP-1 at the mRNA level by constitutive expression of mutant H-*ras* in two independent clones of IEC-18 cells (RAS-3, RAS-4). Right panel—Expression of VEGF and TSP-1 in two independent clones of IEC-18 cells engineered to transiently express mutant H-*ras* under the control of a tetracycline-repressible promoter. Withdrawal of the drug (tet–) for 48 hr leads to expression of VEGF but has little impact on levels of TSP-1 mRNA. (B) Upregulation of VEGF in clones of IEC-18 cells constitutively overexpressing v-*src* (SRC-3, SRC-2). Transformation with v-*src* has little impact on TSP-1 mRNA levels. (D) Relative tumorigenic capacity of H-*ras*- and v-*src*-transformed sublines of IEC-18 cells (on injection of 2×10^6 cells subcutaneously). Parental IEC-18 cells or their nontransformed clones (IEC-18/4A, IEC-18/4B) are nontumorigenic while their counterparts expressing the respective oncoproteins grow rapidly in mice. H-*ras* transfectants appear to be somewhat more tumorigenic, possibly because they do not retain expression of TSP-1.

VEGF mRNA and protein.[26,27] To date, upregulation of VEGF in cells expressing activated Ras oncoprotein has been demonstrated with a remarkable consistency in a large number of diverse experimental models (Table I[28–37]).

Vascular Endothelial Growth Factor as Necessary Mediator of Ras-Dependent Tumor Growth in Vivo

There are reasons to support the contention that endogenous production of VEGF by cancer cells is, at least in some instances, absolutely required for tumor growth and angiogenesis. This notion is supported by several lines of evidence, including (1) the generality of VEGF upregulation by various types of tumor cells,[38] (2) the tumor-suppressing effect of enforced VEGF downregulation by using antisense approaches[39–45] or gene knockout strategies,[46] and (3) the

[26] M. M. Feldkamp, N. Lau, J. Rak, R. S. Kerbel, and A. Guha, *Int. J. Cancer* **81,** 118 (1999).

[27] M. M. Feldkamp, N. Lau, and A. Guha, *Oncogene* **18,** 7514 (1999).

[28] F. Larcher, A. I. Robles, H. Duran, R. Murillas, M. Quintanilla, A. Cano, C. J. Conti, and J. L. Jorcano, *Cancer Res.* **56,** 5391 (1996).

[29] N. M. Mazure, E. Y. Chen, P. Yeh, K. R. Laderoute, and A. J. Giaccia, *Cancer Res.* **56,** 3436 (1996).

[30] J. L. Arbiser, M. A. Moses, C. A. Fernandez, N. Ghiso, Y. Cao, N. Klauber, D. Frank, M. Brownlee, E. Flynn, S. Parangi, H. R. Byers, and J. Folkman, *Proc. Natl. Acad. Sci. U.S.A.* **94,** 861 (1997).

[31] O.V. Volpert, K. M. Dameron, and N. Bouck, *Oncogene* **14,** 1495 (1997).

[32] B. Enholm, K. Paavonen, A. Ristimaki, V. Kumar, Y. Gunji, J. Klefstrom, L. Kivinen, M. Laiho, B. Olofsson, V. Joukov, U. Eriksson, and K. Alitalo, *Oncogene* **14,** 2475 (1997).

[33] F. C. White, A. Benehacene, J. S. Scheele, and M. Kamps, *Growth Factors* **14,** 199 (1997).

[34] K. L. Tober, R. E. Cannon, J. W. Spalding, T. M. Oberyszyn, M. L. Parrett, A. I. Rackoff, A. S. Oberyszyn, R. W. Tennant, and F. M. Robertson, *Biochem. Biophys. Res. Commun.* **247,** 644 (1998).

[35] M. Przybyszewska, J. Miloszewska, and P. Janik, *Cancer Lett.* **131,** 157 (1998).

[36] W. Feleszko, E. Z. Balkowiec, E. Sieberth, M. Marczak, A. Dabrowska, A. Giermasz, A. Czajka, and M. Jakobisiak, *Int. J. Cancer* **81,** 560 (1999).

[37] S. Charvat, M. Duchesne, P. Parvaz, M. C. Chignol, D. Schmitt, and M. Serres, *Anticancer Res.* **19,** 557 (1999).

[38] L. F. Brown, M. Detmar, K. Claffey, J. A. Nagy, D. Feng, A. M. Dvorak, and H. F. Dvorak, *in* "Regulation of Angiogenesis" (I. D. Goldberg and E. M. Rosen, eds.), p. 233. Birkhauser Verlag, Basel, Switzerland 1997.

[39] M. Saleh, S. A. Stacker, and A. F. Wilks, *Cancer Res.* **56,** 393 (1996).

[40] L. M. Ellis, W. Liu, and M. Wilson, *Surgery* **120,** 871 (1996).

[41] S. Y. Cheng, H. J. Huang, M. Nagane, X. D. Ji, D. Wang, C. C. Shih, W. Arap, C. M. Huang, and W. K. Cavenee, *Proc. Natl. Acad. Sci. U.S.A.* **93,** 8502 (1996).

[42] K. P. Claffey, L. F. Brown, L. F. del Aguila, K. Tognazzi, K.-T. Yeo, E. J. Manseau, and H. F. Dvorak, *Cancer Res.* **56,** 172 (1996).

[43] T. Oku, J. G. Tjuvajev, T. Miyagawa, T. Sasajima, A. Joshi, R. Joshi, R. Finn, K. P. Claffey, and R. G. Blasberg, *Cancer Res.* **58,** 4185 (1998).

[44] J. T. Nguyen, P. Wu, M. E. Clouse, L. Hlatky, and E. F. Terwilliger, *Cancer Res.* **58,** 5673 (1998).

[45] S. A. Im, C. Gomez-Manzano, J. Fueyo, T. J. Liu, L. D. Ke, J. S. Kim, H. Y. Lee, P. A. Steck, A. P. Kyritsis, and W. K. Yung, *Cancer Res.* **59,** 895 (1999).

[46] N. Ferrara, K. Carver-Moore, H. Chen, M. Dowd, L. Lu, K. S. O'Shea, L. Powell-Braxton, K. J. Hillan, and M. W. Moore, *Nature (London)* **380,** 439 (1996).

TABLE I
LINKAGE BETWEEN EXPRESSION OF MUTANT *ras* ONCOGENE AND UPREGULATION OF VEGF

Cells/system	Level of analysis	Ref.
H- and K-*ras*-transformed intestinal epithelial cells	mRNA, protein, bioactivity	Rak *et al.* (1995)[19]
v-*ras*-transformed NIH 3T3 cells	mRNA, bioactivity	Grugel *et al.* (1995)[25]
H-*ras* in mouse squamous cell carcinomas	mRNA	Larcher *et al.* (1996)[28]
H-*ras* in NIH 3T3 cells	mRNA, VEGF promoter activity	Mazure *et al.* (1996)[29]
H-*ras*-transformed mouse endothelial cells	mRNA	Arbiser *et al.* (1997)[30]
H-*ras* in Li-Fraumeni $p53^{-/-}$ human fibroblasts	mRNA, protein	Volpert *et al.* (1997)[31]
v-H-*ras* in IMR-90 human lung fibroblasts	mRNA	Enholm *et al.* (1997)[32]
v-H-*ras* in NIH 3T3 cells	mRNA level and stability	White *et al.* (1997)[33]
v-H-*ras* transgene in Tg.AC mice	mRNA level and splicing	Tober *et al.* (1998)[34]
v-H-*ras* in HCV-29 human urothelial cells	mRNA, protein	Przybyszewska *et al.* (1998)[35]
v-H-*ras* in NIH 3T3 cells	Protein	Feleszko *et al.* (1999)[36]
V12G mutant of c-H-*ras* in HaCaT human keratinocytes	Protein	Charvat *et al.* (1999)[37]
V12G mutant of c-H-*ras*-inducible transgene in $INK4a^{-/-}$ mouse melanocytes	Protein	Chin *et al.* (1999)[7]
Constitutively activated wild-type *ras* in human astrocytoma cells	mRNA, protein	Feldkamp *et al.* (1999)[29]

therapeutic efficacy of anti-VEGF antibodies,[47–49] soluble VEGF receptors,[49–52] or VEGF receptor inhibitors.[49,53–57]

In this context it is of interest to know whether VEGF is also indispensable for tumor formation under the influence of mutant *ras*. To address this issue two independent, mutant K-*ras*-positive human colorectal carcinoma cell lines (DLD-1 and HCT116) were subjected to transfection with the $VEGF_{121}$ antisense cDNA.

[47] K. J. Kim, B. Li, J. Winer, M. Armanini, N. Gillett, H. S. Phillips, and N. Ferrera, *Nature (London)* **362,** 841 (1993).

[48] R. S. Warren, H. Yuan, M. R. Mati, N. A. Gillett, and N. Ferrara, *J. Clin. Invest.* **95,** 1789 (1995).

[49] N. Ferrara and T. Davis-Smyth, *Endocr. Rev.* **18,** 4 (1997).

[50] C. K. Goldman, R. L. Kendall, G. Cabrera, L. Soroceanu, Y. Heike, G. Y. Gillespie, G. P. Siegal, X. Mao, A. J. Bett, W. R. Huckle, K. A. Thomas, and D. T. Curiel, *Proc. Natl. Acad. Sci. U.S.A.* **95,** 8795 (1998).

[51] P. Lin, S. Sankar, S. Shan, M. W. Dewhirst, P. J. Polverini, T. Q. Quinn, and K. G. Peters, *Cell Growth Differ.* **9,** 49 (1998).

[52] G. Siemeister, M. Schirner, K. Weindel, P. Reusch, A. Menrad, D. Marme, and G. Martiny-Baron, *Cancer Res.* **59,** 3185 (1999).

TABLE II
EVIDENCE SUPPORTING UPREGULATION OF VEGF AS NECESSARY COMPONENT FOR *ras*-DEPENDENT TUMOR FORMATION *in Vivo*

Approach	Observation	Ref.
VEGF antisense transfection	Downregulation of VEGF abrogates K-*ras*-dependent tumorigenicity of human colorectal cancer cells DLD-1 and HCT116	Okada *et al.* (1998)[58]
VEGF gene knockout	In VEGF$^{-/-}$ ES cells transfection of mutant *ras* does not rescue the tumorigenic phenotype	Shi and Ferrara (1999)[59]
	v-H-*ras*-transformed VEGF$^{-/-}$ mouse embryo fibroblasts are tumorigenically deficient	Grunstein *et al.* (1999)[60]

The resulting 3- to 4-fold reduction in VEGF levels was similar in magnitude to what had previously been achieved through disruption of the single mutant K-*ras* allele itself, in the same cellular background.[19] Such VEGF downregulation resulted in a near complete obliteration of the tumor-forming potential of all clones analyzed.[58] This result, along with at least two other independent studies (Table II[59,60]), suggests that VEGF expression is often (but not always) required for tumor formation in the context of mutant *ras* oncogene.

Molecular Mediators of ras-Dependent Vascular Endothelial Growth Factor Upregulation

Realization that VEGF can contribute to *ras*-dependent tumor formation naturally leads to the idea that signal transduction inhibitors may possess (indirect) antiangiogenic activity.[19,61] In this regard, putative Ras inhibitors such as FTIs can

[53] B. Millauer, M. P. Longhi, K. H. Plate, L. K. Shawver, W. Risau, A. Ullrich, and L. M. Strawn, *Cancer Res.* **56,** 1615 (1996).

[54] B. Millauer, L. K. Shawver, K. H. Plate, W. Risau, and A. Ullrich, *Nature* (*London*) **367,** 576 (1994).

[55] L. Witte, D. J. Hicklin, Z. Zhu, B. Pytowski, H. Kotanides, P. Rockwell, and P. Bohlen, *Cancer Metastasis Rev.* **17,** 155 (1998).

[56] R. M. Shaheen, D. W. Davis, W. Liu, B. K. Zebrowski, M. R. Wilson, C. D. Bucana, D. J. McConkey, G. McMahon, and L. M. Ellis, *Cancer Res.* **59,** 5412 (1999).

[57] T. A. Fong, L. K. Shawver, L. Sun, C. Tang, H. App, T. J. Powell, Y. H. Kim, R. Schreck, X. Wang, W. Risau, A. Ullrich, K. P. Hirth, and G. McMahon, *Cancer Res.* **59,** 99 (1999).

[58] F. Okada, J. Rak, B. St. Croix, B. Lieubeau, M. Kaya, L. Roncari, S. Sasazuki, and R. S. Kerbel, *Proc. Natl. Acad. Sci. U.S.A.* **95,** 3609 (1998).

[59] Y. P. Shi and N. Ferrara, *Biochem. Biophys. Res. Commun.* **254,** 480 (1999).

[60] J. Grunstein, W. G. Roberts, O. Mathieu-Costello, D. Hanahan, and R. S. Johnson, *Cancer Res.* **59,** 1592 (1999).

[61] R. S. Kerbel, A. M. Viloria-Petit, F. Okada, and J. W. Rak, *Mol. Med.* **4,** 286 (1998).

serve as an interesting paradigm in that they suppress VEGF production[19,26,27,37] in various experimental settings. However, it is noteworthy that FTIs may act in a somewhat indirect manner, that is, through inhibition of a small GTPase known as RhoB, which in fibroblasts, is one of the putative mediators of *ras* transformation.[62] This example shows that it may be worthwhile to delineate more precisely the pathway(s) participating in *ras*-dependent upregulation of VEGF, as such an analysis could perhaps point to more practical and/or universal molecular targets for indirect inhibition of tumor angiogenesis.

In general, regulation of VEGF expression and activity is complex and can be influenced at multiple levels.[49] This includes regulation of the gene transcription,[63–67] mRNA stability,[68–71] the rate of translation,[72–75] or posttranslational events.[76] All these modes of regulation can be affected by mutant *ras,* at least in theory.

The main difficulty in establishing the exact mode of VEGF regulation in *ras*-transformed cells lies in the complexity of Ras signaling.[11] There are large numbers of targets and mediators of Ras signaling that need to be considered. In addition, some of the phenotypic changes attributed to *ras* transformation are induced indirectly, for example, by activation of autocrine growth factor circuitry driven by (among other factors) members of the transforming growth factor α

[62] P. F. Lebowitz and G. C. Prendergast, *Oncogene* **17,** 1439 (1998).

[63] E. Tischer, R. Mitchell, T. Hartman, M. Silva, D. Gospodarowicz, J. C. Fiddes, and J. Abrahan, *J. Biol. Chem.* **266,** 11947 (1991).

[64] L. V. Beerepoot, D. T. Shima, M. Kuroki, K.-T. Yeo, and E. E. Voest, *Cancer Res.* **56,** 3747 (1996).

[65] J. A. Forsythe, B. H. Jiang, N. V. Iyer, F. Agani, S. W. Leung, R. D. Koos, and G. L. Semenza, *Mol. Cell. Biol.* **16,** 4604 (1996).

[66] N. M. Mazure, E. Y. Chen, K. R. Laderoute, and A. J. Giaccia, *Blood* **90,** 3322 (1997).

[67] D. Mukhopadhyay, L. Tsiokas, and V. P. Sukhatme, *Cancer Res.* **55,** 6161 (1995).

[68] A. P. Levy, N. S. Levy, and M. A. Goldberg, *J. Biol. Chem.* **271,** 2746 (1996).

[69] I. Stein, M. Neeman, D. Shweiki, A. Itin, and E. Keshet, *Mol. Cell. Biol.* **15,** 5363 (1995).

[70] A. P. Levy, N. S. Levy, O. Iliopoulos, C. Jiang, W. G. Kaplin, Jr., and M. A. Goldberg, *Kidney Int.* **51,** 575 (1997).

[71] J. A. Dibbens, D. L. Miller, A. Damert, W. Risau, M. A. Vadas, and G. J. Goodall, *Mol. Biol. Cell* **10,** 907 (1999).

[72] C. G. Kevil, A. De Benedetti, D. K. Payne, L. L. Coe, F. S. Laroux, and J. S. Alexander, *Int. J. Cancer* **65,** 785 (1996).

[73] I. Stein, A. Itin, P. Einat, R. Skaliter, Z. Grossman, and E. Keshet, *Mol. Cell. Biol.* **18,** 3112 (1998).

[74] G. Akiri, D. Nahari, Y. Finkelstein, S. Y. Le, O. Elroy-Stein, and B. Z. Levi, *Oncogene* **17,** 227 (1998).

[75] P. A. Scott, K. Smith, R. Poulsom, A. De Benedetti, R. Bicknell, and A. L. Harris, *Br. J. Cancer* **77,** 2120 (1998).

[76] J. DiSalvo, M. L. Bayne, G. Conn, P. W. Kwok, P. G. Trivedi, D. D. Soderman, T. M. Palisi, K. A. Sullivan, and K. A. Thomas, *J. Biol. Chem.* **270,** 7717 (1995).

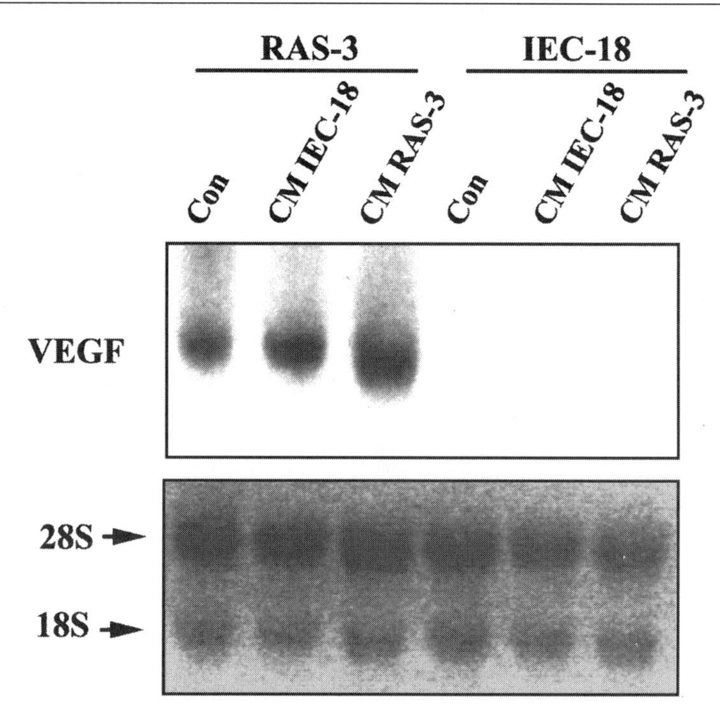

FIG. 2. Inability of to mimic *ras*-dependent upregulation of VEGF by conditioned medium transfer. IEC-18 or their H-*ras*-transformed counterparts were grown to near confluence and the medium from respective cultures was conditioned for 24 hr. The conditioned (IEC-18CM, RAS-3CM) or control medium was subsequently incubated with cell cultures of either type (IEC-18, RAS-3) for 24 hr, after which mRNA was isolated and assayed for VEGF. Ribosomal RNA was used as a loading control.

(TGF-α) family.[77–80] In this regard, our own studies of IEC-18 intestinal epithelial cells have indicated that the relationship between expression of mutant *ras* and induction of VEGF is not mediated by autocrine growth factors. For example, conditioned medium from oncogenic *ras*-transformed IEC-18 cells (RAS-3) did not induce VEGF production in their nontransformed counterparts[18] (Fig. 2).

Studies of signaling pathways that may directly connect activation of *ras* to the molecular machinery regulating VEGF expression have revealed considerable

[77] A. A. Dlugosz, L. Hansen, C. Cheng, N. Alexander, M. F. Denning, D. W. Threadgill, T. Magnuson, R. J. J. Coffey, and S. H. Yuspa, *Cancer Res.* **57,** 3180 (1997).

[78] L. M. Gangarosa, N. Sizemore, R. Graves-Deal, S. M. Oldham, C. J. Der, and R. J. Coffey, *J. Biol. Chem.* **272,** 18926 (1997).

[79] S. M. Oldham, G. J. Clark, L. M. Gangarosa, Jr., R. J. Coffey, and C. J. Der, *Proc. Natl. Acad. Sci. U.S.A.* **93,** 6924 (1996).

[80] J. Filmus, W. Shi, and T. Spencer, *Oncogene* **8,** 1017 (1993).

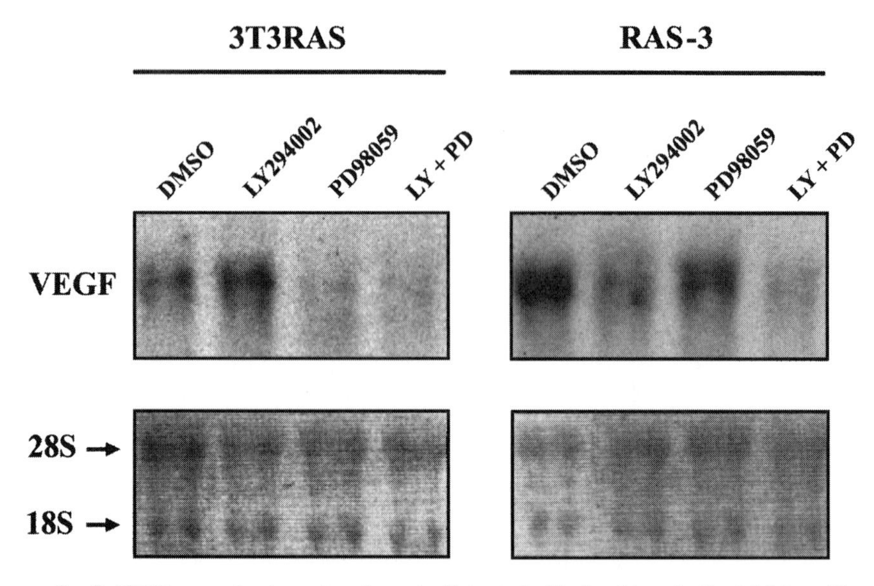

FIG. 3. VEGF expression in *ras*-transformed cells treated with signal transduction inhibitors. The changes observed suggest cell type-specific involvement of two major Ras effectors (PI3K and MEK) in regulation of VEGF expression by mutant *ras*. Sparse cultures of RAS-3 cells or 3T3ras cells were washed with the assay medium [1% (v/v) FBS] and subsequently incubated for 9 hr with either DMSO (vehicle), MEK-1 inhibitor (PD98059; 50 μM), PI3K inhibitor (LY294002; 20 μM), or a combination of the two. Extracted mRNA was analyzed for VEGF expression.

diversity among different model systems. Thus, the Raf–MEK [MAPK (mitogen-activated protein kinase)/ERK (extracellular signal-regulated kinase) Kinase]–MAPK cascade appears to be the predominant signaling module connecting mutant *ras* to VEGF production in cells of fibroblastic origin.[18,25,81] This is based on the observation that expression of the activated form of *raf*[25] or MEK-1[18,81] in NIH 3T3 cells, or hamster CCL39 fibroblasts, is sufficient to substitute for mutant *ras* insofar as upregulation of VEGF is concerned. Also, treatment of *ras*-transformed fibroblasts with a MEK-1 inhibitor (PD98059) normalized their VEGF expression.[18,81] On the other hand, in H-*ras*-transformed epithelial cells (RAS-3), treatment with PD98059 produced only a marginal effect. In those cells a significant decrease in VEGF mRNA and protein expression was achieved in the presence of the powerful inhibitor of phosphatidylinositol 3-kinase (PI3K) known as LY294002.[18] A combined treatment with both MEK-1 and PI3K inhibitors was of even greater potency, suggesting that both of these effector pathways contributed in some way to upregulation of VEGF in *ras*-transformed epithelial cells (Fig. 3). Collectively,

[81] J. Milanini, F. Vinals, J. Pouyssegur, and G. Pages, *J. Biol. Chem.* **273,** 18165 (1998).

these results point to a complex and cell type-specific mode of VEGF regulation by mutant *ras* oncogene.[18]

In addition to causing a global increase in levels of VEGF protein and mRNA,[19,25] expression of mutant ras in various types of cells was also shown to influence VEGF promoter activity,[29] VEGF mRNA stability,[33] and the profile of VEGF splice isoforms expressed.[34]

Coregulation of Vascular Endothelial Growth Factor Expression by ras and Epigenetic Factors

Analysis of *ras*-dependent upregulation of VEGF under different conditions provides an informative example of how genetic and epigenetic influences can cooperate in triggering tumor angiogenesis. Thus in NIH 3T3 fibroblasts expression of mutant H-*ras* was shown to synergize with hypoxia in stimulating the activity of the VEGF promoter via a PI3K-dependent activation of hypoxia-inducible factor 1 (HIF-1).[66] A similar conclusion was reached from an earlier study of an H-*ras*-transformed murine endothelial cell line.[82] In contrast, treatment of mutant K-*ras*-positive HCT116 colorectal carcinoma cells with a hypoxia mimetic, cobalt chloride ($CoCl_2$), failed to produce an increase in VEGF production, and only a weak effect was observed in the case of human DLD-1 or rat RAS-3 cells[58] (Rak *et al.,* unpublished observation). Genetic studies in mice suggest that altered responsiveness to hypoxia may be a feature of gut epithelium.[83]

Another interesting epigenetic influence that appears to influence (i. e., upregulate) VEGF expression in tumor cells is growth under high cell density conditions.[84] Malignant transformation appears to be absolutely required for such VEGF stimulation. Parental, nontransformed IEC-18 cells do not upregulate VEGF in confluent cultures whereas such upregulation does occur readily in their *ras* oncogene-transformed counterparts (RAS-3 cells).[18] Interestingly, RAS-3 cells grown in high-density cultures also become less responsive to treatment with LY294002 (a PI3K inhibitor) in terms of reducing their VEGF production.[18]

Pleiotropic Effect of Ras Transformation on Expression of Multiple Angiogenesis Regulators

Severe angiogenesis deficiency induced in mouse embryos by disruption of even one VEGF allele leads to the notion that this growth factor is central and

[82] J. L. Arbiser, M. A. Moses, C. A. Fernandez, N. Ghiso, Y. Cao, N. Klauber, D. Frank, M. Brownlee, E. Flynn, S. Parangi, H. R. Byers, and J. Folkman, *Proc. Natl. Acad. Sci. U.S.A.* **94,** 861 (1997).

[83] E. Maltepe, J. V. Schmidt, D. Baunoch, C. A. Bradfield, and M. C. Simon, *Nature* (*London*) **386,** 403 (1997).

[84] A. N. Koura, W. Liu, Y. Kitadai, R. K. Singh, R. Radinsky, and L. M. Ellis, *Cancer Res.* **56,** 3891 (1996).

indispensable for all processes involving blood vessel formation.[46,49,85] However, in some instances elevated endogenous production of VEGF by tumor cells may not be essential for tumor growth and angiogenesis.[7,86,87] There are at least two reasons for this. First, it is well established that mutant *ras* may affect expression of a large number of genes relevant for tumor angiogenesis (Table III[88–97]). For example, as shown previously,[18,96,98] various types of *ras*-transformed cells downregulate expression of a potent angiogenesis inhibitor, thrombospondin 1 (TSP-1). Interestingly, this change may be executed via an indirect mechanism because in IEC-18 cells, in which mutant H-*ras* is expressed transiently, under the control of a tetracycline-repressible promoter, TSP-1 mRNA levels remain high regardless of whether the drug is present or not (Fig. 1). Second, another possible circumstance that may diminish the impact of tumor cell-derived VEGF on angiogenesis is production of this growth factor by tumor-infiltrating host stromal cells.[86,99,100] In this regard, we observed that injection of v-H-*ras*-transformed, adult mouse dermal fibroblasts into SCID mice resulted in tumor formation regardless of whether the cells were of VEGF$^{+/+}$ or VEGF$^{-/-}$ genotype, which was accompanied by appreciable infiltration of host stromal cells (Rak *et al.*, unpublished observations).

[85] P. Carmeliet, V. Ferreira, G. Breier, S. Pollefeyt, L. Kieckens, M. Gertsenstein, M. Fahrig, A. Vandenhoeck, K. Harpal, C. Eberhardt, C. Declercq, J. Pawling, L. Moons, D. Collen, W. Risau, and A. Nagy, *Nature (London)* **380,** 435 (1996).

[86] D. Fukumura, R. Xavier, T. Sugiura, Y. Chen, E. C. Park, N. Lu, M. Selig, G. Nielsen, T. Taksir, R. K. Jain, and B. Seed, *Cell* **94,** 715 (1998).

[87] H. Yoshiji, S. R. Harris, and U. P. Thorgeirsson, *Cancer Res.* **57,** 3924 (1997).

[88] N. Iberg, S. Rogelj, P. Fanning, and M. Klagsbrun, *J. Biol. Chem.* **264,** 19951 (1989).

[89] C. J. Marshall, K. Vousden, and B. Ozanne, *Proc. R. Soc. Lond. B. Biol. Sci.* **226,** 99 (1985).

[90] A. B. Glick, M. B. Sporn, and S. H. Yuspa, *Mol. Carcinog.* **4,** 210 (1991).

[91] C. Castelli, M. Sensi, R. Lupetti, R. Mortarini, P. Panceri, A. Anichini, and G. Parmiani, *Cancer Res.* **54,** 4785 (1994).

[92] G. D. Demetri, T. J. Ernst, E. S. Pratt II, B. W. Zenzle, J. G. Rheinwald, and J. D. Griffin, *J. Clin. Invest.* **86,** 1261 (1990).

[93] R. B. Dickson, A. Kasid, K. K. Huff, S. E. Bates, C. Knabbe, D. Bronzert, E. P. Gelmann, and M. E. Lippman, *Proc. Natl. Acad. Sci. U.S.A.* **84,** 837 (1987).

[94] D. F. Bowen-Pope, A. Vogel, and R. Ross, *Proc. Natl. Acad. Sci. U.S.A.* **81,** 2396 (1984).

[95] A. M. Craig, M. Nemir, B. B. Mukherjee, A. F. Chambers, and D. T. Denhardt, *Biochem. Biophys. Res. Commun.* **157,** 166 (1988).

[96] V. Zabrenetzky, C. C. Harris, P. S. Steeg, and D. D. Roberts, *Int. J. Cancer* **59,** 191 (1994).

[97] L. E. Heasley, S. Thaler, M. Nicks, B. Price, K. Skorecki, and R. A. Nemenoff, *J. Biol. Chem.* **272,** 14501 (1997).

[98] N. Sheibani and W. A. Frazier, *Cancer Lett.* **107,** 45 (1996).

[99] L. Hlatky, C. Tsionou, P. Hahnfeldt, and C. N. Coleman, *Cancer Res.* **54,** 6083 (1994).

[100] L. M. Coussens, W. W. Raymond, G. Bergers, M. Laig-Webster, O. Behrendtsen, Z. Werb, G. H. Caughey, and D. Hanahan, *Genes Dev.* **13,** 1382 (1999).

TABLE III
IMPACT OF MUTANT *ras* ON EXPRESSION OF VARIOUS ANGIOGENESIS REGULATORS

Entity[a]	Bioactivity	Angioactivity	Impact of *ras*	Ref.
VEGF	Growth factor	Stimulator	Upregulation	Rak *et al.* (1995)[19] Grugel *et al.* (1995)[25]
bFGF	Growth factor	Stimulator	Upregulation	Iberg *et al.* (1989)[88]
TGF-α	Growth factor	Stimulator	Upregulation	Marshall *et al.* (1985)[89]
TGF-β	Growth factor	Stimulator	Upregulation	Glick *et al.* (1991)[90]
TNF-α	Cytokine	Stimulator (at low conc.)	Upregulation	Castelli *et al.* (1991)[91]
G-CSF	Cytokine	Stimulator	Upregulation	Demetri *et al.* (1990)[92]
IGF-I	Growth factor	Stimulator	Upregulation	Kasid *et al.* (1987)[93]
PDGF	Growth factor	Stimulator	Upregulation	Bowen-Pope *et al.* (1984)[94]
OPN	ECM component	Stimulator	Upregulation	Craig *et al.* (1988)[95]
TSP-1	ECM component	Inhibitor	Downregulation	Zabrenetzky *et al.* (1994)[96]
TIMP	MMP inhibitor	Inhibitor	Downregulation	Chambers *et al.* (1993)[12]
PGE	Prostaglandin	Stimulator	Upregulation	Heasley *et al.* (1985)[97]

[a] TNF-α, tumor necrosis factor α; G-CSF, granulocyte colony-stimulating factor; IGF-I, insulin-like growth factor type I; PDGF, platelet-derived growth factor; OPN, osteopontin; TIMP, tissue inhibitor of metalloproteinase; PGE, prostaglandin E.

Mutant Ras-Dependent Changes in Cell Survival and Mitogenesis: Relationship to Tumor Angiogenesis

Regardless of its molecular nature, the angiogenic phenotype would be expected to confer only a relatively weak selective growth advantage on tumor cells.[101] This is because the presence of a blood vessel capillary network within a given tumor microdomain would be unlikely to favor growth of one specific cancer cell population (clone) over the other, regardless of whether such cells (clone) express a proangiogenic phenotype or not.[102] An indication as to how such a phenotype can be selected for among evolving tumor cell populations may be implied from the following experiments: IEC-18 cells were selected in spheroid culture for resistance to a form of programmed cell death known as "anoikis".[20,103] Without any preexposure to *in vivo* (angiogenesis-dependent) growth conditions, the resulting variant cells lines spontaneously developed tumorigenic and angiogenic

[101] J. Jouanneau, G. Moens, Y. Bourgeois, M. F. Poupon, and J. P. Thiery, *Proc. Natl. Acad. Sci. U.S.A.* **91,** 286 (1994).

[102] F. R. Miller and G. H. Heppner, *Cancer Metastasis Rev.* **9,** 21 (1990).

[103] J. Rak, Y. Mitsuhashi, V. Erdos, S.-N. Huang, J. Filmus, and R. S. Kerbel, *J. Cell Biol.* **131,** 1587 (1995).

properties, including spontaneous upregulation of VEGF.[20] There is a striking similarity between this spontaneous conversion and enforced expression of mutant *ras* oncogene in IEC-18 cells in that the angiogenic phenotype is, in both cases, coexpressed with such potentially highly selectable traits as enhanced mitogenic activity and survival capacity.[104] Hence, *ras* transformation may operate by induction of a cluster phenotype, which may facilitate a collateral selection of angiogenically proficient cells simply because the same cells would also be intrinsically more mitogenic and apoptosis resistant.[20]

Growth factor independence and apoptosis resistance associated with expression of mutant *ras* (and many other transforming genes as well) in tumor cells can, in theory, be expected to alter their relative dependence on the proximity to blood vessels.[14] This is apparent in the case of two tumorigenic clones of IEC-18 cells expressing either high (RAS-7) or low (RAS-3) levels of H-Ras oncoprotein as well as corresponding degrees of cellular transformation.[103] Thus, in mixed tumors composed of RAS-7 and RAS-3 populations the latter (less malignant) cell type is ultimately restricted to areas immediately adjacent (proximal) to the tumor vasculature whereas RAS-7 cells can be found throughout the tumor including its most hypoxic (distal) regions.[22] The ability of RAS-7 cells to grow and survive in areas distant from blood vessels is consistent with their overall greater survival capacity and mitotic activity under a variety of stress conditions. For example, RAS-7 cells have a greater capacity to form micro- and macrometastases in lungs of tumor-bearing mice,[22] proliferate more readily in spheroid culture,[103] and are more resistant to treatment with cisplatin[*cis*-platinum(II) diammine dichloride] (Mitsuhashi *et al.*, unpublished) as compared with their less transformed and more "vascular-dependent" RAS-3 counterparts.

Role of Ras Proteins in Responses of Endothelial Cells to Angiogenic Stimuli

A well-controlled activity of endogenous Ras protein is required for proper function of endothelial cells. Mouse embryos lacking rasGAP (Ras GTPase-activating protein), which restricts Ras activity, leads to impairment in blood vessel formation and death *in utero* by embryonic day 10.5.[15] On the other hand, responses of endothelial cells to angiogenic growth factors such as bFGF or VEGF often lead to activation of various elements of the Ras pathway.[105] Interestingly, Ras FTIs have been shown to possess antiangiogenic activity *in vivo,* in part because of a direct effect on endothelial cell functions[106] (Table IV).

[104] J. Rak, J. Filmus, G. Finkenzeller, S. Grugel, D. Marme, and R. S. Kerbel, *Cancer Metastasis Rev.* **14,** 263 (1995).

[105] G. D'Angelo, I. Struman, J. Martial, and R. I. Weiner, *Proc. Natl. Acad. Sci. U.S.A.* **92,** 6374 (1995).

[106] W. Z. Gu, S. K. Tahir, Y. C. Wang, H. C. Zhang, S. P. Cherian, S. O'Connor, J. A. Leal, S. H. Rosenberg, and S. C. Ng, *Eur. J. Cancer* **35,** 1394 (1999).

TABLE IV
ANTIANGIOGENIC ACTIVITY OF Ras FARNESYLTRANSFERASE INHIBITORS

Compound	Effect	Ref.
L-739,749	Downregulation of VEGF in ras transformed intestinal epithelial cells	Rak *et al.* (1995)[19]
N-Acetyl-*S*-*trans*farnesyl-L-cysteine (AFC)	Downregulation of VEGF in *ras*-transformed cell lines	White *et al.* (1997)[33]
RPR 115135	Downregulation of VEGF in *ras*-transformed keratinocytes	Charvat *et al.* (1999)[37]
L-744,832	Downregulation of VEGF in human astrocytoma	Feldkamp *et al.* (1999)[26,27]
A-170634	Downregulation of VEGF in human colon cancer cells; inhibition of tumor angiogenesis *in vivo;* direct inhibition of endothelial cells *in vitro* and *in vivo*	Gu *et al.* (1999)[106]

Summary

Given the multifaceted role of Ras in tumor angiogenesis, pharmacologic targeting of such proteins may bring about at least three important consequences: (1) partial obliteration of the angiogenic competence of tumor cells, (2) an increase in vascular dependence and sensitization to apoptosis, and (3) a direct inhibition of endothelial cell responses to proangiogenic stimuli. Exploration of some of these possibilities, using various pharmacological compounds and antibodies, has already begun.[18,19,23,26,37,106] An intriguing possibility is that Ras antagonists and signal transduction inhibitors may synergize with a number of other antiangiogenic modalities such as direct acting antiangiogenic agents (e.g., endostatin) or antivascular regimens involving low-dose continuous chemotherapy as a vasculature-targeting strategy.[107,108]

Acknowledgments

This work was supported by a grant from the Medical Research Council of Canada to R.S.K. We gratefully acknowledge the excellent secretarial help of Ms. Lynda Woodcock and Mrs. Cassandra Cheng.

[107] G. Klement, S. Baruchel, J. Rak, S. Man, C. Clark, D. Hicklin, P. Bohlen, and R. S. Kerbel, *J. Clin. Invest.* **105,** R15 (2000).

[108] T. Browder, C. E. Butterfield, B. M. Kraling, B. Marshall, M. S. O'Reilly, and J. Folkman, *Cancer Res.* **60,** 1878 (2000).

[25] Ras Regulation of Radioresistance in Cell Culture

By ANJALI K. GUPTA, VINCENT J. BAKANAUSKAS, W. GILLIES MCKENNA, ERIC J. BERNHARD, and RUTH J. MUSCHEL

Introduction

Some tumor cells have intrinsic resistance to killing by ionizing radiation[1] and this may limit the effectiveness of radiation in cancer treatment. As many as 20% of patients who present with localized disease fail because of uncontrolled disease at the primary site, without signs of disseminated disease. Thus reducing the risk of local failure after radiotherapy could translate into large differences in overall survival.[2–4] One factor known to increase tumor cell resistance to radiation is the presence of activated oncogenes. Transfection with *ras* oncogenes has been shown to increase radioresistance in certain rodent[5,6] and human cells,[7,8] although increased radioresistance was not seen in all cell types after *ras* transfection. We have shown that in rodent cell lines the addition of a second cooperating oncogene such as v-*myc* has a synergistic effect on H-*ras*-mediated radiation resistance, although v-*myc* alone has no effect.[5]

Ras proteins are processed in a series of reactions that result in farnesylation or geranylgeranylation by farnesyltransferase or geranylgeranyltransferase, respectively (reviewed in Ref. 9). This is essential for the attachment of Ras to the inner surface of the plasma membrane and for activity.[10–13] The farnesyl or geranylgeranyl group is appended to the C-terminal cysteine residue that appears in a CAAX

[1] E. P. Malaise and P. Lambin, M. C. Joiner, *Radiat. Res.* **138,** S25 (1994).

[2] S. A. Leibel and Z. Fuks, *Int. J. Radiat. Oncol. Biol. Phys.* **24,** 377 (1992).

[3] E. D. Yorke, Z. Fuks, L. Norton, W. Whitmore, and C. C. Ling, *Cancer Res.* **53,** 2987 (1993).

[4] H. D. Suit, *Int. J. Radiat. Oncol. Biol. Phys.* **23,** 653 (1992).

[5] W. G. McKenna, M. A. Weiss, V. J. Bakanauskas, H. Sandler, M. Kelsten, J. Biaglow, B. Endlich, C. Ling, and R. J. Muschel, *Int. J. Radiat. Oncol. Biol. Phys.* **18,** 849 (1990).

[6] C. C. Ling and B. Endlich, *Radiat. Res.* **120,** 267 (1989).

[7] A. C. Miller, K. Kariko, C. E. Myers, E. P. Clark, and D. Samid, *Int. J. Cancer* **53,** 302 (1993).

[8] A. Hermens and P. Bentvelzen, *Cancer Res.* **52,** 3073 (1992).

[9] J. A. Glomset and C. C. Farnsworth, *Annu. Rev. Cell Biol.* **10,** 181 (1994).

[10] K. Kato, A. D. Cox, M. M. Hisaka, S. M. Graham, J. E. Buss, and C. J. Der, *Proc. Natl. Acad. Sci. U.S.A.* **89,** 6403 (1992).

[11] B. M. Willumsen, K. Norris, A. G. Papageorge, N. L. Hubbert, and D. R. Lowy, *EMBO J.* **3,** 2581 (1984).

[12] B. M. Willumsen, A. Christensen, N. L. Hubbert, A. G. Papageorge, and D. R. Lowy, *Nature (London)* **310,** 583 (1984).

[13] J. H. Jackson, C. G. Cockran, J. R. Bourne, P. A. Solski, J. E. Buss, and C. Der, *Proc. Natl. Acad. Sci. U.S.A.* **87,** 3042 (1990).

0076-6879/00 $35.00

motif (where A is any aliphatic amino acid).[14–16] Some CAAX peptidomimetics and tetrapeptides are competitive inhibitors of the enzyme activity[14] and some are quite selective in that they do not inhibit or compete with geranylgeranyltransferases.

Confirming the role of Ras in radioresistance, we have demonstrated that inhibition of H-Ras farnesylation, using the farnesyltransferase inhibitor FTI-277,[17] increases radiosensitivity in H-*ras*-transformed cells and in human cell lines expressing naturally occurring H-*ras* mutations.[18] Although the reversal of H-Ras action by the CAAX peptidomimetics is relatively effective, the blockade of K-Ras prenylation is more resistant to this class of drugs. This may be both because the affinity of the farnesyltransferase for K-Ras is significantly greater than its affinity for H-Ras and because K-Ras can be prenylated by geranylgeranyltransferases after inhibition of farnesyltransferases.[18] A number of geranylgeranyltransferase inhibitors (GGTIs) have also been developed.[19,20] We found that a combination of FTI and GGTI effectively inhibits K-Ras prenylation and increases the radiosensitivity of a number of human tumor cell lines.[21] The effect is specific in that it is seen only in cells containing activated Ras. These experiments demonstrate that not only does *ras* activation increase the radioresistance of tumor cells, but it also demonstrates that the altered radiosensitivity is a direct effect of expression of the activated oncogene. The radiosensitization of tumors by FTI also occurs *in vivo* using human tumors grown as xenografts that express H-*ras*.[21a]

We have explored possible mechanisms by which *ras* activation results in radiation resistance. Irradiation of eukaryotic cells results in a division delay in all cell types that have been studied and a significant portion of that delay occurs in the G_2 phase of the cell cycle. However, the length of this delay varies depending on the cell type. The radioresistant cells that contained the H-*ras* oncogene showed a much longer G_2 delay than the sensitive cells without activated *ras*.[22] This increased G_2 delay also exerts an antiapoptotic effect.[23] Furthermore, the extent of the

[14] J. Gibbs, A. Oliff, and N. Kohl, *Cell* **77,** 175 (1994).

[15] J. E. Buss, P. A. Solski, J. P. Schaeffer, M. J. MacDonald, and C. J. Der, *Science* **243,** 1600 (1989).

[16] S. Clarke, *Annu. Rev. Biochem.* **61,** 355 (1992).

[17] E. Lerner, Y. Qian, M. Blaskovich, R. D. Fossum, A. Vogt, J. Sun, A. Cox, C. Der, A. Hamilton, and S. Sebti, *J. Biol. Chem.* **270,** 26802 (1995).

[18] E. J. Bernhard, G. Kao, A. D. Cox, S. M. Sebti, A. D. Hamilton, R. J. Muschel, and W. G. McKenna, *Cancer Res.* **56,** 1727 (1996).

[19] E. Lerner, Y. Qian, A. Hamilton, and S. Sebti, *J. Biol. Chem.* **270,** 26770 (1995).

[20] T. F. McGuire, Y. Qian, A. Vogt, A. D. Hamilton, and S. M. Sebti, *J. Biol. Chem.* **271,** 27402 (1996).

[21] E. J. Bernhard, W. G. McKenna, A. D. Hamilton, S. M. Sebti, Y. Qian, J. M. Wu, and R. J. Muschel, *Cancer Res.* **58,** 1754 (1998).

[21a] E. J. Bernhard *et al.*, submitted (2000).

[22] W. G. McKenna, G. Iliakis, M. C. Weiss, E. J. Bernhard, and R. J. Muschel, *Radiat. Res.* **125,** 283 (1991).

[23] W. G. McKenna, E. J. Bernhard, D. A. Markiewicz, M. S. Rudoltz, A. Maity, and R. J. Muschel, *Oncogene* **12,** 237 (1996).

radiation-induced G_2 delay correlates with a depression in cyclin B1 accumulation,[24] which is in a large part due to decreased stability of cyclin B1 mRNA.[25]

There is now ample evidence that *ras* mutations contribute to radiation resistance in human cell lines. Activating mutations of *ras* can be seen in 30% of all human tumors. In certain cancers, such as pancreatic cancer, mutations of *ras* are seen in 75–95% of patients.[26] In addition, *ras* activity can be upregulated by overexpression in the absence of activating mutations.[27,28] Many of these tumors are treated with radiation therapy. Abrogating the effects of *ras* activation could result in clinically improved local control of tumors and thus improve the overall survival of patients.

Methods

Radiation Survival Determination

The key to examining radiation survival in cell culture is a good clonogenic survival assay. Other measures of cell death such as apoptosis counts may not accurately reflect the degree of radiosensitization.[29] For example, data from our laboratory show that 48 hr after 10 Gy of irradiation, primary rat embryo fibroblasts (REFs) show 0–3% apoptosis as compared with 50–60% in REFs transfected with v-*myc*. However, their survival characteristics using full colony formation assays such as clonogenic survival are virtually identical.[22,23] After radiation, apoptosis occurs within 24 hr in cells derived from hematologic or germ cell origin.[30,31] However, for the vast majority of tumor cells in culture, apoptosis is delayed and occurs after cell division and often multiple cell divisions.[32] In addition, many cell lines cease proliferating without undergoing apoptosis.[33] Clonogenic radiation survival thus reflects a combination of early apoptosis, delayed apoptosis, and terminal differentiation or blocked proliferation.[34]

Culture conditions are important for this assay. Cells are maintained in the appropriate medium. When comparisons are performed, the cells being compared

[24] R. J. Muschel, H. B. Zhang, and W. G. McKenna, *Cancer Res.* **5,** 1128 (1993).

[25] A. Maity, W. G. McKenna, and R. J. Muschel, *EMBO J.* **14,** 603 (1995).

[26] G. Capella, S. Cronauer-Mitra, M. A. Pienado, and M. Perucho, *Environ. Health Perspect.* **93,** 125 (1991).

[27] R. Ben-Levy, H. F. Paterson, C. J Marshall, and Y. Yarden, *EMBO J.* **13,** 3302 (1994).

[28] J. E. DeClue, A. G. Papageorge, J. A Fletcher, S. Diehl, N. Ratner, W. C. Vass, and D. L. Lowy, *Cell.* **69,** 265 (1992).

[29] I. R. Radford, T. K. Murphy, J. M. Radley, and S. L. Ellis, *Int. J. Radiat. Biol.* **65,** 217 (1994).

[30] A. R. Clarke, C. A. Purdie, D. J. Harrison, R. G. Morris, C. C. Bird, M. L. Hooper, and A. H. Wyllie, *Nature (London)* **362,** 849 (1993).

[31] J. H. Hendry, A. Adeeko, C. S. Potten, and I. D. Morris, *Int. J. Radiat. Biol.* **70,** 677 (1996).

[32] J. M. Brown and B. G. Wouters, *Cancer Res.* **59,** 1391 (1999).

[33] W. Sinclair, *Radiat. Res.* **21,** 585 (1964).

[34] R. J. Muschel, D. E. Soto, W. G. McKenna, and E. J. Bernhard, *Oncogene* **17,** 3359 (1998).

must be grown under similar conditions. They must be either exponentially growing or growth arrested at the same point in the cell cycle at the time of the assay. This is particularly important because survival varies throughout the cell cycle, so populations being compared need to have equivalent proportions in each phase for valid comparison.[35] Cells are harvested by aspirating off the medium, washing with 10 ml of phosphate-buffered saline, adding 3 ml of trypsin (1 : 250), incubating at 37° for 2–3 min, and then resuspending in fresh medium. It is important to trypsinize completely and pipette cells repeatedly to assure a good single-cell suspension. Proceed with the assay only if the cells are in a single-cell suspension. Cells are counted with a hemocytometer and serially diluted to appropriate concentrations. Cells are then plated in 5 ml of fresh medium in 60-mm dishes or in 10 ml in 100-mm dishes. Because of inherent differences in plating efficiency and radiosensitivity, each cell line has a unique cell plating profile. At first a wide range of cell numbers is plated for each dose point. Only 150 colonies, or fewer, per 60-mm dish can be counted without overlap. Thus adequate dilutions must be allowed for colonies to be counted, but also for enough colonies for statistical significance. The typical initial range for plating cells in a clonogenic assay is 100–400 cells at 0 and 1 Gy, 200–1000 cells at 2 and 3 Gy, 500–2000 cells at 4 Gy, 1000–5000 cells at 6 Gy, and 5000 to 50,000 cells at 8 Gy. In subsequent experiments, cell numbers plated can be narrowed to a range known to yield good data. At each dose, six dishes are plated. Clonogenic survival curves are repeated at least three times per cell line.

Cells are irradiated. We use a Mark I cesium irradiator (J. L. Shepherd, San Fernando, CA) at a dose rate of 1.52 Gy/min. Ten to 14 days after irradiation the cells are stained by aspirating off the medium and staining the cells with crystal violet in methanol (2.5 g/liter). The dishes are allowed to air dry and colonies of 50 or more cells are counted. If satellite colonies are seen, this may be due to shaking of the incubator or loose cells having been detached to form new colonies. This is a frequent cause of erroneous results and can be avoided by not disturbing the plates during the time of incubation. The surviving fraction is calculated by dividing the number of colonies formed by the number of cells plated times plating efficiency. The plating efficiency is the average number of colonies counted on the 0-Gy control dishes. If the plating efficiency is lower than 5–10%, the medium needs to be supplemented or feeder layers used. If it is greater than 100% or even 90–100%, it is likely that the incubator was shaken or the counts are inaccurate. The surviving fraction of cells is then plotted versus dose of radiation.

When doing survival curves in the presence of inhibitors, the above described procedure is slightly modified. When the cells are plated, it is critical that the volume of the medium be kept constant because drug is later added to begin treatment. We have found certain drugs to interfere with the attachment of cells. We

[35] C. C. Ling, M. Guo, C. H. Chen, and T. Deloherey, *Cancer Res.* **55,** 5207 (1995).

therefore let cells attach for 2 hr prior to the addition of the drug. The time frame of pretreatment of cells with the drug is predetermined by appropriate assays performed to determine drug action. Because most drugs are solubilized in dimethyl sulfoxide (DMSO), the control plates are treated with an equal amount of DMSO. Cells can be cultured in noncytotoxic drugs or can be kept in the drug for 24 hr after irradiation, at which time the drug concentration is diluted or the plates are refed to remove drug.

SF2 Determination

Surviving fraction after 2 Gy (SF2) is of special interest because clinical doses are often given as 2 Gy. It is difficult to accurately determine SF2 by the clonogenic assay, as the amount of killing is often less than 50% and the variation of the assay makes statistically significant data difficult to obtain. SF2 as determined by limiting dilution analysis[36] is a more accurate measure of radiosensitivity. With this technique, smaller differences in radiation sensitivity between cell lines or with different drugs can be determined with statistically significant results. As with the clonogenic survival, cells need to be in good condition and matched for cell cycle. For each SF2 determination, 96-well microtest plates are plated in duplicate for 1, 2, 4, 6, or 8 cells per well on average and no irradiation or 2 Gy irradiation. Cells are harvested and counted with a hemocytometer. The first dilution of cells is at 40,000 cells/ml of medium. This is serially diluted to give 400, 300, 200, 100, and 50 cells/ml. Using a multichannel pipettor, 20 μl of the appropriate dilution is plated into each well of the 96-well plates (e.g., 20 μl of the 400-cells/ml dilution into each well yields a final density of 8 cells/well). An additional 180 μl of medium is added to each well. Alternatively, the cells can be diluted so that 200 μl of diluted cells can be plated into each well. The plates are irradiated, wrapped with Parafilm to prevent evaporation of the medium, and incubated for 1–3 weeks. With drug treatment, additional medium to dilute out the drug can be added at later times, often after 24 h. Wells without viable colonies are scored under a microscope. The natural log of negative wells versus cells plated per well is plotted and SF2 calculated by dividing the slope of the irradiated cells by the slope of the nonirradiated cells.

Cell Growth Curves

Growth curves are performed by plating 2.5×10^4 cells per plate in ten 100-mm tissue plates. Two or three plates are harvested each day, and counts are performed manually with a hemocytometer. The data are plotted semilogarithmically, and curve fitting is performed to obtain an equation from which the doubling time can be determined.

[36] I. Lefkovits, "Immunological Methods," p. 355. Academic Press, New York, 1979.

Synchronization

To control for cell cycle perturbations, we often synchronize cells so that the majority of the cells will be in the same phase at the time of irradiation. Our method for synchronization of cells has been previously published by Heintz *et al.*[37] The cells are synchronized by applying a double block of thymidine followed by aphidicolin. The timing of the two blocks may vary slightly from cell line to cell line because of their differences in doubling times and cell cycle distributions. The goal of the first (thymidine) block is to arrest cells that are in S phase at the time the block is instituted and to arrest cells that are in other phases when they reach the G_1/S border. The second (aphidicolin) block follows an interval of release from the first block such that cells initially arrested at the G_1/S border have exited S phase, but cells initially blocked in late S phase have not yet passed the G_1/S border. A general outline of a typical synchrony protocol for a cell line having a doubling time of 20 hr is as follows.

1. Cells (2×10^5) are plated in 9 ml of medium and incubated.
2. Synchronization of these cells is initiated by adding 1 ml of 20 m*M* thymidine (10×) to each dish.
3. After 12 hr of 2 m*M* thymidine treatment, the block is removed by washing the cells two times using with Dulbecco's modified Eagle's medium (DMEM) plus 2.5% (v/v) fetal bovine serum. The cells are then refed medium containing 25 μ*M* thymidine and 25 μ*M* deoxycytidine and allowed to progress through the cell cycle for 18 hr.
4. The second block of 2 μ*M* aphidicolin is then initiated.
5. After 12 hr of the aphidicolin block the cells are washed three times with DMEM plus 2.5% (v/v) fetal bovine serum and refed normal medium to release the block. At this point the cells are synchronized at the G_1/S border. A high degree of synchrony (as high as 85%) can be achieved by this protocol.
6. The cells synchronized at the G_1/S border are then allowed to progress through the cell cycle.

By harvesting the cells, synchronized as described above, at various time points from the release, a population of cells that are predominantly in the S or G_2/M phase can be collected. The time for harvesting varies with different cell lines. Generally, cells in S phase are collected at 2 hr and cells at the G_2/M border are collected 10–12 hr after release. These populations must be monitored by flow cytometry for DNA content, using propidium iodide staining after RNase digestion according to the procedure described by Vindelov *et al.*[38]

[37] N. Cheong, Y. Wang, and G. Illiakis, *Int. J. Radiat. Biol.* **63,** 623 (1993).
[38] L. L. Vindelov, I. J. Christensen, and N. I. Nissen, *Cytometry* **3,** 323 (1983).

1. Start with a minimum of 2×10^5 cells in approximately 0.2 ml of citrate buffer.

2. To these cells 1.8 ml of trypsin solution is added and shaken for 10 min.

3. Then, 1.5 ml of trypsin inhibitor is added and the preparation is again shaken for 10 min.

4. Finally, 1.5 ml of an ice-cold propidium iodide solution is added. The preparation is shaken for 10 min, spun at 1200 rpm for 10 min at 4°, and all but 0.5 ml of the supernatant is removed. From this point forward, these samples must be kept ice cold and protected from light.

5. The remaining suspension is filtered and the DNA content is determined by running these nuclei through a Becton Dickinson (San Jose, CA) FACScan flow cytometer using CellQuest software. Ten thousand events are collected per sample and the data are analyzed using ModFit LT version 2.0.

Conclusion

Using a simple set of procedures including clonogenic survival curves, SF2 determination, and cell synchronization, it is possible to examine the effects of radiation on cells. We have focused on *ras* mutations and their effects on radiation sensitivity. Certainly these procedures can be used to look at almost any oncogene. Although these procedures are straightforward and relatively simple, they require tedious attention to detail and manipulation of the parameters, such as plating numbers for kill curves or times of harvest with synchronization. We have successfully applied these techniques to many cell lines and have found these procedures to be irreplaceable in studying *ras*-mediated radioresistance.

[26] Paired Human Fibrosarcoma Cell Lines That Possess or Lack Endogenous Mutant N-*ras* Alleles as Experimental Model for Ras Signaling Pathways

By SWATI GUPTA and ERIC J. STANBRIDGE

Introduction

Ras proteins and their various relatives regulate many aspects of cell growth, differentiation, control of cytoskeletal architecture, and information flow in eukaryotic cells. The Ras proteins are members of an extended family of GTPases, which cycle between an activated GTP-bound form and an inactive GDP-bound form.[1] The GTP-bound form is slowly converted to the GDP-bound form via an

0076-6879/00 $35.00

intrinsic capacity to hydrolyze GTP much more than GDP. This process is accelerated by GTPase-activating proteins (GAPs). GAP proteins are also implicated in neoplastic progression via mutations that lead to their loss of function, for example, the NF-1 neurofibromatosis gene that encodes a Ras GAP protein.[2] The replacement of GDP with GTP is mediated by guanine nucleotide exchange factors (GEFs), another critical component of the signal transduction cascade. Mutations in the *ras* gene at codons 12, 13, and 61 (those most commonly found in human cancers) result in the protein remaining constitutively active in the GTP-bound state.

The history of the study of the involvement of the *ras* oncogenes in neoplastic progression illustrates how the "activated" oncogene came to be considered as "dominantly acting." The mutant *ras* oncogenes were first identified by the mouse NIH 3T3 cell transformation assay.[3] On the basis of this assay, it was concluded that a single dominantly acting oncogene was capable of malignantly transforming cells. Shortly thereafter, it was appreciated that aneuploid, immortalized 3T3 cells were well down the pathway to malignant conversion and that, when primary cultures of rodent cells were used, combinations of two "cooperating" oncogenes, for example, *ras* and *myc,* were necessary for malignant transformation.[4] The frequent involvement of two oncogenes in human cancer (e.g., the combination of mutated *ras* and deregulated c-*myc* expression) further supported this viewpoint.

Much of the experimentation that led to the notion of "dominantly acting" oncogenes was performed with rodent cell assay systems. In large measure, the extrapolation of dominance was extended to human cancer without supporting experimental evidence. In fact, in those relatively few studies undertaken to determine the functional role of oncogenes, such as *ras,* in human cell transformation assays, it was found that transfection of cellular oncogenes, for example, *myc* and *ras,* singly or in combination, into normal human diploid cells had no discernible phenotypic effects on normal growth behavior and did not result in immortalization or neoplastic transformation.[5,6] Isolated reports of immortalization of normal human cells by the retroviral counterparts of *ras* and *myc,* that is, v-Ha-*ras* and v-*myc,*[7,8] may be reflective of extremely rare immortalizing events that may or

[1] M. S. Boguski and F. McCormick, *Nature (London)* **366,** 643 (1993).

[2] R. M. Cawthon, R. Weiss, G. Xu, D. Viskochil, M. Culver, J. Stevens, M. Robertson, D. Dunn, R. Gesteland, P. O'Connel, and R. White, *Cell* **62,** 193 (1990).

[3] C. Shih, L. C. Padhy, H. Murray, and R. A. Weinbert, *Nature (London)* **290,** 261 (1981).

[4] L. F. Parada, H. Land, R. A. Weinberg, D. Wolf, and W. Rotter, *Nature (London)* **312,** 649 (1984).

[5] R. Sager, *Cancer Cells* **2,** 487 (1984).

[6] E. J. Stanbridge, *in* "Cell Transformation and Radiation-Induced Cancer" (D. Chadwick, C. Seymour, and B. Barnhart, Eds.), pp. 1–10. Adam Hilger, New York, 1989.

[7] G. H. Yoakum, J. F. Lechner, M. G. Gabrielson, A. M. Shamsuddin, B. F. Trump, and C. C. Harris, *Science* **227,** 1176 (1985).

[8] T. C. Morgan, D. J. Yang, D. G. Fry, P. J. Hurlin, S. K. Kohler, V. M. Maher, and J. J. McCormick, *Exp. Cell Res.* **197,** 125 (1991).

may not be influenced by the expression of these retroviral oncogenes. More recently, published reports of neoplastic transformation of normal human diploid cells have documented the need for multiple genetic alterations, including mutant *ras,* inactivated tumor suppressor gene function, and expression of telomerase.[9] The induction of telomerase appears to be a prerequisite for cellular immortalization in most, but not all, experimental model systems,[10,11] and most cancer cells examined possess telomerase activity.[12]

With the advent of recognition that loss of function of tumor suppressor genes plays a critical role in neoplastic progression, more attention was paid to human cell model systems. Molecular characterization of preneoplastic precursors of human malignancies, particularly emphasized in the elegant studies of benign adenomas and carcinomas of the colon by Vogelstein and many other investigators,[13,14] indicates that *ras* mutations frequently are present in preneoplastic tissues and, in and of themselves, are not sufficient to render the cell malignant.

Careful analysis of *in vivo* mouse models of epithelial carcinogenesis, as well as transgenic models of neoplasia involving tissue-specific expression of the *ras* oncogene, indicates that expression of the activated oncogene contributes to neoplastic progression but is not sufficient to convert the cell to a malignant phenotype.[15,16]

How activated *ras* oncogene expression contributes to neoplastic progression is unclear. Much of the analysis of the involvement of Ras in signal transduction and neoplastic transformation has been undertaken with rodent models, both *in vitro* cultured cells and transgenic mouse models. In most of these studies the experimental paradigm has involved overexpression of the *ras* gene in question.

We and others[17,18] have addressed the role of *ras* in human cancer by studying human cancer cells that possess endogenous mutant *ras* alleles whose expression results in constitutively activated Ras-dependent signal transduction pathways. Variants of these cells in which the mutant *ras* allele has been deleted have been isolated.[19] Loss of the mutant allele resulted in dramatic changes in

[9] W. C. Hahn, C. M. Counter, A. S. Lundberg, R. L. Beijersbergen, M. W. Brooks, and R. A. Weinberg, *Nature (London)* **400,** 464 (1999).

[10] S. E. Holt and J. W. Shay, *J. Cell Physiol.* **180,** 10 (1999).

[11] L. M. Colgin and R. R. Reddel, *Curr. Opin. Genet. Dev.* **9,** 97 (1999).

[12] C. W. Greider, *Trends Genet.* **15,** 109 (1999).

[13] C. Lengauer, K. W. Kinzler, and B. Vogelstein, *Nature (London)* **396,** 643 (1998).

[14] M. Ilyas, J. Straub, I. P. Tomlinson, and W. F. Bodmer, *Eur. J. Cancer* **35,** 1986 (1999).

[15] S. Frame, R. Crombie, J. Liddel, D. Stuart, S. Linardopoulos, H. Nagase, G. Portella, K. Brown, A. Street, R. Akhurst, and A. Balmain, *Philos. Trans. R. Soc. Lond. B Biol. Sci.* **353,** 839 (1998).

[16] K. F. Macleod and T. Jacks, *J. Pathol.* **187,** 43 (1999).

[17] R. Plattner, S. Gupta, R. Khosravi-Far, K. Y. Sato, M. Perucho, C. J. Der, and E. J. Stanbridge, *Oncogene* **18,** 1807 (1999).

[18] S. Gupta, R. Plattner, C. J. Der, and E. J. Stanbridge, submitted (2000).

[19] R. Plattner, M. J. Anderson, K. Y. Sato, C. L. Fashing, C. J. Der, and E. J. Stanbridge, *Proc. Natl. Acad. Sci. U.S.A.* **93,** 6665 (1996).

TABLE I
CHARACTERISTICS OF CELLS USED IN THIS STUDY

HT1080 6TG (fibrosarcoma)	MCH603c8
One mutated N-*ras* allele	Wild-type N-*ras* alleles
Round cells	Flat cells
Disorganized actin fibers	Organized actin fibers
Membrane ruffling	Lack of membrane ruffling
Anchorage-independent growth	Anchorage-dependent growth
Constitutively high Raf–MEK–MAPK activity	Low Raf–MEK–MAPK activity

expression of *in vitro*-transformed phenotypes, for example, adhesion, actin stress fiber organization and anchorage-independent growth, and *in vivo* tumorigenic potential.

One cell line in particular, the HT1080 human fibrosarcoma cell line (mutant N-*ras* allele) and its derivative, MCH603c8 (wild-type N-*ras* alleles only), have proved to be an extremely useful experimental model with which to investigate the role of mutant N-*ras* in inducing transformed and tumorigenic phenotypes. A summary of the properties of these two cell lines is presented in Table I. The parental HT1080 cells exhibit typical features of a transformed cell line in culture, including poor adherence, anchorage-independent growth and disorganized actin, and aggressive tumor formation. The MCH603c8 variants have more normal growth characteristics, including a flat adherent morphology, anchorage-dependent growth, and well-organized actin microfilaments. The cells are weakly tumorigenic: tumors are formed in all animals inoculated with the cells, but they grow significantly more slowly than HT1080 cells.[19] Examination of the Ras signal transduction pathways in HT1080 cells showed that downstream members of all pathways examined (see Fig. 1), Raf dependent and Raf independent, have high constitutive activity.[17,18] Conversely, MCH603c8 cells showed only low basal activity except for constitutively activated PKB/Akt and p38. These latter constitutive activities are most likely due to the fact that both HT1080 and MCH603c8 cells secrete platelet-derived growth factor (PDGF), which binds to and activates its cognate receptor (R. Plattner and S. Gupta, unpublished observation, 1999), followed by activation of phosphatidylinositol 3-kinase (PI3K). Utilizing either dominant-negative or constitutively active mutant cDNAs of members of the Raf, Rac, and RhoA pathways, we have downregulated or upregulated individual arms of the Ras signal transduction pathways in HT1080 and MCH603c8 cells, respectively. Distinct alterations in *in vitro* and *in vivo* phenotypic traits are seen and provide evidence of a possible novel signaling pathway that is required for the aggressive tumorigenic phenotype.

Thus, in many (but not all) respects MCH603c8 cells may be considered the human counterparts of mouse NIH 3T3 cells, which have proved to be so useful in

A. Factors that are constitutively active in HT1080 cell line

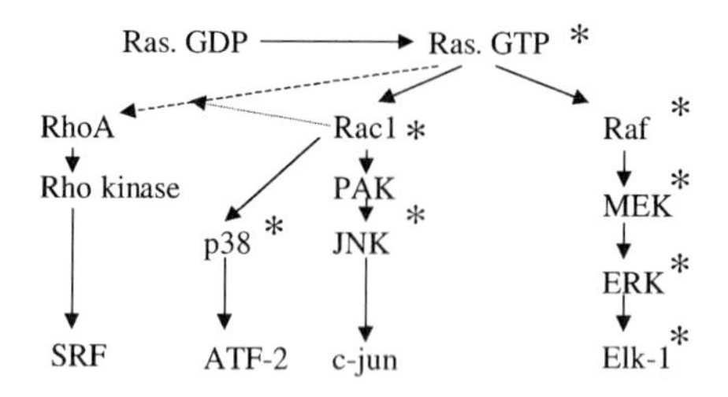

B. Factors that are constitutively active in MCH603c8 cell line

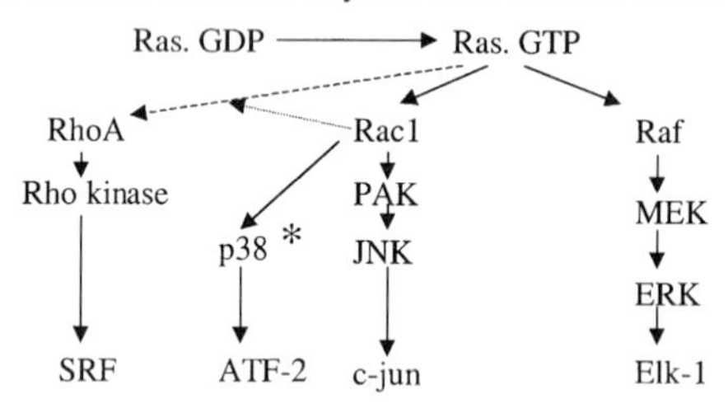

FIG. 1. Schematic diagrams of the constitutively active downstream members of Ras-signaling pathways in HT1080 (A) and MCH603c8 (B) cells.*, Constitutively active.

the study of Ras functions. In this chapter we describe protocols that allow modulation of the individual Ras signaling pathways in both HT1080 and MCH603c8 cells, using dominant-negative (DN) and constitutively active mutant members of the Ras-dependent pathways, e.g., Raf, MEK [MAPK (mitogen-activated protein kinase)/ERK (extracellular signal-regulated kinase) kinase], Rac, RhoA, and PI3K. These modulations in turn, allow examination of the biochemical consequences of downregulating or activating individual arms of the multiple Ras signaling pathways, investigation of cross-talk among pathways, and examination of the biological consequences of such modulations.

Molecular Constructs

To modulate the constitutive levels of activation of downstream members of the Ras signaling pathways, it is necessary to construct mutant cDNAs that function as dominant negatives or constitutively active mutants. Examples of both types of mutants are given in Table II. To check the *in vivo* and *in vitro* phenotypic consequences of expression of a transfected dominant-negative or activating mutant cDNA, it is necessary to establish cells stably expressing the gene of interest. The optimal way of doing that is to use an expression vector that contains the gene of interest plus a selectable marker. The two most commonly used selectable markers

TABLE II
DOMINANT NEGATIVES AND ACTIVATED MUTANTS USED IN THIS STUDY

Mutants	Mutation	Mechanism	Ref.
Dominant Negative			
RafC	Lacks Ras-binding regions, contains kinase domain and MEK-binding domain, encodes C-terminal fragment of human Raf-1	It inhibits Ras transforming activity by forming a complex with endogenous MEK	Brtva *et al.* (1995)[21]
Rac1(17N)	Codon 17 (serine) mutated to asparagine	Forms catalytically inactive complexes	Khosravi-Far *et al.* (1995)[22]
RhoA(19N)	Codon 19 (serine) mutated to asparagine	Forms catalytically inactive complexes	Khosravi-Far *et al.* (1995)[22]
P110CAAX	Encodes the chimeric protein that contains the C*aa*X domain to the catalytic domain of PI3K	Makes it membrane associated, rendering it constitutively active	Wu *et al.* (1996)[24]
Activating			
Raf22W	N-terminal end of Raf1 is truncated	Forms catalytically active Raf1	Yen *et al.* (1994)[20]
Rac1(115I)	Codon 115 (asparagine) mutated to isoleucine	Forms catalytically active Rac1	Khosravi-Far *et al.* (1995)[22]
RhoA(63L)	Codon 63 (glutamine) mutated to leucine	Forms catalytically active RhoA	Khosravi-Far *et al.* (1995)[22]
PTEN(wt)	Encodes a lipid phosphatase	It removes phosphate from the 3′ position of 3-phosphoinositides, thus inhibiting downstream pathway of PI3K	Rodriguez-Viciana *et al.* (1996)[23]

are the genes encoding aminoglycoside phosphotransferase and hygromycin B phosphotransferase, which render the cell expressing the relevant gene resistant to neomycin or hygromycin, respectively. These modified cDNAs, when expressed in the respective cell lines, will make the downstream members of Ras pathway(s) constitutively active, in which case they will be capable of transmitting continuous active signals, or will downregulate the factors downstream of Ras, resulting in the loss of constitutively active function. We have tested a number of dominant negatives and activated mutants of Ras signaling pathways (Table II). The expression construct pCMV(hyg)Raf(22W) encodes an NH_2-terminal truncated human Raf-1 that is catalytically active.[20] The construct pCGN(hyg)RafC encodes a truncated

[20] A. Yen, M. Williams, J. D. Platko, C. Der, and M. Hisaka, *Eur. J. Cell. Biol.* **65,** 103 (1994).

Raf that lacks the Ras-binding sequences but does bind to MEK.[21] This results in a catalytically inactive complex. The pCMV(Neo)Rac1(115I) vector encodes a mutated Rac, in which codon 115 (asparagine) has been mutated to isoleucine, rendering it catalytically active.[22] pCMV(hyg)Rac(17N) encodes a mutated Rac, in which codon 17 (serine) has been changed to asparagine, resulting in a catalytically inactive protein.[22] The pCMV(hyg)RhoA(63L) vector encodes RhoA; in this vector codon 63 has been changed from glutamine to lysine, rendering the protein constitutively active.[22] In pCMV(neo)RhoA(19N), encoding RhoA, codon 19 has been changed from serine to asparagine, and the protein is catalytically inactive.[22] pCMV(hyg)P110CAAX encodes a chimeric protein that contains the plasma membrane-targeting sequence, C*aa*X (where *a* is any aliphatic amino acid), fused to the catalytic domain of PI3K, rendering it membrane associated and hence constitutively active.[23] pcDNA3(neo)PTEN(wt) encodes a lipid phosphatase that removes phosphate from the 3′ position of 3′-phosphoinositides, thus inhibiting the downstream pathway of PI3K. Among its properties is dephosphorylation of Akt, a protein that is downstream of PI3K. The phospho-Akt is the activated form that functions as an antiapoptotic survival factor.[24] Thus, PTEN effectively deactivates phospho-Akt.

Establishment of Stable Transfected Cell Lines

The parental HT1080 cell line is maintained in Dulbecco's modified Eagle's medium (DMEM) supplemented with 10% (v/v) fetal calf serum (FCS; Life Technologies, Rockville, MD). The HT1080 variant, MCH603c8, lacks the activated N-*ras* allele. HT1080 and MCH603c8 cells transfected with the various mutant cDNAs are maintained in their respective antibiotic selection media prior to experimentation.

Subconfluent (70%) 100-mm dishes of MCH603c8 cells or HT1080 cells are transfected for 6 hr with 5 μg of linearized DNA or vector control DNA, using 30 μl of Lipofectin (GIBCO-BRL, Gaithersburg, MD) in Opti-MEM medium (Gibco-BRL). After 6 hr of transfection the medium is aspirated and normal growth medium [DMEM + 10% (v/v) FCS] is added to the cells and kept overnight in an incubator at 37°. The cells are then subcultured (1 : 10) into 100-mm dishes in normal growth medium. Twenty-four hours later growth medium is changed to

[21] T. R. Brtva, J. K. Drugan, S. Ghosh, R. S. Terrell, S. Campbell-Burk, R. M. Bell, and C. J. Der, *J. Biol. Chem.* **270,** 9809 (1995).

[22] R. Khosravi-Far, P. A. Solski, G. J. Clark, M. S. Kinch, and C. J. Der, *Mol. Cell. Biol.* **15,** 6443 (1995).

[23] P. Rodriguez-Viciana, P. H. Warne, B. Vanhaesebroeck, M. D. Waterfield, and J. Downward, *EMBO J.* **15,** 2442 (1996).

[24] X. Wu, K. Senechal, M. S. Neshat, Y. E. Whang, and C. L. Sawyers, *Proc. Natl. Acad. Sci. U.S.A.* **95,** 15587 (1998).

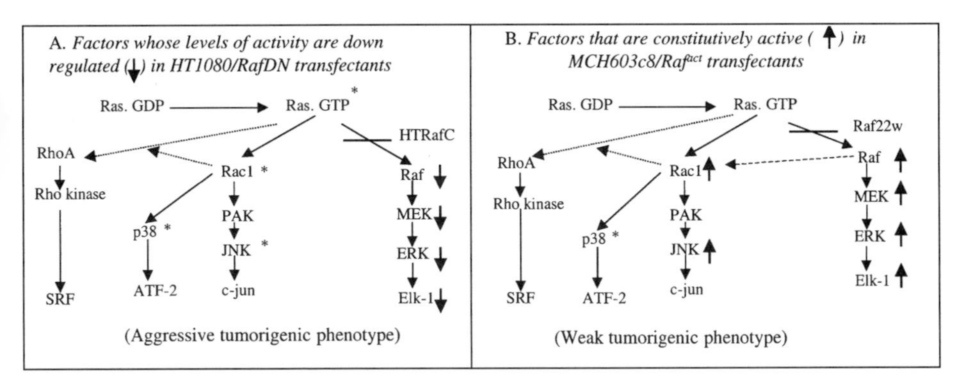

FIG. 2. Schematic diagrams of the downregulated (↓) and constitutively active (↑) downstream members of Ras-signaling pathways in HT1080/RafDN (A) and MCH603c8/Raf^{act} (B) cells.

medium supplemented with the appropriate antibiotic, an 800-μg/ml concentration of geneticin (G418; GIBCO-BRL) for Neo^R cells or 360 U of hygromycin B (Calbiochem, La Jolla, CA) for Hyg^R cells. G418 is made from powder in DMEM as a 50-mg/ml solution and stored at 4°. Hygromycin B is purchased as a solution in phosphate-buffered saline (PBS) and stored at 4°. Parallel transfection with only the vector is used as a control in each case.

The transfected cells are fed with antibiotic-supplemented medium every 3 to 4 days. Cell death is not readily apparent until days 4–5. By day 10–12 the antibiotic-resistant colonies become microscopically visible. Once the clones become more prominent, individual clones are picked and transferred into 96-well plates. Continuous selective pressure should be provided to eliminate the untransfected antibiotic-sensitive cells. After expression of the transfected cell populations by serial passaging, expression of the gene of interest should be estimated by immunoblotting with the respective antibody. Examples of positive and negative clones are shown in Fig. 2 for Raf transfections.

Biochemical Analysis of Stable Transfectants

Measurement of total protein levels derived from the transfected expression vectors does not allow determination of whether the protein in question is functioning efficiently as a dominant-negative or constitutively active protein. To determine this, immunoblotting with antibodies specific for the phosphorylated active forms of the proteins or GTP-binding pulldown assays are performed, in conjunction with kinase assays. The analysis includes examination of the transfected protein in question and, in addition, downstream members of the linear signaling pathway, as well as other Ras-dependent signaling pathways. In this way both linear pathway interactions and "cross-talk" are monitored.

Immunoblot Analysis and Coupled Affinity Precipitation/Immunoblot Analysis

Subconfluent cells are serum starved by incubation in DMEM containing 0.25% (v/v) FCS for 18 hr, and then are lysed directly with 300 μl of 1× cell lysis buffer (New England BioLabs, Beverly, MA) after rinsing the cells with PBS. The lysis buffer contains 20 m*M* Tris (pH 7.4), 1% (v/v) Triton, 2.5 m*M* sodium pyrophosphate, 150 m*M* sodium chloride, 1 m*M* EGTA, 1 m*M* Na_2EDTA, 1 m*M* β-glycerophosphate, 1 m*M* sodium orthovanadate, and phosphatase inhibitors [1 m*M* phenylmethylsulfonyl fluoride (PMSF), and aprotinin, leupeptin, and pepstatin (10 μg/ml each)]. To check with respective antibodies, the lysates are required to be resolved by migration through sodium dodecyl sulfate (SDS)–polyacrylamide gels.

Preparation of Sodium Dodecyl Sulfate–Polyacrylamide Gel. For better resolution, we always use a 130-mm-length plate and 1-mm spacers (CBS Scientific, Del Mar, CA). This system takes approximately 10 ml of separating gel and 5 ml of stacking gel solution. The separating gel solution for a 10% (w/v) gel is made by mixing the following:

Polyacrylamide–bisacrylamide (37.5 : 1, w/w)	5 ml
Tris (pH 8.8), 1.5 *M*	3.75 ml
SDS (10%, w/v)	150 μl
N,N,N′,N′-Tetramethylenediamine (TEMED)	6 μl
Ammonium persulfate (10%, w/v)	150 μl
Distilled water	5.95 ml

After pouring the separating gel, distilled water is carefully overlaid on top of the gel to prevent bubbles. Once the separating gel has polymerized, the water overlay is poured off and an 8% (w/v) stacking gel solution is poured on top of the separating gel. It is made by mixing the following:

Polyacrylamide–bisacrylamide (37.5 : 1, w/w)	1.7 ml
Tris (pH 6.8), 1 *M*	1.25 ml
SDS (10%, w/v)	100 μl
TEMED	10 μl
Ammonium persulfate (10%, w/v)	100 μl
Distilled water	6.8 ml

After pouring the stacking gel the comb is placed in the gel. Running buffer is made by mixing 43.5 g of glycine, 9 g of Trizma base, and 30 ml of 10% (w/v) SDS and the volume is made up to 3 liters with distilled water. The pH of the running buffer should be pH 8.3–8.5. After the gel is polymerized, the gel is set on the gel-running apparatus and the running buffer is added to the top and bottom chambers of the gel apparatus.

Sample Preparation. After checking the protein concentration, a definite concentration of the total protein, for example, 20 μg, of each cell extract is mixed with sample buffer.

Sample buffer (2×):	
SDS (10%, w/v)	4 ml
$NaPO_4$ (pH 7.0), 0.5 *M*	0.4 ml
Glycerol	2 ml
2-Mercaptoethanol	1 ml
Dithiothreitol (DTT), 1 *M*	2 ml
Bromphenol blue (0.4%, v/v)	

The samples are then boiled, centrifuged, and loaded onto the gel along with a molecular weight marker.

Running of Gel and Transferring It to Solid Support. The proteins are resolved on a 10% (w/v) SDS–Polyacrylamide gel at constant current (25 mA) at 100 V through the stacking gel and at 160 V through the separating gel. After electrophoresis the gel is processed further and transferred to the solid support (Immobilon membrane; Millipore, Bedford, MA).[25] The gel is transferred into transfer solution. (9 g of Tris base, 43.5 g of glycine, 600 ml of methanol) and the volume is made up to 3 liters with distilled water. The membrane is incubated sequentially with primary and secondary antibodies, and the antigen–antibody complexes are visualized by enzyme–substrate reactions that produce chemiluminescence.

Western blot analysis is performed as described above, using phospho-Akt or total Akt antibody (New England BioLabs) as the primary antibody and anti-rabbit IgG–horseradish peroxidase (HRP) conjugate as the secondary antibody (Santa Cruz Biotechnology, Santa Cruz, CA). A chemiluminescence detection system (Pierce, Rockford, IL) is used for detection of the antigen–antibody complexes.

To do affinity precipitation for estimating RasGTP, RacGTP, or Cdc42GTP, subconfluent cells are lysed with 1× Mg^{2+} lysis buffer containing 125 m*M* HEPES (pH 7.5), 750 m*M* NaCl, 5% (v/v) Igepal CA-630 (Sigma, St. Louis, MO), 50 m*M* $MgCl_2$, 5 m*M* EDTA, and 10% (v/v) glycerol (Ras and Rac/Cdc42 activation assay kits; Upstate Biotechnology, Lake Placid, NY). Each cell lysate (500 μg) is affinity precipitated with 10 μl of Raf-1 RBD– or PAK-1 PBD–agarose conjugates at 4° overnight. This pulls out the GTP-bound proteins, namely RasGTP and RacGTP/Cdc42GTP, respectively. The agarose beads are collected, washed, and resuspended in 6× Laemmli sample buffer. Western blot analysis is done as described above with a 1-μg/ml concentration of mouse monoclonal anti-Ras or anti-Rac as the primary antibody. Horseradish peroxidase-conjugated anti-mouse IgG (Santa Cruz Biotechnology) is used as the secondary antibody. A chemiluminescence detection system (Pierce) is used for detection of the relevant proteins.

[25] W. N. Burnett, *Anal. Biochem.* **112,** 195 (1981).

Kinase Assays

Raf, MEK, MAPK, and Jun N-terminal Kinase (JNK) kinase assays are performed by precipitating the respective proteins from serum-starved [0.25% (v/v) FCS for 18 hr] cell lysates. Raf, MEK, and MAPK proteins are immunoprecipitated with the relevant antibodies. A Raf-1 polyclonal antibody (C-12; Santa Cruz Biotechnology) is used to immunoprecipitate total Raf-1 from the cell lysates. A polyclonal phospho antibody to MEK 1/2 (Ser-217/221) or a monoclonal phospho antibody to P44/42 MAPK (Thr-202 and Tyr-204) is used to selectively immunoprecipitate active MEK1/2 kinase or active MAPK from the cell lysates, respectively. Stress-activated protein kinase (SAPK)/JNK is selectively "pulled down" from the cell lysates, by using an N-terminal c-Jun[26,27] fusion protein bound to glutathione–Sepharose beads (New England BioLabs). The Raf-1 immunocomplex assay is carried out in a coupled assay, using MEK and MAPK protein as an intermediate and [γ-^{32}P]ATP.[28] This allows immunoprecipitated active Raf to phosphorylate MEK, which in turn will phosphorylate MAPK. The activated MEK assay is carried out by incubating immunoprecipitated phospho-MEK with MAPK protein and unlabeled ATP (New England BioLabs, MEK1/2 kinase assay kit), which allows immunoprecipitated active MEK kinase to phosphorylate MAPK.[29,30] The activated MAPK assay is carried out by incubating immunoprecipitated phospho-MAPK with Elk-1 fusion protein and unlabeled ATP (New England BioLabs, p44/p42 MAPK assay kit), which allows immunoprecipitated active MAPK to phosphorylate Elk.[31] The JNK assays are carried out by incubating the JNK–c-Jun fusion protein complex with unlabeled ATP (New England BioLabs JNK/SAPK assay kit), which results in phosphorylation of c-Jun. All the kinase reactions are performed at 30° for 30 min in the kinase reaction mix of 25 m*M* Tris (pH 7.5), 5 m*M* β-glycerophosphate, 2 m*M* DTT, 0.1 m*M* sodium orthovanadate, and 10 m*M* $MgCl_2$. The reactions are stopped by addition of 6$\times$ Laemmli sample buffer and proteins are separated on a 10% (w/v) SDS–polyacrylamide gel. For the Raf assay, the [γ-^{32}P]MAPK proteins in the gel are visualized by autoradiography. For MEK, MAPK, and JNK assays the relevant gel is transferred onto an Immobilon membrane and Western blot analysis is performed as described above. The blots are performed with phospho-MAPK (Thr-202/Tyr-204) monoclonal antibody for the MEK assay, phospho-Elk-1 (Ser-383) polyclonal antibody for the

[26] B. Derijard, M. Hibi, I. H. Wu, T. Barrett, B. Su, T. Deng, M. Karin, and R. J. Davis, *Cell* **76,** 1025 (1994).

[27] T. Dai, E. Rubie, C. C. Franklin, A. Kraft, D. A. Gillespie, J. Avruch, J. M. Kyriakis, and J. R. Woodgett, *Oncogene* **10,** 849 (1995).

[28] C. W. Reuter, A. D. Catling, T. Jelinek, and M. J. Weber, *J. Biol. Chem.* **270,** 7644 (1995).

[29] J. H. Her, S. Lakhani, K. Zu, J. Vila, P. Dent, T. W. Sturgill, and M. J. Weber, *Biochem. J.* **296,** 25 (1993).

[30] J. H. Her, J. Wu, T. B. Rall, T. W. Sturgill, and M. J. Weber, *Nucleic Acids Res.* **19,** 3743 (1991).

[31] R. Janknecht, W. H. Ernst, V. Pingoud, and A. Nordheim, *EMBO J.* **12,** 5097 (1993).

MAPK assay, and phospho-c-Jun (Ser-63) polyclonal antibody for the JNK assay. To determine the total Raf, MEK, MAPK, and JNK levels, immunoblots are performed, using the respective antibodies that recognize total protein.

Dual Luciferase Reporter Assays. A reporter system [dual luciferase reporter assay kit (Promega, Madison, WI)] is used to study Elk activation in the cell lysates. Dual reporters are used for simultaneous expression and measurement of two individual reporter enzymes within a single system, a technique commonly used to improve experimental accuracy. The "experimental reporter," firefly luciferase, is correlated with the effect of specific experimental conditions, while the activity of the cotransfected "control reporter," *Renilla* luciferase, serves as an internal control. Normalizing the activity of the firefly reporter to the activity of the internal control minimizes experimental variability caused by differences in cell viability during the transfection procedure. Cells are transiently transfected by the liposome-mediated transfection technique (Lipofectin; GIBCO-BRL) as described above, with 2.5 μg of the 5× Gal-luc reporter, and 0.25 μg of the pMMLV-Gal-Elk expression construct. The pRL-CMV *Renilla* luciferase (0.02 μg) is used as the internal control reporter vector. After transfection, cells are serum starved by incubation in DMEM containing 0.25% (v/v) FCS and are lysed 18 hr later in 1× passive cell lysis buffer (Promega). Cells are harvested immediately after the addition of lysis buffer by scraping the cells with a disposable cell lifter. The plate is tilted and the lysates are pipetted up and down to obtain a homogeneous suspension. The lysate is then transferred into a vial for further handling. The cell lysate is subjected to one or two freeze–thaw cycles to accomplish complete lysis of cells. Clearing of the lysate samples is done by centrifugation for 30 sec in a refrigerated microcentrifuge. The cleared lysate is transferred to a fresh tube prior to reporter enzyme analyses. Luciferase assay reagent (LARII; Promega) is prepared by resuspending the provided lyophilized luciferase assay substrate in 10 ml of the supplied luciferase assay buffer II (Promega). Stop & Glo solvent (200 μl; Promega) is added to the Stop & Glo substrate to make it 50× Stop & Glo substrate. For 10 assays, the Stop & Glo reagent is made by mixing 20 μl of 50× Stop & Glo substrate to 1 ml of Stop & Glo buffer. The lysate is analyzed for luciferase activity, using a Moonlight 2010 luminometer (Analytical Luminescence Laboratory, San Diego, CA). LARII (100 μl) is dispensed into the luminometer tube containing 20 μl of the cell lysate and is analyzed in the luminometer for the first measurement. For the second measurement the sample tube is removed from the luminometer and 100 μl of Stop & Glo reagent is added and analyzed in the luminometer. By normalizing the activity of the firefly reporter (first reading) to the activity of the internal control (second reading), the Elk activity is calculated.

In a typical experiment, expression of each of the dominant-negative mutants shows evidence of downregulation of activity of downstream members of the linear pathway (data not shown). Results from a representative analysis (Raf transfections, HT/RafDN) are shown schematically in Fig. 2. Moreover, expression of each of the activating mutants (Raf, MEK, Rac, Rho, and PI3K) resulted in potent activation of the relevant protein, respective downstream members, and some cross-talking members of Ras-dependent pathways (data not shown). Results from a representative analysis (Raf transfections, MCH603/Raf^{act}) are shown schematically in Fig. 2.

Analysis of Growth Rate, Saturation Density, and Serum Dependence

Stably transfected clones from above are seeded at 1×10^4 cells per 25-cm^2 flask in normal growth medium [DMEM + 10% (v/v) FCS], and incubated at 37° (designated as day 0). At 2-day intervals, for a total of 14 days, the cells are counted in triplicate for each point. This is performed by harvesting the cells by adding trypsin, diluting the cells with normal growth medium to a fixed volume, and then using a Coulter (Hialeah, FL) Counter to count the number of cells per milliliter. The growth medium is changed to fresh growth medium on day 4 and thereafter as necessary. Growth rates are determined from the slope of the logarithmic curve during exponential growth and the saturation densities are the cell densities that are reached during the plateau phase of growth. MCH603c8 cells display increased doubling time and decreased saturation densities, in comparison with HT1080 cells, which show higher doubling time and increased saturation densities.[19]

For serum dependence assays cells (1×10^5) are plated in DMEM–10% (v/v) FCS in triplicate 25-cm^2 flasks. The next day the flasks are washed three times with DMEM with no additives and then refed with serum-free medium that is supplemented with 0.1, 0.2, 0.5, 2, 5, or 10% (v/v) FCS. The dishes are maintained in the designated serum concentration for 2 weeks, with a change of the same medium formulation on day 4 and thereafter as necessary. The colonies are stained with Giemsa stain to visualize, washed with Burr's buffer, and then counted. Whereas untransformed cells require 10% (v/v) serum for their optimal growth, the transformed cells will survive and proliferate in a low serum concentration.[32] It is seen that HT1080 cells are able to grow under serum-free conditions, whereas MCH603c8 cells are able to divide in serum-free medium but grow at a slower rate and achieve a much lower maximum cell density.[19]

Analysis of Anchorage-Independent Growth

Whereas MCH603c8 cells (wild-type N-Ras) need to adhere to a solid substratum in order to proliferate, HT1080 cells (mutant N-ras) lost this adherent characteristic and can readily proliferate in suspension culture in semisolid medium.

[32] S. Shirahata, C. Rawson, D. Loo, Y. J. Chang, and D. Barnes, *J. Cell Physiol.* **144,** 69 (1990).

Growth in soft agar is the most commonly used assay for neoplastically transformed cells.

Reagents

Bacto-agar (5%, w/v): Prepare a 10× stock in distilled water; autoclave
DMEM (2×): Make from powdered DMEM (GIBCO-BRL)
FCS (GIBCO-BRL)

Assay

The 5% (w/v) Bacto-agar is melted and kept in a 45° water bath. To make an agar–medium mixture, 1 ml of 2× DMEM is mixed with 1 ml of FCS, 6.9 ml of 1× DMEM, 100 μl of penicillin–streptomycin, and 1 ml of 5% (w/v) Bacto-agar. A concentration of 0.5% (w/v) is used for the bottom layer. The agar–medium mixture (7 ml) is poured into 60-mm dishes. The layer is allowed to solidify at room temperature and the dishes are stored in the incubator until they are ready for use. The cells are washed, trypsinized, and counted in a Coulter counter and 1×10^4 cells/2 ml of DMEM are added to a tube that is prewarmed to 37°. To make a 0.3% (w/v) top agar layer, 4 ml of the prewarmed agar–medium is added to the above cell suspension. After mixing, 1.5 ml of the mixture is swiftly overlaid onto the solidified agar dishes in triplicate. The dishes are incubated in a humidified 37° incubator with 10% CO_2. The cells are fed weekly with growth medium (0.5 ml/plate). In general, the untransformed cells do not form colonies in soft agar, whereas *ras*-transformed cells form colonies within 1 week (Fig. 3E and F). After 3 weeks, colonies (>0.1 mm in size) are scored by counting under a microscope. The colony-forming efficiency of MCH603c8 cells is more than two orders of magnitude less than that of HT1080 cells. MCH603c8 colonies are extremely small (<0.025 mm) and are most likely abortive growth. This result suggests that loss of the mutant N-*ras* allele results in reversion to an anchorage-dependent phenotype.

Morphology and Actin Cytoskeleton Staining

The HT1080 cells have a rounded, more refractile appearance on light microscopic examination, whereas MCH603c8 cells show an elongated and flat morphology (Fig. 3A and B).

One of the characteristics of fibroblasts is the presence of organized actin stress fibers. This is another *in vitro* phenotype that has been used to distinguish between transformed and untransformed cells. Actin stress fibers are visualized by staining cells with Oregon Green-488 phalloidin (Molecular Probes, Engene, OR). Two days after plating, cells are fixed in 3.7% (w/v) paraformaldehyde, treated with 0.1% (v/v) Triton-X solution, and then stained with phalloidin (0.005 U/μl) for 20 min at room temperature and mounted in ProLong fade antifade (Molecular

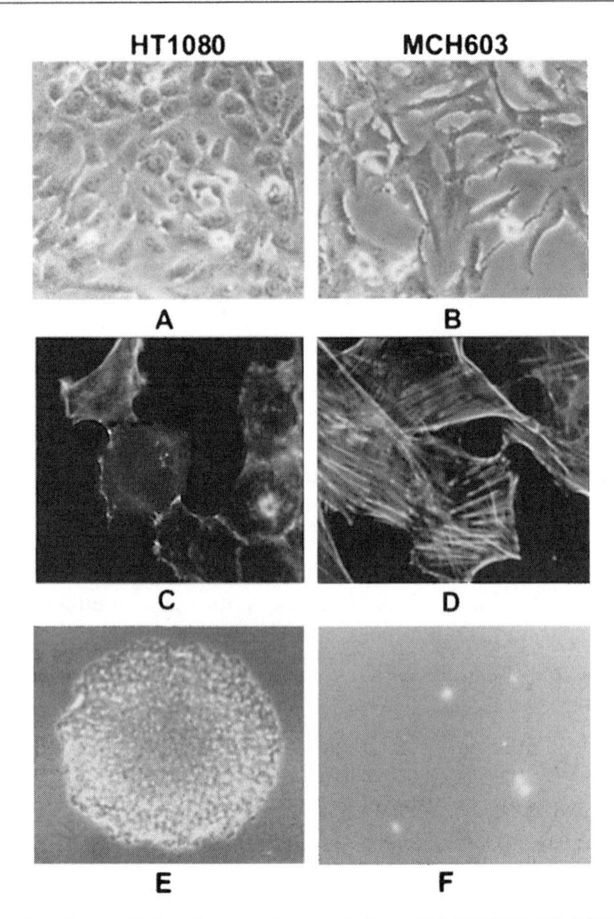

FIG. 3. *In vitro* characteristics. Comparative morphology: (A) HT1080, (B) MCH603c8. Actin stress fiber organization is illustrated by fluorescein isothiocyanate-conjugated phalloidin staining of F-actin: (C) HT1080, (D) MCH603c8. Anchorage-independent growth: (E) HT1080, (F) MCH603c8.

Probes). It is found that HT1080 cells (mutant N-*ras* allele) indeed do have disorganized actin stress fibers. In contrast, MCH603c8 cells have extremely well-organized actin stress fibers. Thus, loss of the mutant N-*ras* allele appears to correlate with reorganization of actin stress fibers (Fig. 3C and D).

Tumorigenicity Assays

The ability to form tumors in athymic (*nu/nu*) nude mice is a reliable *in vivo* technique to determine the tumorigenic phenotype of a cell line. The frequency (number of animals that grow tumors per number of animals inoculated), size

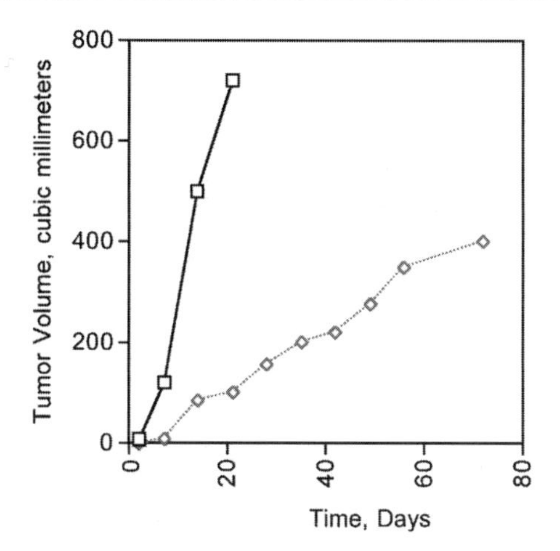

FIG. 4. *In vivo* growth kinetics of HT1080 (□) and MCH603c8(◇) cells.

(volume, mm^3), and time of appearance of the tumors may be used as a measure of the degree of aggressive tumor formation.

Cells in the log phase of growth are harvested by trypsinization, diluted in DMEM medium, and counted. The cells are then resuspended at a concentration of 1×10^7 cells/0.2 ml of DMEM. A volume of cell suspension (0.2 ml) containing 1×10^7 cells is then inoculated into the left and right flanks of 4- to 6-week-old athymic (nude) mice. It is recommended that a 21-gauge needle be used. At least three mice are injected per cell line to be assayed. The mice are monitored at weekly intervals. Tumor sizes are measured as a diameter taken at three positions with vernier calipers over the skin of the animal. It is recommended that the tumor diameter not exceed 1 cm, in order to avoid discomfort to the mice.

HT1080 cells form aggressive tumors with no latency period, as do the vector controls in HT1080 cells, whereas MCH603c8 cells, which have lost the mutant N-*ras* allele, also form tumors, but they grow at a slower rate than HT1080 cells (Fig. 4).

Summary

We present here a human cell model for examination of mutant N-*ras* function. The HT1080 human fibrosarcoma cell line is pseudodiploid and contains a single endogenous mutant N-*ras* allele. MCH603c8 cells are a variant of HT1080 cells, in which the mutant allele has been deleted. The two cell lines differ dramatically in the constitutive levels of activation of downstream members of the Ras signaling pathways, and in biological features of transformation and tumorigenicity. Downregulation or activation of individual Ras-dependent pathways can be accomplished via transfection of dominant negatives or activated mutant cDNAs into HT1080 and MCH603c8 cells, respectively. The biochemical and biological consequences of expression of these mutant cDNAs can be assessed. There are dramatic effects on both the transformed and tumorigenic phenotype, depending on the cell line and mutant cDNA that is transfected.

Acknowledgment

Our research is supported by grants from the National Institutes of Health to E.J.S. (CA69515).

[27] Orally Bioavailable Farnesyltransferase Inhibitors as Anticancer Agents in Transgenic and Xenograft Models

By MING LIU, W. ROBERT BISHOP, LORETTA L. NIELSEN, MATTHEW S. BRYANT, and PAUL KIRSCHMEIER

Introduction

This chapter describes our experience in the evaluation of the preclinical efficacy of inhibitors of farnesyl protein transferase (FPT) as anticancer agents. Using the described methodologies, we have evaluated the bioavailability and efficacy of lead FPT inhibitors (FTIs) discovered using the enzyme and cell assays described previously.[1,2] FTIs with appropriate enzyme and cell potency were first evaluated for pharmacokinetics after oral dosing in mice to ensure adequate exposure and acceptable stability in the host animal systems, as well as tumor models, following various dosing schedules (twice a day or four times a day). Compounds with acceptable mouse pharmacokinetic parameters [i.e., oral bioavailability, half-life, area under the curve (AUC), C_{max}, and C_{min}] were further evaluated for pharmacokinetics in rats and monkeys to confirm consistent oral bioavailability in various animal species, so that sufficient information could be gathered in preparation for toxicological studies (Fig. 1). FPT inhibitors with satisfactory pharmacologic parameters in the mouse and other animal species were then evaluated in *in vivo* tumor models. Transgenic (Wap-*ras*) models and xenograft models were used

[1] W. R. Bishop, R. Bond, J. Petrin, L. Wang, R. Patton, R. Doll, F. G. Njoroge, J. J. Catino, J. Schwartz, W. Windsor, R. Syto, D. Carr, L. James, and P. Kirschmeier, *J. Biol. Chem.* **270,** 30611 (1995).

[2] D. B. Whyte, P. Kirschmeier, T. N. Hockenberry, I. Nunez-Oliva, L. James, J. J. Catino, W. R. Bishop, and J.-K. Pai, *J. Biol. Chem.* **272,** 14459 (1997).

0076-6879/00 $35.00

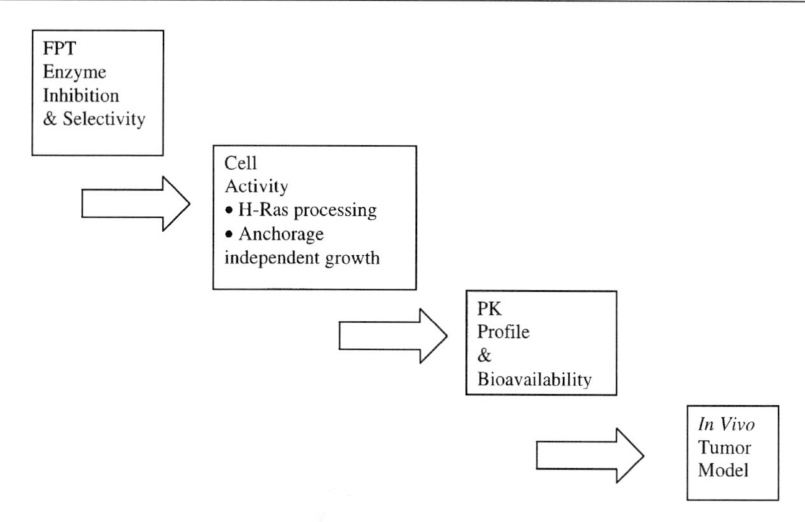

FIG. 1. Pathway to the selection of farnesyl protein transferase inhibitors.

for *in vivo* efficacy evaluation. The purpose of this evaluation was to compare the efficacy of a series of FTIs to identify potential clinical candidates.

All animal studies described in this chapter were carried out in the animal facility of Schering-Plough Research Institute (Kenilworth, NJ) in accordance with institutional guidelines. Animals were maintained in accordance with the National Institutes of Health *Guide for the Care and Use of Laboratory Animals.* Experimental protocols were reviewed, and the experimental progress was supervised, by the Schering-Plough Animal Care and Use Committee.

Pharmacokinetic Studies: Serum and Tumor Drug Levels

The first important step in our efficacy evaluation process is to determine the pharmacokinetic profile of candidate FTIs in the host animal (i.e., mouse) of the tumor models. To demonstrate the efficacy of an oral agent, a reasonable half-life and an appropriate plasma level should be maintained after oral administration. To study the pharmacokinetics and oral bioavailability of FTIs, nude mice (Crl : Nu/Nu-nu Br; Charles River Laboratories, Wilmington, MA) are used. The FTI is routinely dissolved in 20% (w/v) hydroxypropyl-β-cyclodextrin (HPβCD), if necessary, with the aid of three to six pulses of brief (20- to 30-sec) micronization, using a Vibra Cell probe sonicator (Sonics & Materials, Danbury, CT) at 80- to 100-W output. Oral administration of the FTI solution is conducted with 20-gauge sterile disposable animal feeding needles (Popper and Sons, New Hyde Park, NY) connected to sterile tuberculin syringes. Blood samples are collected at nine time points (2 min, 5 min, 15 min, 30 min, 1 hr, 2 hr, 4 hr, 7 hr, and

24 hr) after a single oral or intravenous (trail vein) dose of FTI (25 mg/kg). Three mice are used for each time point and samples are collected by cardiac puncture after euthanasia with carbon dioxide. After clotting on ice, serum is isolated by centrifugation and stored frozen at -20° until analysis. A 40-μl aliquot of serum is added to a polypropylene microcentrifuge tube and subjected to protein precipitation with 100 μl of acetonitrile–methanol (90 : 10, v/v) containing a 1-ng/μl concentration of the internal standard (e.g., SCH 56580). After vortexing for 30 sec and centrifugation at 12,000 *g* for 8 min at 4°, the supernatant is transferred into high-performance liquid chromatography (HPLC) injection vials. Quantitation of FTI serum levels is achieved by using HPLC–atmospheric pressure chemical ionization (APCI) tandem mass spectrometry as described in detail earlier.[3–5] To compare bioavailability in different vehicles, FTI is also suspended in 0.4% (w/v) methylcellulose and dosed orally for pharmacokinetic analysis.

In many of the efficacy studies, both serum and tumor samples are collected from mice at various times after the final dosing. FTI is quantified in serum as described above. Quantitation of the FTI in tumor samples requires additional processing steps including pulverizing the frozen tissue, homogenization, and protein precipitation.[3–5]

Initial compounds in the tricyclic FTI series, such as SCH 44342, display less than optimal pharmacokinetic properties in the athymic nude mouse, including a rapid oxidative metabolism.[1,4] Blocking the susceptible metabolic sites on SCH 44342 greatly improves the pharmacokinetic properties of compounds in this series. SCH 66336, an 11-piperidinyl trihalogenated analog, is one of the lead compounds that has emerged from these efforts (Table I). This compound possesses not only excellent bioavailability in both the mouse and monkey but it also displays improved metabolic stability after intravenous administration ($t_{1/2}$ of 1.4 hr in the mouse and 3 hr in the monkey). After a single oral dose of 25 mg/kg, a peak serum concentration (C_{max}) of 8.8 μ*M* is achieved. Because of these improvements, SCH 66336 persists in mouse serum at a concentration ≥ 1 μ*M* for more than 7 hr after a single oral dose of 25 mg/kg. On the basis of the mechanism of this compound, our working hypothesis is that the trough serum concentration (C_{min}) achieved between doses is a critical pharmacokinetic parameter. If serum levels fall below

[3] M. S. Bryant, W. A. Korfmacher, S. Wang, C. Nardo, A. A. Nomeir, and C.-C. Lin, *J. Chromatogr. A* **777,** 61 (1997).

[4] M. Liu, M. S. Bryant, J. Chen, S. Lee, B. Yaremko, P. Lipari, M. Malkowski, E. Ferrari, L. L. Nielsen, N. Prioli, J. Dell, D. Sinha, J. Syed, W. A. Korfmacher, A. A. Nomeir, C.-C. Lin, L. Wang, A. G. Taveras, R. J. Doll, F. G. Njoroge, A. K. Mallams, S. Remiszewski, J. J. Catino, V. M. Girijavallabhan, P. Kirschmeier, and W. R. Bishop, *Cancer Res.* **58,** 4947 (1998).

[5] M. Liu, M. S. Bryant, J. Chen, S. Lee, B. Yaremko, Z. Li, J. Dell, P. Lipari, M. Malkowski, N. Prioli, R. R. Rossman, W. A. Korfmacher, A. A. Nomeir, C.-C. Lin, A. K. Mallams, P. Kirschmeier, R. J. Doll, J. J. Catino, V. M. Girijavallabhan, and W. R. Bishop, *Cancer Chemother. Pharmacol.* **43,** 50 (1999).

TABLE I
PHARMACOKINETIC PROFILE OF SCH 66336 IN THE NUDE MOUSE

Route[a]	Vehicle	C_{max} (μM)	AUC (0–24 hr) (μg · hr/ml)	Bioavailability (%)	Half-life (hr)
A. After a single dose of 25 mg/kg					
Intravenous	20% HPβCD	—	31.8	—	1.4
Per os	20% HPβCD	8.84	24.1	75.8	—
Per os	0.4% methylcellulose	16.63	41.45	100	—

B. Correlation of concentration in serum and tumor xenograft after multiple doses

Dose[a] (mg/kg)	C_{max} SCH 66336 Serum (μM)	C_{max} SCH 66336 Tumor (μmol/kg)	*In vivo* tumor growth inhibition (%)
10	1.1	0.88	32
50	31.5	26.7	76

[a] Per os, four times daily.

a critical threshold, protein farnesylation will occur, replenishing the pool of farnesylated Ras and other proteins. In addition to the improved pharmacokinetics of SCH 66336, the intrinsic potency (IC_{50} of 1.9 nM) is also substantially improved compared with earlier compounds in this series. The enhanced potency and pharmacokinetic properties of SCH 66336 are critical for its antitumor efficacy when administered orally on either a four times daily or twice daily schedule.[1,4–7]

In a human colon carcinoma DLD1 xenograft study in the nude mouse, the concentration of SCH 66336 in the tumor tissue after the final dosing is similar to the steady state concentration achieved in the serum (Table IB). Although this cell line is relatively resistant to the growth inhibitory effects of SCH 66336 in soft agar (IC_{50} of 2.5 μM), these high tumor levels support antitumor efficacy. At a tumor drug level of 0.88 μmol/kg, 32% tumor growth inhibition is achieved, while a tumor drug level of 26.7 μmol/kg results in 76% inhibition. In a second tumor model utilizing HTB177 human lung carcinoma cells, tumor levels of SCH 66336 range from 14 to 33% of that in the serum.[4] However, because this cell line

[6] F. G. Njoroge, A. G. Taveras, J. Kelly, S. Remiszewski, A. K. Mallams, R. Wolin, A. Afonso, A. B. Cooper, D. F. Rane, Y.-T. Liu, J. Wang, B. Vibulbhan, P. Pinto, J. Deskus, C. S. Alvarez, J. del Rosario, M. Connolly, J. Wang, J. Desai, R. R. Rossman, W. R. Bishop, R. Patton, L. Wang, P. Kirschmeier, M. S. Bryant, A. A. Nomeier, C.-C. Lin, M. Liu, A. T. McPhail, R. J. Doll, V. M. Girijavallabhan, and A. K. Ganguly, *J. Med. Chem.* **41,** 4890 (1998).

[7] F. G. Njoroge, B. Vibulbhan, P. Pinto, W. R. Bishop, M. S. Bryant, A. A. Nomeir, C.-C. Lin, M. Liu, R. J. Doll, V. Girijavallabhan, and A. K. Ganguly, *J. Med. Chem.* **41,** 1561 (1998).

displays much greater sensitivity to soft agar growth inhibition by SCH 66336, these lower tumor drug levels also support antitumor efficacy. The reason for differences in tumor drug levels between the two xenograft models is unknown but may reflect, in part, differences in the extent of tumor vascularization. The observed pharmacokinetic properties described above support the use of SCH 66336 as an oral agent and indicate that SCH 66336 can readily reach the target tumor tissue.

In Vivo Efficacy Studies Using Wap–*ras* Transgenic Mouse Models

To evaluate SCH 66336 efficacy in a model with Ras mutation, spontaneous tumor occurrence, and intact host immunity, Wap-*ras* transgenic mice carrying an activated Ha-*ras* oncogene are used. *ras* expression is driven by the whey acidic protein promoter in this model. Because the Ha-*ras* transgene is carried on the Y chromosome, only male mice develop tumors (mammary and salivary). The founder mice are obtained from A. C. Andres[8] and sublines (69-2 and 69-2F) have been developed in house.[9]Subline 69-2 mice spontaneously develop tumors between 3 and 6 months of age, while the tumor onset time in 69-2F mice (Wap-*ras* transgene bred into the FVB/N strain[10]) is between 6 and 9 weeks. These tumors express the human p21 Ras protein at high levels and display a high mitotic index. Neoplastic tissue has a high mitotic index and tumor-bearing animals have an ongoing immune system response as evidenced by immune cell infiltration of the affected tissue.[11]

Differences in the time of tumor onset allow us to use 69-2F mice for testing prophylactic efficacy and 69-2 mice for evaluating therapeutic efficacy. For prophylactic studies, 69-2F mice are enrolled when they are 35 days old, before tumor onset (42–63 days). For therapeutic studies, 69-2 mice are enrolled on development of a palpable tumor. For both studies, vehicle or FTI treatment lasts for 4 weeks on a four times daily schedule with 10 mice per treatment group. The mice in study are monitored for general health conditions at least once a day and palpated to determine tumor onset/tumor growth twice per week. The FTI is dissolved in 20% (w/v) HPβCD. Vehicle controls receive 20% (w/v) HPβCD. Vehicle or drug solution (0.1 ml) is administered by oral gavage, as described above, every 6 hr (four times daily) for 20 or 21 days. A no-treatment control is always included along with the vehicle control to evaluate the influence of vehicle and of the four times daily gavage treatment. Once palpable, tumor volume is measured in three dimensions twice weekly and calculated with the formula $V = 1/6\ (\pi LWT)$, where

[8] A.-C. Andres, C.-A. Schonenberger, B. Groner, L. Hennighausen, M. LeMeur, and P. Gerlinger, *Proc. Natl. Acad. Sci. U.S.A.* **84,** 1299 (1987).

[9] L. L. Nielsen , M. Gurnani, J. J. Catino, and R. D. Tyler, *Anticancer Res.* **15,** 385 (1995).

[10] L. L. Nielsen, M. Gurnani, and R. D. Tyler, *Cancer Res.* **52,** 3733 (1992).

[11] L. L. Nielsen, C. M. Discafani, M. Gurnani, and R. D. Tyler, *Cancer Res.* **51,** 3762 (1991).

L, *W*, and *T* represent length, width, and thickness, respectively.[4,5] *T/C* value in percent is calculated where *T* and *C* are the mean tumor volume of the treated and control groups, respectively, at the end of each experiment. Average inhibition is used to compare the efficacy of various treatments and is derived by subtracting the *T/C* values of each treatment from 100. A single-tailed Student *t* test is used for statistical analysis.

When the efficacy of SCH 66336 is evaluated, clear antitumor activity is demonstrated when dosing is initiated prior to tumor onset (at 35 days of age) and continued for 4 weeks (to 63 days of age) in the 69-2F strain. SCH 66336 treatment delays tumor onset, reduces the average number of tumors per mouse, and reduces the average tumor weight per animal.[4] More significant antitumor effects are seen at the high (40 mg/kg) dose level, where animals remained tumor free throughout the dosing period. These animals remain tumor free for a minimum of another 20 days after treatment is terminated, at which time tumors do develop in some mice.

The Wap-*ras* model is also utilized in a therapeutic mode in which treatment is initiated after mice have developed palpable tumors. Strain 69-2 mice are treated with SCH 66336 at 2.5, 10, 20, or 30 mg/kg (four times daily) for 4 weeks. Growth curves for the various treatment groups are shown in Fig. 2. The mean tumor size at the start of dosing (day 0) is 200 mm^3. In the vehicle-treated group, tumors grow throughout the course of the experiment to a volume of >1500 mm^3 by the end of the study. SCH 66336 at 2.5 or 10 mg/kg significantly slows the rate of tumor growth. SCH 66336 at the 20- or 30-mg/kg dose level results in significant and complete tumor regression. Similar results are obtained in strain 69-2F mice. Significant and rapid tumor regressions are also observed in 69-2 mice when SCH 66336 treatment is initiated after tumors have achieved sizes in the 2000-mm^3 range (data not shown).

In cross-over studies, mice initially assigned to the vehicle control or low-dose SCH 66336 groups (2.5 or 10 mg/kg are subsequently treated with SCH 66336 at 30 mg/kg for 3 more weeks. As soon as the high-dose treatments are initiated, all tumors regress in a comparable fashion (Fig. 3). Such cross-over efficacy is demonstrated in both the 69-2 and 69-2F Wap-*ras* strains. The regressed tumor reappears and continues to grow 14 to 21 days after FTI treatment is withdrawn. However, every recurrent tumor can be induced into regression again by retreatment with SCH 66336 (data not shown).

In Vivo Efficacy Studies Using *ras*-Transformed Rodent Fibroblasts and Human Tumor Xenografts

After 1 week of acclimation, 5- to 7-week-old female nude mice are subcutaneously inoculated with various cell lines on day 0. The construction, cloning, and selection of NIH 3T3-CVLS cells transfected with activated Ha-Ras containing its native C-terminal sequence CVLS have been described previously.[1,5] The number

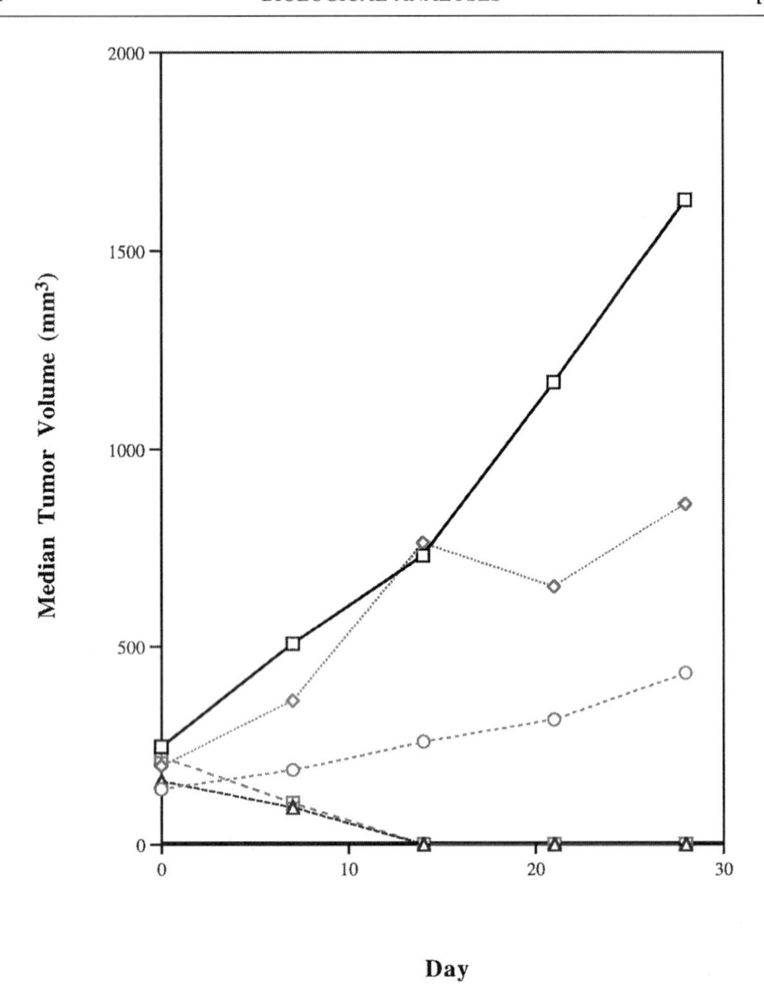

FIG. 2. SCH 66336 induces tumor growth inhibition and tumor regression when dosed therapeutically to Wap-*ras* transgenic mice. Wap-*ras* transgenic mice (strain 69-2) were used to evaluate SCH 66336 in a therapeutic mode in which treatment was initiated after mice had developed palpable tumors (>100 mm^3). Mice were orally treated every 6 hr with escalating doses [(◇) 2.5, (○) 10, (△) 20, or (⊞) 30 mg/kg] of SCH 66336 for 4 weeks. The median tumor size measured on days 0, 7, 14, 21, and 28 is plotted for each group. (□) 20% HPBCD.

of cells inoculated are as follows: 3.0×10^5 for NIH 3T3 Ki-Ras-CVIM; 2×10^6 for NIH 3T3 Ha-Ras-CVLL and -CVLS; 3×10^6 for MSV-3T3; and 5×10^6 for PT-24. All the cell lines are cultured and passaged routinely, as a monolayer, in Dulbecco's modified Eagle's medium (DMEM) with nonessential amino acids plus 10% (v/v) fetal bovine serum at 37° and 5% CO_2. Subconfluent (90–99%)

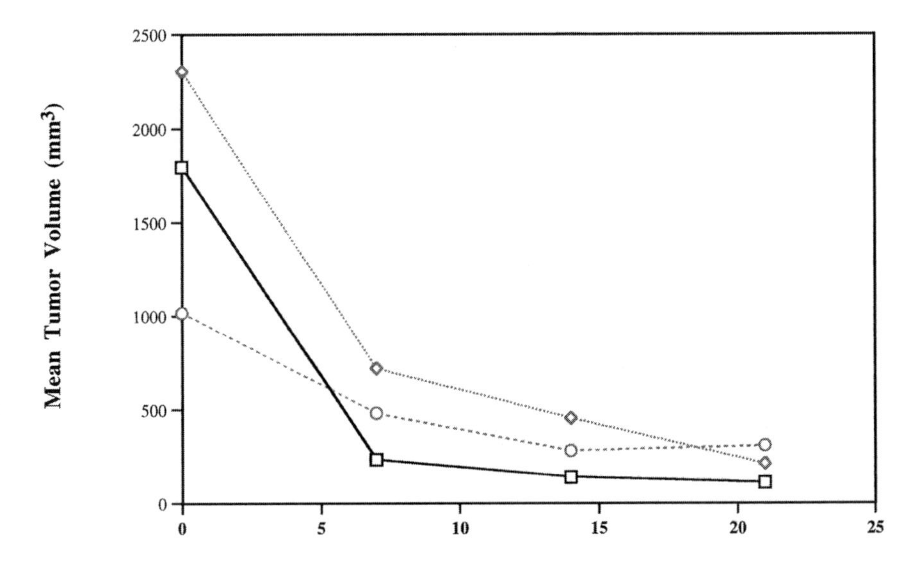

FIG. 3. Wap-*ras* 69-2 tumors, previously treated with no or low doses [(□) 0, (◇) 2.5, or (○) 10 mg/kg] of SCH 66336 for 4 weeks, regress in similar fashion when treated with a high dose (30 mg/kg) of SCH 66336. After the experiment described in Fig. 2 was completed, four mice of each no- or low-dose group were randomly selected and enrolled into the cross-over study. Each mouse in the cross-over study received gavage of SCH 66336 at 30 mg/kg every 6 hr for an additional 3 weeks. The mean tumor size measured on days 0, 7, 14, and 21 is plotted.

cells are trypsinized, washed twice in phosphate-buffered saline (PBS), and enumerated before being diluted to the appropriate concentration for inoculation. All the xenograft tumor models used for our FTI evaluations have a 100% take rate and grow consistently in every animal. Animals are randomly assigned to control and treatment groups (10 animals per group) before the first treatment. Drug treatment at either 10 or 50 mg/kg is initiated on day 1.

In human tumor xenograft models, nude mice are subcutaneously inoculated with various cell lines on day 0. Human cancer cell lines used are as follows: lung, A-549, HTB-177; pancreas, AsPc-1, HPAF-II, Hs700T, MIA PaCa; colon; HCT-116, DLD-1, prostate, DU-145; urinary bladder, EJ. All cell line stocks (except EJ) are obtained from the American Type Culture Collection (Rockville, MD). The numbers of cells inoculated are as follows: 3.0×10^6 for HTB-177; 5.0×10^6 for DLD-1, DU-145, HCT-116, HPAF-II, and MIA PaCa; 6.0×10^6 for A-549, AsPc-1, and Hs700T. Animals are randomly assigned to control and treatment

TABLE II
In Vivo EFFICACY OF SCH 66336 IN XENOGRAFT MODELS

		Average growth inhibition (%)		
Tumor (site)	Mutation status	2.5 mg/kg, four times daily	10 mg/kg, four times daily	40 mg/kg, four times daily
A 549 (lung)	K-Ras	42	61	70
HTB-177 (lung)	K-Ras	40	57	83
AsPc-1 (pancreas)	K-Ras	29	33	72
HPAF-II (pancreas)	K-Ras	9	19	67
Hs700T (pancreas)	Not detected	40	61	78
MIA PaCa (pancreas)	K-Ras	46	48	72
DU-145 (prostate)	Not detected	29	59	86
HCT-116 (colon)	K-Ras	38	67	84
EJ[a] (bladder)	H-Ras	59	80	100[b]
NIH 3T3-CVLS[a]	H-Ras	16	55	100[b]
LOX[a] (melanoma)	Not detected		57	97[c]
DLD-1[d] (colon)	K-Ras		32	76

[a] Result of a 5, 20, and 80 mg/kg, twice daily treatment.
[b] Ten of 10 mice were tumor free at the end of the experiment.
[c] Five of 10 mice were tumor free at the end of the experiment.
[d] Result of a 10 and 50 mg/kg, four times daily treatment.

groups (10 animals per group) before the first treatment. Drug treatment at 0.625, 2.5, 10, or 40 mg/kg, four times a day, is initiated on day 1. In the xenograft models, the making of FTI dosing solution, the use of vehicle controls and no-treatment controls, the methodology used to measure the tumor volume, and the methods to determine the significance of efficacy of different treatment groups are all as described above in the section on transgenic mouse models. In some experiments, vehicle-treated animals experience 5–10% weight loss, while no additional weight loss is observed in FTI-treated animals.

When NIH 3T3 cells transformed with activated Ha-Ras-CVLS are used to evaluate the antitumor effect of SCH 66336, dose-dependent inhibition of tumor growth (Table II) is observed, with 16, 55, and 100% tumor growth inhibition at the 5-, 20-, and 80-mg/kg dose levels, respectively. This inhibition is significant at $p < 0.005$ for the 20- and 80-mg/kg treatment groups. The efficacy of SCH 66336 observed in this model is consistent with the results observed with an earlier tricyclic analog (e.g., SCH 59228), which selectively inhibits, growth of tumors from cells transformed with a farneslylated versus geranylgeranylated form of Ha-Ras.[4,5]

When further evaluated in a panel of human tumor xenograft models, SCH 66336 demonstrates significant *in vivo* antitumor activity in a variety of models when dosed orally in nude mice (Table II). In most of these studies, tumor cells are implanted subcutaneously on day 0 and oral gavage is administered four times a day beginning on day 1 and continuing throughout the study. Typically, four times daily doses of 2.5, 10, and 40 mg/kg are administered by oral gavage with using 20%(v/v) HPβCD as vehicle. Using this paradigm, antitumor activity is observed in xenograft models of lung carcinoma, pancreatic carcinoma, colon carcinoma, and prostate carcinoma. Average inhibition of tumor volume in drug-treated versus vehicle control groups at the end of the experiment ranges from 67 to 86% at the 40-mg/kg, four times daily level ($p < 0.01$ in each model). Significant antitumor activity is also observed at lower dose levels (10 and 2.5 mg/kg, four times daily). Antitumor activity is also observed when SCH 66336 is administered by a twice a day schedule in the HTB-177 lung carcinoma model. Twice daily doses of 80 and 20 mg/kg produce equivalent tumor growth inhibition (87 and 59%, respectively) compared with four times daily doses of 40 and 10 mg/kg. This result is anticipated on the basis of the known pharmacokinetic properties of SCH 66336 in mice. Using the twice daily dosing schedule in a human bladder carcinoma (EJ) model, a human melanoma (LOX) model, and a *ras*-transfected rodent fibroblast (NIH 3T3-CVLS) model, dose-dependent tumor growth inhibition is observed with doses of 5 and 20 mg/kg, respectively. Complete inhibition is achieved with a dose of 80 mg/kg. These data, along with the promising pharmacokinetic properties of SCH 66336 in primates, suggest that a twice daily dosing schedule may be achievable clinically.

Combination of Farnesyltransferase Inhibitor with Cytotoxic Chemotherapy

HTB-177 human lung carcinoma cells are also used to evaluate the combination efficacy of SCH 66336 with cytotoxic agents. Cytotoxic agents cyclophosphamide, 5-fluorouracil (5-FU), vincristine, and paclitaxel are obtained from Sigma (St. Louis, MO). HTB-177 cells are inoculated into nude mice and efficacy is evaluated according to the methodology described above. In one study, nude mice (10 mice per group) inoculated with HTB 177 cells on day 0 are treated with either SCH 66336 alone (40 mg/kg, four times daily, day 1 to day 26), or cyclophosphamide alone (200 mg/kg once on day 13, intraperitoneal injection), 5-fluorouracil alone (50 mg/kg once on day 13, intraperitoneal injection), or vincristine alone (1 mg/kg once on day 13, intraperitoneal injection). Some animals are treated with the combination of SCH 66336 and each of the cytotoxic agents.[4]

SCH 66336 alone results in 60% tumor growth inhibition. Cytotoxic agents alone yield 9, 28, and 7% tumor growth inhibition for cyclophosphamide (200 mg/kg), 5-FU (50 mg/kg), and vincristine (1 mg/kg), respectively. When SCH 66336 is used in combination with these agents, tumor growth inhibition of 81, 80, and 80% is observed for the combination with cyclophosphamide (Cytoxan), 5-FU, and vincristine, respectively. These results are significant compared with the SCH 66336 alone result, with $p < 0.05$. The greater efficacy observed with the combinations indicates that there is no antagonism on combining SCH 66336 with these agents.

In another study, paclitaxel is also evaluated in combination with SCH 66336. HTB-177-bearing nude mice are treated either with SCH 66336 alone (20 mg/kg twice a day from day 1 to day 14), paclitaxel alone (5 mg/kg once a day from day 4 to day 7, intraperitoneal injection), or the combination of SCH 66336 and paclitaxel. The combination results in significant efficacy compared with SCH 66336 or paclitaxel alone.[12] This combination of SCH 66336 and paclitaxel demonstrates either synergistic or additive activity over a broad panel of human tumor cell lines, except for one breast cancer cell line, against which the combination demonstrates antagonism.

Interestingly, while cyclophosphamide at 200 mg/kg or SCH 66336 at 10 mg/kg, administered singly, as does not result in tumor regression in the Wap-*ras* transgenic tumors, the combination of these treatments results in significant regression.[4] This result corresponds well with the enhanced efficacy observed with the combination of SCH 66336 and cyclophophamide in the HTB-177 xenograft model and supports the possibility that SCH 66336 may act in an additive or greater than additive manner when combined with standard cytotoxic cancer therapy. Similarly, combination of SCH 66336 with paclitaxel also results in enhanced efficacy in Wap-*ras* transgenic tumors.[12]

Analysis of Apoptosis, Proliferation, and Other Surrogates in Tumor Specimens

Ten mammary tumor-bearing strain 69-2 Wap-*ras* mice are treated with SCH 66336 at 40 mg/kg, four times daily, for five consecutive days. Every 24 hr after the initiation of treatment, two mice are killed, and tumors are removed and preserved in 10% (v/v) buffered formalin for histological study. Formalin-fixed and paraffin-embedded tissues are cut into 5-μm sections, using a Leitz (Wetzlar, Germany) model 1512 microtome, and are analyzed for the number of apoptotic cells, using the terminal deoxynucleotidyltransferase-mediated dUTP-biotin nick end-labeling (TUNEL) method as previously described.[13] The assay is performed with reagents obtained from Oncor (Gaithersburg, MD). Cells undergoing apoptosis are evaluated by light microscopy. Positively stained cells within a 6.5×9 mm grid in the eyepiece are counted over 10 different $\times 200$ magnification fields for

[12] B. Shi, B. Yaremko, G. Hajian, G. Terracina, W. R. Bishop, M. Liu, and L. L. Nielsen, *Cancer Chemother. Pharmacol.* **46,** 387 (2000).

[13] R. E. Barrington, M. A. Subler, E. Rands, C. A. Omer, P. J. Miller, J. E. Hundley, S. K. Koester, D. A. Troyer, D. J. Bearss, M. W. Conner, J. B. Gibbs, K. Hamilton, K. S. Koblan, S. D. Mosser, T. J. O'Neill, M. D. Schaber, E. T. Senderak, J. J. Windle, A. Oliff, and N. E. Kohl, *Mol. Cell. Biol.* **18,** 85 (1998).

each tumor section. The numbers are recorded and the mean and standard deviation are calculated. The Student *t* test is used to compare the significance between treatments.

Inhibition of cellular proliferation *in vivo* is assessed by bromodeoxyuridine (BrdU) labeling. After 5 days of treatment with SCH 66336 (80 mg/kg, twice daily), two strain 69-2 Wap-*ras* mice are given intraperitoneal injections of 0.1 ml of BrdU solution (10 mg/kg) 1 hr after the final SCH 66336 dose. Four hours after BrdU injection, the mice are killed and mammary tumors are removed and preserved for immunohistochemical evaluation. Proliferating tumor cells are detected with antibodies directed against BrdU incorporated into the DNA.[14] The assay is performed with reagents obtained from Boehringer Mannheim (Indianapolis, IN). Anti-BrdU-stained cells are quantified by light microscopy as described above.

When hematoxylin-eosin-stained tissue sections of tumors from these transgenic animals are examined microscopically, a marked increase in apoptosis and a decrease in mitotic figures are observed on SCH 66336 treatment. A marked increase in apoptosis is observed as early as 24 hr after the initiation of SCH 66336 treatment in the Wap-*ras* tumors. Similar results are seen in the EJ xenograft model.[4] In the Wap-*ras* model, maximal apoptosis is observed at 24 hr; however, the apoptotic index remains, elevated for at least 5 days. A significant dose-dependent decrease in BrdU labeling is also detected on day 5 after dosing was initiated in both wap-*ras* transgenic tumors and EJ xenograft tumors, indicating a significant reduction in cell proliferation.[4]

While the FTIs clearly function through inhibition of farnesyl transferase activity, it is unclear which farnesylated protein(s) contribute to their antiproliferative activities. We have been working to identify intracellular farnesylated proteins that would be useful indicators of farnesyltransferase inhibition *in vivo*. To date we have evaluated a number of proteins including H-Ras, PxF (HK33), dna J, and prelamin A. dna J[15] has proved to be a useful surrogate marker and we have observed unfarnesylated dna J to increase in a dose-dependent fashion in xenograft tumors after treatment with various FTIs (W. R. Bishop *et al.*, unpublished data). The prelamin system has been shown successfully to be a surrogate marker with which quantify the preclinical and clinical efficacy of SCH 66336.[16]

Summary

The *in vivo* evaluation process described here was instrumental in the identification of SCH 66336 as a clinical candidate. Our lead FTI, SCH 66336, and

[14] H. G. Gratzner, *Science* **218,** 474 (1982).

[15] D. A. Andres, H. Shao, D. C. Crick, and B. S. Finlin, *Arch. Biochem. Biophys.* **346,** 113 (1997).

[16] A. A. Adjei, C. Erlichman, J. N. Davis, D. Cutler, J. A. Sloan, R. Marks, L. H. Hanson, P. A. Svingen, W. R. Bishop, P. Kirschmeier, and S. H. Kaufmann, *Cancer Res.* **60,** 1871 (2000).

several other FTIs are being evaluated in early-phase clinical trials to establish proof-of-principle for farnesyl transferase inhibition in human patients.[17] The pre-clinical studies described here suggest that FTIs may have utility against a wide array of human cancers as a single agent and may, at least in some cases, lead to tumor regression. In addition, the results to date in combination with cytotoxic chemotherapeutic agents in animal models indicate that these combinations may enhance the clinical efficacy of FPT inhibitors. Further preclinical studies should help to guide the clinical development of this class of novel antitumor agents.

[17] E. K. Rowinsky, J. J. Windle, and D. D. Von Hoff, *J. Clin. Oncol.* **17,** 3631 (1999).

[28] Animal Models for Ras-Induced Metastasis

By CRAIG P. WEBB and GEORGE F. VANDE WOUDE

Introduction

Tumor metastasis, the process by which cancer spreads from the site of origin to secondary tissues, is responsible for the majority of mortalities among cancer patients. As such, a greater understanding of the fundamental processes involved in tumor metastasis is key to the development of potential therapeutics. In this chapter, we describe an experimental system for investigating the complex cascade of events during Ras-mediated metastasis *in vivo*.

Metastasis is a multistage process during which cells originating from a primary tumor mass migrate to and colonize at distant sites, forming secondary metastases. A number of key cellular events occur during metastasis, including cell proliferation, migration, invasion, survival, and angiogenesis.[1] These events are intricately regulated by host–tumor cell interactions that ultimately determine the fate of a cancer. The first stages of metastasis include the formation of a primary tumor, tumor vascularization (angiogenesis), and the escape of cells into the host circulation ("intravasation"). Circulating tumor cells must overcome a number of barriers prior to the development of metastases, including evasion of the immune response and survival from apoptosis in the absence of substrate attachment. On adhesion to the vasculature, tumor cells then migrate and invade through the endothelium and surrounding stromal tissue at a secondary location, a process known as "extravasation." After extravasation, individual tumor cells proliferate to initially form microscopic nodules, and ultimately macroscopic metastases. The conversion

[1] E. Ruoslahti, *Adv. Cancer Res.* **76,** 1 (1999).

0076-6879/00 $35.00

of micrometastases to macrometastases appears to represent a major rate-limiting step during metastasis.[2,3] Indeed, the majority of tumor cells that successfully exit the vasculature fail to form macrometastases, and may lay dormant for many years.[2]

Although a number of key events have been identified during the formation of metastases, the detailed molecular events governing this process remain to be elucidated. The Ras superfamily of GTP-binding proteins has long been of interest in the field of cancer research. Activating mutations in *ras* genes are commonly found in a variety of human tumors,[3a] and such oncogenic *ras* mutants confer both tumorigenic and metastatic properties to recipient cell lines in culture.[4] Thus, oncogenic Ras has served as a prototype for the study of cellular transformation, tumor growth, and metastasis. In its active GTP-bound configuration, Ras mediates intracellular signaling events leading to alterations in cell behavior. Ras activation is controlled by a series of proteins including GTPase-activating proteins (GAPs), guanine nucleotide exchange factors (GEFs), and guanine nucleotide dissociation stimulators/inhibitors, which mediate the interconversion between active GTP-bound and inactive GDP-bound Ras.[5] A region of the Ras protein known as the effector domain (amino acids 32–40 in H-Ras) has been identified as an essential mediator of the interaction between GTP–Ras and downstream effectors, which in mammalian cells include Raf-1, phosphatidylinositol 3-kinase (PI3K), Ras-GAPs, Ral guanine nucleotide dissociation stimulator (RalGDS), MEKK1 [MAPK (mitogen-activating protein kinase)/ERK (extracellular signal-regulated kinase) kinase kinase 1] KSR (kinase suppressor of Ras), Rin1, and AF6/Rsb1.[5–7] Point mutations within the effector domain of oncogenic Ras generate mutant proteins deficient in specific effector function and therefore activation of specific down-stream signaling pathways.[8,9] We made use of three of these Ras effector domain mutants, V12S35 H-Ras, V12G37 H-Ras, and V12C40 H-Ras, to determine the signaling pathways contributing to Ras-mediated metastasis.[10] The Ras effector

[2] A. F. Chambers, I. C. MacDonald, E. E. Schmidt, V. L. Morris, and A. C. Groom, *Cancer Metastasis Rev.* **17,** 263 (1999).

[3] V. L. Morris, E. E. Schmidt, I. C. MacDonald, A. C. Groom, and A. F. Chambers, *Cancer Metastasis Rev.* **17,** 263 (1997).

[3a] J. L. Bos, *Cancer Res.* **49,** 4682 (1989).

[4] A. F. Chambers and A. B. Tuck, *Crit. Rev. Oncol.* **4,** 95 (1993).

[5] S. L. Campbell, R. Khosravi-Far, K. L. Rossman, G. J. Clark, and C. J. Der, *Oncogene* **17,** 1395 (1998).

[6] T. Hunter, *Cell* **88,** 333 (1997).

[7] L. Van Aelst, M. A. White, and M. H. Wigler, *Cold Spring Harbor Symp. Quant. Biol.* **59,** 181 (1994).

[8] M. A. White, C. Nicolette, A. Minden, P. Polverino, L. Van Aelst, M. Karin, and M. H. Wigler, *Cell* **80,** 533 (1995).

[9] R. Khosravi-Far, M. A. White, J. K. Westwick, P. A. Solski, M. Chrzanowska-Wodnicka, L. Van Aelst, M. H. Wigler, and C. J. Der, *Mol. Cell. Biol.* **16,** 3923 (1996).

domain mutant molecules have been previously characterized on the basis of their respective abilities to mediate a variety of signaling events and induce subsequent cellular phenotypes *in vitro,* including cellular transformation, cell cycle progression, apoptosis, and cell migration.[5,11,12] As such, it has been shown that V12S35 H-Ras, V12G37 H-Ras, and V12C40 H-Ras mediate their cellular effects predominantly through activation of either Raf, RalGDS, or PI3K, respectively.[8,9,11] Although these data provide valuable insights into the mechanisms of Ras-mediated events *in vitro,* they are limited with regard to the study of tumor formation and progression *in vivo.*

In this chapter we describe the application of the Ras effector domain mutants to study tumor growth and metastasis *in vivo*. The nature of these mutants allows the investigator to study the relative contribution of downstream signaling pathways to Ras-mediated events *in vivo,* and provides an excellent isogeneic system to investigate the detailed molecular mechanisms of tumor growth and metastasis.

Characterization of Stable NIH 3T3 Cell Lines Expressing *ras* Effector Domain Mutants

Principles

The three mammalian *ras* genes, H-, K-, and N-*ras,* are constitutively activated by naturally occurring oncogenic mutations at amino acid positions 12, 13, or 61, due to impaired GTPase activity, thereby resulting in the accumulation of Ras in the GTP-bound or active state.[13] Thus, an oncogenic variant containing a glycine-to-valine mutation at position 12 (G12V) of H-*ras* (termed throughout as V12 H-*ras*) transforms a variety of rodent cell lines, including mouse NIH 3T3 cells.[4] When additional mutations are introduced into V12 H-Ras within its effector domain (amino acids 32–40), the ability of activated Ras to interact with, and hence activate, distinct downstream effectors is severely impaired.[8,9] These effector domain mutants can therefore be used to investigate the contribution of individual Ras effectors and signaling pathways to various Ras-mediated phenotypes both *in vitro* and *in vivo*. For *in vivo* studies, it is essential that stable cell lines be generated that express the different V12 H-Ras effector domain mutants to allow for the study of long-term tumor growth and metastasis *in vivo*. Prior to administration of these cells into mice, the cell lines expressing the different Ras proteins should be carefully characterized for expression and activation of signaling pathways downstream of Ras, to ensure that the chosen cell line behaves

[10] C. P. Webb, L. Van Aelst, M. H. Wigler, and G. F. Vande Woude, *Proc. Natl. Acad. Sci. U.S.A.* **95,** 8773 (1998).

[11] T. Jonesone, and D. Bar-Sagi, *J. Mol. Med.* **75,** 587 (1997).

[12] J.-J. Yang, J.-S. Kang, and R. S. Krauss, *Mol. Cell. Biol.* **18,** 2586 (1998).

[13] M. Barbacid, *Annu. Rev. Biochem.* **56,** 779 (1987).

like those previously characterized using this system. The ability of these V12 H-Ras effector domain mutants to activate downstream signaling pathways likely differs between different cell types (C. P. Webb, unpublished observations, 1999). We routinely use a subclone of NIH 3T3 cells [NIH 3T3 (490)] that displays a low rate of spontaneous transformation, and serves as an excellent recipient for ectopic gene expression. After gene transfection and selection for cells expressing the gene(s) of interest, we use pooled populations of cells as apposed to individual cell clones. This avoids problems of clonal variation frequently observed when comparing individual cell clones. Moreover, metastasis is clonal in origin, and only a small subset of a population of tumor cells eventually form macrometastases.[2] By injecting a heterogeneous population of transfected cells and obtaining tumor-derived explants (see Culturing of Tumor Explants, below), we are better able to characterize the molecular basis of metastasis.

Plasmids

The effector domain mutants of V12 H-Ras were identified after randomized mutagenesis.[8,9] V12S35 H-Ras contains a threonine-to-serine mutation at amino acid position 35 (T35S), V12G37 H-Ras contains a glutamic acid-to-glycine mutation at position 37 (EG37), and V12C40 H-Ras contains a tyrosine-to-cysteine mutation at amino acid position 40 (Y40C). V12 H-*ras* and the effector domain mutants of V12-H-*ras* are expressed with the pDCR mammalian expression vector. pDCR contains a cytomegalovirus (CMV) promoter followed by unique *Sal*I and *Bam*HI sites and the rabbit β-globin terminator and splice sequence. This region and the *neo* (neomycin resistance) gene under the control of the simian virus 40 (SV40) promoter are flanked by Moloney murine leukemia retrovirus 5′ and 3′ long terminal repeats (LTRs). The mutant *ras* genes were cloned into pDCR, using the *Sal*I and *Bam*HI restriction sites, and were tagged at the amino terminus with an hemagglutinin (HA) sequence to allow easy detection of ectopically expressed Ras proteins.[9,10]

Cell Lines and Transfections

NIH 3T3 (490) cells, a subclone of the murine NIH 3T3 fibroblast cell line, display a low level of spontaneous transformation. NIH 3T3 (490) cells are grown in Dulbecco's modified Eagle's medium (DMEM; GIBCO-BRL, Gaithersburg, MD) supplemented with 10% (v/v) fetal bovine serum (GIBCO-BRL) and are maintained at 37° in a humidified atmosphere of 5% CO_2.

V12 H-*ras* and V12 H-*ras* effector domain mutant cDNAs encoded within the pDCR plasmids are transfected into NIH 3T3 cells with the cationic liposome reagent 1,2-dioleoyl-3-trimethylammoniumpropane (DOTAP; Roche, Nutley, NJ). Cells are grown to approximately 50% confluence in a 100-mm tissue culture dish, and the medium is replaced with fresh growth medium 24 hr prior to transfection.

Ten micrograms of plasmid DNA is mixed with the DOTAP reagent as per the manufacturer instructions, and after incubation this is mixed with fresh growth medium and applied directly to the cells for 8 hr. After this time, the cells are washed and grown for 24 hr in fresh growth medium. The medium is then replaced with growth medium containing geneticin (G418, 400 μg/ml; GIBCO-BRL), which allows for the selection of cells expressing the *neo* gene (contained within the pDCR plasmid) and therefore the *ras* genes of interest. After approximately 1–2 weeks, discrete cell colonies should be apparent, and these are pooled as a mixed population. As discussed above, we typically use pooled populations of cells to avoid possible differences due to clonal variation.

Analyzing Expression of Ras Proteins

After transfection of the *ras*cDNA, we analyze the expression of the Ras proteins to check for differences that may exist between cell lines harboring the different Ras mutants. This is readily achieved by sodium dodecyl sulfate–polyacrylamide gel electrophoresis (SDS–PAGE) followed by immunoblotting with an antibody against the HA epitope. Stably transfected cells are grown in 100-mm culture dishes, and washed twice with ice-cold phosphate-buffered saline (PBS), and solubilized in 0.5–1 ml of lysis buffer [20 m*M* piperazine-*N,N′*-bis(2-ethanesulfonic acid) (PIPES, pH 7.4), 150 m*M* NaCl, 1 m*M* EGTA, 1.5 m*M* $MgCl_2$, 1% (v/v) Triton X-100, aprotinin (10 μg/ml), leupeptin (10 μg/ml), 1 m*M* phenylmethylsulfonyl fluoride (PMSF), 1 m*M* sodium orthovanadate, and 0.1% (w/v) SDS]. Cell lysates are clarified by centrifugation at 15,000 *g* for 5 min at 4° to remove insoluble material, and the amount of protein per sample is determined by a standard protein assay (Pierce, Rockford, IL).

For SDS–PAGE, 10 μg of protein sample is resuspended 3 : 1 into 4× Laemmli buffer [0.4 *M* Tris-HCl (pH 6.8), 0.4 *M* dithiothreitol (DTT), 8% (w/v) SDS, 39% (v/v) glycerol, 0.04% (w/v) bromphenol blue], boiled for 5 min, and loaded into a well of an SDS–polyacrylamide gel (Novex, San Diego, CA) alongside appropriate molecular mass markers. We typically use 14% (w/v) gels to adequately resolve the Ras proteins that have an approximate molecular mass of 21 kDa. Proteins are then transferred to nitrocellulose (Hybond-ECL; Amersham, Arlington Heights, IL) solid support overnight, using a typical electroblotting transfer apparatus. After transfer, blots are reversibly stained with Ponceau S (Sigma, St. Louis, MO) to visualize proteins prior to immunodetection. This ensures correct transfer from the SDS–polyacrylamide gel to nitrocellulose, and serves as an additional verification for equal protein loading between lanes. Blots are blocked with 5% (w/v) bovine serum albumin (BSA) in Tris-buffered saline plus 0.1% (v/v) Tween 20 (TBS-T, pH 7.4) to prevent nonspecific binding, and then probed initially with polyclonal antiserum against the HA epitope (clone Y-11, 2 μg/ml; Santa Cruz Biotechnology, Santa Cruz, CA), followed by a 1 : 15,000 dilution of a goat anti-rabbit peroxidase-conjugated antiserum (Sigma). This allows detection of the HA-tagged Ras proteins, using enhanced chemiluminescence (ECL) reagents (Amersham), and subsequent X-ray film exposure and development. A typical X-ray film showing HA-tagged Ras protein expression in stable pools of NIH 3T3 (490) cells transfected with the different Ras effector domain mutants is shown in Fig. 1A.

Activation of Extracellular Signal-Regulated Kinase 1/2 Pathway by Ras Effector Domain Mutants

The Ras effector domain mutants used in this study were originally characterized on the basis of their differential abilities to interact with the Raf-1 kinase in a yeast two-hybrid system.[8,9] The Raf kinase is a well-characterized effector of active GTP-bound Ras that lies upstream in the classic MAP kinase (ERK1/2) pathway. In this pathway, activation of Ras leads to the sequential activation of a series of intracellular kinases (Raf, MEK, ERK1/2), culminating in alterations in gene transcription and cell behavior.[5] The V12S35 H-Ras mutant retains Raf-1 association, whereas both V12G37 H-Ras and V12C40 H-Ras mutants are defective in this function.[8,9] It is important to determine the ability of the Ras mutants to activate the Raf–MEK–ERK1/2 pathway, as well as other signaling pathways of interest (such as RalGDS and PI3K) in the recipient cell because variations in the extent of activation of these pathways are observed in different cell types (C. P. Webb, unpublished observations, 1999).[8,10,14] Activation of the Raf–MEK–ERK1/2 pathway can be determined at multiple levels. The physical association between Ras and Raf can be determined by coimmunoprecipitation analysis, while the activation of either MEK or ERK1/2 can be determined either by *in vitro* kinase assays or by using antibodies specific for the activated (phosphorylated) forms of the respective proteins.

We are able to detect complexes between the V12 H-Ras proteins and Raf-1 by immunoprecipitation of the HA-tagged Ras protein, followed by Western blotting with an antibody against Raf-1.[10] In brief, cell lysates are produced and clarified as described above. Protein lysates (500 μg) are then incubated overnight with 20 μg of anti-HA antiserum at 4°. Immune complexes consisting of HA-tagged Ras and associated proteins are then bound to protein A–agarose beads, which can be subsequently separated from the noncomplexed proteins by a series of centrifugations and washing steps. Immune-complexed proteins can then be separated from the protein A–agarose by boiling in 2× Laemmli buffer (see above), and are resolved by 8% (w/v) SDS–PAGE (Raf-1 has an approximate molecular mass of 74 kDa). After transfer to nitrocellulose, any Raf-1 complexed to HA-tagged

[14] P. Rodriguez-Viciana, P. H. Warne, A. Khwaja, B. M. Marte, D. Pappin, P. Das, M. D. Waterfield, A. Ridley, and J. Downward, *Cell* **89,** 457 (1997).

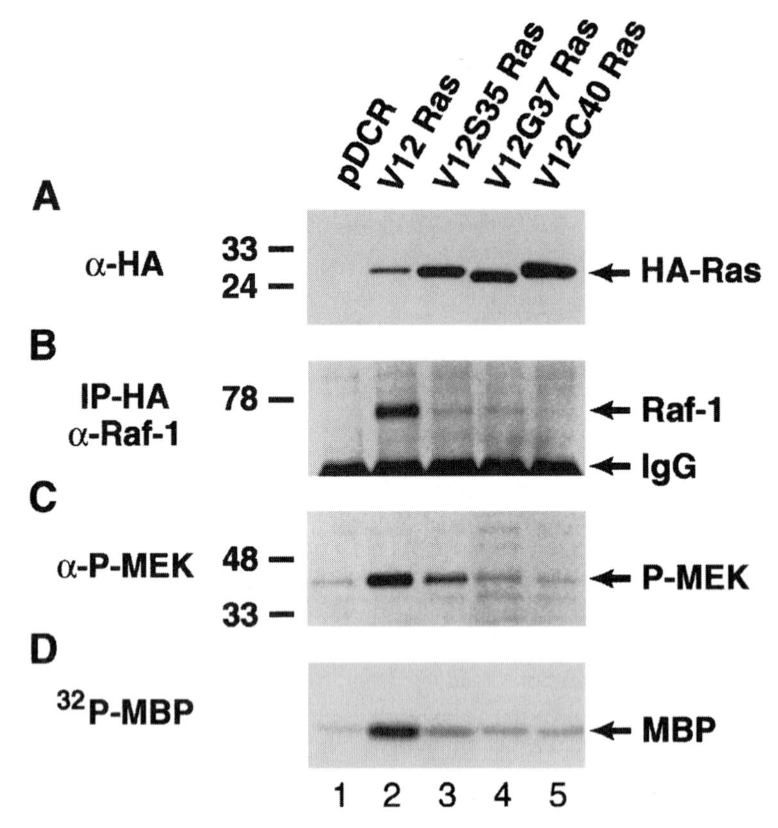

FIG. 1. Characterization of the Raf–MEK–ERK1/2 pathway in NIH 3T3 (490) cells transfected with vector alone (lane 1), V12 H-Ras (lane 2), V12S35 H-Ras (lane 3), V12G37 H-Ras (lane 4), or V12C40 H-Ras (lane 5). (A) Western blot analysis of HA-tagged Ras oncoproteins, showing comparable levels of expression in cell pools stably expressing the different Ras effector domain mutants. The positions of the HA-tagged Ras proteins are shown. (B) Ras–Raf-1 association (coimmunoprecipitation) in cells expressing the different Ras effector domain mutants. Cell lysates are immunoprecipitated with an HA antiserum, followed by anti-Raf-1 immunoblotting. The positions of $p74^{Raf-1}$ and immunoglobulin heavy chains are indicated. (C) The levels of MEK phosphorylation (activation) in cells expressing the different Ras effector domain mutants can be determined by immunoblotting, using an antibody against only phosphorylated MEK1/2. The positions of phospho-MEK1/2 are shown. (D) ERK1/2 activity in cells expressing the different Ras effector domain mutants. ERK1/2 immunoprecipitates can be used in an *in vitro* kinase assay with myelin basic protein as a substrate in the presence of [γ-^{32}P]ATP. The position of phosphorylated myelin basic protein is indicated. The positions of the relative molecular mass markers are shown throughout. [Reproduced from Webb *et al., Proc. Natl. Acad. Sci. U.S.A.* **95,** 8773 (1998).]

Ras proteins can be visualized with an antiserum against Raf-1 (clone C12; Santa Cruz Biotechnology) and ECL detection as described above.

Activation of the MEK and ERK kinases is concomitant with their phosphorylation and antibodies that recognize the phosphorylated conformation of either MEK or ERK1/2 can be used to determine their activation. Protein lysates (10 μg) are resolved by 8% (w/v) SDS–PAGE as described above, and immunoblots are probed with antibodies reactive to the phosphorylated forms of MEK1/2 (New England BioLabs, Beverly, MA) or ERK1/2 (Promega, Madison, WI) ECL and X-ray film exposure are completed as described above.

In vitro kinase assays can be used to analyze the degree of ERK1/2 kinase activity as described previously.[9,10] In brief, ERK1 and ERK2 are immunoprecipitated from cell lysates as described above, using a mouse monoclonal anti-ERK1/2 antibody (clone ERK-7D8; Zymed, South San Francisco, CA). Immune complexes are then washed three times with lysis buffer and twice with kinase buffer [25 m*M* HEPES (pH 7.2), 100 m*M* NaCl, 5 m*M* $MgCl_2$, 1 m*M* DTT, 0.1 m*M* sodium vanadate, 10 μ*M* ATP, 0.1% (w/v) BSA, 0.1% (v/v) Triton X-100]. Each kinase reaction is carried out in the presence of 4 μg of myelin basic protein and 10 μCi of [γ-^{32}P]ATP at 30° for 30 min. The reactions are stopped by the addition of a one-third volume of 4× Laemmli buffer, and proteins are separated by 14% (w/v) SDS–PAGE. The SDS–polyacrylamide gels are then dried and visualized by autoradiography.

As shown, V12 H-Ras strongly activates the Raf–MEK–ERK1/2 pathway in NIH 3T3 (490) cells (Fig. 1B–D). V12S35 H-Ras induces a partial activation of this signaling pathway, whereas the V12G37 H-Ras and V12C40 H-Ras effector domain mutants are unable to activate the Raf–MEK–ERK1/2 pathway.

Determining Tumorigenic and Metastatic Properties of *ras*-Transformed Cell Lines *in Vivo*

Principles

The complex multifactorial nature of metastasis dictates the need for experimental systems that more accurately mimic the series of events in the metastatic cascade. The most prominent feature of tumor growth and metastasis that is lacking from most *in vitro* studies is the intricate interaction between the "host" tissue and the tumor cells. In 1889, Paget proposed the "seed and soil" concept for metastatic spread to selective tissues, in which it was suggested that a tumor cell (the seed) could grow only within an environment (the soil) that supported its growth.[15] It is now established that the microenvironment of the host plays a key role in determining the fate of metastases. More recently, it has been observed that many tumor

[15] S. Paget, *Lancet* **1,** 571 (1889).

cells undergo the process of extravasation independently of their ultimate abilities to form macrometastases.[3] The study of the cellular and genetic events involved in these later stages of metastatic growth is therefore of utmost importance.

In the mouse, several experimental assays are utilized to model metastatic spread *in vivo*.[16] We have extensively used the "experimental metastasis" assay, in which cells are injected directly into the tail vein of immunocompromised (athymic nude) mice. Metastatic cells form multiple lesions predominantly (although not exclusively) within the lung, which can readily be counted, sized, and further analyzed by a number of means. It should be noted that this assay has its limitations, as well as advantages. Because cells are injected directly into the host vasculature, this approach does not assay for the earlier events in the metastatic cascade prior to extravasation (such as primary tumor growth, intravasation). However, this assay is both rapid and avoids what has been repeatedly observed in many primary tumors produced in mice, i.e., a failure to metastasize to secondary sites. Moreover, findings demonstrating that events occurring postextravasation determine the ultimate fate of metastatic growth suggest that many aspects of the experimental assay are reasonable for the study of metastasis *in vivo*. The future use of transgenic and knockout mice that develop primary tumors that endogenously metastasize represents an exciting means to study the progressive stages of tumor development.

Subcutaneous Tumor Growth

The ability of cells to form tumors in mice is frequently determined at the subcutaenous region on the backs of mice. The mice used for these studies are typically female athymic (*nu/nu*) mice of approximately 4–6 weeks of age. The autosomal recessive nude gene in homozygous (*nu/nu*) mice is responsible for the lack of fur and an abnormal thymus. The subsequent deficiency in T cell function allows athymic mice to accept and grow xenografts as well as allografts of normal and malignant tissues. Animal care is provided in strict accordance with the procedures outlined in the *Guide for the Care and Use of Laboratory Animals*.[17] We routinely inject 10^5 cells per site in a volume of 0.1 ml of HEPES-buffered saline solution (HBSS). Cells in culture are gently trypsinized, counted, and resuspended at a density of 10^6 cells/ml in HBSS. A 0.1-ml volume of this even cell suspension is then injected under the skin of each mouse, using a sterile 25-gauge needle. Mice are monitored daily and developing tumors are measured with calipers. It is useful to measure the forming tumors in three dimensions (length, width, and thickness) such that accurate tumor volumes may be calculated over time. Using this assay,

[16] A. I. McClatchey, *Oncogene* **18,** 5334 (1999).

[17] National Institutes of Health (NIH), "Guide for the Care and Use of Laboratory Animals." NIH Publication No. 86–23. U.S. Government Printing Office, Washington, D.C., 1985.

TABLE I
TUMORIGENIC AND METASTATIC PROPERTIES OF NIH 3T3 (490) CELLS EXPRESSING Ras EFFECTOR DOMAIN MUTANTS[a]

Transfected gene	Tumorigenicity[b]	Metastasis[c]
Vector alone (pDCR)	1/6 (8)	0/7
V12 H-*ras*	6/6 (2)	9/9 (3–4)
V12S35 H-*ras*	6/6 (3)	7/7 (6–7)
V12G37 H-*ras*	7/8 (4)	0/11
V12C40 H-*ras*	8/8 (3)	0/11

[a] Adapted from C. P. Webb, L. Yan Aelst, M. H. Higler, and G. F. Vande Wonde, *Proc. Natl. Acad. Sci. U.S.A.* **95,** 8773 (1998).

[b] Number of mice with tumors per total number of mice injected subcutaneously after 9 weeks. Elapsed times (in weeks) after injection before visible tumors appear is shown in parentheses.

[c] Number of mice with lung metastases per total number of mice injected intravenously after 14 weeks. Elapsed times (in weeks) after injection before mice display signs of metastatic lung disease (labored breathing, loss of weight) is shown in parentheses. The presence or absence of lung metastases is verified by pathology.

we determined that all the Ras effector domain mutants induced comparable tumor growth when expressed in NIH 3T3 cells, but none as effectively as the V12 H-*ras* oncogene (Table I).[10] This demonstrates that multiple downstream effectors of Ras can mediate tumorigenesis, although the full complement of signaling pathways is required for maximal effect.[9]

Experimental Metastasis Assay

On injection of metastatic cells into the tail vein of mice, macroscopic lung metastases will reproducibly form at a relatively high frequency. Cells should be prepared as described above for subcutaneous injections, such that 0.1 ml of cells (representing 10^5 cells) is injected directly into the vasculature via the tail vein, using a 27-gauge sterile needle. Unlike subcutaneous tumor growth, which is readily observed throughout the duration of the experiment, there are no visual means to determine the growth rate of lung metastases. As such, this assay is strictly an end-point assay, that is, at the end of the experiment the extent of metastatic growth is determined. The extent of lung metastasis depends on a number of variables including the cell type, cell number, the length of time postinjection, as well as the transforming gene. Typically, tail vein injection of 10^5 NIH 3T3 (490) cells expressing V12 H-Ras will result in extensive lung metastasis in approximately 4 weeks. It is essential to observe mice on a regular basis because they will become moribund and display breathing difficulties and weight loss as signs of metastatic disease. After sacrifice, the lungs of the mice should be inflated with 10% (v/v) buffered neutral formalin to readily allow visualization of the macroscopic metastases. An example is shown in Fig. 2. The abilities of the Ras effector domain mutants to induce the metastatic phenotype in NIH 3T3 (490) cells are dramatically different (Table I), such that only the V12S35 H-*ras* mutant confers metastatic propensity to NIH 3T3 cells. Cells transformed by either V12G37 H-*ras* or V12C40 H-*ras* fail to produce macroscopic experimental lung metastases, despite forming subcutaneous tumors (Table I). This implicates the Raf–MEK–ERK1/2 pathway in the formation of metastases in this system.[10] It remains to be seen whether the Ras effector domain mutants behave similarly when expressed in different recipient cell lines.

Culturing of Tumor Explants

Metastasis is believed to be clonal in origin, whereby subpopulations of tumor cells acquire the ability to complete all steps of the metastatic cascade, to yield macroscopic metastases. It is therefore valuable to analyze the metastatic lesions, and compare these either with the primary tumor giving rise to the metastases, or in the case of the experimental metastasis assay, with the population of cells injected into the tail vein. We determined that certain genetic events are associated with the acquisition of the metastatic phenotype.[10] For this purpose, after the mice have been killed, the lungs can be dissected and teased apart into culture medium within a 100-mm tissue culture dish, using a sterile scalpel and forceps. After 24 hr, the medium containing host- and tumor-derived lung tissue should be replaced with growth medium containing geneticin (400 μg/ml). This ensures that only tumor-derived cells (containing the *neo* gene) survive, thereby eliminating host-derived cells. In this manner, metastasis-derived cell lines can be obtained for further analysis. Using this approach, we have shown that metastasis-derived cells expressing the V12S35 H-Ras protein display higher levels of activation of the Raf–MEK–ERK1/2 pathway relative to the pooled population of cells injected into mice,[10] further suggesting a key role for this signaling pathway during Ras-mediated metastasis.

Pathological Examination of Tumors and Metastasis

The presence of visible macroscopic metastases in the lung is a clear sign of metastatic potential of the injected cell line. In the absence of these lesions, it cannot be concluded that the injected cell line fails to metastasize unless the lung is pathologically examined for micrometastases. Micrometastases are difficult to detect by simple observation of the lung. The difference between the detection of micro- and macrometastases may relate to subtle differences in the respective growth rates within the lung, and therefore important information regarding the metastatic potential of a chosen cell line could be overlooked in the absence of pathological examination. We examine all lungs. Lungs are inflated with formalin

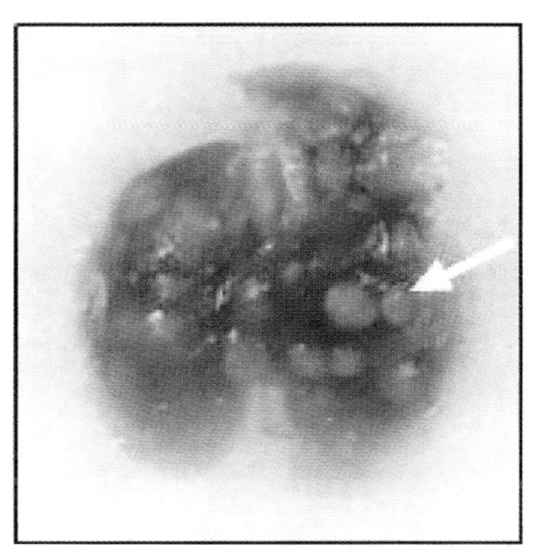

FIG. 2. Photograph of a whole lung possessing numerous metastatic lesions after tail vein injection of V12 H-*ras*-transformed cells. Macrometastases are apparent approximately 4 weeks after administration of 10^5 cells. The arrow highlights a typical macroscopic lesion that can be observed. [Modified from Webb *et al.*, *Proc. Natl. Acad. Sci. U.S.A.* **95,** 8773 (1998).]

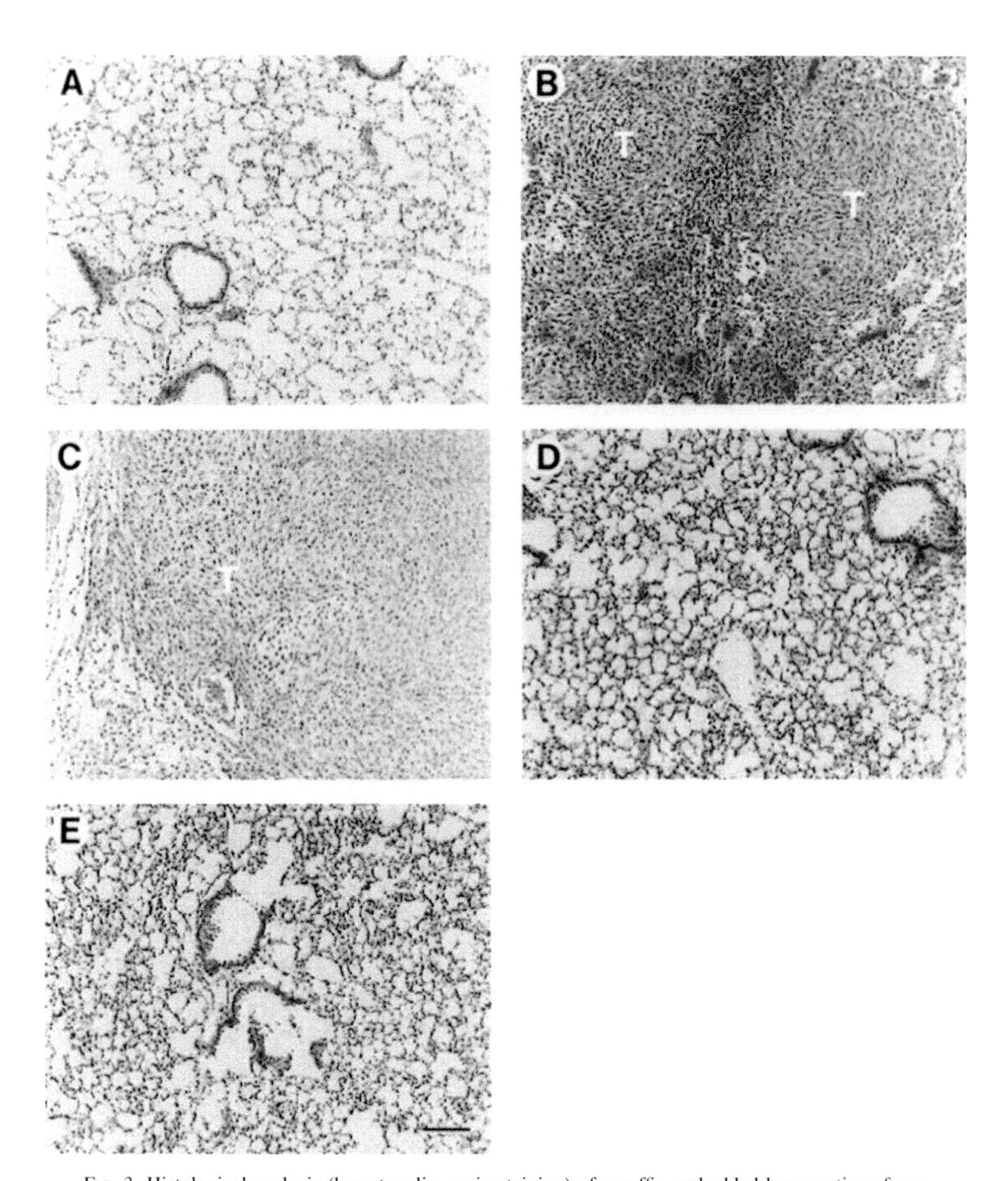

FIG. 3. Histological analysis (hematoxylin–eosin staining) of paraffin-embedded lung sections from mice after intravenous injection of NIH 3T3 (490) cells expressing the different Ras effector domain mutants. Diffuse metastatic fibrosarcoma tumor tissue (T) can be observed in the lungs of mice injected with cells expressing either V12 H-Ras (*B*) or V12S35 H-Ras (*C*) but not in those injected with cells expressing either control vector (*A*), V12G37 H-Ras (*D*), or V12C40 H-Ras (*E*), the lungs from which were completely free of metastatic lesions 14 weeks postinjection. Bar: 100 μm. [Reproduced from Webb *et al.*, *Proc. Natl. Acad. Sci. U.S.A.* **95,** 8773 (1998).]

as described above, embedded in paraffin, sectioned at 5 μm, and stained with hematoxylin and eosin (H&E). Slides are then examined microscopically to determine the extent of metastases. NIH 3T3 cells transformed by either V12G37 H-*ras* or V12C40 H-*ras* failed to produce microscopic metastases, whereas cells transformed by either V12 H-*ras* or V12S35 H-*ras* produce clearly visible macroscopic lesions when analyzed by H&E staining (Fig. 3). Thus the Ras mutants display a clear difference in metastatic propensity, which does not appear to relate to subtle differences in growth rate within the lung.

Summary and the Future

We have described an isogeneic system for studying the cellular and genetic events in experimental metastasis, using effector domain mutants of the V12 H-*ras* oncogene. On expression in NIH 3T3 (490) cells, the V12S35 H-Ras mutant that signals predominantly through activation of the Raf–MEK–ERK1/2 pathway is able to mediate both subcutaneous tumor growth and experimental metastasis to the lung. Conversely, neither the V12G37 H-Ras nor V12C40 H-Ras mutant, which activate the RalGDS and PI3K pathway, respectively, are able to induce the metastatic phenotype in NIH 3T3 cells despite inducing comparable tumorigenesis in athymic nude mice at the subcutaneous site. This isogeneic system can be extremely useful for studying the signaling pathways downstream of Ras as well as subsequent genetic events that are key to tumorigenesis and metastasis.

Acknowledgments

We thank Linda Van Aelst and Mike Wigler for providing the Ras mutants used in these studies, and for providing excellent collaborative support. Thanks also to Tracy Webb for critical review of this manuscript, and Michelle Reed for help in preparing the manuscript.

Section III

Regulation of Guanine Nucleotide Association

[29] Nonradioactive Determination of Ras–GTP levels Using Activated Ras Interaction Assay

By STEPHEN J. TAYLOR, ROSS J. RESNICK, and DAVID SHALLOWAY

Introduction

In their activated, GTP-bound forms the Ras proteins interact with a variety of intracellular effector proteins to initiate signaling pathways that contribute to cell proliferation, differentiation, or death, depending on cellular context. The level of Ras activation (the fraction of total Ras molecules in the GTP-bound form) and the duration of Ras activation (the length of time a given fraction is in the GTP-bound form) are controlled by the opposing actions of exchange factors that stimulate GDP for GTP exchange and GTPase-activating proteins (GAPs), which stimulate the intrinsic GTPase activity of Ras. In most normal circumstances Ras is switched on by a net increase in exchange factor activity, for example, after mitogenic stimulation, and then switched off, after an appropriate delay, by a net increase in GAP activity. Both the level and duration of Ras activation probably contribute to downstream signal strength and selection. Proliferative responses appear to involve transient Ras activation, while more prolonged activation can lead to differentiation. Sustained and maximal Ras activation by mutations that render Ras unresponsive to GAPs can have profound consequences. In most normal cells this leads to senescence or death, but in the context of mutation or misexpression of other genes, sustained Ras activation can result in cellular transformation. To fully understand the roles that Ras proteins play in normal and deregulated cellular function it has been essential to be able to measure the level, the timing and the duration, of cellular Ras activation.

The first generation of Ras–GTP assays directly measured the ratio of GTP to GDP bound to Ras extracted from cells.[1–4] This was achieved by incubating tissue culture cells with $^{32}P_i$, immunoprecipitating Ras proteins with a monoclonal antibody (which at the same time inhibited GAP-stimulated GTP hydrolysis) and measuring the ratio of bound radiolabeled GTP to GDP by thin-layer chromatography. The assay was used to confirm that oncogenic mutants of Ras did indeed have much higher levels of bound GTP in cells than wild-type Ras and to demonstrate that growth factors thought to act via Ras actually increased Ras–GTP levels

[1] J. B. Gibbs, M. D. Schaber, M. S. Marshall, E. M. Skolnick, and I. S. Sigal, *J. Biol. Chem.* **262,** 10426 (1987).

[2] T. Satoh, M. Endo, S. Nakamura, and Y. Kaziro, *FEBS Lett.* **236,** 185 (1988).

[3] J. B. Gibbs, *Methods Enzymol.* **255,** 118 (1995).

[4] T. Satoh and Y. Kaziro, *Methods Enzymol.* **255,** 149 (1995).

0076-6879/00 $35.00

in cells. Although this assay has been useful in studies of Ras function, it suffers from some drawbacks. First, the experiments require the use of millicuries of $^{32}P_i$, necessitating extensive radiation safety procedures and precluding use in many laboratories. Second, the success of the assay is based in part on the ability of the anti-Ras antibody Y13-259 to block GAP interaction with Ras and hence to inhibit GTP hydrolysis on Ras during the experimental procedure. Because Y13-259 recognizes all Ras isoforms, it cannot be used to measure GTP binding to individual isoforms of Ras. Moreover, because other antibodies with the appropriate binding/blocking characteristics are not available, it has been difficult to efficiently extrapolate the procedure to measure activation of related GTPases. Third, attaining adequately high specific activities of labeled nucleotides requires culturing cells in phosphate-depleted medium, possibly placing the cells under metabolic stress. Fourth, and perhaps most significantly, treatment of cells in tissue culture with even low levels of radioisotopes can result in rapid cell cycle arrest or apoptosis, often by activation of p53-dependent checkpoints.[5,6] Because Ras is intimately involved with the regulation of cell cycle progression and apoptosis or cell survival it is possible that, at least in some situations, the requirement for labeling with $^{32}P_i$ may result in the experimental procedure itself influencing the outcome of the experiment. The requirement for radiolabeling was eliminated by the development of a method for measuring amounts of GTP and GDP bound to immunoprecipitated Ras using enzyme-linked photometric assays.[7] An advantage of this assay was its ability to measure Ras–GTP levels in human tissue biopsies.[8] However, like the radiolabeling assay, this assay requires the use of the "neutralizing" antibody Y13-259, limiting its use in measuring activation of different isoforms or related GTPases. The enzyme-linked assay has not found widespread use in different laboratories, making assessment of its usefulness and versatility difficult.

To overcome these various problems a new type of assay for measuring Ras–GTP levels in cells was developed.[9,10] The assay exploits the specificity with which Ras–GTP interacts with the Ras-binding domain (RBD) of one of the major effectors of Ras in cells, the Raf-1 protein kinase. Specifically, Ras–GTP binds to Raf RBD with an affinity approximately three orders of magnitude higher than Ras–GDP.[11] The Raf RBD is expressed as a glutathione *S*-transferase

[5] R. Dover, Y. Jayaram, K. Patel, and R. Chinery, *J. Cell Sci.* **107,** 1181 (1994).

[6] J. Yeargin and M. Haas, *Curr. Biol.* **5,** 423 (1995).

[7] J. S. Scheele, J. M. Rhee, and G. R. Boss, *Proc. Natl. Acad. Sci. U.S.A.* **92,** 1097 (1995).

[8] A. Guha, N. Lau, I. Huvar, D. Gutmann, J. Provias, T. Pawson, and G. Boss, *Oncogene* **12,** 2121 (1996).

[9] S. Taylor and D. Shalloway, *Curr. Biol.* **6,** 1621 (1996).

[10] J. de Rooij and J. L. Bos, *Oncogene* **14,** 623 (1997).

[11] C. Herrmann, G. A. Martin, and A. Wittinghofer, *J. Biol. Chem.* **270,** 2901 (1995).

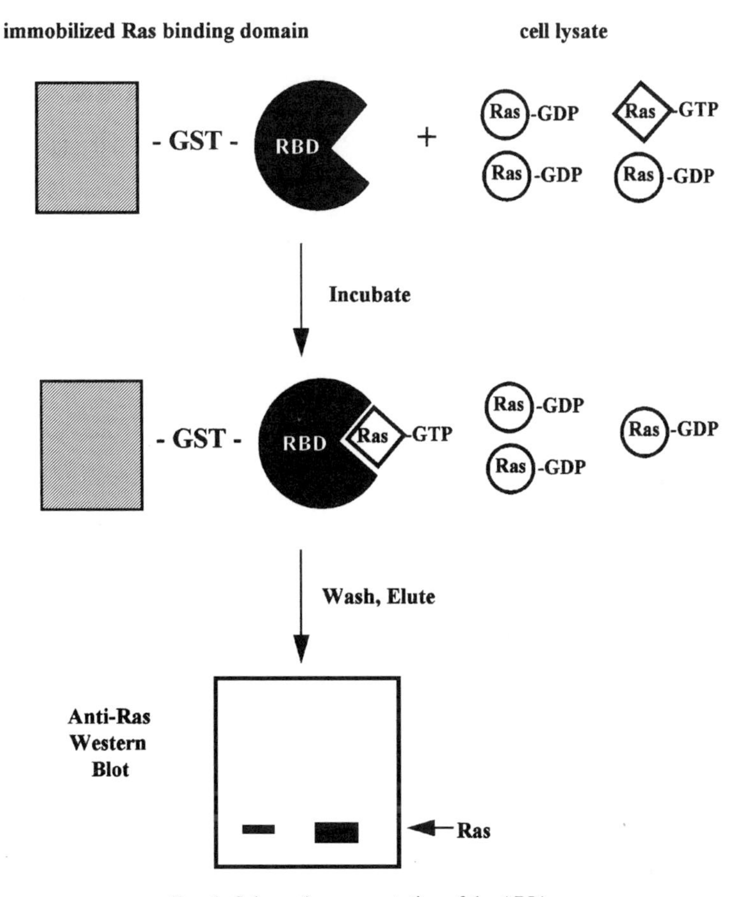

FIG. 1. Schematic representation of the ARIA.

(GST) fusion protein, immobilized on glutathione-Sepharose beads, and then used to affinity precipitate Ras–GTP from cell lysates. The affinity-precipitated Ras is then detected by immunoblotting with specific anti-Ras antibodies (Fig. 1).

Activated Ras Interaction Assay Method: ARIA

Below we describe the methods used in our laboratory to prepare the GST–RBD protein and to use it for Ras–GTP affinity precipitation from tissue culture cells. We have tried to point out alternative procedures, and also to indicate which procedures and conditions appear to be particularly important for success.

Preparation of GST–RBD

We use the bacterial expression vector pGEX-RBD, which encodes amino acids 1–149 of c-Raf-1, including the Ras-binding domain, fused to GST. A construct encoding amino acids 51–131 of Raf-1 has also been used successfully.[10] *Escherichia coli* DH5α (other strains such as BL21 have been used successfully) are transformed with pGEX-RBD and stored as a 20% (v/v) glycerol stock at −70°. Liquid cultures can be started directly from the frozen stock or from a colony from a streaked plate. A 5- to 10-ml culture in "super broth" [3.2% (w/v) tryptone, 2.5% (w/v) yeast extract, 0.5% (w/v) NaCl, pH 7; other enriched media can be used] containing ampicillin at 0.1 mg/ml is started and grown overnight with shaking at 37°. The starter culture is added to ~350 ml of super broth plus ampicillin and grown for a further 4–6 hr or until the OD_{600} is 0.6–0.8. GST–RBD expression is induced with 1 m*M* isopropyl-β-D-thiogalactopyranoside (IPTG) for 3–4 hr at 37°. The bacteria are pelleted by centrifugation (15,000 rpm for 15 min at 4° in a Sorvall GS-3 rotor or equivalent), resuspended in ~160 ml of cold HBS [25 m*M* HEPES (pH 7.5), 150 m*M* NaCl; it is easiest first to resuspend the pellet in a few milliliters and then make up to volume] and then aliquoted into four 50-ml conical tubes and pelleted by centrifugation for 20–30 min at 4°. The bacterial pellets are then frozen in liquid N_2 and stored at −70° until use. For most purposes one pellet will provide more than enough fusion protein for an experiment.

Prior to an experiment a pellet is resuspended in 10 ml of RBD lysis buffer [RLB: 20 m*M* HEPES (pH 7.5), 120 m*M* NaCl, 10% (v/v) glycerol, 2 m*M* EDTA, 1 m*M* dithiothreitol (DTT), leupeptin (10 μg/ml), aprotinin (10 μg/ml)] and sonicated on ice twice (15 sec each) at a medium setting. Some researchers have found that oversonication can adversely affect performance of the GST–RBD in binding assays; if this appears to be a problem it may be worthwhile trying different output settings to determine the lowest level necessary for GST–RBD recovery. As an alternative, the bacteria can be lysed by adding lysozyme (~0.2 mg/ml) to the suspension and incubating with rocking for 15 min at 4°; if viscosity becomes a problem, DNase and 5 m*M* $MgCl_2$ can be added. Other researchers have successfully used the B-PER reagent from Pierce (Rockford, IL) to lyse the bacteria. The lysates are centrifuged at 15,000 rpm in a Sorvall SS-34 rotor, or equivalent, and the supernatant is collected. Nonidet P-40 (NP-40; 10%, v/v) is added to 0.5% (v/v) final concentration and glutathione–Sepharose beads (Amersham Pharmacia, Piscataway, NJ) (~300–500 μl of packed beads, washed and equilibrated with HBS) are added. The lysates and beads are incubated with rocking for 30–60 min in a cold room. The beads are then washed six to eight times with ~0.6 ml of RLB containing 0.5% (v/v) NP-40. To check yield and purity, serial dilutions of the GST–RBD beads can be run out on a sodium dodecyl sulfate (SDS)–polyacrylamide gel alongside known amounts of a pure protein, for example, bovine serum albumin, and stained with Coomassie blue. GST–RBD runs at

~42 kDa on an SDS–polyacrylamide gel and should represent >80% of the total purified protein. Contaminating proteins (in preparations from DH5α, prominent bands at ~68 and 15–25 kDa are seen) should not interfere with the assay. The beads can also be used directly for protein assay.

In our laboratory we have found that the stability of GST–RBD beads, that is, their ability to bind Ras–GTP efficiently with time, varies between different batches. However, we and many other researchers have found that there is some loss in Ras–GTP binding capacity after a few days of storage at 4°. It is therefore strongly recommended that a new batch of GST–RBD beads be made either on the day of an experiment or 1 or 2 days before. Because multiple bacteria pellets can be made at the same time and stored frozen, and because the GST–RBD beads can be prepared within about 2 hr, we have found this to be convenient. If performing the assay frequently with few samples per experiment, it may be advisable to scale down the GST–RBD preparation to avoid wasting glutathione beads.

Affinity Precipitation of Ras–GTP

The amount of starting material, that is, the number of cells, required to obtain a readily detectable Ras signal will depend on many factors, most notably the expression level of Ras proteins in the cells under study and the level of activation of Ras in response to the stimuli under study. If using a cell type not previously assayed for Ras activation by this method we suggest using one or two 100-mm plates, 70–80% confluent, with or without a known or expected Ras activator, preferably alongside positive controls as described below. For NIH 3T3 or HeLa cells one 100-mm tissue culture plate or 200–400 μg of protein per precipitation is sufficient (scale up if probing multiple immunoblots). After treatment of cells the plates are washed twice with ice-cold HBS or phosphate-buffered saline (PBS) and 0.4–0.8 ml per 100-mm plate of ice-cold Mg-containing lysis buffer [MLB: 25 m*M* HEPES (pH 7.5), 150 m*M* NaCl, 1% (v/v) NP-40, 0.25% (w/v) sodium deoxycholate, 10% (v/v) glycerol, 10 m*M* $MgCl_2$, 1 m*M* EDTA, leupeptin (10 μg/ml) aprotinin (10 μg/ml)] is added. The phosphatase inhibitors NaF (25 m*M*) and sodium vanadate (1 m*M*) can be included in MLB if protein phosphorylation is to be simultaneously examined; sodium pyrophosphate should not be used. The plates are placed on a rocker in a cold room for 15 min, or the cells can be scraped in the MLB into an Eppendorf tube and lysed with rocking for 15 min. At this point the unclarified lysates can be snap-frozen in liquid N_2 and stored at −70° until use; it is not recommended to freeze cells prior to lysis. The lysates are centrifuged at 4° for 10 min at 15,000 rpm in a Sorvall SM-24 rotor or equivalent; centrifugation for 10 min at full speed in a microcentrifuge in the cold is also probably sufficient.

If lysates were prepared from paired plates, for example, equal numbers of cells plus or minus drug, equal volumes of lysate can be used for affinity precipitation (AP). Otherwise protein assays should be performed to allow adjustment to equal amounts of protein in each sample. It is important to work quickly and to keep lysates on ice, as the Ras–GTP will be susceptible to GAP activity until incubation with GST–RBD; the DC protein assay (Pierce) can be carried out in less than 30 min. It is generally not necessary to quantitate the amount of GST–RBD per bead volume used in the AP. However, it is advisable to determine the approximate yield of GST–RBD when preparing for the first time or after changing preparation conditions; thereafter yields should be fairly constant. In general 10–15 μl of packed beads (usually 20–50 μg of protein) is used per AP; add the beads to the lysate from a 30–50% slurry, using a pipette tip with the end cut off. Cleared lysates are incubated with GST–RBD beads on a rocker in a cold room for 30 min. The GST–RBD beads are washed three times with 0.6 ml of MLB per wash. Again, it is important to work quickly; 10-sec spins in a microcentrifuge and resuspension by inversion should allow washing multiple samples in ~5 min. Proteins are eluted by boiling in SDS–PAGE sample buffer and the samples, along with lysates and prestained molecular weight markers, are boiled in sample buffer for a few minutes and run on an 11% (w/v) acrylamide gel. Ras proteins run at ~21 kDa on SDS-PAGE; however, different isoforms and mutants have slightly different mobilities. If using prestained markers from New England BioLabs (Beverly, MA), all Ras isoforms run between the 16.5- and 25-kDa markers. By cutting the gel at the 32.5-kDa marker it is possible to transfer the Ras proteins (lower) and check recovery and amount [by running a protein standard, e.g., bovine serum albumin (BSA)] of the GST–RBD by staining the upper part. The Ras proteins are transferred to a polyvinylidine difluoride (PVDF) membrane; transfer conditions will vary between different transfer units but Ras requires less time and current than larger proteins. The blots are blocked with 2.5% (w/v) BSA in 25 m*M* Tris (pH 7.5), 150 m*M* NaCl, 0.1% (v/v) Tween 20 (TTBS) and then probed with anti-Ras antibodies (see below), followed by the appropriate secondary antibody coupled to peroxidase. Proteins are detected by enhanced chemiluminescence or chemifluorescence.

Anti-Ras Antibodies for Ras Detection

Anti-pan Ras antibodies used successfully for immunoblotting include the mouse monoclonal antibodies clone 18 (Transduction Laboratories, Lexington, KY) and RAS 10 (Calbiochem, La Jolla, CA; Upstate Biotechnology, Lake Placid, NY). Isoform-specific antibodies can be used to specifically assay activation of the individual isoforms; these include H-Ras F235 (Calbiochem; Santa Cruz Biotechnology, Santa Cruz, CA), K-Ras F234 (Calbiochem, Santa Cruz Biotechnology), and N-Ras F155 (Santa Cruz Biotechnology) monoclonal antibodies. Figure 2 shows immunoblots of whole cell lysates (lanes 1–4, 13, and 14), GST–RBD APs (lanes 5–8, 15, and 16), and Y13-259 IPs (lanes 9–12, 17, and 18) from some human cancer cell lines probed with anti-pan Ras, anti-K-Ras, and anti-N-Ras. Anti-pan Ras detects two bands, which in these cells are primarily K-Ras and N-Ras; N-Ras

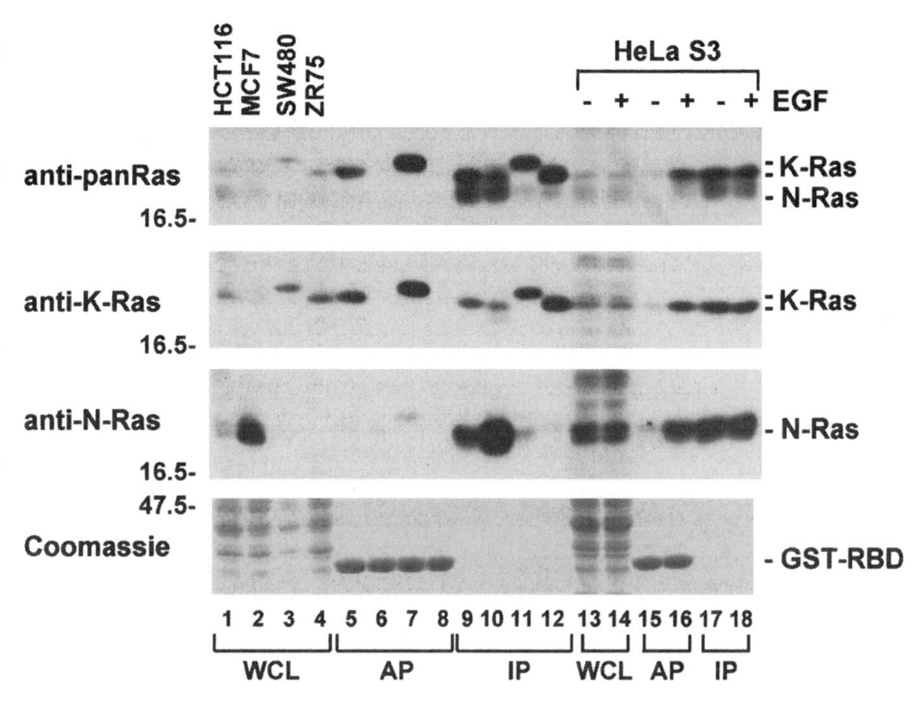

FIG. 2. Detection of Ras activation in human cancer cell lines. The cancer cell lines HCT116 (lanes 1, 5, and 9), MCF7 (lanes 2, 6, and 10), SW480 (lanes 3, 7, and 11), ZR75 (lanes 4, 8, and 12), and HeLa S3 (lanes 13–18) were lysed and subjected to affinity precipitation and immunoprecipitation as described in text. HeLa S3 cells were placed into 0.1% serum for 24 hr and then treated without (odd-numbered lanes) or with (even-numbered lanes) EGF (200 ng/ml) for 10 min before lysis. Affinity precipitation was with ~30 μg of GST–RBD bound to ~15 μl of beads and was from 1.0 (HCT116, MCF7, and ZR75), 0.5 (SW480), or 1.5 (HeLa S3) mg of lysate protein; 20% of each AP was loaded per gel lane. Immunoprecipitation was with 1 μg of Y13-259 and was from 40% of the AP input. Twenty percent of the APs and IPs were loaded per lane. Whole cell lysates represent 4% of the total AP protein input, that is, the equivalent of half the IP lane input. Three identical gels were run, transferred, and probed with anti-pan Ras (clone 18; Transduction Laboratories), anti-K-Ras (F234; Santa Cruz Biotechnology), or anti-N-Ras (F155; Santa Cruz Biotechnology). The upper part of one gel was stained with Coomassie blue to visualize GST–RBD. The positions of K-Ras and N-Ras as well as prestained molecular weight markers (New England BioLabs) are indicated.

runs with faster gel mobility and more diffusely than K-Ras. The rat monoclonal antibody Y13-259 [available from Calbiochem and Santa Cruz Biotechnology; hybridoma cells available from the American Type Culture Collection (ATCC, Manassas, VA); CRL-1742] works well for immunoprecipitation (see Fig. 2). The rat monoclonal Y13-238 (Calbiochem, Santa Cruz Biotechnology) also immunoprecipitates well but does not recognize N-Ras. These antibodies can also be used for immunoblotting but require an anti-rat secondary antibody.

Controls

Nonspecific precipitation of GDP-bound Ras from cell lysates is generally not a problem. However, if signal to background is low it may be advisable to perform a parallel affinity precipitation with GST alone or with an irrelevant GST fusion protein. An additional control for specificity entails preincubation of lysates with anti-Ras antibodies. The rat monoclonal Y13-259 binds close to the Ras effector-binding region and blocks GST–RBD binding, preventing affinity precipitation of Ras–GTP. Another rat monoclonal, Y13-238, does not block precipitation.[9]

It is recommended that an aliquot of the lysate be run on the gel, usually 10–20% of that used for the AP. However, depending on the level of Ras expression and the anti-Ras antibody being used it is possible that signal to background may be poor in lysates. In fact, we have found that signal to noise is generally proportionately weaker in lysate lanes than in immunoprecipitate lanes (see Fig. 2, lanes 1–4 compared with lanes 9–12 and lanes 13 and 14 compared with lanes 17 and 18). It is therefore a good idea also to immunoprecipitate (IP) an aliquot of lysate alongside the APs. Anti-Ras Y13-259 is excellent for IP and recognizes all isoforms. (Run the antibody alone when using it for the first time, to establish the gel mobility of Ras vs. the IgG signal.) For this purpose, incubate 10–50% of the AP input with 0.5–1 μg of Y13-259 on ice for 1 hr, and add 10–20 μl of packed protein G–Sepharose beads (Gammabind Plus; Amersham Pharmacia); rock for 30 min, wash three times with MLB, and resuspend in sample buffer.

To establish that the assay is working, activated epitope-tagged Ras mutants, for example, V12 or L61, can be transiently expressed in cells and recovery of Ras in APs can be compared with the wild-type or dominant-negative, N17, forms. Alternatively, established cancer cell lines harboring activating *ras* mutations can be used; the colorectal cancer cell lines SW480 (ATCC CCL-228) and HCT116 (ATCC CCL-247) harbor activating mutations at codons 12 and 13, respectively, in K-*ras*. As shown in Fig. 2, at least 50% of total K-Ras can be detected in APs from these cell lines, while N-Ras from these cells or K- and N-Ras from the breast cancer cell lines MCF-7 and ZR75 (in which *ras* is not mutated) are not detectably affinity precipitated (Fig. 2, lanes 5–8). Note that the mutant Ras in SW480 cells runs with a slower gel mobility than normal Ras and mutant Ras from HCT116 cells. SW480 and HCT116 cells can be grown in Dulbecco's Modified Eagle's Medium (DMEM) containing 10% (v/v) fetal calf serum. Figure 2 also illustrates the large variation in expression levels of different Ras isoforms that can occur between cell lines; the breast cancer cell lines MCF7 and ZR75 express differentially high levels of N-Ras and K-Ras (lanes 2, 4, 10, and 12).

If trying to detect a low level of Ras activation it may be necessary to optimize assay conditions using normally activated, that is, nonmutated, Ras. Treatment of a wide variety of serum-starved cells with serum or growth factors such as epidermal growth factor (EGF) or platelet-derived growth factor (PDGF) results in Ras activation that can be readily detected by activated Ras interaction assay (ARIA). For example, treatment of 24-hr serum-starved HeLa cells with EGF (50–200 ng/ml) for 10 min will give strong Ras activation. As shown in Fig. 2, up to 50% of total Ras can be recovered in EGF-treated APs with anti-pan Ras, anti-K-Ras, or anti-N-Ras antibodies whereas Ras is almost undetectable in APs from untreated cells (lanes 15 and 16). If NIH 3T3 fibroblasts are available, the strongest signal can be obtained by treating serum-depleted NIH 3T3 cells for 4–6 hr with 20% (v/v) calf serum.[9] Note that NIH 3T3 cells cannot be placed in DMEM alone, as they will round up and die; the cells are washed once with DMEM and then refed with DMEM containing 0.2–0.5% (v/v) serum. When comparing lysate protein concentrations between serum-starved and serum-treated cells it is important to bear in mind that serum proteins, particularly albumin, will stick to plates or cells, even after washing, and increase protein concentration. In this type of experiment it is advisable to use equal volumes for the APs and to check lysate protein levels by staining the blot.

Applications of ARIA

The ARIA method was first used to demonstrate biphasic activation of Ras during G_1 phase progression in HeLa and NIH 3T3 cells; early activation was associated with mitogen-activated protein (MAP) kinase activation but mid-G_1 activation occurred in the absence of MAP kinase activation.[9] In fact, the assay was developed in our laboratory to enable cell cycle analysis of Ras–GTP levels without cell cycle progression being compromised by radiolabeling of the cells. de Rooij and Bos independently developed the same assay and used it to demonstrate activation of Ras by insulin in A14 cells and by glial cell line-derived neurotrophic factor (GDNF) in SK-P2 cells.[10] By performing the assay in parallel with *in vivo* labeling assays they showed that the nonradioactive assay had a greater level of sensitivity. The assay has since been used to demonstrate modulation of Ras–GTP levels in a wide variety of cells types in response to many different conditions. Some examples of established cell lines that have been used successfully include the T cell line Jurkat,[12] the B cell line Daudi,[13] Madin–Darby Canine Kidney (MDCK) cells,[14] $Rb^{-/-}$ and $Rb^{+/+}$ mouse embryonic fibroblasts,[15] Chinese hamster ovary (CHO) cells,[16] and many cancer cell lines, while primary cell cultures

[12] K. E. Wilson, Z. Li, M. Kara, K. L. Gardner, and D. D. Roberts, *J. Immunol.* **163,** 3621 (1999).

[13] X. Li and R. H. Carter, *J. Immunol.* **161,** 5901 (1998).

[14] B. Gay, S. Suarez, C. Weber, J. Rahuel, D. Fabbro, P. Furet, G. Caravatti, and J. Schoepfer, *J. Biol. Chem.* **274,** 23311 (1999).

[15] K. Y. Lee, M. H. Ladha, C. McMahon, and M. E. Ewen, *Mol. Cell. Biol.* **19,** 7724 (1999).

[16] R. V. Fucini, S. Okada, and J. E. Pessin, *J. Biol. Chem.* **274,** 18651 (1999).

used include neonatal cardiac myocytes,[17] primary T lymphocytes,[18] primary embryonic striatal neurons,[19] primary glomerular mesangial cells,[20] and myotubes.[21] It is reasonable to expect that the assay should be applicable to any cell type that expresses sufficient levels of Ras protein for detection. It is also likely that the assay could be successfully used to measure Ras–GTP levels in whole tissues, perhaps even whole organisms. In the case of assaying Ras–GTP in whole tissues, factors of particular importance would be rapid homogenization and lysis to minimize hydrolysis or dissociation of Ras-bound GTP before detergent solubilization and dilution.

The assay has been modified to enable detection of other GTPases in their GTP-bound forms. Franke *et al.*[22] used the RBD of another Ras effector, RalGDS, to affinity precipitate the GTP-bound form of the closely related GTPase Rap1. Detection of activation of other GTPase families has also been achieved. Manser *et al.*[23] and Bagrodia *et al.*[24] used the p21-binding domains (PBD) from Pak1 and Pak3 to measure activation of transiently expressed, epitope-tagged forms of the Rho family GTPases Rac and Cdc42 by coexpressed exchange factors. (An important technical point to note is that, although the phosphatase inhibitors NaF and sodium vanadate can be included in the lysis buffer for ARIA without affecting the assay outcome, we have found that inclusion of one or both of these inhibitors in the lysis buffer for activated Rac/Cdc42 assays results in greatly increased recovery of nonactivated Rac and Cdc42 in GST–PBD APs. This is probably due to the ability of fluoride to stabilize the interaction of these GTPases in their GDP-bound forms with Pak, a situation that does not occur in the case of Ras interaction with Raf.) A similar assay was developed for Rho, using the Rho-binding domain of a Rho effector[25] and it is reasonable to expect that the affinity precipitation principle will be successfully applied to measure the activation level of many other GTPases.

Acknowledgments

We thank our many colleagues for helpful comments on their experience with the ARIA assay and Michael Dehn for help with preparation of the manuscript.

[17] C. Montessuit and A. Thorburn, *J. Biol. Chem.* **274,** 9006 (1999).
[18] V. Lafont, F. Ottones, J. Liautard, and J. Favero, *J. Biol. Chem.* **274,** 25743 (1999).
[19] R. Maue, Personal communication, 1999.
[20] M. Foschi, S. Chari, M. J. Dunn, and A. Sorokin, *EMBO J.* **16,** 6439 (1997).
[21] T. Tsakiridis, A. Bergman, R. Somwar, C. Taha, K. Aktories, T. F. Cruz, A. Klip, and G. P. Downey, *J. Biol. Chem.* **273,** 28322 (1998).
[22] B. Franke, J. W. Akkerman, and J. L. Bos, *EMBO J.* **16,** 252 (1997).
[23] E. Manser, T. H. Loo, C. G. Koh, Z. S. Zhao, X. Q. Chen, L. Tan, I. Tan, T. Leung, and L. Lim, *Mol. Cell.* **1,** 183 (1998).
[24] S. Bagrodia, S. J. Taylor, K. A. Jordan, L. Van Aelst, and R. Cerione, *J. Biol. Chem.* **273,** 23633 (1998).
[25] X. D. Ren, W. B. Kiosses, and M. A. Schwartz, *EMBO J.* **18,** 578 (1999).

[30] Measurement of GTP-Bound Ras-Like GTPases by Activation-Specific Probes

By MIRANDA VAN TRIEST, JOHAN DE ROOIJ, and JOHANNES L. BOS

Introduction

Small GTPases are activated by the conversion of the GDP-bound conformation into the GTP-bound conformation and inactivated by GTP hydrolysis. Until recently the common method to measure the active GTP-bound form of Ras was by immunoprecipitation of ^{32}P-labeled Ras followed by separation of the labeled GDP/GTP bound to Ras. This method allows the determination of the percentage of total GDP/GTP-bound Ras that is in the GTP-bound form. This method still is the "gold standard," but it is technically demanding and needs relatively high levels of radioactivity. Furthermore, side effects due to radiation-induced DNA damage cannot be excluded. For several other small GTPases an additional problem was that suitable antisera to precipitate the GTPase were unavailable. For instance, for Rap1, despite many attempts, a satisfactory precipitating antiserum was not obtained. Therefore, to study activation of endogenous Rap1 it was essential to develop a novel procedure.[1] This method is based on the high difference in affinity of the GTP- versus GDP-bound form for specific binding domains of effector proteins *in vitro*. By using glutathione *S*-transferase (GST) fusion proteins containing these binding domains, the GTP-bound form of the GTPase can be precipitated from cell lysates. In principle, this method can be used for all small GTPases and has proved to function for Ras, Rap1, Rap2, R-ras, and Ral.[2–6] Here we describe a general procedure to measure the GTP-bound form of Ras-like GTPases.

Expression and Isolation of Activation-Specific Probes

As activation-specific probes we use for Ras and R-ras the ras-binding domain (RBD) [amino acids (aa) 51–131] of Raf1,[4,6] for Rap1 (and Rap2) the RBD of RalGDS (97 aa),[1] and for Ral the Ral-binding domain (aa 397–518) of

[1] B. Franke, J. W. N. Akkerman, and J. L. Bos, *EMBO J.* **16,** 252 (1997).

[2] Z. Luo, B. Diaz, M. S. Marshall, and J. Avruch, *Mol. Cell. Biol.* **17,** 46 (1997).

[3] S. J. Taylor and D. Shalloway, *Curr. Biol.* **6,** 1621 (1996).

[4] J. de Rooij and J. L. Bos, *Oncogene* 14, 623 (1997).

[5] R. M. F. Wolthuis, B. Franke, M. van Triest, B. Bauer, R. H. Cool, J. H. Camonis, J. W. N. Akkerman, and J. L. Bos, *Mol. Cell. Biol.* **18,** 2486 (1998).

[6] J. de Rooij, N. M. Boenink, M. van Triest, R. H. Cool, A. Wittinghofer, and J. L. Bos, *J. Biol. Chem.* **274,** 38125 (1999).

0076-6879/00 $35.00

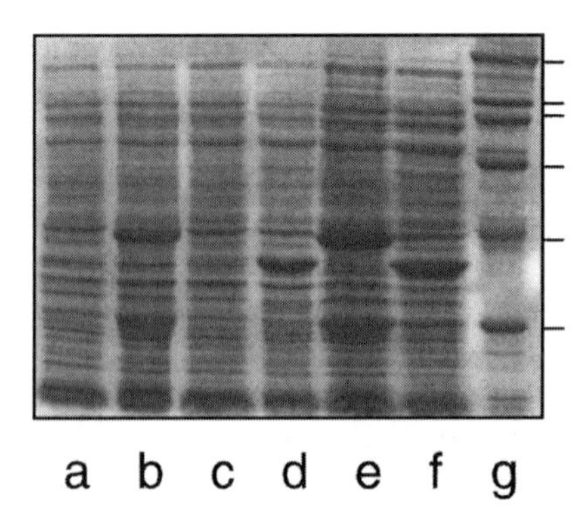

FIG. 1. Coomassie-stained gel containing the following samples: (a and b) bacteria lysed in sample buffer with GST–Raf–RBD prior to (a) and after (b) induction with IPTG; (c and d) bacteria lysed in sample buffer with GST–RalGDS–RBD prior to (c) and after (d) induction with IPTG; (e) bacterial lysate used in the assay containing GST–RalBD; (f) bacterial lysate used in the assay containing GST–RalGDS–RBD; (g) molecular mass marker (205, 116, 97.5, 66, 45, and 29 KDa).

RLIP76 (RalBP).[5] Each of the domains was cloned in pGEX vectors (Pharmacia, Piscataway, NJ), to obtain bacterial expression vectors that produce GST fusion proteins with the GST part at the N terminus. Cloning and characterization of each of the probes have been described previously.[1,4,5]

The activation-specific probes are isolated as follows.

1. Transform *Escherichia coli* BL21 with GST fusion constructs. Incubate the bacteria in 50 ml of LB medium, containing ampicillin (75 μg/ml), overnight at 37° with stirring.

2. Dilute the culture 1 : 50 in 1 liter of LB medium containing ampicillin (75 μg/ml) and 0.4% (w/v) D-glucose and let the bacteria multiply to an OD_{600} of 0.6–0.7. Add isopropyl-β-D-thiogalactopyranoside (IPTG) to a final concentration of 0.1 m*M* and incubate the bacteria for an additional 2–3 hr at 37°. Spin the bacteria down (7700 *g* at 4° for 10 min). At this point the pellet can be stored overnight at −80° before checking whether the GST fusion protein is induced (Fig. 1).

3. Resuspend the pellet in 25 ml of cold 1× bacterial lysis buffer [20% (w/v) sucrose, 10% (v/v) glycerol, 50 m*M* Tris (pH 8.0), 2 m*M* dithiothreitol (DTT), 2 m*M* $MgCl_2$, 1 m*M* phenylmethylsulfonyl fluoride (PMSF), leupeptin (1 μg/ml), and aprotinin (2 μg/ml)] and sonicate 10 times (30 sec each) on ice.

4. Centrifuge for 1 hr at 12,000 g at 4° to remove bacterial debris. Collect the supernatant and store aliquots at −80°. The presence of GST fusion protein in the clear lysate can be checked by sodium dodecyl sulfate–polyacrylamide gel electrophoresis (SDS–PAGE) (see Fig. 1).The bacterial lysates are stable for about 1 month at −80°. However, multiple freeze–thawing should be avoided.

Escherichia coli BL21 did not give good yields of GST–RalBD. We therefore tested several bacterial strains and found that the protease-negative *E. coli* AD202[7] gave much better yields. Because we were satisfied with the yield of GST–Raf–RBD and GST–RalGDS–RBD we have not tried the AD202 protocol.

1. Transform AD202 with the GST–RalBD construct. Incubate the bacteria in 50 ml of LB medium containing both ampicillin (75 μg/ml) and kanamycin (25 μg/ml) overnight at 37° with stirring.
2. Dilute the culture 1 : 50 in 1 liter of LB medium containing ampicillin (75 μg/ml), kanamycin (25 μg/ml), 0.4% (w/v) D-glucose, and 10% (v/v) brain heart infusion medium (Difco, Detroit, MI) and let the bacteria multiply to OD_{600} of 0.6-0.7. Add IPTG to a final concentration of 0.1 m*M* and incubate the bacteria overnight at room temperature.

Indentification of GTP-Bound GTPases with Activation-Specific Probes

The GTP-bound GTPases are precipitated from clear cell lysate, using activation-specific probes that are precoupled to glutathione–agarose beads. Detection occurs by gel electrophoresis and Western blotting using antibodies specific for a small GTPase.

1. Wash 75 μl of a 10% (v/v) glutathione–agarose suspension two times with 1× lysis buffer [1% (v/v) Nonidet P-40, 10% (v/v) glycerol, 50 m*M* Tris-HCl (pH 7.4), 200 m*M* NaCl, 2.5 m*M* $MgCl_2$, 1 m*M* PMSF, 2 m*M* sodium orthovanadate, leupeptin (1 μg/ml), aprotinin (2 μg/ml), trypsin inhibitor (10 μg/ml), and NaF]. Add 100 μl of bacterial lysate containing the activation-specific probes to the washed glutathione–agarose beads for 30–60 min on a tumbler at 4°. Wash the beads four times with 1× lysis buffer. (The amount of bacterial lysate can be adjusted depending on the expression of the GST fusion protein, but in general we use a vast excess.)
2. Transfer the culture dishes (usually 9-cm dishes) containing the cells for analysis on ice and wash two times with ice-cold phosphate-buffered saline (PBS).
3. Add 1 ml of cold 1× lysis buffer. Scrape the cells with a rubber policeman and transfer the cell lysate to an Eppendorf tube. Spin the lysate for 10 min at 14,000 rpm in an Eppendorf centrifuge at 4°.
4. Add the supernatant to the activation-specific probe precoupled to glutathione–agarose beads (step 1) and incubate for 45 min on a tumbler at 4°. Wash the beads four times with 1× lysis buffer. After the final wash remove the

[7] H. Nakano, T. Yamazaki, M. Ikeda, H. Masai, S. Miyatake, and T. Saito, *Nucleic Acids Res.* **22,** 543 (1994).

remaining fluid carefully with an insulin syringe and collect the beads in 20 μl of Laemmli sample buffer. Heat the samples at 95° for 5 min.

5. Separate the proteins on a 12.5% (w/v) SDS–polyacrylamide gel. Transfer the proteins to polyvinylidene difluoride (PVDF) membrane by Western blotting. Stain the membrane with Ponceau S to check for equal distribution of GST–probe over the different samples.

6. Block the membrane for 1 hr at room temperature in blocking buffer [2% (w/v) nonfat dry milk, 0.5% (w/v) bovine serum albumin in PBS–0.1% (v/v) Tween 20]. After this incubation add to the membranes 10% (v/v) blocking buffer in PBS–0.1% (v/v) Tween 20 containing the monoclonal anti-Ras (1 : 1000), anti-Rap1 (1 : 500), anti-Ral A (1 : 8000) (all from Transduction Laboratories, Lexington, KY), anti-Rap2 (1 : 2500), or anti-R-ras (1 : 500) (Santa Cruz Biotechnology, Santa Cruz, CA) for the corresponding GTPase and incubate overnight at 4°.

7. Remove the primary antibody and wash three times with PBS–0.1% (v/v) Tween 20, 10 min each, at 4°. Incubate for 1 hr with goat anti-mouse or goat anti-rabbit secondary antibody conjugated to horseradish peroxidase (HRP) diluted (1 : 8000) in PBS–0.1% (v/v) Tween 20 at 4°. Wash three times with PBS–0.1% (v/v) Tween 20, 5 min each, at 4°. Visualize the bound secondary antibody by enhanced chemiluminescence (New England Nuclear, Boston, MA) according to the manufacturer protocol.

Figures 2 and 3 show typical examples of the activation-specific probe assay for Ras, Rap1, Ral, and R-ras. To measure Ras, Rap1, and Ral in the GTP-bound form, A14 cells were stimulated with different stimuli and the activated GTPases were precipitated with the different specific GST–probes. Figure 2 (panel 1) shows a clear increase in GTP-bound Ras after stimulation with insulin (5 μg/ml) for 5 min. Stimulation of the A14 cells with 100 n*M* endothelin for 3 min or with 12-O-tetradecanoylphorbol-13-acetate (TPA, 100 ng/ml) for 15 min does not affect Ras–GTP levels in these cells. As a control the amount of Ras in total lysate was measured. No difference in Ras protein levels was observed (Fig. 2, panel 2). Figure 2 (panel 3) shows that in contrast to Ras, Rap is activated by stimulation with TPA and endothelin, while insulin fails to do so. Ral was activated by stimulation of A14 cells with insulin, and to a minor extent by endothelin (Fig. 2, panel 5). As shown in Fig. 3 GST–Raf1–RBD can also be used to measure GTP-bound R-ras. When human blood platelets were stimulated with α-thrombin (1 U/ml) an increase in R-rasGTP could be detected. However, despite many attempts we were unable to find activation of R-ras after growth factor stimulation of several other cell types tested. Thus far we have been unable to demonstrate activation of Rap2 after growth factor stimulation using GST–RalGDS–RBD as probe, presumably because in the cell types we have tested Rap2 is predominantly in the GTP-bound from.[5] However, when a guanine nucleotide exchange factor for Rap2 is cotransfected with Rap2, we do observe a small increase, indicating that the method in principle can be used for Rap2 as well.

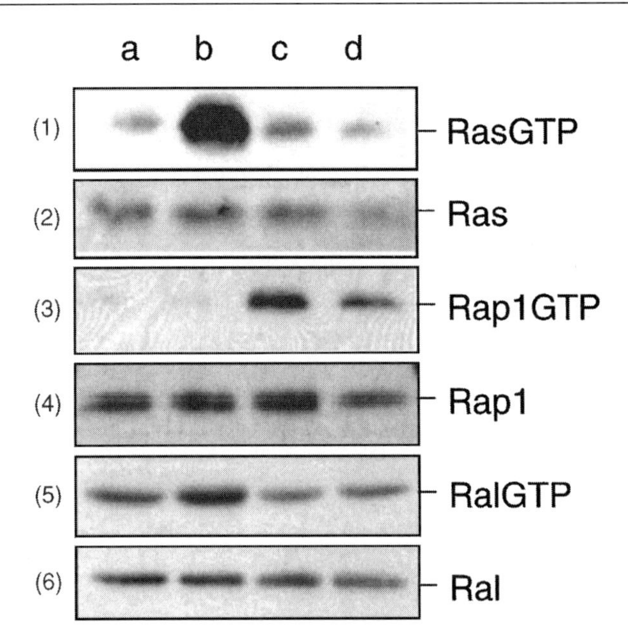

FIG. 2. Activation of Ras, Rap1, and Ral. A14 cells serum starved for 16 hr were unstimulated (a) or stimulated with insulin (5 μg/ml, 5 min), (b) TPA (100 ng/ml, 15 min) (c), or endothelin (100 n*M*, 3 min) (d). GTP-bound forms of Ras (panel 1), Rap (panel 3), and Ral (panel 5) were precipitated with the activation-specific probes and analyzed by SDS–PAGE followed by Western blotting. In panels 2, 4, and 6 Western blots of the corresponding total lysates are shown.

The above-described assay with activation-specific probes is rapid and easy, and allows handling of many samples in a short period of time. It has great advantages over the assay using radiolabeled extracts. In particular, it is nonradioactive and it does not require highly specific immunoprecipitating antibodies. However, the method determines only the relative increase in the GTP-bound conformation of the GTPase. This increase can be determined rather accurately by a serial dilution strategy, but the total percentage of the GTPase in the GTP-bound form can only be guessed. A more serious problem could be that if a GTPase is strongly bound to

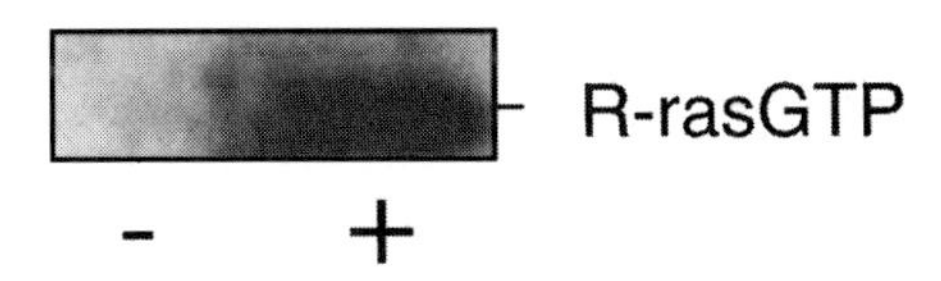

FIG. 3. Activation of R-ras. Human blood platelets are stimulated for 30 sec with 1 U/ml α-thrombin. GST–Raf1–RBD was used as an activation specific probe to precipitate R-rasGTP. Samples were analyzed by SDS-PAGE followed by Western blotting using anti-R-ras polyclonal antibody.

an effector, the GST–probe may not bind and, thus, the measured increase may be an underestimate. However, in all cases in which we have compared the activation-specific probe assay with the assay using radiolabeled cells, we observed a good correlation, indicating that the fraction of active GTPase that is a stable complex may be relatively low, or that there is sufficient probe available to compete for that activation. A final possible problem could arise from GTPase-activating proteins (GAPs) that hydrolyze the GTP after lysis. However, we observed GAP effects only when the time between lysis of the cells and addition of the probe was too long or when the incubations were too long and not performed in the cold.

Acknowledgments

The work in our laboratory is supported by the Dutch Cancer Society, the Netherlands Heart Foundation, and the Council of Earth and Life Sciences of the Netherlands Organization for Scientific Research (NWO-ALW).

[31] Immunocytochemical Assay for Ras Activity

By LARRY S. SHERMAN and NANCY RATNER

Introduction

Approximately 30% of all human tumors contain point mutations in H-, K-, or N-*ras* genes.[1] These activating *ras* mutations cause loss of intrinsic Ras GTPase activity or an increased rate of guanine nucleotide exchange, both resulting in a higher proportion of active (GTP-bound) Ras. Mutations in *ras* genes that result in constitutively high Ras activity are likely to be early events in the progression of numerous cancers.[2] Increased Ras expression, presumably leading to increased Ras activity, has also been linked to malignant progression in some tumors.[3] Alterations in Ras mediators can also result in elevated Ras activity in affected cells (see Fig. 1). Notably, tumors and cell lines from patients with neurofibromatosis type 1 lack neurofibromin, a Ras GTPase, and have elevated Ras activity.[4]

At present, efforts have been focused on developing reverse transcription-polymerase chain reaction (RT-PCR) techniques that identify activating *ras*

[1] J. L. Bos, *Cancer Res.* **49,** 4682 (1989).
[2] E. R. Fearon and B. Vogelstein, *Cell* **61,** 759 (1990).
[3] B. Gulbis and P. Galand, *Hum. Pathol.* **24,** 1271 (1993).
[4] B. Weiss, G. Bollag, and K. Shannon, *Am. J. Med. Genet.* **89,** 14 (1999).

0076-6879/00 $35.00

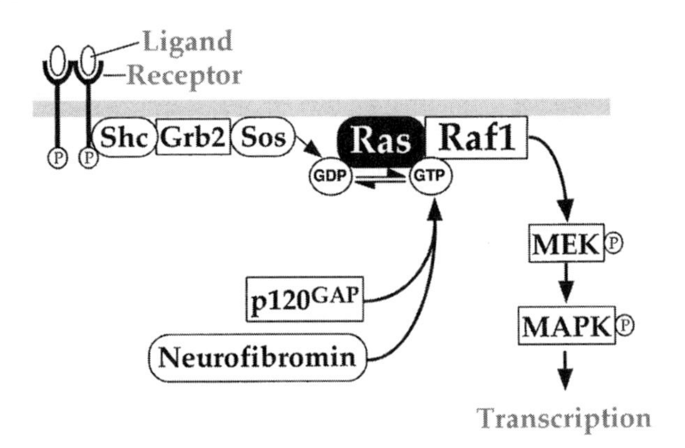

FIG. 1. Activation of Raf1 by native Ras proteins. There are several known mechanisms leading to Ras activation. One of the best understood mechanisms involves activation of receptor tyrosine kinases by their extracellular ligands (e.g., epidermal growth factor and the EGF receptor), followed by activation of guanine exchange factors such as Sos (the *Drosophila* homolog of son-of-sevenless) through the actions of particular adapter proteins. Guanine exchange factors convert Ras–GDP (the inactive form) into Ras–GTP (the active form). Raf1 binds Ras–GTP, becomes active, and phosphorylates a number of substrates, including MEK. GTPase-activating proteins (such as p120^{GAP} and/or neurofibromin) convert Ras back into its inactive, GDP-bound state.

mutations in tissues and other samples (e.g., Villa *et al.*[5]). While these assays may have predictive diagnostic and prognostic value, they do not measure Ras activity and cannot be used to identify which cells in a sample are affected. More importantly, these assays fail to identify cells with elevated Ras activity due to increased Ras expression or mutations in Ras mediators. The detection of increased Ras expression may also have predictive diagnostic and prognostic value.[3] Immunohistochemistry with Ras antibodies, however, cannot demonstrate whether elevated Ras protein expression is accompanied by increased Ras activity. The shortcomings of these assays could be overcome by evaluating Ras activity directly in cells and tissues. An advantage of such an assay would be that measurements could be made without having to know which *ras* gene or which Ras mediator is involved.

Ras-Binding Domain of Raf1 Used as Probe for Ras Activity

The original assay for Ras activity involves culturing cells in the presence of ortho[^{32}P]phosphate, and then determining the ratio of GTP : GDP bound to

[5] E. Villa, A. Dugani, A. Rebecchi, A. Vignoli, A. Grottola, P. Buttafoco, L. Losi, M. Perini, P. Trande, A. Merighi, R. Lerose, and F. Manenti, *Gastroenterology* **110,** 1346 (1996).

Ras.[6–8] This technique, however, is technically demanding and is not useful for most clinical applications. An enzymatic assay has also been used to measure Ras activity in tumor and cell lysates.[9,10] This method, however, cannot detect altered Ras activity in samples in which only a small number of cells are affected, as is often the case in preneoplastic or early neoplastic tissues.

The Ras-binding domain of Raf1 (RBD; amino acids 51–131) binds to Ras–GTP with high affinity (K_d 20 nM) while its affinity for inactive, Ras–GDP is three orders of magnitude lower.[11–13] Raf1 is a major effector of active Ras (see Fig. 1). de Rooij and Bos[14] used Raf1–RBD to measure the relative levels of activated Ras by incubating cell lysates with glutathione *S*-transferase (GST)-tagged Raf1–RBD immobilized on glutathione–agarose, and then examining the amount of activated Ras (e.g., Ras proteins bound to Raf1–RBD) by Western blotting with a Ras antibody (*see* Chapter 30, this volume). This technique is sensitive enough to measure small changes in Ras activity. The drawback of this approach is that it cannot evaluate Ras activity in individual cells and therefore cannot be used to determine which cells in a sample have elevated Ras activity.

We have adapted Raf1–RBD–GST as an immunocytochemical probe to identify cells with elevated Ras *in vitro*. This approach has all the advantages of other Ras activity assays, with the added benefits of (1) allowing evaluations to be made on small samples, (2) preserving samples so that the identity of cells with elevated Ras can be determined, and (3) requiring little or no specialized equipment or methodologies. The major drawbacks of this technique are that it can be used only to demonstrate qualitative and not quantitative differences in Ras activity, and it has limited sensitivity. It is also unclear whether RBD–GST recognizes other molecules, such as activated Rap1, although previous findings suggest that this association does not occur at detectable levels.[14] This assay, however, can be used when other Ras activity assays are not feasible, and should ultimately be adaptable for clinical diagnostic laboratories.

[6] J. Downward, J. D. Graves, P. H. Warne, S. Rayter, and D. Cantrell, *Nature (London)* **346,** 719 (1990).

[7] T. Satoh, M. Endo, M. Nakafuku, S. Nakamura, and Y. Kaziro, *Proc. Natl. Acad. Sci. U.S.A.* **87,** 5993 (1990).

[8] J. B. Gibbs, M. S. Marshall, E. M. Scolnick, R. A. Dixon, and U. S. Vogel, *J. Biol. Chem.* **265,** 20437 (1990).

[9] A. Guha, N. Lau, I. Huvar, D. Gutmann, J. Provias, T. Pawson, and G. Boss, *Oncogene* **12,** 507 (1996).

[10] J. S. Scheele, J. M. Rhee, and G. Boss, *Proc. Natl. Acad. Sci. U.S.A.* **92,** 1097 (1995).

[11] C. Herrmann, G. Martin, and A. Wittinghofer, *J. Biol. Chem.* **270,** 2901 (1995).

[12] N. Nassar, G. Horn, C. Herrmann, A. Scherer, F. McCormick, and A. Wittinghofer, *Nature (London)* **375,** 554 (1995).

[13] J. R. Sydor, M. Engelhard, A. Wittinghofer, R. S. Goody, and C. Herrmann, *Biochemistry* **37,** 14292 (1998).

[14] J. de Rooij and J. L. Bos, *Oncogene* **14,** 623 (1997).

RBD–GST Binding to Cells with Constitutively Activated v-H-Ras

Our strategy for detecting active Ras *in situ* is to treat fixed, permeabilized cells with RBD–GST, and then use anti-GST antibodies to detect RBD–GST that has complexed to Ras–GTP (Fig. 2a). To accomplish this, cells are fixed in ice-cold

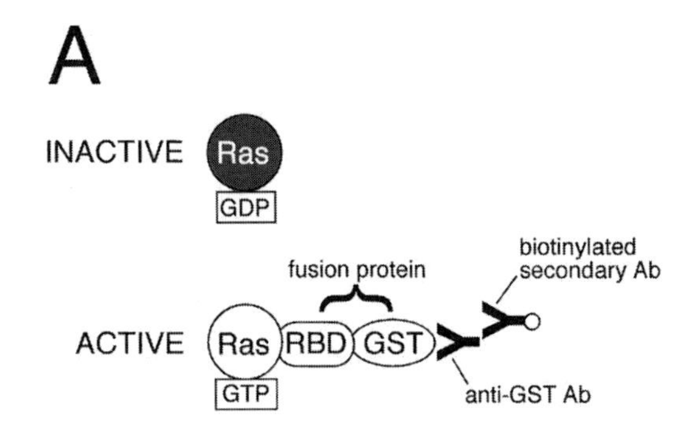

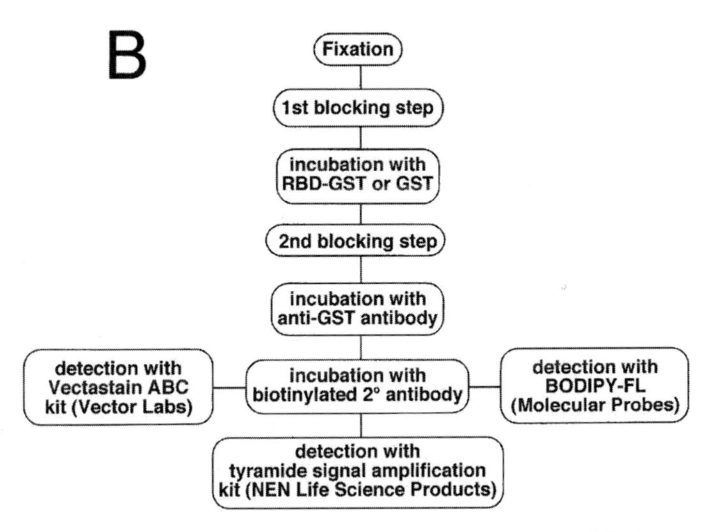

FIG. 2. Description of activated Ras detection assay. (A) Principle of Ras–GTP detection mechanism. RBD–GST binds to Ras–GTP (the active form of Ras) but has significantly lower affinity for Ras–GDP (the inactive form). RBD–GST bound to activated Ras can then be visualized with antibodies (Ab) recognizing GST, followed by immunocytochemistry. (B) Outline of experimental protocol. Cells are fixed and permeabilized, blocked with a buffer that prevents nonspecific binding of GST, and then incubated with RBD–GST for 3–5 hr. To control for nonspecific binding by the GST fusion protein, GST alone is added to parallel cultures. After a second blocking step and incubation with an anti-GST antibody, RBD–GST is detected by one of several methods described in text.

ethanol–acetic acid, blocked with a buffer containing 0.5% (w/v) bovine serum albumin (BSA) and 5% (v/v) normal goat serum, and then incubated with either GST alone (as a negative control) or GST–RBD. We have obtained a Raf1–RBD–GST construct (a generous gift from G. Bollag and F. McCormick, Onyx Pharmaceuticals, Richmond, CA) and purified RBD–GST protein, using a standard glutathione–agarose chromatography protocol (see below). Cells are then processed for GST immunocytochemistry (Fig. 2B).

We have established the conditions under which RBD–GST can be utilized as a probe for activated Ras, using stable clones of NIH 3T3 cells transfected with v-H-*ras* (61L mutant, which is constitutively in the active, GTP-bound state) as compared with cells transfected with empty vector. Using avidin BODIPY-FL (Molecular Probes, Eugene, OR) to detect signal, we have found that RBD–GST signal accumulates significantly in v-H-*ras*-transformed cells, while vector controls have little or no signal (compare Fig. 3A and B with Fig. 3C and D). Activated Ras appears to be highly localized in a membrane-associated compartment in the center of these overexpressing cells (Fig. 3E). This localization has not been analyzed further. If cells are incubated with GST in place of RBD–GST, no signal is detected (Fig. 3F). Similar results are obtained with avidin–horseradish peroxidase (HRP) followed by 3,3′-diaminobenzidine (DAB) for detection, indicating that the assay can be performed using bright-field microscopy (data not shown). However, development with DAB does not allow for the distinct intracellular localization of activated Ras. We have also used a tyramide signal amplification kit (NEN Life Science Products, Boston, MA), which we found significantly enhances the sensitivity of the assay for the detection of endogenous Ras–GTP.[15]

Methods and Procedures

Production and Purification of RBD–GST

The Raf1–RBD–GST protein is synthesized and purified by glutathione–agarose chromatography as previously described, with minor changes from the published protocol.[11] Pellets of bacteria transformed with pGEX 2T-RBD (generously provided by G. Bollag and F. McCormick, Onyx Pharmaceuticals) and induced with 0.1 m*M* isopropyl-β-D-thiogalactopyranoside (IPTG) for 6 hr are suspended in ice-cold phosphate-buffered saline (PBS) with 2 m*M* dithiothreitol, 1 μ*M* aprotinin, 1 μ*M* leupeptin, 1 μ*M* pepstatin, and 20 μ*M* Pefabloc and sonicated with a probe at maximum power on ice for 1–2 min with 10-sec pulses. Triton X-100 is added to a final concentration of 1% (v/v). After gentle stirring for 30 min, lysates are centrifuged at 10,000 rpm for 20 min at 4° and the supernatants are mixed with an ice-cold 50% (v/v) slurry of glutathione–Sepharose (Pharmacia,

[15] L. S. Sherman, R. Atit, T. Rosenbaum, A. D. Cox, and N. Ratner, *J. Biol. Chem.* **275,** 30740 (2000).

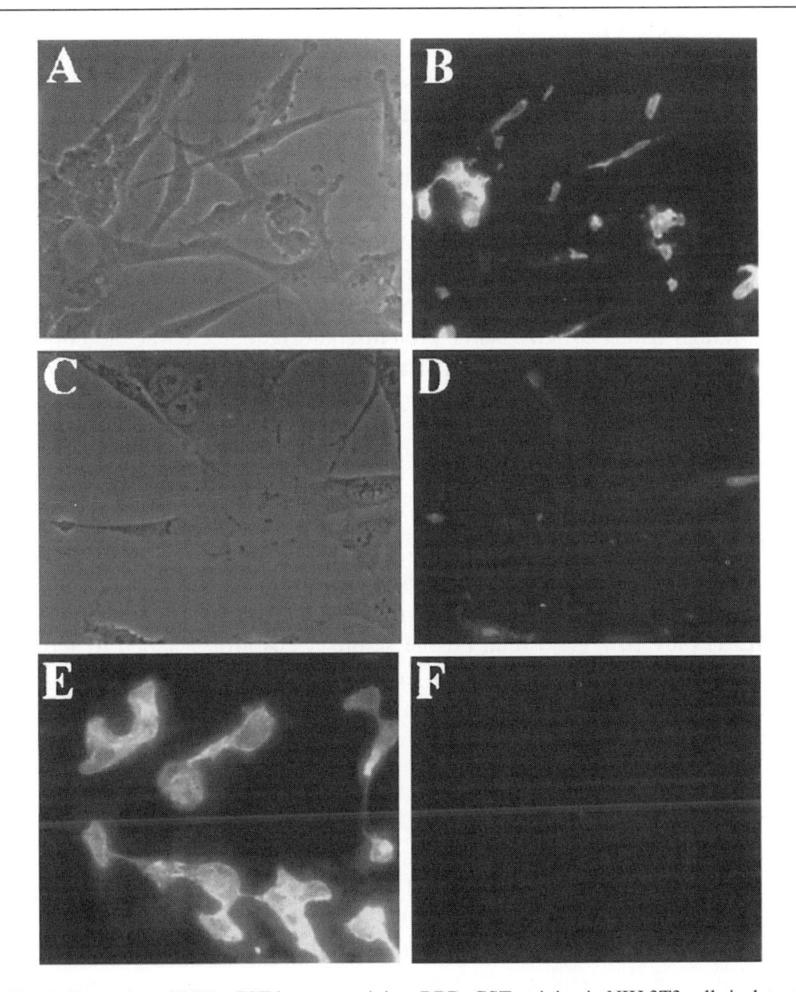

FIG. 3. Examples of RBD–GST immunostaining. RBD–GST staining in NIH 3T3 cells is detected by the protocol described in text, followed by a biotinylated secondary antibody to recognize anti-GST and avidin-BODIPY-FL. (A and B) NIH 3T3 cells stably transfected with v-H-*ras*. (C and D) NIH 3T3 cells stably transfected with empty vector. (A and C) Phase-contrast photomicrographs of the cells shown in (B) and (D), respectively. (E) High-power photomicrograph of RBD–GST localization in v-H-*ras*-transformed cells. (F) High-power photomicrograph of v-H-*ras*-transformed cells incubated with GST in place of RBD–GST.

Piscataway, NJ) in PBS and incubated with rotation for 1 hr at 4°. Beads are then washed three times with 4 volumes of ice-cold PBS, and packed into an Econo-Pac column (Bio-Rad, Hercules, CA). Protein is eluted in 1-ml aliquots, using 50 m*M* sodium bicarbonate at pH 10.8. Aliquots are neutralized with 87 μl of 1 *M* sodium citrate, pH 5.0. To determine which fractions contain RBD–GST, 5 μl of each fraction is examined by sodium dodecyl sulfate–polyacrylamide gel electrophoresis (SDS–PAGE) and Coomassie blue staining. Positive fractions are then purified by a second round of glutathione–Sepharose purification. Positive aliquots are pooled, concentrated to 10 mg/ml, and stored at −80° in 50-μl aliquots for up to 2 weeks. GST is also prepared by this protocol, and is stored for up to several months.

Under these conditions, we have found that RBD–GST stability is highly variable. Some batches can be used for longer than 2 weeks, while others are stable for only 3–5 days. We also have batches that did not work at all in this assay, even though bands of RBD–GST protein could be detected at the correct size by Western blotting with an anti-GST antibody. We therefore recommend testing an aliquot from each batch, using control cells (e.g., parental NIH 3T3 cells and v-Ha-*ras*-transformed NIH 3T3 cells), before performing any experiments.

Cell Culture

NIH 3T3 cells stably transfected with v-Ha-*ras* and empty vector-transfected controls are plated onto eight-well chamber slides (Fisher Scientific, Pittsburgh, PA) at various densities and grown in Dulbecco's modified Eagle's medium (DMEM) plus 10% (v/v) fetal bovine serum. Cells are switched to serum-free medium for 24–48 hr before fixation to reduce background levels of Ras activity. We have also performed this assay with primary cultures of human fibroblasts and Schwann cells, using similar culture protocols.

Cell Fixation

The fixation procedure that works best for this assay is to gently aspirate medium from tissue culture wells, and then directly replace it with ice-cold acid alcohol fixative [90% (v/v) ethanol, 5% (v/v) glacial acetic acid, 5% (v/v) distilled water] for 20 min. We have attempted to use several other fixation reagents, including ice-cold methanol (for 20 min), Bouin's fixative (aqueous; for 15–20 min), alcoholic Bouin's fixative (Brasil–Duboscq fluid; for 15–20 min), and 2 and 4% (w/v) paraformaldehyde (for 20 min). After applying some of these fixation conditions, we have also permeabilized cells with 0.05–2% (v/v) Triton X-100. Other detergent–fixative combinations have not been attempted. Under none of these conditions have we detected adequate signal, using the procedures described below. We therefore have adopted acid–alcohol fixation for this assay.

RBD–GST Incubation and Glutathione S-Transferase Immunocytochemistry

After fixation, cells are washed twice with PBS, and then incubated in blocking buffer A [ice-cold PBS with 5% (v/v) normal goat serum and 0.5% (w/v) bovine serum albumin] for 1 hr at 4°. This buffer substantially reduces background GST binding, and is made fresh for each assay. The levels of background staining increase if older buffer preparations are used. We have attempted to use several other blocking buffers, including different concentrations (1–10%) of different sera (goat, fetal bovine, mouse), but have found the goat serum–albumin combination to be most effective. Cells are then incubated with a 10-μg/ml concentration of RBD–GST or GST (as a negative control) in buffer A for 3 hr at 4°, washed three times in blocking buffer without albumin (buffer B), and then incubated for 1 hr with mouse anti-GST monoclonal antibody (1 μg/ml) in buffer B at room temperature. We have used two separate anti-GST mouse monoclonal antibodies, both of which perform equally well: one from Clontech (Palo Alto, CA), the other from Santa Cruz Biotechnology (Santa Cruz, CA). Cells are then washed three times with buffer B.

Signal Detection

We have successfully used three different signal detection kits to visualize bound RBD–GST. For detecting Ras–GTP levels in cells expressing only endogenous Ras proteins or activated *ras* alleles, we use a tyramide signal amplification kit (NEN Life Science Products). Cells are incubated in the kit blocking buffer for 30 min. The blocking buffer is drained and replaced with wash buffer [0.1 *M* Tris-HCl (pH 7.5), 0.15 *M* NaCl, 0.05% (v/v) Tween 20] and then incubated for 1 hr with a 1 : 1000 dilution of goat anti-mouse horseradish peroxidase (HRP)-conjugated secondary antibody (Bio-Rad). Cells are then washed three times for 5 min each in wash buffer and processed for tyramide fluorophore immunofluorescence according to the kit manufacturer instructions. Labeled cells are mounted and examined by fluorescence microscopy.

For detecting Ras–GTP in cells expressing activated *ras* alleles (e.g., cells transformed with v-Ha-*ras*) by fluorescence microscopy, cells are incubated with a 1 : 2500 dilution of a biotinylated goat anti-mouse IgG (Vector Laboratories, Burlingame, CA) for 30 min at room temperature. Cells are then washed three times with buffer B and incubated for an additional 30 min with a 1 : 1000 dilution of avidin-conjugated BODIPY-FL (Molecular Probes). For detection by bright-field microscopy, avidin-BODYPY-FL is replaced with avidin–HRP from a Vector Laboratories Vectastain Elite kit according to the manufacturer instructions. Signal is detected with 3,3′-diaminobenzidine (DAB) as substrate.

Acknowledgments

This work was sponsored by the NIH (NINDS-28840) and the DOD Program on Neurofibromatosis.

CHAPTER 1, TABLE II*

Mutant Shc Proteins[a]	Ref.
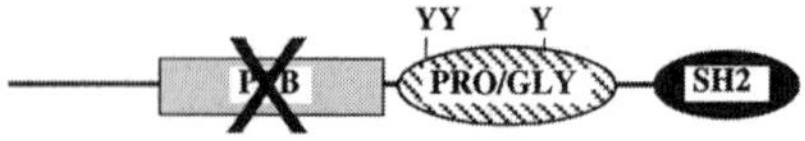 PTB (X) — YY Y PRO/GLY — SH2 [b]	23-25
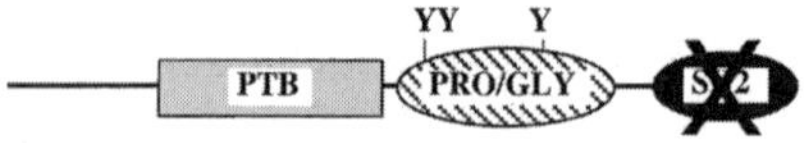 PTB — YY Y PRO/GLY — SH2 (X) [c]	24, 26
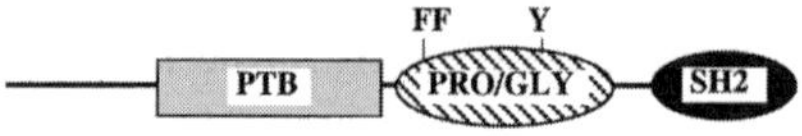 PTB — FF Y PRO/GLY — SH2	23-25, 27, 27a, 38
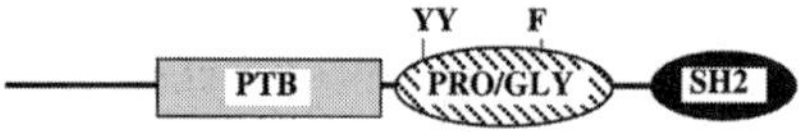 PTB — YY F PRO/GLY — SH2	23-27a, 37-39
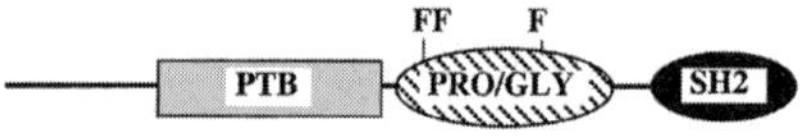 PTB — FF F PRO/GLY — SH2	23-25, 33, 38
Y RO/GLY — SH2 [d]	29
PTB — YY Y PRO/GLY [e]	34
PTB — YY F PRO/GLY	26, 36
PTB — YY — SH2 [f]	30
 PTB	23, 25, 31-34, 38
 SH2	22, 26, 31-36, 38

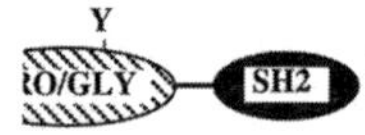

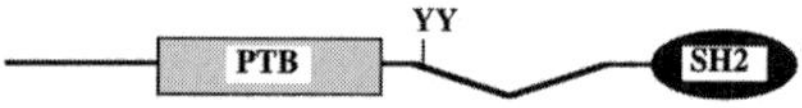

[a] The mutant version of Shc were in the p52^{ShcA}. The truncated versions were of either ShcA or ShcC.

[b] Mutations abolish recognition of phosphotyrosine ligands.

[c] Mutations abolish recognition of phosphotyrosine ligands.

[d] Truncation of the PTB domain and the part of the CH1 domain encompassing the phosphorylation sites at tyrosines 239 and 240.

[e] Truncation of all or part of the SH2 domain.

[f] Deletion of amino acids 250-379 of the CH1 region which removes the phosphorylation site at tyrosine 317.

*See page 12, this volume.

Author Index

A

Aaronson, S. A., 203, 204(7)
Abate, C., 45
Abbott, D. W., 37
Abe, J., 89
Abe, K., 98
Abo, A., 35, 89
Abraham, J. A., 72
Abrahan, J., 276
Adeeko, A., 286
Adjei, A. A., 317
Adnane, J., 139
Advani, S., 28
Aebersold, R., 151
Afonso, A., 309
Agani, F., 276
Aguer, M., 139, 140(4; 5), 149(4)
Aguirre-Ghiso, J. A., 108
Ahmad, M. F., 45
Ahn, N. G., 56, 61, 63(2), 192, 268, 270(18), 271(18), 277(18), 278(18), 279(18), 280(18), 283(18)
Aidoudi, S., 169
Aizawa, S., 247
Akerman, M., 106
Akhurst, R., 292
Akira, S., 139, 140(7–10), 141, 149, 149(7)
Akiri, G., 276
Akiyama, S. K., 165
Akkerman, J. W. N., 342, 343, 344(1; 5), 346(5)
Aktas, H., 130
Aktories, K., 342
Alahari, S., 151
al-Alawi, N., 200
Albanese, C., 74, 75, 116, 116(1), 117, 118, 118(8; 9), 119, 119(9), 121, 126, 126(12), 130, 131(24), 146, 148(38), 149
Alcorta, D. A., 45
Alemá, S., 232
Alessi, D. R., 156, 196
Alexander, J. S., 276
Alexander, N., 277
Alitalo, K., 271(32), 273
Allgayer, H., 105, 106, 107, 109(33), 111
Allis, C. D., 45
Alonso, D. F., 115
Alper, S. L., 124
Alvarez, C. S., 309
Alvegard, T., 106
Amanatullah, D. F., 116, 127
Ambegaokar, A., 108
Ambrose, D. M., 56, 58(6)
Amir, C., 29, 30(12), 34(12)
Amrani, Y., 49
Anders, R. A., 132
Andersen, P., 16
Anderson, M. J., 292, 293(19), 302(19)
Anderson, R. G., 173, 174, 176(21), 179
Anderson, W. B., 28, 35(10)
Andjelkovic, M., 46, 53(17)
Ando, A., 217
Ando, H., 217
Andrabi, K., 53
Andreasen, P. A., 105
Andres, A.-C., 310
Andres, D. A., 317
Andrews, P. C., 90
Anichini, A., 280, 281(91)
Anisowicz, A., 173
Anthony, N. J., 267
Aono, A., 141
Aplin, A. E., 151, 152, 153(1), 156, 156(1), 157, 158(12)
Apolloni, A., 169, 174(5)
Apone, S., 237
App, H., 275
Arand, G., 265
Arap, W., 273
Arbiser, J. L., 271(30), 273, 279
Arkona, C., 98
Armanini, M., 274
Arnold, A., 116(1), 117, 118(8), 127, 130, 131(24)
Arnold, R. S., 89
Aronheim, A., 200

Asano, T., 187
Ashendel, C. L., 232, 233, 233(5), 236
Ashmun, R. A., 130, 247
Ashworth, A., 88, 156, 196
Askari, A., 30(17), 34
Aslakson, C. J., 270
Atkinson, S., 98, 99(24)
Auer, K. L., 28, 29(1), 32(1), 35, 37(1)
Augenlicht, L. H., 127
Aukhil, I., 156
Ausubel, F. M., 238
Avantaggiati, M., 117
Avruch, J., 45, 46, 53, 53(21), 300, 343
Axelrod, J. H., 106, 109(30), 112(30)
Ayusawa, D., 3, 6(4)
Azam, M., 141
Azzam, H., 265

B

Babic, R., 106
Bach, E. A., 139, 140(4), 149(4)
Back, R., 182
Bacon, C. M., 147
Bae, S. N., 265
Baeuerle, P. A., 89, 90
Bagg, A., 37
Bagrodia, S., 56, 58(6), 342
Bakanauskas, V. J., 284
Baker, E. K., 165
Baldari, C. T., 12
Baldwin, A. S., Jr., 73, 74, 82(7; 8; 12), 84(13), 85(6), 88, 89, 119
Balkowiec, E. Z., 271(36), 273
Ballin, M., 98
Balmain, A., 292
Baltimore, D., 7, 90, 173, 251
Banerjee, P., 45
Bankston, L., 45
Barbacid, M., 247, 320
Bardeesy, N., 267, 274(7), 280(7)
Bardelli, A., 115
Barker, K. T., 203
Barleon, B., 271, 274(25), 278(25), 279(25), 281(25)
Barnard, D., 56
Barnes, C. A., 217
Barnes, D., 302
Barradas, M., 130, 248, 250(14), 255(14)
Barrett, T., 88, 122, 300
Barrington, R. E., 316
Bar-Sagi, D., 15, 58, 73, 320
Bartek, J., 128, 131, 132(41)
Bartkova, J., 131, 132(41)
Baruchel, S., 283
Bassand, P., 45
Bates, S. E., 280, 281(93)
Batlle, D., 117, 126(12)
Batzer, A., 7
Bauer, B., 343, 344(5), 346(5)
Bauer, R., 111
Baunoch, D., 279
Baxter, J. D., 119, 121(21)
Bayko, L., 268, 269(19), 271(19), 274(19), 275(19), 276(19), 279(19), 281(19), 283(19)
Bayne, M. L., 276
Beach, D., 204, 247, 248, 250(11), 251(11), 255(11)
Beach, D. H., 75, 129, 130, 131(18; 22)
Bearss, D. J., 316
Beavis, A. J., 124
Becker, J. M., 229
Becker, L. B., 118
Becker, S., 151
Bedner, S., 139, 140(13)
Beerepoot, L. V., 276
Beg, A. A., 74, 82(8)
Behn, M., 131
Behre, G., 70
Behrendt, N., 105
Behrendtsen, O., 280
Behrens-Jung, U., 124
Behrmann, I., 149
Beijersbergen, R. L., 292
Beinhauer, J., 219
Bell, R. M., 295(21), 296
Bell, S. M., 106
Benehacene, A., 271(33), 273, 279(33), 283(33)
Ben-Levy, R., 286
Bentvelzen, P., 284
Ben-Ze'ev, A., 121
Berchtold, S., 149
Bergers, G., 280
Berglund, E. B., 237
Bergman, A., 342
Berk, B. C., 89
Bernards, R., 131

Bernhard, E. J., 96, 98, 99(17), 108, 284, 285, 285(22; 23), 286
Berta, P., 141
Besser, D., 114, 115, 116(53), 149
Bessereau, J. L., 237
Bett, A. J., 274
Betts, J. C., 75
Bi, C., 187, 189(4), 191(4), 193(4), 194(4), 197(4), 199(4), 200(4), 202(4)
Biaglow, J., 284
Bicknell, R., 276
Binetruy, B., 97
Bird, C. C., 286
Birnbaum, D., 12
Birrer, M., 28, 29(1), 32(1), 37(1), 109, 114(38), 115(38)
Bishop, J. M., 204, 248, 250(13), 255(13)
Bishop, W. R., 306, 308, 308(1), 309, 309(1; 4; 5), 311(1; 4; 5), 314(4; 5), 315(4), 316, 317, 317(4)
Bist, A., 174
Blaikie, P., 5, 12, 15(14)
Blasberg, R. G., 273
Blasi, F., 105, 106, 112, 112(18; 20; 22), 114(20; 22)
Blaskovich, M., 285
Blau, H. M., 236
Blenis, J., 45, 46, 53, 53(20), 54, 54(16)
Boast, S., 106, 112(18)
Bocquel, T., 123
Bodmer, W. F., 292
Boenink, N. M., 343
Bogenberger, J., 12
Bogoyevitch, M. A., 156
Boguski, M. S., 88, 290(1), 291
Bohlen, P., 275, 283
Bokoch, G. M., 56, 58(6), 60, 202
Bollag, G. E., 40, 348
Bond, R., 306, 308(1), 309(1), 311(1)
Bonfini, L., 3
Borg, J. P., 12
Bortner, D. M., 268
Bos, J. L., 201, 319, 334, 336(10), 341(10), 342, 343, 344(1; 5; 5), 346(5), 348, 350
Bosch, E., 250
Boss, G., 267, 271(9), 334, 350
Bouck, N., 271(31), 273
Boulton, T. G., 147
Bourgeois, Y., 281
Bourne, J. R., 284
Bours, V., 90
Bouzahzah, B., 75, 117, 118(9), 119(9)
Bowen-Pope, D. F., 280, 281(94)
Bower, G. D., 29, 30(11), 34(11)
Bowers, G., 28, 32(9)
Bowman, T., 139, 149
Boyd, D., 106, 107, 108, 109, 109(21; 33), 110(41), 111, 112, 112(19), 114(19; 38), 115(19; 21; 38)
Bradfield, C. A., 279
Bradshaw, R. A., 12(32; 37), 13
Brasher, P., 98
Brass, L. F., 165
Brattain, M. G., 109
Braungart, E., 105
Bravo, R., 90
Breier, G., 280
Brennan, T. J., 237
Brenner, D. A., 226
Brent, R., 238
Brenz-Verca, S., 35
Briane, D., 97
Bridges, A. J., 263
Bromberg, J., 138, 146, 148, 148(38), 149
Bronzert, D., 280, 281(93)
Brooks, M. W., 292
Browder, T., 283
Brown, C., 149
Brown, J. M., 286
Brown, K., 89, 90, 292
Brown, L. F., 273
Brown, M., 131, 139, 140(13)
Brownlee, M., 271(30), 273, 279
Brtva, T. R., 295(21), 296
Brugge, J., 7
Brundage, R., 105
Brunet, A., 118, 130
Brunner, G., 106, 107(31)
Brusselbach, S., 131
Bryant, M. S., 306, 308, 309, 309(4; 5), 311(4; 5), 314(4; 5), 315(4), 317(4)
Bucana, C. D., 275
Buch, M. B., 53
Buckle, R. S., 131
Buick, R. N., 268, 271(17)
Bujard, H., 128
Bull, P., 90
Burakoff, S. J., 4
Burd, P. R., 90
Burgering, B. M. T., 44, 46

Burk, R., 173
Burnett, W. N., 299
Burridge, K., 153, 154(6), 203
Buskin, J. N., 237
Buss, J. E., 284, 285
Bustelo, X. R., 203, 204, 204(4), 208(10), 210(4), 211(4)
Butler, G., 98, 99(24)
Buttafoco, P., 349
Butterfield, C. E., 283
Byers, H. R., 271(30), 273, 279

C

Cabrera, G., 274
Cai, H., 130
Cajot, J. F., 106
Calaycay, J., 27
Caldenhoven, E., 149
Caldwell, G. A., 229
Camonis, J. H., 343, 344(5), 346(5)
Campbell, D. G., 46, 53(18), 139, 140(4), 149(4), 268, 276(11)
Campbell, S. L., 55, 73, 187, 203, 204(5), 208(5), 217, 319
Campbell-Burk, S., 295(21), 296
Candia, J. M., 192
Canete-Soler, R., 96
Cannon, R. E., 271(34), 273, 279(34)
Cano, A., 271(28), 273
Cantley, L., 3, 6(5), 13(5)
Cantrell, D., 350
Cao, Y., 271(30), 273, 279
Capella, G., 286
Caracciolo, A., 115
Caravatti, G., 341
Carboni, J. M., 203, 204(4), 210(4), 211(4)
Cardarelli, P. M., 169
Carmeliet, P., 280
Carr, D., 306, 308(1), 309(1), 311(1)
Carracciolo, A., 106, 112(22), 114(22)
Carson, R. T., 139, 140(11), 149(11)
Carter, R. H., 341
Carter, S., 28, 29(1), 32(1), 37(1)
Carver-Moore, K., 139, 140(4), 149(4), 273, 280(46)
Casalino, L., 115
Casey, P. J., 218
Casnellie, J. E., 16, 24(5)
Castelli, C., 280, 281(91)
Castro, A. F., 187, 189(4), 191(4), 193(4), 194(4), 197(4), 199(4), 200(4), 202(4)
Catino, J. J., 306, 308, 308(1), 309(1; 4; 5), 310, 311(1; 4; 5), 314(4; 5), 315(4), 317(4)
Catlett-Falcone, R., 142, 149(28)
Catling, A. D., 300
Cato, A. C. B., 119
Caudwell, F. B., 46, 53(21)
Caughey, G. H., 280
Cavenee, W. K., 273
Cawthon, R. M., 291
Cayrol, C., 124
Celis, J. E., 217
Cepko, C. L., 191, 204, 250, 251(19)
Ceresa, B. P., 5, 15(14)
Cerione, R., 342
Cerwinski, H., 250
Cesareni, G., 3, 6(5), 7(6)
Chait, B. T., 146, 148(37), 150(37)
Chakrabarty, S., 109
Chakraborty, A., 142, 149(27)
Chambers, A. F., 97, 268, 280, 281(12; 95), 319, 326(3)
Chan, A. M., 163, 164(5), 169(5), 187, 203, 204(7)
Chan, T. O., 163, 164(5), 169(5)
Chang, E., 28, 35(10)
Chang, Y. J., 302
Changeux, J. P., 237
Chao, J.-R., 116(3), 117, 130
Chao, L., 203, 204(7)
Chaplin, D. J., 271
Chapman, R. S., 139, 140(9)
Chari, S., 342
Charter, C., 35
Charvat, S., 271(37), 273, 276(37), 283(37)
Cheatham, B., 46
Cheatham, L., 54
Chen, A., 55
Chen, C. H., 90, 287
Chen, E. Y., 271(29), 273, 276, 279(29; 66)
Chen, H., 198, 199(16), 273, 280(46)
Chen, J., 88, 308, 309(4; 5), 311(4; 5), 314(4; 5), 315(4), 317(4)
Chen, J. D., 123
Chen, J. H., 108
Chen, K., 146
Chen, P. B., 28
Chen, Q., 153, 154, 154(6), 156, 156(7)

Chen, R. H., 45
Chen, X., 129, 131(18), 138, 139
Chen, X. Q., 342
Chen, Y., 280
Chen, Z., 56
Cheng, C., 277
Cheng, J. J., 89
Cheng, M., 117, 129
Cheng, S. Y., 273
Cheong, N., 289
Cheresh, D. A., 12, 169
Cherian, S. P., 282, 283(106)
Chernoff, J., 56, 58(6), 60
Cheung, P., 45
Chida, K., 12
Chignol, M. C., 271(37), 273, 276(37), 283(37)
Chilvers, E. R., 49
Chin, L., 130, 247, 267, 274(7), 280(7)
Chinali, G., 175
Chinery, R., 334
Chiu, C. P., 236
Chiu, V. K., 169
Chmura, S. J., 28
Cho, S. S., 147
Choong, P. F. M., 106
Chou, H. S., 129
Chou, M. M., 45, 46, 54, 54(16)
Chow, K. L., 237
Choy, E., 169
Christensen, A., 284
Christensen, I. J., 289
Chrzanowska-Wodnicka, M., 319, 320(9), 321(9), 325(9), 327(9)
Chung, A. B., 89
Chung, C. D., 141
Chung, J., 45, 46
Ciliberto, G., 142, 149(28)
Cirillo, G., 115
Cisek, L. J., 16
Claffey, K. P., 273
Clark, A. R., 139, 140(9), 286
Clark, C. J., 217, 226(8), 283, 295(22), 296
Clark, E. P., 284
Clark, G., 7, 55
Clark, G. J., 73, 117, 118(11), 187, 188, 190(7), 192(7), 203, 204(2), 206, 207(17), 209(2), 210(2), 211(2), 217, 218(7), 223, 225(7), 231, 231(7), 277, 319
Clark, K., 229
Clark, R., 40, 139, 140(4), 149(4), 202
Clarke, S., 285
Clouse, M. E., 273
Coats, S. R., 130
Cobb, M. H., 56, 61, 226
Cobb, R. R., 169
Cockran, C. G., 284
Coe, L. L., 276
Coffer, P. J., 44, 46, 56
Coffey, R. J. J., 277
Cogswell, P. C., 74, 84(13), 88
Cohen, C., 247
Cohen, P., 16, 46, 53(18; 21), 156, 196
Cohen, S. L., 146, 148(37), 150, 150(37)
Colburn, N., 97
Coleman, C. N., 280
Colgin, L. M., 292
Collen, D., 280
Collier, I. E., 99
Collins, L. R., 12
Collins, T., 75
Colman, M. S., 62, 68(6), 206
Colombo, L. L., 130
Colussi, P. A., 105, 112(6)
Comoglio, P., 115
Conklin, E., 27
Conn, G., 276
Conner, M. W., 267, 316
Connolly, D. C., 106, 309
Contessa, J. N., 28, 29, 30(11; 12), 34(11; 12), 35
Conti, C. J., 117, 130, 271(28), 273
Cool, R. H., 343, 344(5), 346(5)
Cooper, A. B., 309
Cooper, G. M., 12, 35, 130
Cooper, J. A., 203, 204(2), 209(2), 210(2), 211(2), 212
Copeland, J., 159
Coppenrath, E., 105
Corden, J. L., 16
Cordon-Cardo, C., 28, 130, 247, 267, 274(7), 280(7)
Cornish, A. L., 132, 138(49)
Cortez, D., 74, 82(12)
Corthals, G. L., 151
Couchman, J. R., 265
Counter, C. M., 292
Courtneidge, S. A., 12(34), 13, 15(34)
Coussens, L. M., 97, 280
Coutavas, E., 204
Cowburn, D., 142

Cowell, S., 98, 99(24)
Cowie, A., 122
Cowley, S., 156, 196
Cox, A. D., 7, 14, 188, 190(7), 192(7), 203, 204(2–4), 206, 206(3), 207(17), 208(3), 209(2), 210(2; 4), 211(2; 4), 218, 223, 226, 261, 284, 285
Craig, A. M., 280, 281(95)
Cramer, R., 89
Crespo, P., 204, 208(10)
Crews, C. M., 45
Crick, D. C., 317
Crombie, R., 292
Crompton, M. R., 203
Cronauer-Mitra, S., 286
Crosby, S., 4
Crosio, C., 45
Cruz, T. F., 342
Culver, M., 291
Cunningham, M., 165
Curiel, D. T., 274
Cutler, A. J., 220
Cutler, D., 317
Czajka, A., 271(36), 273
Czernik, A. J., 16

D

Dabrowska, A., 271(36), 273
Dai, T., 300
Dalby, K. N., 46, 53(21)
Dallas, P., 118
Dalton, W. S., 142, 149(28)
Daly, R., 7
Dameron, K. M., 271(31), 273
Damert, A., 276
D'Amico, M., 75, 117, 118(9), 119(9), 121
D'Angelo, G., 282
Dano, K., 105
Darnell, J. E., Jr., 138, 139, 141(17; 18), 142, 142(21), 146, 147, 148, 148(1; 21; 37; 38), 149, 149(1; 30), 150(37)
Das, P., 56, 58(2), 88, 323
Davey, H. W., 139, 140(14)
Davide, J. P., 267
Davis, D. W., 275
Davis, J. N., 317
Davis, R. J., 28, 88, 115, 122, 300
Davis, R. L., 233, 236(12), 237
Davis-Smyth, T., 274, 276(49), 280(49)
De Benedetti, A., 276
De Cesare, D., 106, 112(22), 114(22), 115
de Chasseval, R., 123
Decker, S. J., 263
Declercq, C., 280
DeClue, J. E., 286
Degen, J. L., 106
Degitz, K., 105
De Giuseppe, A., 3, 6(5), 7(6)
Degregori, J., 131
De Groot, R. P., 149
Degryse, E., 35
Dejong, V., 97
de Kier Joffe, E. B., 108, 115
Dekker, G. A., 105
del Aguila, L. F., 273
DeLeo, F. R., 89
Dell, J., 308, 309(4; 5), 311(4; 5), 314(4; 5), 315(4), 317(4)
del Rosario, J., 309
Demetri, G. D., 280, 281(92)
Deng, T., 88, 122, 300
Denhardt, D. T., 280, 281(95)
Denning, M. F., 277
Dennis, P. B., 46, 53, 53(17)
Dent, P., 28, 29, 29(1), 30(11; 12), 32(1; 9), 34(11; 12), 35, 35(10), 37, 37(1), 300
Dente, L., 3, 6(5), 7(6)
DePinho, R. A., 130, 247, 267, 274(7), 280(7)
Der, C. J., 3, 6(5), 7, 12(31), 13, 13(5), 14, 55, 62, 65, 68, 73, 74, 75, 82(8), 84(13), 88, 89, 117, 118(9–11), 119(9), 154, 156(7), 187, 188, 189(4), 190(7), 191(4), 192(7), 193(4), 194(4), 197(4), 199(4), 200(4), 202, 202(4), 203, 204(2–5), 206, 206(3), 207(17), 208(3; 5), 209(2), 210, 210(2; 4), 211(2; 4), 217, 218, 218(7), 223, 225(7), 226, 231, 231(7), 261, 267, 268, 276(11), 277, 284, 285, 292, 293(17–19), 295, 295(21; 22), 296, 302(19), 319, 320(9), 321(9), 325(9), 327(9)
Derbin, C., 97
Derecka, K., 123
Dérijard, B., 88, 115, 122, 300
de Rooij, J., 201, 334, 336(10), 341(10), 343, 344(4), 350
Desai, J., 309
Desai, S., 27
Deskus, J., 309

De Smet, C., 187
deSolms, S. J., 58, 267
Detmar, M., 273
Dettmar, P., 105, 106
D'Eustachio, P., 203, 204, 204(3), 206(3), 208(3)
Devgan, G., 146, 148(38)
de Villarty, J. P., 123
Dewhirst, M. W., 274
Dhand, R., 115
Di, S. M. M., 12
Diaz, B., 56, 343
Diaz-Meco, M. T., 97
Dibbens, J. A., 276
Dickens, M., 28
Dickson, R. B., 280, 281(93)
Didichenko, S., 115
Diehl, J. A., 117, 129
Diehl, S., 286
Diekmann, D., 89
Dieterle, A., 35
Di Fiore, P. P., 5, 15(13)
Dighe, A. S., 139, 140(4; 5), 149(4)
Dignam, J. D., 114
Dilks, D. W., 96
Dillon, P. J., 90
Dilworth, S. M., 12
Ding, L., 97
Di Nocera, P., 112
DiRenzo, J., 131
DiSalvo, J., 276
Discafani, C. M., 310
Dixon, R. A., 350
Djeu, J. Y., 139
Dleloherey, T., 287
Dlugosz, A. A., 277
Dobrzanski, P., 90
Doh, K., 98
Doherty, P. C., 139, 140(11), 149(11)
Doll, R. J., 306, 308, 308(1), 309, 309(1; 4; 5), 311(1; 4; 5), 314(4; 5), 315(4), 317(4)
Dong, C., 88
Dong, H.-J., 60
Donoghue, M., 237
D'Orazio, D., 114
Douwes, K., 105
Dover, R., 334
Dowd, M., 273, 280(46)
Dowdle, E. B., 100
Dowdy, S. F., 128, 131
Downey, G. P., 342
Downing, J. R., 130, 247
Downward, J., 37, 38, 40(3), 41(5), 43, 43(6), 56, 58(2), 73, 88, 115, 117, 121, 149, 163, 164(2), 295(23), 296, 323, 350
Draetta, G., 129
Draznin, B., 15
Drenning, S. D., 142, 149(27)
Dritschilo, A., 37
Drivas, G. T., 203, 204, 204(3), 206(3), 208(3)
Drobes, B. L., 232, 233(2), 236(2)
Drugan, J. K., 203, 204(5), 208(5), 295(21), 296
Du, X. P., 165
Du Bois, R. N., 139, 140(4), 149(4)
Duchesne, M., 271(37), 273, 276(37), 283(37)
Duckett, C. S., 90
Dudley, D. T., 263
Duffy, M. J., 105
Dufner, A., 46, 53(17)
Dugani, A., 349
Duncan, K. G., 119
Dunn, D., 202, 291
Dunn, M. J., 342
Duran, H., 271(28), 273
Durand, R. E., 271
Durbin, J. E., 139, 140(3), 149(3)
Dvorak, A. M., 182, 273
Dvorak, H. F., 273
Dwarki, V. J., 237
Dynlacht, B., 128

E

Eberhardt, C., 280
Eblen, S. T., 132
Eckstein, J., 129
Edwards, D. R., 98
Egawa, S., 97
Ehrhardt, G. R., 187
Eichten, A., 74
Eklund, N., 117, 118(8), 130
Ellerbroek, S. M., 105
Ellis, C., 217, 226(8)
Ellis, L. M., 273, 275, 279
Ellis, S. L., 286
Ellis, V., 105
Elroy-Stein, O., 276
Emerson, C. P., Jr., 232, 233(8), 236(8)
Endlich, B., 284
Endo, M., 333, 350

Endo, T., 187
Engelberg, D., 200
Engelhard, M., 350
Engelman, J. A., 173
Engert, J. C., 237
Engstorm, L., 16, 27(7)
Enholm, B., 271(32), 273
Erdos, V., 281, 282(103)
Erhardt, P., 30(16), 34
Erikson, E., 45, 46
Erikson, R. L., 45
Eriksson, U., 271(32), 273
Erkell, L. J., 106, 107(31)
Erlichman, C., 317
Ernsberger, D., 257
Ernst, T. J., 280, 281(92)
Ernst, W. H., 300
Erwin, R. A., 147
Esumi, H., 267
Eszterhas, A. J., 49
Etheridge, M., 169, 174(5)
Evan, G., 56
Evans, R. M., 123
Evans, T., 40
Ewen, M. E., 131, 132(51), 133, 341
Ezhevsky, S. A., 128

F

Faasen, A. E., 265
Fabbro, D., 341
Fahrig, M., 280
Fan, Z., 149
Fanning, P., 280, 281(88)
Farias, E. F., 108, 115
Farnsworth, J., 28, 284
Farrar, W. L., 147
Fashing, C. L., 292, 293(19), 302(19)
Fattaey, A., 129
Fautsch, M. P., 132
Favero, J., 342
Fearon, E. R., 89, 267, 348
Feig, L. A., 12, 198
Feldkamp, M. M., 267, 271(9), 273, 276(26; 27), 283(26; 27)
Feleszko, W., 271(36), 273
Feliciello, I., 175
Felzien, L. K., 75
Feng, D., 273
Fenrich, M., 30
Feoktisitov, M., 169
Fernandez, C. A., 271(30), 273, 279
Fernandez-Luna, J. L., 142, 149(28)
Ferno, M., 106
Fero, M., 129
Ferrans, V. J., 88
Ferrara, N., 273, 274, 275, 276(49), 280(46; 49)
Ferrari, E., 308, 309(4), 311(4), 314(4), 315(4), 317(4)
Ferrari, S., 45
Ferreira, V., 280
Fesik, S. W., 4
Fiddes, J. C., 276
Field, J., 55, 56, 57(7), 58, 58(7)
Fielder, T., 219
Fielding, C. J., 173, 174
Fielding, P. E., 173, 174
Figueroa, C., 203
Filmus, J., 130, 268, 269(19), 270, 270(18), 271(17; 18; 22), 274(19), 275(19), 276(19), 277, 277(18), 278(18), 279(18; 19), 280(18), 281, 281(19), 282, 282(22; 103), 283(18)
Finbloom, D., 147
Finco, T. S., 74, 82(8), 85(6)
Finkel, T., 88, 89
Finkelstein, Y., 276
Finkenzeller, G., 271, 274(25), 278(25), 279(25), 281(25), 282
Finlin, B. S., 317
Finn, R., 273
Fischer, K., 105, 116
Fisher, P. B., 28, 29(1), 32(1), 37(1), 123
Fisher, T. L., 46, 53(20)
Fishman, D. A., 105
Flavell, R. A., 97
Flemington, E. K., 124
Fletcher, J. A., 286
Flier, J. S., 30(15), 34
Flimus, J., 269, 281(20), 282(20)
Flint, A., 123
Flint, D. J., 139, 140(9)
Florenes, V. A., 269, 281(20), 282(20)
Flores, P., 106
Flynn, E., 271(30), 273, 279
Foehr, E. D., 12(32), 13
Folch, 42
Folk, W. R., 106
Folkman, J., 267, 271(5; 30), 273, 279, 283

Fong, A., 257, 259(4)
Fong, T. A., 275
Foos, G., 61, 209
Forsyth, J., 163, 164(2; 3)
Forsyth, P. A., 98
Forsythe, J. A., 276
Foschi, M., 342
Fossum, R. D., 285
Foster, D. A., 108
Fourest-Lieuvin, A., 229
Fournier, E., 12
Fra, A. M., 173
Frackelton, A. R., 12(36), 13
Frame, S., 292
Frandsen, T. L., 265
Frangioni, J. V., 124
Frank, D. A., 139, 271(30), 273, 279
Franke, B., 342, 343, 344(1; 5), 346(5)
Frankel, P., 108
Franklin, C. C., 300
Franze, A., 106, 112(18)
Franzoso, G., 89
Frazier, W. A., 280
Freake, H., 257
French, B. A., 237
Frevert, E. U., 30(16), 34
Friedman, J. M., 146
Frisch, S. M., 99
Fritsch, E. R., 119, 125(19), 132, 134(47), 190
Frodin, M., 53
Frost, J. A., 56
Fry, D. G., 291
Fry, M. J., 37, 115
Fu, J., 53
Fu, M., 75, 116, 117, 118(9), 119(9)
Fu, X. Y., 148
Fuall, R. J., 165
Fucini, R. V., 341
Fueyo, J., 273
Fujimoto, N., 98
Fujiyama, A., 229
Fukasawa, K., 192
Fuks, Z., 28, 284
Fukuda, T., 149
Fukumura, D., 280
Furcht, L. T., 259
Furet, P., 341
Furuchi, T., 174, 176(21)
Furuse, M., 267, 271(6)

G

Gabbay, R. A., 30(15), 34
Gabrielson, M. G., 291
Galaktionov, K., 129, 131(18)
Galand, P., 348, 349(3)
Galang, C. K., 61, 62, 68, 209, 210
Galbiati, F., 173
Galeffi, P., 112
Gammeltoft, S., 53
Gangarosa, L. M., 277
Gangarosa, L. M., Jr., 277
Ganguly, A. K., 309
Gantzer, M., 35
Garbisa, S., 97, 105
Garcia, R., 149
Garcia-Ramirez, J., 68
Gardner, K. L., 341
Garrett, L., 139, 140(12)
Garrett, M. D., 129
Gartmann, C., 149
Gavrilovic, J., 98, 99(24)
Gay, B., 341
Gazit, A., 16
Gearing, A. J., 89
Gebel, S., 119
Gelmann, E. P., 280, 281(93)
Georgio, M., 3, 6(5), 7(6)
Gerlinger, P., 310
Gerritsen, M. E., 75
Gertsenstein, M., 280
Gesteland, R., 202, 291
Geurrini, L., 106
Ghiso, J. A., 115, 279
Ghiso, N., 271(30), 273
Ghitescu, L., 182
Ghosh, S., 74, 75, 90, 295(21), 296
Giaccia, A. J., 271(29), 273, 276, 279(29; 66)
Giancotti, F. G., 12(38), 13, 152
Gibbs, J. B., 56, 58, 267, 285, 316, 333, 350
Gibson, A. W., 98
Giermasz, A., 271(36), 273
Gietz, D., 220
Gilbert, C., 56
Gill, G., 122
Gillespie, D. A., 300
Gillespie, G. Y., 274
Gillett, N., 274
Gilman, M. Z., 147
Gilmer, T. M., 203

Gineitis, D., 159
Ginsberg, M. H., 163, 164(2; 3), 165, 169
Ginty, D. D., 45
Giorgio, M., 3, 5(2), 15(2)
Girijavallabhan, V. M., 308, 309, 309(4; 5), 311(4; 5), 314(4; 5), 315(4), 317(4)
Gish, G. D., 4, 12(30; 34), 13, 15(34)
Gishizky, M. L., 12
Giuli, S., 3, 6(5), 7(6), 12
Giuliani, E. A., 58, 267
Gius, D. R., 128
Glass, A. A., 165
Glass, C. K., 123
Glick, A. B., 116(4), 117, 130, 280, 281(90)
Glomset, J. A., 284
Gloss, B., 123
Gnudi, L., 30(15–17), 34
Gofraind, J. M., 16
Goh, K. C., 139
Gokhale, P. C., 37
Goldberg, G. I., 99
Goldberg, M. A., 276
Goldfarb, R., 105
Goldman, C. K., 274
Goldschmidt-Clermont, P. J., 88, 89, 89(8)
Goldstein, N., 270, 283(23)
Gomez, D. E., 98, 115, 267
Gomez-Manzano, C., 273
Gonias, S. L., 116
Goodall, G. J., 276
Goodman, L. E., 229
Goody, R. S., 350
Gordon, D. F., 120
Gordon, J. A., 8
Goretzki, L., 105
Gosparowicz, D., 276
Gossen, M., 128
Goto, K., 217
Gotoh, N., 12
Gotoh, T., 199
Goueli, B. S., 16, 17
Goueli, S. A., 16, 17, 26(10), 27(10)
Gouilleux, F., 149
Gouilleux-Gruart, V., 149
Gout, I., 4, 115
Graeff, H., 105, 106, 112(25), 115(25), 116
Graham, K. R., 187, 189(4), 191(4), 193(4), 194(4), 197(4), 199(4), 200(4), 202(4)
Graham, S. L., 58, 267
Graham, S. M., 7, 188, 190(7), 192(7), 203, 204(2–5), 206, 206(3), 207(17), 208(3; 5), 209(2), 210(2; 4), 211(2; 4), 223, 284
Grammer, T., 46
Grana, X., 130
Grandis, J. R., 142, 149(27)
Granner, D. K., 30(15), 34
Grant, S., 37
Grasso, L., 187
Gratzner, H. G., 317
Graves, J. D., 350
Graves-Deal, R., 277
Green, D., 28
Green, H., 248, 249(18)
Greenberg, M. E., 28, 45
Greengard, P., 16
Greenlund, A. C., 139, 140(4), 149(4)
Greider, C. W., 292
Griendling, K. K., 89
Griffin, J. D., 280, 281(92)
Griffiths, B., 131
Griffiths, G., 180, 182, 182(29)
Grommoll, J., 123
Gromov, P. S., 217
Grondahl-Hansen, J., 105
Gronemeyer, H., 123
Groner, B., 149, 151, 310
Groom, A. C., 319, 326(3)
Grossman, Z., 276
Grosveld, G., 130, 139, 140(11; 13), 149(11), 247
Grottola, A., 349
Grove, J. R., 45, 53
Gruber, S. B., 98, 99(17)
Grugel, S., 271, 274(25), 278(25), 279(25), 281(25), 282
Grunstein, J., 275
Grusby, M. J., 139, 140(15), 149(15)
Grützner, K. U., 106
Grynkiewicz, G., 162
Gu, W. Z., 282, 283(106)
Guha, A., 267, 271(9), 273, 276(26; 27), 283(26; 27), 334, 350
Gulbis, B., 348, 349(3)
Guldener, U., 219
Gum, R., 106, 108, 109, 109(21), 112(19), 114(19; 38), 115(19; 21; 38)
Gunji, Y., 271(32), 273
Guo, M., 287
Gupta, A. K., 284

Gupta, S., 290, 292, 293, 293(17; 18)
Gurnani, M., 310
Gustafson, P., 106
Gutkind, S., 108
Gutman, A., 106
Gutmann, D., 334, 350
Guttridge, D. C., 74, 119

H

Haas, M., 334
Hackenmiller, R., 139, 140(3), 149(3)
Hafter, R., 105
Hagen, S. T., 259
Hagner, B., 98
Hahn, W. C., 292
Hahnfeldt, P., 280
Haimovitz-Friedman, A., 28
Hajian, G., 316
Halaas, J. L., 146
Hall, A., 88, 89
Halpern, J., 142
Hamilton, A. D., 217, 218(7), 225(7), 231(7), 285
Hamilton, K., 267, 316
Han, J. W., 53, 163
Hanahan, D., 275, 280
Hanauer, A., 45
Hancock, J. F., 40, 169, 174(5), 178, 182(2), 196, 197(14a)
Handt, L. K., 267
Hani, S., 139, 140(10)
Hansen, K., 128
Hansen, L., 116(4), 117, 130, 277
Hanson, L. H., 317
Hao, Q., 116
Haque, S. J., 139
Harada, H., 247
Harari, I., 150
Harbeck, N., 116
Harding, A., 169, 174(5)
Harlan, J. E., 4
Harlow, E., 128, 129, 134
Harpal, K., 268, 280, 282(15)
Harris, A. L., 276
Harris, C. C., 280, 281(96), 291
Harris, S. R., 280
Harrison, D. J., 286
Hartman, G. D., 267
Hartman, T., 276
Hasegawa, H., 267
Hashimoto, K., 121, 149
Haspel, R. L., 139, 141(17; 18)
Hassel, J. A., 116, 122
Hatase, O., 199
Hato, T., 169
Hattori, S., 12, 199
Hauschka, S. D., 237
Hauser, C. A., 14, 61, 62, 63(4), 68, 68(6), 190, 206, 209, 210
Hayflick, L., 248
He, C., 99
Heaney, M. L., 233
Heasley, L. E., 280, 281(97)
Heck, S., 219
Heder, A., 74
Hegemann, J. H., 219
Heike, Y., 274
Heimann, R., 28
Heinrich, P. C., 149
Heinzel, T., 123
Heiss, M. M., 105, 106, 111
Heiss, P., 116
Helin, K., 128
Hemmings, B. A., 46, 53, 53(17)
Hemmings, H. C., Jr., 16
Hendrik Gille, H., 117
Hendrix, M. J., 97
Hendry, J. H., 286
Henglein, B., 117
Henkel, G., 62, 68(6), 206
Henkel, T., 89, 90
Henkemeyer, M., 4, 268, 282(15)
Hennighausen, L., 139, 140(12), 149, 310
Heppner, G. H., 281
Her, J. H., 300
Hermann, A. S., 192
Hermens, A., 284
Herrlich, P., 119
Herrmann, C., 334, 350
Hershenson, M. B., 118
Herskowitz, I., 12
Herzinger, T., 128
Heussen, C., 100
Heyman, R. A., 123
Hibi, M., 62, 88, 149, 226, 300
Hicklin, D. J., 275, 283
Hieter, P., 228
Higashi, H., 132(52), 133

Higler, M. H., 327
Hijmans, E. M., 131
Hildberg, F., 106
Hill, R. J., 12
Hillan, K. J., 273, 280(46)
Himelstein, B. P., 96, 97
Hinds, P. W., 131
Hinz, M., 74
Hirai, H., 131
Hirano, T., 149
Hirata, M., 141
Hirth, K. P., 275
Hisaka, M. M., 284, 295
Hlatky, L., 273, 280
Hockenberry, T. N., 306
Hodge, J. A., 118
Hoey, T., 148
Hofmann, J., 24
Holash, J., 267, 274(7), 280(7)
Hollenbach, P. W., 12
Hollenberg, A., 126
Hollenberg, S. M., 196, 197(14), 212, 237
Holloway, J. M., 119, 121(21)
Holmyard, D. P., 268, 282(15)
Holt, J., 37
Holt, S. E., 292
Holzman, L. B., 139
Hooper, M. L., 286
Hori, T., 128
Horn, F., 149
Horn, G., 350
Horner, J. W., 247, 267, 274(7), 280(7)
Horvath, C. M., 142, 146, 149
Horwitz, A. F., 263
Hoshiai, H., 98
Houghton, C., 12(30), 13
Houseknecht, K. L., 30(16), 34
Howe, A. K., 118, 151, 152, 153(2), 157(2), 159(2)
Hoxie, J. A., 165
Hsaio, K., 16, 17, 26(10), 27(10)
Hsieh, H. J., 89
Hsuan, J., 4
Hua, J., 98
Huang, C. M., 273
Huang, H. J., 273
Huang, L. H., 142
Huang, S.-N., 281, 282(103)
Huang, Y., 203
Hubbard, M. J., 16
Hubbert, N. L., 284
Huchcroft, S. A., 98
Huchet, M., 237
Huckle, W. R., 274
Huff, K. K., 280, 281(93)
Huff, S. Y., 203, 204(2), 209(2), 210(2), 211(2)
Hughes, E. N., 98
Hughes, P. E., 163, 164(2; 3)
Hume, D. A., 62, 68(6), 114, 206
Hundley, J. E., 316
Hung, M.-C., 12, 270, 283(23)
Hunter, T., 16, 127, 247, 267, 319
Hunter, W., 109
Hunziker, R., 97
Hurlin, P. J., 291
Hussaini, I. M., 116
Huttenlocher, A., 263
Huvar, I., 334, 350
Hvalby, O., 16

I

Iberg, N., 280, 281(88)
Ichiba, M., 149
Igarashi, M., 129
Ihida, K., 182
Ihle, J. N., 139, 140(11; 13), 149(11), 150
Ikeda, M., 132(52), 133, 345
Ikonen, E., 173
Ikuta, K., 148, 149(48)
Iliakis, G., 285, 285(22)
Iliopoulos, O., 276
Illiakis, G., 289
Ilyas, M., 292
Im, S. A., 273
Imai, T., 128
Imamura, T., 12(35), 13
Inagami, T., 116
Inoko, H., 217
Inoue, M., 149
Inoue, Y., 98
Irani, K., 88, 89, 89(8)
Irigoyen, J. P., 115, 116(53)
Isaacs, C. M., 124
Isakoff, S. J., 12(38), 13
Ishihara, H., 12(35), 13, 217, 264
Ishihara, M., 247
Ishiki, M., 12(35), 13
Isselbacher, K. J., 268

Itin, A., 276
Itoh, Y., 99
Ivanov, I. E., 169
Iwasaki, J., 257
Iwase, H., 98
Iwata, H., 98
Iyer, N. V., 276

J

Jacks, T., 268, 282(15), 292
Jackson, J. H., 284
Jacobs, C. W., 218
Jacobson, K., 264
Jacobson, M., 247
Jacobson, N. G., 142
Jacquot, S., 45
Jäger, C., 105, 106, 112(25), 115(25)
Jahne, R., 149
Jain, R. K., 280
Jakobisiak, M., 271(36), 273
Jakoi, L., 131
James, L., 306, 308(1), 309(1), 311(1)
Jameson, J. L., 126
Jänicke, F., 105, 106
Janik, P., 271(35), 273
Janknecht, R., 300
Jankun, J., 106
Janulis, M., 108
Jauch, K. W., 106
Jay, P., 141
Jayaram, Y., 334
Jeffers, M., 106, 112(25), 115(25)
Jelinek, T., 300
Jensen, C. J., 53
Jensen, V., 16
Jeruzalmi, D., 139
Jessus, C., 129
Jhun, B. H., 15
Ji, X. D., 273
Jiang, B. H., 276
Jiang, C., 276
Jiang, H., 108
Jiang, M.-C., 116(3), 117, 130
Jinno, S., 129
Johnsen, M., 106, 112(20; 22), 114(20; 22)
Johnson, B., 264
Johnson, D. I., 218
Johnson, G. L., 88, 115, 156
Johnson, J., 117, 118(8), 130
Johnson, R. S., 275
Johnson, S. E., 232, 233(4; 5)
Johnston, C., 89
Johnston, R. N., 98
Johnstone, M., 89
Joiner, M. C., 284
Jonat, C. H., 119
Jones, F. E., 148
Jones, M., 12(34), 13, 15(34)
Jones, P. A., 232
Jones, S. W., 45
Joneson, T., 58, 320
Jooss, K. U., 131
Jorcano, J. L., 271(28), 273
Jordan, K. A., 342
Joshi, A., 273
Joshi, R., 273
Jouanneau, J., 281
Joukov, V., 271(32), 273
Jove, R., 139, 142, 149, 149(28)
Joyce, D., 75, 117, 118(9), 119(9)
Joyce-Shaikh, B., 12
Juarez, J., 106, 108, 109(21), 115(21)
Juliano, R. L., 151, 152, 153, 153(1; 2), 154, 154(6), 156, 156(1; 7), 157, 157(2; 5), 158(12), 159(2)
Jung, H.-K., 132(52), 133

K

Kahn, B. B., 30(15), 34
Kahn, C. R., 46
Kaisho, T., 139, 140(8)
Kajita, T., 141
Kalebic, T., 97
Kalejta, R. F., 124
Kalo, M. S., 163, 169(6)
Kaluz, S., 123
Kaluzova, M., 123
Kamei, Y., 123
Kamijo, T., 130, 247
Kamps, M., 271(33), 273, 279(33), 283(33)
Kanakura, Y., 121, 149
Kanaoka, Y., 129
Kang, J.-S., 320
Kao, A. W., 5, 15(14)
Kao, G., 285
Kao, M. Y., 120

Kaplan, D. H., 139, 140(4; 5), 149(4)
Kaplan, M. H., 139, 140(15), 149(15)
Kaplan, W. G., Jr., 276
Kara, M., 341
Kariko, K., 284
Karin, M., 28, 62, 74, 82(11), 88, 96, 115, 122, 200, 226, 300, 319, 320(8), 321(8)
Karnezis, A. N., 118
Kartenbeck, J., 174
Kashiwamura, S., 149
Kasid, A., 280, 281(93)
Kasid, U., 28, 35, 35(10), 37
Kaslauskas, A., 46
Kasuga, M., 4
Katagiri, Y., 165
Kato, J.-Y., 129, 132(11; 51; 52), 133
Kato, K., 284
Katz, M. E., 73
Kauffmann-Zeh, A., 56
Kaufmann, S. H., 317
Kaufmann, W. E., 217
Kavanaugh, W. M., 3, 7(3), 28
Kawada, M., 130
Kawashima, T., 148, 149(48)
Kaya, M., 275, 279(58)
Kaziro, Y., 333, 350
Ke, L. D., 273
Keely, P. J., 256, 257, 259(4), 261, 264
Keivens, V. M., 163, 164(2; 3)
Kelly, B. L., 128
Kelly, J., 309
Kelly, K., 90
Kelsten, M., 284
Kemp, B. E., 16, 17(4)
Kendall, R. L., 274
Kerbel, R. S., 267, 268, 269, 269(19), 270, 270(18), 271(18; 19; 22), 273, 274(19), 275, 275(19), 276(19; 26), 277(18), 278(18), 279(18; 19; 58), 280(18), 281, 281(19; 20), 282, 282(14; 20; 22; 103), 283, 283(18; 19; 23; 26)
Kerins, D. M., 116
Kerkhoff, E., 130
Kerr, I. M., 138
Keshet, E., 276
Kessler, H., 116
Kevil, C. G., 276
Khosravi-Far, R., 55, 73, 88, 187, 203, 217, 267, 268, 276(11), 292, 293(17), 295(22), 296, 319, 320(9), 321(9), 325(9), 327(9)
Khwaja, A., 37, 38, 41(5), 56, 58(2), 88, 323
Kido, Y., 4
Kieckens, L., 280
Kiefer, B., 114
Kim, J. S., 273
Kim, K. J., 274
Kim, S.-H., 131
Kim, Y. H., 275
Kimmelman, A. C., 163, 164(5), 169(5), 187
Kimura, M., 217
Kimura, T., 247
Kinch, M. S., 153, 154(6), 203, 217, 218(7), 225(7), 231(7), 295(22), 296
Kinder, M., 263
Kingston, R. E., 238, 250
Kinzler, K. W., 247, 292
Kiosses, W. B., 342
Kirken, R. A., 147
Kirn, D., 88
Kirschmeier, P., 306, 308, 308(1), 309, 309(1; 4; 5), 311(1; 4; 5), 314(4; 5), 315(4), 317, 317(4)
Kirstein, M., 97
Kishimoto, T., 139, 140(7; 8), 141, 149, 149(7)
Kitadai, Y., 279
Kitagawa, M., 132(52), 133, 247
Kitajima, M., 267
Kitamura, T., 121, 148, 149, 149(48)
Kitaoka, T., 149
Kitsis, R. N., 117
Kiuchi, N., 149
Kivinen, L., 271(32), 273
Klagsbrun, M., 280, 281(88)
Klauber, N., 271(30), 273, 279
Klefstrom, J., 271(32), 273
Klein, D. J., 265
Klein, J. U., 117, 118(9), 119(9)
Klein, U. J., 75
Klement, G., 283
Kline, R. L., 237
Klingmuller, U., 149
Klip, A., 342
Klompmaker, R., 131
Klostergaard, J., 112
Kman, P., 16, 27(7)
Knabbe, C., 280, 281(93)
Knauper, V., 98, 99(24)
Knaus, U. G., 56, 58(6), 60
Knudsen, K. A., 263
Kobari, M., 97

Kobayashi, M., 12(35), 13
Kobayashi, S., 98
Koblan, K. S., 267, 316
Koester, S. K., 316
Koh, C. G., 342
Kohalmi, S. E., 220
Kohl, H., 267
Kohl, N. E., 58, 285, 316
Kohler, S. K., 291
Kojima, H., 149
Koleske, A. J., 173
Kolesnik, R., 28
Kometiani, P., 30(17), 34
Kong, Y., 232, 233(4)
Konieczny, S. F., 232, 233, 233(2; 4–6; 8), 236(2; 6; 13), 237
Koos, R. D., 276
Kopp, E. B., 74
Korfmacher, W. A., 308, 309(4; 5), 311(4; 5), 314(4; 5), 315(4), 317(4)
Kornfeld, S., 178
Kortylewski, M., 149
Kossakowska, A. E., 98
Kotanides, H., 275
Koura, A. N., 279
Kovary, K., 15
Kozlosky, C., 45
Kozlowski, M. T., 53
Kozma, S. C., 45, 46, 53, 53(17)
Kraemer, M., 97
Kraft, A., 300
Krag, T. O., 53
Kral, A. M., 267
Kraling, B. M., 283
Krappmann, D., 74
Krauss, R. S., 320
Krestow, J. K., 269, 281(20), 282(20)
Kristensen, P., 105
Krymskaya, V. P., 45
Kufe, D. W., 28
Kuhn, W., 106
Kumar, S., 105, 112(6)
Kumar, V., 123, 271(32), 273
Kunz, C., 114
Kurakawa, R., 123
Kuriyan, J., 139, 146, 148(37), 150(37)
Kuroki, M., 276
Kurzchalia, T. V., 169, 173, 173(6)
Kushner, P. J., 119, 121(21)
Kwok, P. W., 276
Kyriakis, J. M., 53, 118, 300
Kyritsis, A. P., 273
Kyrmskawa, V. P., 49

L

LaBaer, J., 129
Lachance, P., 122
Laderoute, K. R., 271(29), 273, 276, 279(29; 66)
Ladha, M. H., 131, 341
LaFlamme, S. E., 165
Lafont, V., 342
Lai, K.-M. V., 4, 5, 15(13)
Laig-Webster, M., 280
Laiho, M., 271(32), 273
Laing, T. D., 98
Lakhani, S., 300
Lakonishok, M., 263
L'Allemain, G., 118, 130
Lamb, P., 142, 149(30)
Lambert, Q. T., 12(31), 13, 75, 117, 118(10; 11), 231
Lambeth, J. D., 89
Lambin, P., 284
Lamphier, M. S., 247
Land, H., 131, 247, 250, 251, 291
Landowski, T. H., 142, 149(28)
Landt, H., 130
Lane, A., 169
Lane, D., 124
Lanfrancone, L., 3, 5, 5(2), 15(2; 13)
Lange-Carter, C., 88, 115, 156
Langer, S. J., 268
Langstrom, E., 106
Larcher, F., 271(28), 273
Laroux, F. S., 276
Lassar, A. B., 232, 233, 233(3), 236(12), 237
Lassegue, B., 89
Lathem, W. W., 149
Lau, N., 267, 271(9), 273, 276(26; 27), 283(26; 27), 334, 350
Lavoie, J. N., 118, 130
Lawry, D. L., 286
Le, S. Y., 276
Leal, J. A., 282, 283(106)
Leberer, E., 229
Lebovitz, R. M., 114
Lebowitz, P. F., 276
Lechner, J. F., 291

Lee, A. K., 129
Lee, C. S. L., 132, 138(49), 141
Lee, E. J., 97
Lee, F., 187
Lee, H.-W., 130, 247
Lee, H. Y., 273
Lee, J. W., 151, 264
Lee, K. Y., 341
Lee, R. J., 75, 116(1), 117, 118, 118(9; 10), 119(9), 126(12), 130, 131(24)
Lee, S., 308, 309(4; 5), 311(4; 5), 314(4; 5), 315(4), 317(4)
Lees, E., 250
Leeuw, T., 229
Leevers, S. J., 156, 196
Lefkovits, I., 288
Leibel, S. A., 284
LeMeur, M., 310
Lemon, K. P., 46
Lengauer, C., 292
Lengyel, E., 105, 106, 108, 109, 109(21), 112, 112(19; 25), 114(19; 38), 115(19; 21; 25; 38)
Leof, E. B., 132
Leone, G., 131
Lerner, E., 285
Lerose, R., 349
Leslie, K. B., 187
Leung, K., 75
Leung, S. W., 276
Leung, T., 56, 342
Levi, B. Z., 276
Levitt, R. C., 187
Levitzki, A., 16, 142, 149(28)
Levy, A. P., 276
Levy, D. E., 139, 140(3), 149(3)
Levy, N. S., 276
Lewin, D. A., 4
Lewis, T. S., 56, 61, 63(2)
Li, B., 274
Li, J. J., 97, 118
Li, K., 12
Li, L., 237
Li, N., 200
Li, S., 177
Li, W., 7, 163, 164(5), 169(5)
Li, X., 341
Li, Z., 308, 309(5), 311(5), 314(5), 341
Liao, J., 141
Liaudet-Coopman, E., 105
Liautard, J., 342
Liddel, J., 292
Lieubeau, B., 275, 279(58)
Lim, L., 56, 342
Lin, A. W., 88, 115, 130, 131(22), 204, 248, 250(11; 14), 251(11), 255(11; 14)
Lin, B., 202
Lin, C.-C., 308, 309, 309(4; 5), 311(4; 5), 314(4; 5), 315(4), 317(4)
Lin, N., 131
Lin, P., 274
Lin, S.-C., 123
Lin, T. H., 153, 154, 154(6), 156, 156(7)
Linardopoulos, S., 292
Lindsay, M., 169
Ling, C. C., 284, 287
Liotta, L. A., 97, 105
Liou, H. C., 90
Lipari, P., 308, 309(4; 5), 311(4; 5), 314(4; 5), 315(4), 317(4)
Lippman, M. E., 280, 281(93)
Lisanti, M. P., 173, 177
Liu, B., 141
Liu, J.-J., 56, 116(3), 117, 130, 182
Liu, M., 306, 308, 309, 309(4; 5), 311(4; 5), 314(4; 5), 315(4), 316, 317(4)
Liu, P. T., 118
Liu, T. J., 273
Liu, W., 273, 275, 279
Liu, X., 139, 140(12), 149
Liu, Y.-T., 309
Livingston, D. M., 132(51), 133
Lloyd, A. C., 130, 250
Lockshon, D., 237
Loda, M., 129
Loftus, J. C., 165
LoGrasso, P., 27
Logsdon, C., 28, 30, 32(9)
Löhrs, U., 106
Longhi, M. P., 275
Lonzonschi, L., 97
Loo, D., 302
Loo, T. H., 342
Loomans, C. J. M., 131
Lopez, G. N., 119, 121(21)
Lorenzi, M. V., 187
Lorenzo, M. J., 12(30), 13
Losi, L., 349
Louahed, J., 187
Lounsbury, K. M., 73

Lourenco, P. C., 139, 140(9)
Lowe, S. W., 74, 84(13), 88, 130, 131(22), 204, 248, 250(11; 14), 251(11), 255(11; 14)
Lowy, D. R., 284
Lu, L., 273, 280(46)
Lu, N., 280
Lu, Y.-L., 12
Lu, Z., 108
Lubenski, I., 98
Lucibello, F. C., 131
Ludolph, D. C., 233
Luetterforst, R., 169, 174(5), 175
Lukas, J., 128, 131, 132(41)
Lundberg, A. S., 292
Luo, G., 146
Luo, Z., 343
Lupetti, R., 280, 281(91)
Luther, V., 116
Lutz, V., 105, 116
Lynch, M. J., 203, 204(4), 210(4), 211(4)

M

Ma, J., 65
Macara, I. G., 73
MacDonald, I. C., 319, 326(3)
MacDonald, M. J., 65, 285
MacDougall, J. R., 98
MacIntyri, I., 257
Mackiewicz, A., 149
Macleod, K. F., 292
Madrid, L. V., 74
Madsen, P., 217
Magdolen, V., 105, 116
Magnuson, T., 277
Maher, M., 90
Maher, V. M., 106, 291
Mai, M., 267
Maihle, N. J., 106
Mainiero, F., 12(38), 13
Maity, A., 285, 285(23), 286
Mak, T. W., 247
Maki, R. A., 62, 68(6), 206
Malaise, E. P., 284
Malkowski, M., 308, 309(4; 5), 311(4; 5), 314 (4; 5), 315(4), 317(4)
Mallams, A. K., 308, 309, 309(4; 5), 311(4; 5), 314(4; 5), 315(4), 317(4)
Maller, J. L., 45, 46
Maltepe, E., 279
Man, S., 283
Manenti, F., 349
Maniatis, T., 119, 125(19), 132, 134(47), 190
Manseau, E. J., 273
Manser, E., 56, 342
Mansour, S. J., 192, 268, 270(18), 271(18), 277(18), 278(18), 279(18), 280(18), 283(18)
Mao, X., 274
Marais, R., 62, 194
Marcantonio, E. E., 12(38), 13
Marcocci, C., 257
Marcy, A., 27
Marczak, M., 271(36), 273
Margolis, B., 5, 12, 15(14)
Markiewicz, D. A., 285, 285(23)
Marks, R., 317
Marksitzer, R., 114
Marme, D., 271, 274, 274(25), 278(25), 279(25), 281(25), 282
Marschall, C., 105
Marsh, M., 182
Marshall, B., 283
Marshall, C. J., 37, 156, 169, 182(2), 196, 204, 218, 280, 281(89), 286
Marshall, M. S., 56, 333, 343, 350
Marte, B. M., 37, 38, 43, 43(6), 56, 58(2), 88, 323
Martial, J., 282
Martin, C. B., 187, 189(4), 191(4), 193(4), 194(4), 197(4), 199(4), 200(4), 202(4), 203, 204(5), 208(5)
Martin, G. A., 334, 350
Martin, M. E., 106
Martin, S., 28
Martiny-Baron, G., 274
Marwaha, S., 58
Masai, H., 345
Masaoka, A., 98
Mathieu-Costello, O., 275
Mati, M. R., 274
Matrisian, L. M., 97, 98
Matsuda, M., 199
Matsumoto, K., 115, 187
Matsumoto, M., 139, 140(7), 149, 149(7)
Matsumoto, T., 141
Matsumura, I., 121, 149
Matsuno, S., 97
Matsushime, H., 129, 132(11; 51), 133

Matsuyama, T., 247
Matten, W. T., 192
Mattson, G., 27
Mauceri, H. J., 28
Maue, R., 342
May, M. J., 74
Mayo, M. W., 73, 74, 84(13), 88
Mazure, N. M., 271(29), 273, 276, 279(29; 66)
Mbamalu, G., 7, 268, 282(15)
McCance, D. J., 113
McCarthy, J. B., 259, 265
McCarthy, K., 105
McCarthy, S. A., 72
McClatchey, A. I., 326
McConkey, D. J., 275
McCormick, F., 73, 88, 202, 290(1), 291, 350
McCormick, J. J., 106, 291
McCurrach, M. E., 130, 131(22), 204, 248, 250(11), 251(11), 255(11)
McDonnell, S., 97
McDonough, M., 58
McGlade, J., 4, 7, 12(34), 13, 15(34)
McGuire, T. F., 285
McIntosh, D. P., 182
McKay, C., 139, 140(13)
McKenna, W. G., 284, 285, 285(22; 23), 286
McKiernan, C., 73
McLoughlin, M., 28
McMahon, C., 341
McMahon, G., 275
McMahon, M., 72, 88, 115, 204, 248, 250, 250(13), 255(13)
McPhail, A. T., 309
McRae, D., 37
Meadows, R. P., 4
Meckler, J., 129
Medcalf, R. L., 106
Mehtali, M., 35
Mele, S., 3, 5, 5(2), 6(5), 7(6), 15(2; 13)
Mellman, I., 4
Mendelsohn, J., 149
Menke, S. L., 232, 233(2), 236(2)
Menrad, A., 274
Meraz, M. A., 139, 140(4), 149(4)
Merighi, A., 349
Messiers, C., 116
Meyerson, M., 134
Miao, W., 56
Michaelson, D., 169
Michalides, R. J. A. M., 131
Migliaccio, E., 3, 5, 5(2), 15(2; 13)
Miki, T., 203, 204(7)
Mikkelsen, R., 28, 29, 29(1), 30(11; 12), 32(1), 34(11; 12), 37(1)
Milanini, J., 278
Milia, E., 12
Millauer, B., 275
Miller, A. C., 284
Miller, A. D., 251
Miller, B. E., 270
Miller, D. B., 97
Miller, D. L., 276
Miller, F. R., 270, 281
Miller, K. M., 89
Miller, P. J., 267, 316
Miller, S. J., 131
Milocco, L. H., 142, 149(30)
Miloszewska, J., 271(35), 273
Miltenyi, S., 124
Minami, M., 149
Minamoto, S., 141
Minamoto, T., 267
Minden, A., 88, 96, 115, 319, 320(8), 321(8)
Minetti, C., 173
Miranti, C. K., 45
Misawa, K., 148, 149(48)
Miskin, R., 106, 109(30), 112(30)
Misra, R. P., 45
Mitchell, R., 276
Mitin, N., 232
Mitsuhashi, Y., 268, 269, 269(19), 270(18), 271(18), 274(19), 275(19), 276(19), 277(18), 278(18), 279(18; 19), 280(18), 281, 281(19; 20), 282(20; 103), 283(18)
Miura, H., 139, 140(10)
Miyagawa, T., 273
Miyata, S., 217
Miyatake, S., 345
Mizuki, N., 217
Mizuno, S., 130
Mizzen, C. A., 45
Moarefi, I., 146, 148(37), 150(37)
Moens, G., 281
Moghal, N., 124
Monaco, L., 45
Moncalvo, S., 89
Monfar, M., 54

Monia, B. P., 37
Montesano, R., 115
Montessuit, C., 342
Montreau, N., 97
Moons, L., 280
Moore, D. D., 238, 250
Moore, M. W., 273, 280(46)
Moore, S., 75
Moran, E., 118
Morgan, T. C., 291
Morgensen, S., 27
Morgersten, J. P., 251
Mori, N., 3, 6(4), 7
Moriggl, R. M., 149
Morimoto, T., 169
Morioka, H., 12(35), 13
Moroni, M. C., 128
Morrice, N., 46, 53(21)
Morris, I. D., 286
Morris, R. G., 286
Morris, S., 60
Morris, V. L., 319, 326(3)
Mortarini, R., 280, 281(91)
Moscat, J., 97
Moscinski, L., 142, 149(28)
Moses, M. A., 271(30), 273, 279
Mosser, S. D., 58, 267, 316
Motokura, T., 127
Movilla, N., 204, 208(10)
Mueller, E., 97
Mui, A. L., 148, 149(48)
Mukai, M., 267
Mukherjee, B. B., 280, 281(95)
Mukhopadhyay, D., 276
Mukhopadhyay, N. K., 53
Muller, C. W., 151
Muller, M., 120
Müller, R., 118, 130, 131
Muller, W., 124
Mulligan, R. C., 191, 204
Murakami, Y., 130
Murano, G., 109, 110(41)
Muraoka, S., 7
Murata, M., 173
Murillas, R., 271(28), 273
Muroya, K., 12
Murphy, G., 98, 99(24)
Murphy, K. M., 142
Murphy, T. K., 286
Murray, H., 291
Muschel, R. J., 96, 97, 98, 99(17), 108, 284, 285, 285(22; 23), 286
Musgrove, E. A., 132, 138(49)
Myers, C. E., 284

N

Nabel, G. J., 75, 90
Nagahara, H., 128
Nagamine, Y., 114, 115, 116(53)
Nagane, M., 273
Nagase, H., 99, 292
Nagata, A., 129
Nagy, A., 280
Nagy, J. A., 273
Nahari, D., 276
Naider, F., 229
Nairn, A. C., 16
Naka, T., 141
Nakae, K., 149
Nakafuku, M., 350
Nakajima, K., 149
Nakamura, S., 12, 199, 333, 350
Nakamura, T., 3, 6(4), 7, 115
Nakanishi, K., 149
Nakano, H., 345
Narazaki, M., 141
Nardo, C., 308
Nassar, N., 350
Nathans, D., 146, 217
Navre, M., 97
Nead, M. A., 113
Neel, B. G., 124
Neeman, M., 276
Negro, A., 97
Neish, A. S., 75
Nemenoff, R. A., 280, 281(97)
Nemerow, G. R., 169
Nemir, M., 280, 281(95)
Nerlov, C., 106, 109, 112(20; 22), 114(20; 22; 38), 115(38)
Neshat, M. S., 122, 295(24), 296
Neuman, E., 131
Nevins, J. R., 122, 131
Neznanov, N. N., 68
Ng, S. C., 282, 283(106)
Ng, S.-Y., 116(3), 117, 130

Nguyen, D. H. D., 116
Nguyen, J. T., 273
Nicke, B., 28, 30
Nicks, M., 280, 281(97)
Nicolaides, N. C., 187
Nicolette, C., 96, 319, 320(8), 321(8)
Niedbala, M., 109
Nielander, G., 27
Nielsen, G., 280
Nielsen, L. L., 306, 308, 309(4), 310, 311(4), 314(4), 315(4), 316, 317(4)
Nielsen, L. S., 105
Niino, Y., 199
Nishiguchi, T., 106, 116
Nishimoto, N., 141
Nishimura, S., 132(52), 133
Nissen, N. I., 289
Njoroge, F. G., 306, 308, 308(1), 309, 309(1; 4), 311(1; 4), 314(4), 315(4), 317(4)
Nobutoh, T., 105
Noda, K., 98
Nodzenski, E., 28
Nogami, N., 217
Noguchi, K., 139, 140(7), 149(7)
Nojima, H., 129
Nolan, G. P., 7, 90, 251
Nomeir, A. A., 308, 309, 309(4; 5), 311(4; 5), 314(4; 5), 315(4), 317(4)
Nordeen, S., 121
Nordheim, A., 300
Norris, J. L., 73, 74, 82(8)
Norris, K., 284
Norton, L., 284
Nosaka, T., 148, 149(48)
Nunez, G., 142, 149(28)
Nunez-Oliva, I., 306
Nyormoi, O., 111

O

Obata, K., 98
Oberyszyn, A. S., 271(34), 273, 279(34)
Oberyszyn, T. M., 271(34), 273, 279(34)
O'Brien, J. M., 218
O'Brien, R. M., 30(15), 34
O'Bryan, J. P., 3, 6(5), 12(31), 13, 13(5)
O'Connel, P., 291
O'Connor, S., 282, 283(106)
Oegema, T. R., 265
Oertli, B., 163
Ogata, Y., 99
Oh, P., 182
O'Hagen, R., 267, 274(7), 280(7)
Ohmura, G., 98
Ohno, S., 217
Oishi, M., 3, 6(4)
Okabayashi, Y., 4
Okada, F., 275, 279(58)
Okada, S., 5, 15(14), 341
Okada, Y., 98
Okamoto, T., 173, 177
Okayama, H., 129
Oku, T., 273
Okumura, K., 217
Old, L. J., 139, 140(5)
Oldham, S. M., 203, 204(5), 208(5), 277
Olefsky, J. M., 12(33), 13, 15, 15(33)
Olejniczak, E. T., 4
Oliff, A., 58, 267, 285, 316
Olive, P. L., 271
Olivier, J. P., 4
Olivier, P., 129
Olofsson, B., 271(32), 273
Olsen, A., 108
Olson, E. N., 237
Olson, M. F., 88
Omer, C. A., 267, 316
O'Neill, T. J., 267, 316
Ono, A., 217
Oppelt, P., 105
Oppenheim, H., 97
Orei, L., 115
O'Reilly, M. S., 283
O'Rourke, E. C., 202
Orsini, M. J., 49
Osada, M., 163, 164(5), 169(5), 187
O'Shea, J. J., 147
O'Shea, K. S., 273, 280(46)
Oshima, R. G., 68
Oshiro, M. M., 142, 149(28)
Ossowski, L., 109
Ostrowski, M. C., 62, 68(6), 206, 268
O'Toole, T. E., 165
Otsuka-Murakami, H., 120
Ottones, F., 342
Overall, R. W., 232, 233(3)
Oxford, G., 264
Ozanne, B., 280, 281(89)

P

Paavonen, K., 271(32), 273
Pache, L., 105, 106
Padhy, L. C., 291
Page, K., 118
Pages, G., 278
Paget, S., 325
Pahk, A., 55
Pai, J.-K., 306
Palade, G. E., 182
Palisi, T. M., 276
Palmero, I., 247, 248
Pampori, N., 169
Panceri, P., 280, 281(91)
Pandolfi, P. P., 3, 5(2), 6(5), 7(6), 15(2)
Panettieri, R. A., 49
Pang, L., 263
Pantginis, J., 267, 274(7), 280(7)
Pantoja, C., 248
Papageorge, A. G., 284, 286
Pappin, D., 37, 56, 58(2), 88, 323
Parada, L. F., 247, 291
Parangi, S., 271(30), 273, 279
Parise, L. V., 261
Park, E. C., 280
Park, J. P., 28, 32(9)
Park, O. K., 146
Park, S., 90
Park, S. H., 139, 140(14)
Parmiani, G., 280, 281(91)
Parrett, M. L., 271(34), 273, 279(34)
Parry, D., 250
Parsons, J. T., 233
Parton, R. G., 169, 173, 173(6), 174, 174(5), 175, 196, 197(14a)
Parvaz, P., 271(37), 273, 276(37), 283(37)
Pasquale, E. B., 163, 169(6)
Patel, K., 334
Paterson, H. F., 169, 182(2), 204, 286
Patriarca, P., 89
Patton, R., 306, 308(1), 309, 309(1), 311(1)
Paulsen, O., 16
Pavasant, P., 265
Pavirani, A., 35
Pawling, J., 280
Pawson, A., 267, 271(9)
Pawson, T., 3, 4, 5, 6(5), 12(30; 34), 13, 13(5), 15(13; 34), 268, 282(15), 334, 350
Payne, D. K., 276
Pear, W. S., 7, 250, 251, 251(19)
Pearson, R. B., 16, 17(4), 53
Pelech, S., 53
Pelicci, G., 3, 5, 5(2), 6(5), 7, 7(6), 12, 15(2; 13)
Pelicci, P. G., 3, 5, 5(2), 6(5), 7, 7(6), 15(2; 13)
Pena, L. A., 28
Pena, P. P., 118
Pendergast, A. M., 74, 82(12)
Penn, R. B., 49
Pepper, M. S., 115
Peranen, J., 173
Perez-Roger, I., 131
Pergola, F., 106, 112(22), 114(22)
Perini, M., 349
Perkins, N. D., 75, 90
Perou, C. M., 229
Perucho, M., 286, 292, 293(17)
Pessin, J. E., 5, 15(14), 341
Pestell, R. G., 74, 75, 116, 116(1), 117, 118, 118(8–11), 119, 119(9), 121, 126, 126(12), 127, 129, 130, 131, 131(24), 146, 148(38), 149, 173, 231
Peter, K., 165
Peters, K. G., 274
Petrin, J., 306, 308(1), 309(1), 311(1)
Petros, A. M., 4
Petry, K., 124
Pfaff, M., 163, 164(2; 3), 165
Pflanz, M., 105
Philips, M. R., 169
Phillips, H. S., 274
Phillips, P. C., 58
Picarella, M. S., 109
Pienado, M. A., 286
Piette, J., 106, 237
Pine, R., 147
Pingoud, V., 300
Pinto, P., 309
Pirola, L., 35
Pitt, A. S., 142, 149(27)
Plate, K. H., 275
Platko, J. D., 295
Plattner, R., 292, 293, 293(17–19), 302(19)
Pledger, W. J., 130
Plevin, R. J., 49
Plow, E. F., 165
Poenie, M., 162
Pohl, J., 106, 107(31)
Pollefeyt, S., 280
Polsky, D., 247

Polverini, P. J., 274
Polverino, A., 96
Polverino, P., 319, 320(8), 321(8)
Pomerantz, J., 247, 267, 274(7), 280(7)
Pompliano, D. L., 58
Ponder, B. A. J., 12(30), 13
Ponta, H., 119
Ponzetto, C., 115
Porfini, E., 40
Portella, G., 292
Potten, C. S., 286
Poulsom, R., 276
Poupon, M. F., 281
Pouysségur, J., 118, 130, 278
Powell, T. J., 275
Powell-Braxton, L., 273, 280(46)
Pozzatti, R., 97
Prall, O. W. J., 132
Pratt, E. S. II, 280, 281(92)
Predescu, D., 182
Prendergast, G. C., 276
Prezioso, V. R., 147
Price, B., 280, 281(97)
Price, D. J., 45, 53
Prioli, N., 308, 309(4; 5), 311(4; 5), 314(4; 5), 315(4), 317(4)
Prior, I., 169
Pritchard, C. A., 72
Provias, J., 334, 350
Przybyzewska, M., 271(35), 273
Ptashne, M., 65, 122
Pullen, N., 46, 53, 53(17)
Purdie, C. A., 286
Puri, M. C., 268, 282(15)
Pytowski, B., 275

Q

Qi, H., 45
Qian, Y., 285
Qiu, R. G., 88
Quaranta, V., 165
Quaroni, A., 268, 271(17)
Quelle, D. E., 129, 130, 132(11), 247
Quelle, F. W., 150
Quilliam, L. A., 62, 63(4), 177, 187, 189(4), 190, 191(4), 193(4), 194(4), 197(4), 198, 199(4; 16), 200(4), 202, 202(4)
Quinn, M. T., 89
Quinn, T. Q., 274
Quinones, S., 97
Quintanilla, M., 271(28), 273
Quintans, J., 28
Qureshi, S. A., 142

R

Raastad, M., 16
Radbruch, A., 124
Radford, I. R., 286
Radinsky, R., 279
Radler-Pohl, A., 106, 107(31)
Radley, J. M., 286
Raffioni, S., 12(32), 13
Rahman, A., 37
Rahmsdorf, K.-K., 119
Rahuel, J., 341
Rainey, W. E., 117, 126(12)
Raingeaud, J., 28, 122
Rak, J., 267, 268, 269, 269(19), 270, 270(18), 271(18; 22), 273, 274(19), 275, 275(19), 276(19; 26), 277(18), 278(18), 279(18; 19; 58), 280(18), 281, 281(19; 20), 282, 282(14; 20; 22; 103), 283, 283(18; 23; 26)
Rall, T. B., 300
Ram, P. A., 139, 140(14)
Ramocki, M. B., 232, 233(5; 6)
Rands, E., 267, 316
Rane, D. F., 309
Rao, X., 142
Rapp, U. R., 130
Ratner, N., 286, 348
Ravichandran, K. S., 4, 12(36), 13
Rawson, C., 302
Ray, L. B., 46
Raymond, W. W., 280
Rayter, S., 350
Reardon, D. B., 28, 29, 30(11; 12), 32(9), 34(11; 12)
Rebecchi, A., 349
Rebhun, J. F., 187, 198, 199(16)
Reboldi, P., 3, 5(2), 15(2)
Reddel, R. R., 292
Reddy, E. P., 130
Reed, J. C., 163, 164(4), 169(4)
Reed, S. I., 128
Rees, R. C., 147
Reich, E., 105

Reich, R., 99, 106, 109(30), 112(30)
Remiszewski, S., 308, 309, 309(4), 311(4), 314(4), 315(4), 317(4)
Ren, X. D., 342
Renauld, J.-C., 187
Renkawitz, R., 120
Renshaw, M. W., 163, 164(2; 3)
Renzing, J., 124
Resnick, R. J., 333
Resnitzky, D., 128
Reuning, U., 105, 106, 116
Reusch, P., 274
Reutens, A. T., 116(1), 117, 118, 129, 130, 131(24)
Reuter, C. W., 300
Reuther, G. W., 74, 82(12)
Reuther, J. Y., 74, 82(12), 119
Rewcastle, N. B., 98
Rhee, J. M., 334, 350
Rheinwald, J. G., 280, 281(92)
Rhodes, S. J., 233, 236(13)
Richards, S. A., 53, 73
Ricketts, W. A., 12, 12(33), 13, 15(33)
Ridgway, E. C., 120
Ridley, A., 37, 56, 58(2), 88, 323
Ried, S., 106, 112(25), 115(25)
Riley, J. K., 139, 140(4), 149(4)
Rincon, M., 97
Risau, W., 275, 276, 280
Ristimaki, A., 271(32), 273
Rivera, V. M., 45
Robbiati, F., 106, 112(18)
Robens, J. M., 45
Roberts, B. E., 191, 204
Roberts, D. D., 280, 281(96), 341
Roberts, J. M., 128, 129
Roberts, W. G., 182, 275
Robertson, F. M., 271(34), 273, 279(34)
Robertson, M, 291
Robinson, G. W., 139, 140(12)
Robles, A. I., 116(4), 117, 130, 271(28), 273
Roche, S., 12(34), 13, 15(34)
Rochefort, H., 105
Rockoff, A. I., 271(34), 273, 279(34)
Rockwell, P., 270, 275, 283(23)
Rodig, S. J., 139, 140(4), 149(4)
Rodriguez, A., 124
Rodriguez-Puebla, M. L., 116(4), 117, 130
Rodriguez-Viciana, P., 37, 38, 40(3), 41(5), 43(6), 56, 58(2), 88, 115, 295(23), 296, 323
Roeder, R. G., 114
Rogelj, S., 280, 281(88)
Rogers-Graham, K. S., 74, 84(13), 88, 203, 217, 218(7), 225(7), 231(7)
Rohde, M., 131, 132(41)
Rohme, D., 248, 259(15)
Rolls, B., 169, 174(5)
Romanelli, A., 53
Romeo, D., 89
Romer, L., 156
Ron, D., 62
Roncari, L., 275, 279(58)
Rong, S., 192
Rorrer, W. K., 28
Rorth, P., 106, 112(20), 114(20)
Rosario, M., 204
Rose, D. M., 169
Rose, D. W., 12(33), 13, 15, 15(33), 123
Rosen, C. A., 90
Rosen, J., 142, 149(30)
Rosen, P. P., 257
Rosenberg, S. H., 282, 283(106)
Rosenfeld, M. G., 123
Rosenthal, N., 237
Rosette, C., 28
Rosner, M. R., 118
Rosnet, O., 12
Ross, R., 280, 281(94)
Rossi, D. J., 268, 282(15)
Rossi, F., 89
Rossman, K. L., 55, 73, 187, 203, 217, 268, 276(11), 319
Rossman, R. R., 308, 309, 309(5), 311(5), 314(5)
Rotter, W., 291
Roussel, M. F., 117, 129, 130, 247
Rowinsky, E. K., 318
Roy, S., 169, 174(5)
Rozakis-Adcock, M., 7
Ruben, S. M., 90
Rubie, E., 300
Rudoltz, M. S., 285, 285(23)
Ruff-Jamison, S., 146
Ruggeri, Z. M., 165
Rui, H., 147
Ruley, H. E., 247
Rundell, K., 118
Ruoslahti, E., 152, 163, 164(4), 169(4; 6), 318
Rusconi, S., 35
Rush, M. G., 203, 204, 204(3), 206(3), 208(3)
Russell, M., 115

Rusyn, E. V., 261
Rutberg, S. E., 97
Rutkowski, J. L., 58
Ruzicka, C., 17, 26(10), 27(10)
Rydholm, A., 106
Ryseck, R. P., 90

S

Saatcioglu, F., 122
Sabichi, A., 109, 114(38), 115(38)
Sadowski, H. B., 147
Sadowski, I., 65
Saelman, E. U. M., 264
Saez, E., 97
Saez, R., 203
Saffiotti, U., 97
Sage, D., 127
Sager, R., 173, 291
Saito, T., 345
Sakaguchi, K., 4
Salcini, A., 5, 15(13)
Salditt-Georgieff, M., 139, 141(17)
Saleh, M., 273
Salihuddin, H., 56
Saltiel, A. R., 15, 263
Sambrook, T., 119, 125(19), 132, 134(47), 190
Samid, D., 284
Samuels, M. L., 72
Sanan, D., 179
Sanchez, T., 73
Sanders, L. K., 217
Sandhu, C., 129
Sandler, H., 284
Sanghera, J., 53
Sangster, M. Y., 139, 140(11), 149(11)
Sankar, S., 274
Sano, S., 139, 140(10)
Sanokawa, R., 3, 6(4), 7
Santana, P., 28
Santoro, S. A., 257, 259(4), 264
Santos, E., 203
Sanz, L., 97
Sarawar, S. R., 139, 140(11), 149(11)
Sarevic, B., 132
Sargiacomo, M., 177
Sasajima, T., 273
Sasaki, Y., 3, 6(4)
Sasaoka, T., 12(35), 13, 15
Sasazuki, S., 275, 279(58)
Sasazuki, T., 107, 109(33), 267, 268, 269(19), 271(6; 19), 274(19), 275(19), 276(19), 279(19), 281(19), 283(19)
Sassone-Corsi, P., 45
Sathyanarayana, B. K., 97
Sato, H., 97, 108
Sato, K. Y., 292, 293(17; 19), 302(19)
Sato, M., 217
Satoh, T., 333, 350
Sattler, I., 218
Sattler, M., 4
Saucan, L., 182
Saus, J., 97
Savage, D., 27
Savino, R., 142, 149(28)
Sawa, T., 12(35), 13
Sawaguchi, K., 267
Sawyers, C. L., 122, 142, 295(24), 296
Schaber, M. D., 267, 316, 333
Schaefer, T. S., 146
Schaeffer, J. P., 285
Schaller, M., 156
Schaufele, F., 119, 121(21)
Scheele, J. S., 271(33), 273, 279(33), 283(33), 334, 350
Scheidereit, C., 74
Scherer, A., 350
Scherer, P. E., 173
Schiestl, R. H., 220
Schindler, C., 139, 140(6; 15), 141, 147, 149(15)
Schirner, M., 274
Schirrmacher, V., 106, 107(31)
Schlechte, W., 109, 110(41)
Schlegel, A., 173
Schlessinger, J., 150, 200
Schleuning, W. D., 106
Schmalfeldt, B., 106
Schmid, R. M., 90
Schmidt, E. E., 319, 326(3)
Schmidt, J. V., 279
Schmidt-Ullrich, R., 28, 29, 29(1), 30(11; 12), 32(1; 9), 34(11; 12), 37(1)
Schmitt, D., 271(37), 273, 276(37), 283(37)
Schmitt, M., 105, 106, 112(25), 115(25), 116
Schniertschauer, U., 149
Schnitzer, J. E., 182
Schoepfer, J., 341
Schonenberger, C.-A., 310
Schottelius, A. J., 74

Schrader, J. W., 187
Schrager, J. A., 265
Schreck, R., 90, 275
Schreiber, R. D., 138, 139, 140(4; 5), 142, 149(4)
Schreiner, R., 173
Schuchman, E. H., 28
Schuetz, E., 139, 140(13)
Schule, R., 120
Schultz, R., 108
Schwartz, J., 306, 308(1), 309(1), 311(1)
Schwartz, M. A., 152, 163, 164(2; 3), 342
Schwartz, R. J., 237
Scolnick, E. M., 58, 350
Scott, M. L., 7, 251
Scott, P. A., 276
Scott, P. H., 49
Sears, R., 131
Sebti, S. M., 139, 217, 218(7), 225(7), 231(7), 285
Seed, B., 280
Seftor, E. A., 97
Segal, A. W., 89
Segall, J. E., 75, 116(1), 117, 118(9), 119(9), 130, 131(24)
Segawa, K., 132(52), 133
Sehgal, G., 98
Seidel, H. M., 142, 149(30)
Seidman, J. G., 238, 250
Seiki, M., 97, 108
Seizinger, B. R., 203, 204(4), 210(4), 211(4)
Sekharam, M., 139
Selbert, S., 139, 140(9)
Selig, M., 280
Sells, M. A., 56, 58(6)
Semenza, G. L., 276
Senderak, E. T., 316
Senechal, K., 122, 295(24), 296
Sensi, M., 280, 281(91)
Serrano, M., 130, 131(22), 204, 247, 248, 250(11; 14), 251(11), 255(11; 14)
Serres, M., 271(37), 273, 276(37), 283(37)
Sethi, T., 163, 164(2)
Sewing, A., 130, 131, 250
Shaheen, R. M., 275
Shalloway, D., 38, 333, 334, 340(9), 341(9), 343
Shambaugh III, G. E., 129
Shamsuddin, A. M., 291
Shan, S., 274
Shankaran, V., 139, 140(5)
Shannon, K., 348
Shao, H., 317
Shao, R., 12
Shapiro, P. S., 56, 61, 63(2)
Shattil, S. J., 165, 169
Shaw, P. E., 56
Shawver, L. K., 275
Shay, J. W., 292
Sheehan, C., 268, 269, 270(18), 271(18), 277(18), 278(18), 279(18), 280(18), 281(20), 282(20), 283(18)
Sheehan, K. C., 139, 140(4), 149(4)
Sheldrick, L., 123
Shen, Q., 267, 274(7), 280(7)
Shen, Y., 156
Sheng, S., 173
Shenk, T., 124
Sherman, L. S., 348, 352
Sherr, C. J., 116(2), 117, 127, 129, 129(1), 130, 131, 132(11; 51), 133, 247
Shi, B., 316
Shi, J., 89
Shi, W., 130, 139, 140(7), 149, 149(7), 277
Shi, Y. P., 75, 275
Shibuya, M., 12, 129, 132(11)
Shiebani, N., 280
Shih, A., 204
Shih, C. C., 273, 291
Shih, T. S., 268, 282(15)
Shima, D. T., 276
Shirahata, S., 302
Shirasawa, S., 107, 109(33), 267, 271(6)
Shoelson, S., 12(33), 13, 15(33)
Short, S. M., 151, 152, 153, 153(1), 156(1), 157(5), 161
Shtutman, M., 121
Shu, I.-W., 118
Shuai, K., 141, 142, 147
Shurtleff, S. A., 129, 132(11)
Shweiki, D., 276
Shyy, Y. J., 89
Sibai, B. M., 105
Sicinski, P., 116(4), 117, 130
Siebenkotten, G., 124
Siebenlist, U., 89, 90
Sieberth, E., 271(36), 273
Siegel, G., 105
Siegmann, M., 45
Siemeister, G., 274
Sigal, I. S., 333
Sikorski, R. S., 228

Silberman, S., 108
Silletti, J., 169
Silva, M., 276
Silverman, R. H., 138
Simcha, I., 121
Simon, A., 105
Simon, M. C., 139, 140(3), 149(3), 279
Simoni, M., 123
Simons, K., 173
Sinclair, W., 286
Singh, B., 109, 114(38), 115(38)
Singh, R. K., 279
Singhal, A., 30, 34(14)
Sinha, C. C., 98
Sinha, D., 308, 309(4), 311(4), 314(4), 315(4), 317(4)
Sizemore, N., 277
Skaliter, R., 276
Skolnick, E. M., 333
Skorecki, K., 280, 281(97)
Skriver, L., 105
Slingerland, J. M., 129
Sloan, J. A., 317
Sluss, H. K., 132(51), 133
Smeal, T., 62
Smiley, S. T., 139, 140(15), 149(15)
Smith, D. P., 12(30), 13
Smith, J. A., 238, 250
Smith, K., 276
Smith, L. T., 70
Smith, R. L., 58
Smoluk, J., 97
Snell, R. G., 139, 140(14)
Snider, L., 237
Soderman, D. D., 276
Sollott, S. J., 89
Solski, P. A., 68, 284, 285, 295(22), 296, 319, 320(9), 321(9), 325(9), 327(9)
Somasundaram, K., 117
Somwar, R., 342
Song, S. K., 177
Songyang, Z., 3, 6(5), 13(5)
Sorbat, B., 106
Sorescu, D., 89
Soroceanu, L., 274
Sorokin, A., 342
Sotgia, F., 173
Sotiropoulos, A., 149, 159
Sotirpoulu, A., 173
Soto, D. E., 286
Spalding, J. W., 271(34), 273, 279(34)
Spencer, T., 277
Spiegelman, B. M., 97
Sporn, M. B., 280, 281(90)
St. Croix, B., 275, 279(58)
St. Jean, A., 220
Staartz, W. D., 264
Stacey, K. J., 62, 68(6), 206
Stack, M. S., 105
Stacker, S. A., 273
Stan, R. V., 182
Stanbridge, E. J., 290, 291, 292, 293(17–19), 302(19)
Stang, E., 169, 174, 174(5), 175
Stanton, H., 98, 99(24)
Stark, G. R., 138
Steck, P. A., 273
Steeg, P. S., 280, 281(96)
Steen, H., 56
Steer, J., 75, 117, 118(9), 119(9)
Stein, I., 276
Stein, R. B., 142, 149(30)
Stenman, G., 173
Stepp, E., 106, 109(21), 112(19), 114(19), 115(19; 21)
Stern, D. F., 148
Stevens, J., 291
Stevenson, L. E., 12(36), 13
Stevenson, L. F., 129
Stockert, E., 139, 140(5)
Stoffel, M., 146
Stone, J. C., 130, 248, 250(14), 255(14)
Stonehouse, T. J., 12(30), 13
Stoppelli, M. P., 112
Storm, J. F., 16
Strauss, M., 74, 131, 132(41)
Stravopodis, D., 139, 140(13)
Strawn, L. M., 275
Street, A., 292
Strehlow, I., 141
Stricker, C., 123
Struab, J., 292
Struhl, K., 238, 250
Struman, I., 282
Stuart, D., 292
Stuart-Tilley, A., 124
Stupack, D. G., 169
Sturgill, T. W., 35, 46, 300
Su, B., 88, 300
Suarez, S., 341

Subler, M. A., 316
Sudarshan, C., 147
Sugden, P. H., 156
Sugimoto, Y., 4
Sugimura, T., 267
Sugiura, T., 280
Suh, Y. A., 89
Suit, H. D., 284
Sukhatme, V. P., 276
Sulciner, D. J., 88
Sullivan, K. A., 276
Sun, H., 56
Sun, J., 285
Sun, L., 275
Sun, Y. L., 148
Sunamura, M., 97
Sundaresan, M., 89
Superti-Furga, G., 5, 15(13)
Sutherland, C., 30(15), 34, 46, 53(18)
Sutherland, G., 98
Sutherland, R. L., 132, 138(49)
Suto, K., 129
Suy, S., 28, 35, 35(10)
Suzuki, K., 130
Suzuki Takahashi, I., 132(52), 133
Svingen, P. A., 317
Swarbrick, A., 132, 138(49)
Sweet, L. J., 45
Swisshelm, K., 173
Sydor, J. R., 350
Syed, J., 308, 309(4), 311(4), 314(4), 315(4), 317(4)
Symons, M., 75, 88, 117, 118(10; 11), 231
Syto, R., 306, 308(1), 309(1), 311(1)
Szabo, S. J., 142

T

Tabancay, A. P., 217, 218(4), 222(4), 229(4), 230
Taga, T., 141
Taguchi, S., 217
Taguchi, T., 217
Taha, C., 342
Tahir, S. K., 282, 283(106)
Takamiya, M., 90
Takata, Y., 12(35), 13
Takeda, J., 139, 140(8; 10)
Takeda, K., 139, 140(7–10), 149, 149(7)
Taksir, T., 280
Talbot, G., 153, 157(5)
Tam, A., 267, 274(7), 280(7)
Tamai, K., 132(52), 133
Tamanoi, F., 202, 217, 218, 218(4), 222(4), 229, 229(4; 13), 230
Tamir, A., 268, 270(18), 271(18), 277(18), 278(18), 279(18), 280(18), 283(18)
Tamura, R., 165
Tan, I., 342
Tan, L., 342
Tanaka, H., 149
Tanaka, K., 202
Tanaka, N., 247
Tanaka, T., 139, 140(7), 149, 149(7)
Tang, B., 150
Tang, C., 275
Tang, H., 116
Tang, W. J., 106
Tang, Y., 55, 56, 57(7), 58, 58(7)
Taniguchi, T., 247
Taparowsky, E. J., 232, 233, 233(2; 4; 5; 6), 236, 236(2)
Tapscott, S., 237
Tarutani, M., 139, 140(10)
Tató, F., 232
Tavassoli, F. A., 257
Taveras, A. G., 308, 309, 309(4), 311(4), 314(4), 315(4), 317(4)
Taya, Y., 132(52), 133
Taylor, S. J., 38, 333, 334, 340(9), 341(9), 342, 343
Taylor, S. M., 232
Tedder, T. F., 124
Teegarden, D., 236
Teglund, S., 139, 140(13)
Telford, J. L., 12
Tempst, P., 90
Tenen, D. G., 70
ten Hoeve, J., 142
Tennant, R. W., 116(4), 117, 130, 271(34), 273, 279(34)
Tennekoon, G. I., 58
Teramoto, T., 267
Tereba, A., 17
Terranova, V., 105
Terrell, R. S., 295(21), 296
Terwilliger, E. F., 273
Testa, J. E., 106
Thaler, S., 280, 281(97)
Thayer, M. J., 232, 233(3)

Thelen, M., 115
Thierfelder, W. E., 139, 140(11), 149(11), 150
Thiery, J. P., 281
Thimmapaya, B., 117, 118
Thomas, D., 12(32; 37), 13
Thomas, D. Y., 229
Thomas, G., 45, 46, 53, 53(17)
Thomas, K. A., 274, 276
Thomas, S., 7
Thompson, E. W., 265
Thompson, L. M., 12(32), 13
Thompson, T. C., 98
Thomssen, C., 105
Thorburn, A., 342
Thorgeirsson, U. P., 98, 280
Threadgill, D. W., 277
Tischer, E., 276
Tjuvajev, J. G., 273
Tober, K. L., 271(34), 273, 279(34)
Todaro, G. J., 248, 249(18)
Todd, D. G., 28, 29, 30(11), 34(11)
Tognazzi, K., 273
Tojo, A., 12
Tokuda, M., 199
Tokuyasu, K. T., 180
Tolkacheva, T., 163, 164(5), 169(5), 187
Tomerup, N., 217
Tomlinson, I. P., 292
Tonner, E., 139, 140(9)
Toomik, R., 16, 27(7)
Torchia, J., 123
Torri, J., 265
Toshizaki, K., 141
Totty, N. F., 4, 45
Tournaire, R., 97
Towers, R. P., 139, 140(14)
Tozer, E. C., 165
Trande, P., 349
Treisman, R., 61, 62, 159, 194
Trempus, C., 116(4), 117, 130
Tripp, R. A., 139, 140(11), 149(11)
Trivedi, P. G., 276
Troyer, D. A., 316
Trump, B. F., 291
Truong, T., 263
Tsakiridis, T., 342
Tseng, M. J., 30
Tsichlis, P. N., 163, 164(5), 169(5)
Tsien, R., 162
Tsiokas, L., 276
Tsionou, C., 280
Tuck, A. B., 268, 281(12), 319
Turkson, J., 139, 142, 149, 149(28)
Tweardy, D. J., 142, 149(27)
Tyler, R. D., 310

U

Udy, G. B., 139, 140(14)
Ueda, M., 267
Uehara, Y., 130
Uht, R. M., 119
Ullrich, A., 275
Ulrich, E., 56
Upton, T. M., 131
Urano, J., 217, 218, 218(4), 222(4), 229(4; 13), 230
Urbanski, S. J., 98
Usui, I., 12(35), 13

V

Vadas, M. A., 276
Vaidya, T. B., 236
Vaisse, C., 146
Valerie, K., 28, 29, 29(1), 30, 30(11; 12), 32(1; 9), 34(11; 12; 14), 37(1)
Vallone, D., 115
van Aeist, L., 96
Van Aelst, L., 58, 96, 117, 118(11), 130, 231, 248, 250(14), 255(14), 319, 320, 320(8; 9), 321(8; 9), 323(10), 325(9; 10), 327, 327(9; 10), 328(10), 342
Van Beveren, C., 62, 68(6), 206
Vandenhoeck, A., 280
Vanden Hoek, T. L., 118
van der Geer, P., 4
van der Sman, J., 131
van Deursen, J. M., 139, 140(11; 13), 149(11)
Vande Woude, G. F., 106, 112(25), 115(25), 192, 318, 320, 323(10), 325(10), 327, 327(10), 328(10)
Vanhaesebroeck, B., 37, 40(3), 115, 295(23), 296
Van Roost, E., 187
van Triest, M., 343, 344(5), 346(5)
Vass, W. C., 286
Vaughan, D. E., 116

Verde, P., 106, 112, 112(18; 22), 114(22), 115
Verducci-Galletti, B., 3, 6(5), 7(6)
Verma, I., 237
Vetriani, C., 3, 6(5), 7(6)
Vibulbhan, B., 309
Vignali, D. A., 139, 140(11), 149(11)
Vignoli, A., 349
Vila, J., 300
Villa, E., 349
Villalobos, J., 90
Viloria-Petit, A. M., 268, 270, 270(18), 271(18), 275, 277(18), 278(18), 279(18), 280(18), 283(18; 23)
Vinals, F., 278
Vindelov, L. L., 289
Vinkemeier, U., 139, 146, 148(37), 150(37)
Viskochil, D., 291
Vlahos, C. J., 46
Vocero-Akbani, A. M., 128
Voest, E. E., 276
Vogel, A., 280, 281(94)
Vogel, U. S., 350
Vogelstein, B., 247, 267, 292, 348
Vogt, A., 285
Vojtek, A. B., 196, 197(14), 203, 204(2), 209(2), 210(2), 211(2), 212
Voll, R. E., 75
Volonte, D., 173
Volpert, O. V., 271(31), 273
Von Hoff, D. D., 318
Vousden, K., 280, 281(89)
Vrana, J. A., 37
Vu, D., 117, 118(8), 130
Vuori, K., 163, 164(4), 169(4)

W

Wade, W. S., 4
Wagner, K. U., 139, 140(12)
Wakao, H., 121, 149
Walker, T. R., 49
Wang, B., 163, 169(6)
Wang, C. Y., 74, 84(13)
Wang, D., 89, 139, 140(13), 273
Wang, H., 106, 107, 109(21; 33), 111, 115(21), 163, 164(4), 169(4), 220
Wang, J., 309
Wang, L., 46, 306, 308, 308(1), 309, 309(1; 4), 311(1; 4), 314(4), 315(4), 317(4)
Wang, Q., 55
Wang, S., 308
Wang, S. H., 229
Wang, X. Z., 62, 275
Wang, Y. C., 282, 283(106), 289
Wang, Y.-Z., 130
Warne, P. H., 37, 38, 40(3), 41(5), 43(6), 56, 58(2), 88, 115, 295(23), 296, 323, 350
Warren, R. S., 274
Wary, K. K., 12(38), 13
Wasylyk, C., 106
Watanabe, G., 117, 118, 118(8), 126(12), 129, 130, 173, 267
Watanabe, S., 217
Waterfield, M. D., 4, 37, 40(3), 56, 58(2), 88, 115, 295(23), 296, 323
Watson, C. J., 139, 140(9)
Watts, C. K. W., 132
Waxman, D. J., 139, 140(14)
Way, M., 175
Webb, C. P., 96, 318, 320, 323(10), 325(10), 327, 327(10), 328(10)
Webb, P., 119, 121(21)
Weber, C., 341
Weber, M. J., 300
Weber-Nordt, R. M., 142
Webster, C., 236
Wei, M. C., 128
Wei, S., 149
Weichel, W., 124
Weichselbaum, R. R., 28
Weinberg, R. A., 116(4), 117, 127, 129(2), 130, 247, 248, 291, 292
Weinbert, R. A., 291
Weindel, K., 271, 274, 274(25), 278(25), 279(25), 281(25)
Weiner, R. I., 282
Weinmann, R., 203, 204(4), 210(4), 211(4)
Weintraub, H., 232, 233, 233(3), 236(12), 237
Weiss, B., 348
Weiss, M. A., 284, 285, 285(22)
Weiss, R., 202, 291
Weissman, B., 65
Welte, T., 148
Wen, Z., 139, 142, 142(21), 146, 147, 148, 148(21)
Weng, Q.-P., 53
Wennström, S., 38, 41(5), 43(6)
Wentworth, B. M., 237
Werb, Z., 97, 280

West, B. L., 119
Westwick, J. K., 62, 63(4), 68, 74, 75, 82(8), 117, 118(9–11), 119(9), 190, 226, 231, 261, 319, 320(9), 321(9), 325(9), 327(9)
Wettenhall, R. E. H., 53
Weyman, C. M., 233, 236
Whang, Y. E., 122, 295(24), 296
White, F. C., 271(33), 273, 279(33), 283(33)
White, J. M., 139, 140(4), 149(4)
White, M. A., 96, 232, 233(5; 6), 319, 320(8; 9), 321(8; 9), 325(9), 327(9)
White, R., 202, 291
Whitehead, I. P., 75, 117, 118(10), 261
Whiteside, T., 35
Whiteway, M. S., 229
Whitmarsh, A. J., 122
Whitmore, W., 284
Whyte, D. B., 306
Wiederanders, B., 98
Wieland, F., 173
Wieler, J. S., 187
Wientjens, E., 131
Wigler, M. H., 96, 319, 320, 320(8; 9), 321(8; 9), 323(10), 325(9; 10), 327(9; 10), 328(10)
Wildmann, C., 187
Wiley, S., 4
Wilhelm, O., 105, 106, 116
Wilhelm, S. M., 99
Wilkins, R. J., 139, 140(14)
Wilks, A. F., 273
Willen, H., 106
Williams, A. J., 75
Williams, B. R., 138, 139
Williams, L. T., 3, 7(3)
Williams, M., 295
Williamson, E., 173
Williamson, N. A., 53
Willumsen, B. M., 284
Wilson, K. E., 341
Wilson, M. R., 273, 275
Windle, J. J., 316, 318
Windsor, W., 306, 308(1), 309(1), 311(1)
Winer, J., 274
Winston, J. T., 130
Wiseman, B., 130, 250
Wisniewski, D., 27
Witte, L., 275
Witthuhn, B. A., 150
Wittinghofer, A., 334, 343, 350
Wolf, D., 291
Wolfe, K. G., 128
Wolfson, M., 233
Wolin, R., 309
Wolthuis, R. M. F., 343, 344(5), 346(5)
Wong, C., 98
Wong, M. J., 130, 267, 274(7), 280(7)
Wood, W. M., 120
Woodgett, J. R., 300
Woods, D., 202, 204, 248, 250, 250(13), 255(13)
Woods, R. A., 220
Worley, P. F., 217
Woude, G. F., 96
Wouters, B. G., 286
Wrzeszczynska, M. H., 146, 148(38)
Wu, C., 229
Wu, I. H., 88, 300
Wu, J., 139, 300
Wu, J. E., 264
Wu, J. M., 285
Wu, P., 273
Wu, S., 263
Wu, X., 122, 295(24), 296
Wun, T. C., 105
Wung, B. S., 89
Wyllie, A. H., 286
Wymann, M., 35
Wynne, J., 62, 194
Wynshaw-Boris, A., 139, 140(12)

X

Xavier, R., 280
Xia, Y., 89
Xia, Z., 28
Xie, Z., 30(17), 34
Xin, J. H., 122
Xing, J., 45
Xu, G. F., 202, 291
Xu, J., 147
Xu, L. Z., 73, 123, 142
Xu, X. A., 89, 148

Y

Yamada, K. M., 165
Yamagata, K., 217
Yamagoe, S., 130

Yamaguchi, Y., 139, 140(10)
Yamamoto, K., 98, 139, 140(11), 149(11)
Yamanaka, Y., 149
Yamashita, H., 147
Yamashita, N., 267
Yamazaki, M., 217
Yamazaki, T., 345
Yan, J., 169
Yan, N., 203, 204(4), 210(4), 211(4)
Yancopoulos, G. D., 147, 267, 274(7), 280(7)
Yang, B.-S., 62, 68(6), 206
Yang, C. Y., 88
Yang, D. J., 291
Yang, J.-J., 320
Yang, W., 217, 218, 218(4), 222(4), 229(4; 13)
Yang, Y., 111
Yang-Yen, H.-F., 116(3), 117, 130
Yaniv, M., 106
Yarden, Y., 150, 286
Yaremko, B., 308, 309(4; 5), 311(4; 5), 314(4; 5), 315(4), 316, 317(4)
Yeargin, J., 334
Yee, W., 217
Yeh, L., 12
Yeh, P., 271(29), 273, 279(29)
Yen, A., 295
Yen, J. J.-Y., 116(3), 117, 130
Yeo, K.-T., 273, 276
Yoakum, G. H., 291
Yokoyama, N., 267, 271(6)
Yorke, E. D., 284
Yoshida, E., 132(52), 133
Yoshida, N., 139, 140(7; 8), 149, 149(7)
Yoshiji, H., 280
Yoshikawa, K., 139, 140(10)
Yoshioka, T., 98
Young, M. R., 97
Yu, J., 56, 57(7), 58(7)
Yu, Z. X., 88
Yuan, H., 274
Yu-Lee, L., 146
Yung, W. K., 273
Yuspa, S. H., 97, 116(4), 117, 130, 277, 280, 281(90)
Yutzey, K. E., 237

Z

Zabrenetzky, V., 280, 281(96)
Zafonte, B. T., 116, 127
Zandi, E., 74, 82(11)
Zebrowski, B. K., 275
Zeng, Q., 142, 149(27)
Zenzle, B. W., 280, 281(92)
Zhang, C.-F., 209
Zhang, F. L., 218
Zhang, H. B., 286
Zhang, H. C., 282, 283(106)
Zhang, K., 202
Zhang, X., 173
Zhang, Y., 139
Zhang, Z., 163, 164(4), 169(4)
Zhao, J., 128
Zhao, Y., 139, 146, 148(38)
Zhao, Z. S., 56, 342
Zheng, C.-F., 61, 73
Zhong, H., 75
Zhong, Z., 139, 142, 142(21), 146, 147, 148(21)
Zhou, H., 267, 274(7), 280(7)
Zhou, M. M., 4
Zhou, M. Y., 142, 149(27)
Zhu, J., 204, 248, 250(13), 255(13)
Zhu, X., 142
Zhu, Z., 275
Zhuo, Y., 55
Zhurinsky, J., 121
Zindy, F., 117, 130, 247
Ziober, B., 109
Zisch, A. H., 163, 169(6)
Zohn, I. E., 203, 204(5), 208(5)
Zon, L., 118
Zong, H., 187
Zorzi, N., 175
Zou, J. X., 163, 169(6)
Zou, Z., 173
Zozulya, S., 12
Zu, K., 300
Zutter, M. M., 257, 259(4), 264
Zweier, J. L., 89
Zwicker, J., 131
Zwijsen, R. M. L., 131

Subject Index

A

Activated Ras interaction assay
  affinity precipitation of Ras–GTP, 337–338
  applications, 341–342
  controls, 340
  GTPase types amenable for assay, 342
  optimization, 340–341
  principle, 334–335
  Raf–glutathione *S*-transferase fusion protein preparation, 336–337
  Western blot analysis
    antibodies, 338–339
    blotting, 338
    electrophoresis, 338
Akt
  phosphatidylinositol 3-kinase signaling, 38, 43
  Ras activation in cells
    COS cell cotransfection assay, 44
    kinases, 43
Angiogenesis, Ras regulation
  cell lines, 268
  cell survival and mitogenesis relationship to tumor angiogenesis, 281–282
  clinical application, 283
  endothelial cell growth/survival assay, 269
  farnesyltransferase inhibitor studies, 282–283
  gene types under Ras control, 280–281
  signaling overview, 267–268, 276–279
  thrombospondin 1 expression assay, 270, 280
  tumor growth model in nude mice, 270
  vascular dependence assay of transformed cell lines, 270–271
  vascular endothelial growth factor induction
    angiogenic switch, 271, 273
    coregulation with epigenetic factors, 279
    expression assay, 270
    levels of regulation, 276, 278–279
    model systems, 273–274
    signal transduction
      inhibitor studies, 270–271, 275–276, 278
      mitogen-activated protein kinase, 278
    tumor growth role and evidence, 273–275, 279–280
Apoptosis
  farnesyltransferase inhibitor induction, 316–317
  H-Ras(V12) induction following inactivation of nuclear factor-κB
    DNA laddering on gels, 84–87
    β-galactosidase expression assay for cell viability, 85
    terminal deoxynucleotidyltransferase-mediated dUTP nick-end labeling assay, 84, 87
ARIA, *see* Activated Ras interaction assay

B

Biotinylated kinase substrate, *see* Protein kinase A; Protein kinase C
Boyden chamber, *see* Metastasis
Breast adenocarcinoma, *see* Metastasis

C

Calcium flux, integrin signaling through G protein-coupled receptors, 161–162
Caveolin
  cholesterol regulation, 173–174
  dominant-negative mutant screening, 174
  functions, 173
  Ras function analysis with dominant-negative mutants
    caveats, 174–175
    cholesterol depletion effects using cyclodextrin, 176
    constructs, 175
    electron microscopy of plasma membranes
      immunogold labeling, 180, 182
      rip-off technique, 179–180

sucrose density gradient centrifugation of lysates, 177–179
transfection, 175–176
Western blot analysis of caveolin, 177, 179
tumor inhibition, 173
types, 173
CDK, *see* Cyclin-dependent kinase
Cholesterol
caveolin regulation, 173–174
membrane depletion using cyclodextrin, 176
Cyclin-dependent kinase
activation, 127
cell cycle regulation, 127–128
cyclin dependence, 127
Ras signaling effects on cyclin complexes
assays
activating kinase assays, 134, 138
applications, 131
cell extract preparation, 132, 134–135
gel electrophoresis, 134, 137–138
immunoprecipitation of cyclin complexes, 133, 135–136
kinase assay, 133–134, 137
kinase substrate preparation, 132–133, 135
materials, 132–134
overview, 130–131
regulators of cyclin complex activity, 129
retinoblastoma protein phosphorylation role, 127–128
Cyclin D1
nuclear factor-κB transcription control, 74–75
phosphorylation and degradation, 117
Ras induction
Dbl protein roles, 118
Rac1 induction, 118–119
transfection analysis
artifacts, 120–122
calcium phosphate transfection, 125
enrichment artifacts, 124
internal control reporter gene artifacts in normalization, 122–124, 126
luciferase assay, 125–126
reporters, 125
vector backbone sequences, 119–120
viral promoter regulation, 122–123
Ras-induced transformation requirement, 116–117, 126–127
transcription factors, 117–118

D

Dominant-negative mutants, *see* Caveolin; Ras N17; Shc proteins

E

Electron microscopy, plasma membranes
immunogold labeling, 180, 182
rip-off technique, 179–180
Electron paramagnetic resonance, reactive oxygen species assay by spin trapping, 92–93
Electrophoretic mobility shift assay (EMSA)
nuclear factor-κB binding to DNA
binding reaction, 80, 94
gel electrophoresis, 80–81, 94–95
nuclear extract isolation, 78–80, 93–94
principle, 77, 93
Rac1 activation study, 93–95
radiolabeled probe, 80
Ras activation study, 81–82
supershift assay, 77–78
signal transducer and activator of transcription activation
affinity of binding, 146–147
binding reaction, 144
electrophoresis, 144–145
extraction of proteins
cell cultures, 145
tissues, 145–146
gel casting, 143
migration of dimers, 142, 146
positive controls, 142–143
probe preparation, 143–144
urokinase transcription factor identification
AP-1/PEA3 site, 114–115
supershift assay, 114
Elk-1
M-Ras/R-Ras3 activation assay
luciferase reporter assay, 195
principle, 193–194
transfection, 194
N-Ras activation in fibrosarcoma cell lines, 301–302

EPR, *see* Electron paramagnetic
resonance

F

Farnesyltransferase inhibitor
angiogenesis inhibition studies,
282–283
cancer treatment prospects, 317–318
cytotoxic chemotherapy combination with
SCH 66336 studies, 315–316
pharmacokinetic studies in mouse
administration of drug, 307
overview, 306–307
sample collection, 307–308
SCH 44342, 308
SCH 66336, 308–310
radioresistance effects, 285
tumor xenograft models in nude mice
human tumors, 313–315
Ras-transformed rodent fibroblasts,
311–313
Wap–*ras* transgenic mouse studies of SCH
66336
animal features, 310
apoptosis assay, 316–317
cell proliferation assay, 317
drug administration, 310
treatment onset effects, 311
tumor characterization, 310–311
FTI, *see* Farnesyltransferase inhibitor

G

GAL4 fusion proteins, *see* Nuclear factor-κB;
Ras
GAP, *see* GTPase-activating protein
Gelatinase zymography
band visualization, 101
cell culture, 99–100
gel casting, 100–101
gel electrophoresis, 101
matrix metalloproteinase identification,
99
GTPase-activating protein
M-Ras/R-Ras3 assay, 202
Ras activation, 333
tumorigenesis role with Ras proteins,
290–291
Guanine nucleotide exchange
M-Ras/R-Ras3
identification of exchange factors
binding assay, 198–199
Ras–glutathione *S*-transferase fusion
protein preparation, 198
Western blot analysis, 199
overview, 187–188, 197–198
in vivo exchange assays
nonradioactive assay, 201–202
radioactive assay, 199–200
Ras–GTP level determination, *see also*
Activated Ras interaction assay
radioassays and limitations, 333–334, 343
Ras-like GTPase–GTP determination with
activation-specific probes
advantages, 347
principle, 343
Ras-binding domain–glutathione
S-transferase fusion protein
preparation, 343–345
troubleshooting, 347–348
Western blot analysis, 345–346
tumorigenesis role with Ras proteins,
290–291

H

H-Ras, *see* Ras

I

Immunohistochemistry
muscle-specific proteins
antibody specificity, 243–244
detection, 244
fixation, 244
reagents, 244
staining and microscopy, 246
Ras
applications, 349
Raf1–glutathione *S*-transferase fusion
protein as probe
advantages and limitations, 350
binding to cells with constitutively
activated v-H-Ras, 351–352,
354–355
cell culture and fixation, 354
preparation of probe, 352–354
rationale, 349–350
signal detection, 355

signal transducer and activator of transcription activation, 147–148
tumor cells in lungs, 104
Integrins
G protein-coupled receptor signaling, calcium flux measurements, 161–162
mitogen-activated protein kinase signaling assays
cell anchoring with anti-integrin antibodies, 152–153
cell lysate preparation, 153–154
controls, 152
mitogen-activated protein kinase immunoprecipitation and kinase assay, 154–155
phospho-specific antibodies for phosphorylated kinase analysis, 157–158
Raf immunoprecipitation and kinase assay, 155–157
transient cotransfections for morphology and signaling analysis
epitope-tagged kinase assays, 160–161
green fluorescent protein constructs, 158
immunofluorescence microscopy, 158–160
Rho GTPase constructs, 158
R-Ras/H-Ras activation
cell adhesion assays of R-Ras activation
absorbance assay, 171
cell lines, 169
fluorescent assay, 169–170
interpretation, 171
cell line specificity, 164
Chinese hamster ovary cells for assays, 165
flow cytometry reagents and applications, 168–169
PAC-1 binding assay
antibody for flow cytometry, 165
cell harvesting and staining, 167–168
flow cytometry and data analysis, 168
transfection, 166–167
signal transduction
functions, 163
types, 151–152
Invasion, *see* Metastasis

K

K-Ras, *see* Ras

L

Lipid raft, association of Ras isoforms, 172–173, 175

M

MAPK, *see* Mitogen-activated protein kinase
Matrix metalloproteinases
gelatinase zymography
band visualization, 101
cell culture, 99–100
gel casting, 100–101
gel electrophoresis, 101
identification of proteinase types, 99
metastasis
collagen degradation, 97–98
MMP-1 role, 98–99
Metastasis
breast adenocarcinoma T47D cell model for Ras and Rho effects
Boyden chamber transwell migration assay
counting of cells, 260–261
inhibitor studies, 261, 263
principle, 259
stable transfectant migration, 259–261
transient transfectant migration, 261
cell characteristics and advantages, 256–257
culture, 257–258
gel invasion assays, 265
time-lapse video microscopy of migration, 264
transfection
stable, 258
transient, 258–259
transwell invasion assay, 264–265
wound assays of migration, 263–264
immunohistochemistry of tumor cells in lungs, 104
matrix metalloproteinases
collagen degradation, 97–98
MMP-1 role, 98–99
nude mouse system, 102–103

proteases, 105
Ras signaling
effector domain mutant studies in cell culture
mitogen-activated protein kinase pathway activation, 323, 325, 329
NIH 3T3 cell transfection, 321–322
plasmids, 321
principles, 320–321
Raf complex immunoprecipitation, 323, 325
types of mutants, 319–320
Western blotting of Ras proteins, 322–323
effector domain mutant studies in mouse
metastasis assay, 327–328
pathological examination, 328–329
principles, 325–326
subcutaneous tumor growth, 326–327
tumor explant culture, 328
overview, 96–97, 319
stages, 318–319
transfected cell visualization in lung sections with fluorescence microscopy, 103–104
urokinase role, 105–108
Migration, *see* Metastasis
Mitogen-activated protein kinase
assay, 300–301
integrin signaling assays
cell anchoring with anti-integrin antibodies, 152–153
cell lysate preparation, 153–154
controls, 152
mitogen-activated protein kinase immunoprecipitation and kinase assay, 154–155
phospho-specific antibodies for phosphorylated kinase analysis, 157–158
Raf immunoprecipitation and kinase assay, 155–157
transient cotransfections for morphology and signaling analysis
epitope-tagged kinase assays, 160–161
green fluorescent protein constructs, 158
immunofluorescence microscopy, 158–160
Rho GTPase constructs, 158
M-Ras/R-Ras3 induction assay, 195–196
Ras effector domain mutant pathway activation, 323, 325, 329
Ras N17 blocking of radiation-induced activation, *see* Ras N17
Rheb signaling, 226
S6 kinase activation, 46
TC21/R-Ras2 induction assay, 210
vascular endothelial growth factor induction by Ras signaling, 278
MMPs, *see* Matrix metalloproteinases
M-Ras/R-Ras3
C2 cell differentiation assay, 193
Elk-1 transcription assay
luciferase reporter assay, 195
principle, 193–194
transfection, 194
gene cloning, 187
GTPase-activating protein assay, 202
guanine nucleotide exchange
identification of exchange factors
binding assay, 198–199
Ras–glutathione *S*-transferase fusion protein preparation, 198
Western blot analysis, 199
overview, 187–188, 197–198
in vivo exchange assays
nonradioactive assay, 201–202
radioactive assay, 199–200
homology with other Ras isoforms, 187
mitogen-activated protein kinase assay, 195–196
transformation assays
NIH 3T3 focus-forming assay
cell culture, 188, 190
interpretation, 191–192
overview, 188
transfection, 190–191
secondary focus-forming assay, 192
soft agar assay, 192–193
yeast two-hybrid analysis
controls, 197
β-galactosidase assay, 197
principle, 196–197
Myc, immortalization of cells with v-Myc, 291–292
Myoblast
5-azacytidine conversion from mouse fibroblasts, 232–233
Ras effects
cell lines for study

23A2 myoblasts, 234, 236
C_2C_{12} myoblasts, 236
media, 233–234, 236
transfected mouse 10T1/2 fibroblasts, 234, 236
immunostaining of muscle-specific proteins
antibody specificity, 243–244
detection, 244
fixation, 244
reagents, 244
staining and microscopy, 246
muscle regulatory factor inhibition, 232
stable transfection
23A2 myoblasts, 243
overview, 242
reagents, 242–233
transient transfection
luciferase assay, 239–240
mouse 10T1/2 fibroblast transfection, 239
overview, 236–238
promoter selection, 238
reagents, 238–239
reporter genes, 237

N

NF-κB, *see* Nuclear factor-κB
Northern blot, urokinase induction by Ras analysis, 111–112
Nuclear factor-κB
cyclin D1 transcription control, 74–75
DNA-binding region, 75, 90
electrophoretic mobility shift assay
binding reaction, 80
gel electrophoresis, 80–81
nuclear extract isolation, 78–80
principle, 77
radiolabeled probe, 80
supershift assay, 77–78
family members, 89–90
Gal4 fusion studies of p65 subunit transactivation, 82–84
inhibitor
apoptosis induction, 84, 86–87
mechanism, 74, 90
types, 90
Rac1-induced activation
electrophoretic mobility shift assay of factor activation, 93–95
reactive oxygen species production assays, 91–93
signaling, 88–89
transient transfection assay of reporter genes, 95–96
Ras activation
apoptosis induction by H-Ras(V12) following inactivation of factor
DNA laddering on gels, 84–87
β-galactosidase expression assay for cell viability, 85
terminal deoxynucleotidyltransferase-mediated dUTP nick-end labeling assay, 84, 87
electrophoretic mobility shift assay of H-Ras activation, 81–82
p65 subunit transactivation, 83–84
signaling pathway, 74, 88
transient transfection assay of H-Ras activation, 77
transient transfection assay for transcriptional activity
NIH 3T3 cell transfection, 76
overview, 75–76
reporter activity assays, 76–77

P

p21-activated kinase, Ras activation assay
cell culture and transfection, 59
extract preparation, 59
immunoprecipitation of kinase, 60
kinase reaction and gel electrophoresis, 60
materials, 58–59
troubleshooting, 60–61
cell lines for study, 56, 58
evidence, 56
PAK, *see* p21-activated kinase
Phosphatidylinositol 3-kinase
Ras activation
activation assay in cells
deacylation and high-performance liquid chromatography analysis, 43
phosphorous-32 labeling and lipid analysis, 42
principle, 41
thin-layer chromatography, 42–43
transfection of COS cells, 41–42
activation assay *in vitro*

incubation conditions and thin-layer chromatography, 40
principle, 39–40
Ras reconstitution in liposomes, 40
Akt signaling, 38, 43
p110 Ras-binding domain interactions
binding reaction, 39
glutathione *S*-transferase fusion protein preparation, 38–39
Ras loading with guanine nucleotides, 39
protein–protein interactions, 37
S6 kinase activation, 46
Phosphocellulose filter, binding assays for protein kinases, 16–17
PKA, *see* Protein kinase A
PKC, *see* Protein kinase C
Plasminogen zymography, urokinase assay, 108–109
Protein kinase A, biotinylated peptide substrate assay
advantages, 26–27
buffers and solutions, 18
calculations, 21–22
extract preparation, 19–20
high-throughput application, 26–27
incubation conditions, 20–21
membrane binding, 21
principle, 17
quantification of radioactivity, 21
response curves, 24–25
substrate sequence and synthesis, 17–18
Protein kinase C, biotinylated peptide substrate assay
advantages, 26–27
buffers and solutions, 18–19
calculations, 23
extract preparation, 19–20, 22
high-throughput application, 26–27
incubation conditions, 22–23
membrane binding, 21
principle, 17
quantification of radioactivity, 21
response curves, 24–25
substrate sequence and synthesis, 17–18

R

Rac
activation assay, 90–91
cyclin D1 induction, 118–119
nuclear factor-κB activation
electrophoretic mobility shift assay of factor activation, 93–95
signaling, 88–89
transient transfection assay of reporter genes, 95–96
reactive oxygen species induction
assays
dichlorofluorescein assay, 92
electron paramagnetic resonance spin trapping, 92–93
overview, 91
Ras signaling, 88–89
Radioresistance, Ras regulation
cell cycle
modulation of radioresistance, 285–286
synchronization for studies, 289–290
cell lines, 284
farnesyltransferase inhibitor effects, 285
geranylgeranyltransferase inhibitor effects, 285
growth curves for cells, 288
survival assay for cells
clonogenic survival, 286
culture, 286–287
inhibitor studies, 287–288
irradiation, 287
surviving fraction after 2 Gy, 288
Raf
assay, 300–301
glutathione *S*-transferase fusion protein
activated Ras interaction assay, 336–337
Ras immunohistochemistry
advantages and limitations, 350
binding to cells with constitutively activated v-H-Ras, 351–352, 354–355
cell culture and fixation, 354
preparation of probe, 352–354
rationale, 349–350
signal detection, 355
Rheb-binding assay
pull-down assay, 227
Raf purification, 227
rheb–glutathione *S*-transferase fusion protein preparation, 226–227
Ras, *see also* M-Ras/R-Ras3; TC21/R-Ras2
Akt activation in cells
COS cell cotransfection assay, 44
kinases, 43

angiogenesis regulation, *see* Angiogenesis, Ras regulation
caveolin dominant-negative mutant studies, *see* Caveolin
cyclin-dependent kinase regulation, *see* Cyclin-dependent kinase
cyclin D1 promoter regulation, *see* Cyclin D1
fibrosarcoma cell lines for N-Ras allele studies
actin cytoskeleton staining, 303–304
activating mutants, 294–295, 306
anchorage-independent growth assay, 302–303
dominant negative mutants, 294–295, 306
Elk activation assay with luciferase reporter, 301–302
growth rate assay, 302
HT1080 cell features, 293, 305
kinase assays, 300–301
MCH603c8 cell features, 293–294, 305
morphology, 303
plasmids, 295–296
saturation density assay, 302
serum dependence assay, 302
stable transfection, 296–297
tumorigenicity assay in nude mice, 304–305
Western blot analysis of signaling proteins
antibody staining, 299
blotting, 299
electrophoresis, 299
gel casting, 298
GTP pull-down, 299
lysate preparation, 298
rationale, 297
GAL4 fusion proteins for transcriptional activation assays
controls, 70–71
Ets2 fusion constructs, 68
luciferase assay, 67–68
NIH 3T3 cell line and culture, 63, 65
plasmids, 65
principle, 62–63
signal transduction monitoring using Ras pathway component fusions, 71–73
transient transfection, 65–67
guanine nucleotide exchange, *see* Guanine nucleotide exchange
homolog enriched in brain, *see* Rheb
immunohistochemistry
applications, 349
Raf1–glutathione *S*-transferase fusion protein as probe
advantages and limitations, 350
binding to cells with constitutively activated v-H-Ras, 351–352, 354–355
cell culture and fixation, 354
preparation of probe, 352–354
rationale, 349–350
signal detection, 355
integrin activation
cell adhesion assays of R-Ras activation
absorbance assay, 171
cell lines, 169
fluorescent assay, 169–170
interpretation, 171
cell line specificity, 164
Chinese hamster ovary cells for assays, 165
flow cytometry reagents and applications, 168–169
PAC-1 binding assay
antibody for flow cytometry, 165
cell harvesting and staining, 167–168
flow cytometry and data analysis, 168
transfection, 166–167
isoforms and effector activation, 172, 267
lipid raft association of isoforms, 172–173, 175
metastasis role, *see* Metastasis
muscle regulation, *see* Myoblast
mutation in tumors, 88, 96–97, 267, 292, 319, 348
nuclear factor-κB activation
apoptosis induction by H-Ras(V12) following inactivation of factor
DNA laddering on gels, 84–87
β-galactosidase expression assay for cell viability, 85
terminal deoxynucleotidyltransferase-mediated dUTP nick-end labeling assay, 84, 87
electrophoretic mobility shift assay of H-Ras activation, 81–82
p65 subunit transactivation, 83–84
Rac1-induced activation
electrophoretic mobility shift assay of factor activation, 93–95
reactive oxygen species production assays, 91–93
signaling, 88–89

transient transfection assay of reporter genes, 95–96
signaling pathway, 74, 88
transient transfection assay of H-Ras activation, 77
oncogenesis and effectors, 247–248
p21-activated kinase activation
assay
cell culture and transfection, 59
extract preparation, 59
immunoprecipitation of kinase, 60
kinase reaction and gel electrophoresis, 60
materials, 58–59
troubleshooting, 60–61
cell lines for study, 56, 58
evidence, 56
phosphatidylinositol 3-kinase activation
activation assay in cells
deacylation and high-performance liquid chromatography analysis, 43
phosphorous-32 labeling and lipid analysis, 42
principle, 41
thin-layer chromatography, 42–43
transfection of COS cells, 41–42
activation assay *in vitro*
incubation conditions and thin-layer chromatography, 40
principle, 39–40
Ras reconstitution in liposomes, 40
Akt signaling, 38, 43
p110 Ras-binding domain interactions
binding reaction, 39
glutathione *S*-transferase fusion protein preparation, 38–39
Ras loading with guanine nucleotides, 39
protein–protein interactions, 37
posttranslational modification, 172–173, 284–285
radioresistance modulation, *see* Radioresistance, Ras regulation
reporter genes for transcriptional activation studies, 61–62
reverse transcription–polymerase chain reaction for transcript quantification, 348–349
senescence induction, *see* Senescence, induction by oncogenic Ras
signaling overview, 55–56, 61, 73, 88
S6 kinase activation, 45–46, 54–55
urokinase induction
electrophoretic mobility shift assay for transcription factor identification
AP-1/PEA3 site, 114–115
supershift assay, 114
evidence, 106–107
laminin degradation mediation by receptor-bound urokinase, 109–110
metastasis role, 107–108
Northern blot analysis, 111–112
Ras transfectants and magnetic cell sorting, 110–111
reporter assay of transactivation using transient transfection, 112–113
signaling pathway, 115–116
Ras N17
mitogen-activated protein kinase, blocking of radiation-induced activation
blocking activity of Ras N17, 35, 37
epidermal growth factor receptor antisense cell generation and culture, 29, 34
immunoprecipitation of lysates, 32
irradiation of cells, 32, 35
kinase assay, 32, 34
overview, 28–29
U0126 kinase inhibitor treatment of cells, 32, 35, 37
Western blot, 34
recombinant adenoviral vectors
generation
Escherichia coli recombination, 30–32
human 293 cell recombination, 30
infection efficiency, 28, 34
infection of epidermal growth factor receptor antisense cells, 32
Reactive oxygen species, Rac induction
assays
dichlorofluorescein assay, 92
electron paramagnetic resonance spin trapping, 92–93
overview, 91
signaling, 88–89
Rheb
farnesylation
*Caa*X motif and mutation, 217–218, 228
enzymes, 227–228
farnesyltransferase deletion effects in yeast, 228–230
radiolabeling, 230

functional overview, 217–218
NIH 3T3 cell studies
focus-forming assay, 224–225
signal transduction assays, 225–226
stable cell line establishment, 225
subcellular localization, 230–231
transfection, 221, 223–224
Raf-binding assay
pull-down assay, 227
Raf purification, 227
rheb–glutathione *S*-transferase fusion protein preparation, 226–227
species distribution and conserved features, 217
yeast function analysis
arginine uptake rates, 221
canavanine sensitivity, 220–221
gene disruption, 219–220
lysine uptake rates, 221
materials, 218–219
thialysine sensitivity, 220–221
Rho
cell migration and invasion effects, *see* Metastasis
S6 kinase activation, 54–55
ROS, *see* Reactive oxygen species
R-Ras, *see* M-Ras/R-Ras3; Ras; TC21/R-Ras2
RSK, *see* S6 kinase

S

SCH 66336, *see* Farnesyltransferase inhibitor
Senescence, induction by oncogenic Ras
analysis
morphology of cells, 254–255
overview of assays, 253–254
senescence-associated β-galactosidase expression, 255–256
thymidine incorporation assay, 254
human diploid fibroblast culture, 249–250
mouse embryo fibroblast preparation, 248–249
overview, 247–248
retroviral transduction of fibroblasts, 250–253
Shc proteins
domains, 3–4
dominant-negative mutant analysis
epidermal growth factor transcriptional activation system, 13–14
mechanisms of signaling inhibition, 12
reporter assay, 14–15
transfection, 13
genes, 3
ShcA protein isoforms, 4–5
ShcB protein isoforms, 6–7
ShcC protein isoforms, 5–6
signaling overview, 3
tyrosine phosphorylation
activation, 3, 15
assay
antibodies, 11–12
cell culture, 7–8
cell lysis, 8–9
immunoprecipitation, 9–10
Western blot, 10–11
sites on protein, 4
Signal transducers and activators of transcription
dimerization, 139
domains, 151
electrophoretic mobility shift assay of activation
affinity of binding, 146–147
binding reaction, 144
electrophoresis, 144–145
extraction of proteins
cell cultures, 145
tissues, 145–146
gel casting, 143
migration of dimers, 142, 146
positive controls, 142–143
probe preparation, 143–144
immunohistochemistry of activation, 147–148
isoforms, 139, 141
ligands leading to activation, 138–140
regulators of activation, 141
STAT1 purification from recombinant baculovirus expression system
cation-exchange chromatography, 150
cell lysis, 150
cysteine alkylation, 150
heparin affinity chromatography, 150
hydrophobic affinity chromatography, 150
vectors, 149
STAT3 purification from recombinant *Escherichia coli*, 151
transfection assays of transcriptional competence, 148–149
types and knockout phenotypes, 139–141
tyrosine phosphorylation

reaction *in vitro*, 150–151
site, 138–139
Western blot analysis of activation, 147
S6 kinase
antibody generation, 46–47
assay
cell extract preparation, 49
filter assays, 50–51, 53
gel electrophoresis, 50
immunoprecipitation, 49–50
kinase reaction, 50
p70S6k activation by Ras and Rho in cells, 54–55
S6–glutathione *S*-transferase fusion protein preparation as substrate, 47–49
Western blot analysis of phosphorylative activation, 53
basal activity of p70S6k, 55
kinases in activation, 45–46
types and functions
p70S6k, 45
RSK, 45
STATs, *see* Signal transducers and activators of transcription

T

TC21/R-Ras2
cell differentiation blocking, 208
glutathione *S*-transferase fusion protein preparation, 210–211
homology with other Ras isoforms, 203
mammalian expression vectors, 204–206
mutation in tumors, 203
prenylation, 203
protein–protein interaction assays
pull-down assay, 211–212
yeast two-hybrid analysis
HIS3 reporter transactivation assessment, 214–215
LacZ reporter transactivation assessment, 215
media, 213–214
plasmids, 212–213
principle, 212
transformation, 214
yeast strain, 213
signal transduction
assays, 208–210
mitogen-activated protein kinase assay, 210
overview, 204
transcription factor activation, 208–210
transformation assays
cell lines, 206–207
NIH 3T3 focus-forming assay, 207
soft agar assay, 207–208
Terminal deoxynucleotidyltransferase-mediated dUTP nick-end labeling assay
farnesyltransferase inhibitor induction of apoptosis, 316–317
H-Ras(V12) effects following inactivation of nuclear factor-κB, 84, 87
Thrombospondin 1, expression assay, 270, 280
TUNEL assay, *see* Terminal deoxynucleotidyltransferase-mediated dUTP nick-end labeling assay

U

Urokinase
metastasis role, 105–106
plasminogen zymography, 108–109
Ras induction
electrophoretic mobility shift assay for transcription factor identification
AP-1/PEA3 site, 114–115
supershift assay, 114
evidence, 106–107
laminin degradation mediation by receptor-bound urokinase, 109–110
metastasis role, 107–108
Northern blot analysis, 111–112
Ras transfectants and magnetic cell sorting, 110–111
reporter assay of transactivation using transient transfection, 112–113
signaling pathway, 115–116
transcription factors, 106, 114

V

Vascular endothelial growth factor, Ras induction
angiogenic switch, 271, 273
coregulation with epigenetic factors, 279
expression assay, 270
levels of regulation, 276, 278–279
model systems, 273–274

signal transduction
inhibitor studies, 270–271, 275–276, 278
mitogen-activated protein kinase, 278
tumor growth role and evidence, 273–275, 279–280
VEGF, *see* Vascular endothelial growth factor

W

Western blot
activated Ras interaction assay
antibodies, 338–339
blotting, 338
electrophoresis, 338
caveolin in sucrose gradients, 177, 179
Ras-like GTPase–GTP determination with activation-specific probes, 345–346
Ras mutants, 322–323
Shc protein phosphorylation, 10–11
signal transducer and activator of transcription activation, 147

Y

Yeast two-hybrid system
M-Ras/R-Ras3–effector interactions
controls, 197
β-galactosidase assay, 197
principle, 196–197
TC21/R-Ras2–effector interactions
HIS3 reporter transactivation assessment, 214–215
LacZ reporter transactivation assessment, 215
media, 213–214
plasmids, 212–213
principle, 212
transformation, 214
yeast strain, 213

Z

Zymography, *see* Gelatinase zymography; Plasminogen zymography

ISBN 0-12-182234-6